Oxidative Stress and Antioxidant Defenses in Biology

Oxidative Stress and Antioxidant Defenses in Biology

edited by
Sami Ahmad

CHAPMAN & HALL

I(T)P An International Thomson Publishing Company

New York • Albany • Bonn • Boston • Cincinnati
• Detroit • London • Madrid • Melbourne • Mexico City
• Pacific Grove • Paris • San Francisco • Singapore
• Tokyo • Toronto • Washington

Copyright © 1995
Softcover reprint of the hardcover 1st edition   1995
By Chapman & Hall
A division of International Thomson Publishing Inc.
I(T)P The ITP logo is a trademark under license

For more information, contact:

Chapman & Hall
One Penn Plaza
New York, NY 10119

Chapman & Hall
2-6 Boundary Row
London SE1 8HN

International Thomson Publishing
Berkshire House 168-173
High Holborn
London WC1V 7AA
England

International Thomson Editores
Campos Eliseos 385, Piso 7
Col. Polanco
11560 Mexico D.F. Mexico

Thomas Nelson Australia
102 Dodds Street
South Merlbourne, 3205
Victoria, Australia

International Thomson Publishing Gmbh
Königwinterer Strasse 418
53228 Bonn
Germany

Nelson Canada
1120 Birchmount Road
Scarborough, Ontario
Canada, M1K 5G4

International Thomson Publishing Asia
221 Henderson Road
#05-10 Henderson Building
Singapore 0315

International Thomson Publishing-Japan
Hirakawacho-cho Kyowa Building, 3F
1-2-1 Hirakawacho-cho
Chiyoda-ku, 102 Tokyo
Japan

1 2 3 4 5 6 7 8 9 10 XXX 01 00 99 97 96 95

Library of Congress Cataloging-in-Publication Data

Oxidant-Induced stress and antioxidant defenses in Biology / editor, Sami Ahmad.
    p. cm.
  Includes bibliographical references and index.
  ISBN-13: 978-1-4615-9691-2   e-ISBN-13: 978-1-4615-9689-9
  DOI: 10.1007/978-1-4615-9689-9
   1. Active oxygen—Pathophysiology. 2. Antioxidants. I. Ahamd, Sami.
R8170.094   1995
574.2'19214—dc20
                           94-13342
                           CIP

Contents

1

Mechanisms of oxygen activation and reactive oxygen species
detoxification
Enrique Cadenas

2

Pathophysiology and reactive oxygen metabolites
Yan Chen, Allen M. Miles and Mathew B. Grisham

3

Free radical mechanisms of oxidative modification of
low density lipoprotein (or the rancidity of body fat)
Balaraman Kalyanaraman

4

Synthetic pro-oxidants: drugs, pesticides and other
environmental pollutants
Sidney J. Stohs

5

Metabolic detoxification of plant pro-oxidants
May R. Berenbaum

6

Antioxidant mechanisms of secondary natural products
Richard A. Larson

7

Antioxidant mechanisms of enzymes and proteins
Sami Ahmad

8

Antioxidant defenses of *Escherichia coli* and *Salmonella typhimurium*
Richard D. Cunningham and Holly Ahern

9

Antioxidant defenses of plants and fungi
David A. Dalton

10

Antioxidant defenses of vertebrates and invertebrates
Gary W. Felton

11

Genetic regulation of antioxidant defenses in
Escherichia coli and *Salmonella typhimurium*
Holly Ahern and Richard P. Cunningham

Contributors

HOLLY AHERN
Department of Biological Sciences
State University of New York
  at Albany
New York 12222

SAMI AHMAD
Department of Biochemistry
University of Nevada
Reno, Nevada 89557-0014

MAY R. BERENBAUM
Department of Entomology
University of Illinois
Urbana, Illinois 61801-3795

ENRIQUE CADENAS
Institute for Toxicology and
  Department of Molecular
  Pharmacology & Toxicology
University of Southern California
Los Angeles, California 90033

YAN CHEN
Department of Physiology and
  Biophysics
Louisiana State University
  Medical Center
Shreveport, Louisiana 71130

RICHARD P. CUNNINGHAM
Department of Biological Sciences
State University of New York
  at Albany
New York 12222

DAVID A. DALTON
Biology Department,
Reed College
Portland, Oregon 97202

GARY W. FELTON
Department of Entomology
University of Arkansas
Fayetteville, Arkansas 72703

MATHEW B. GRISHAM
Department of Physiology and
  Biophysics
Louisiana State University
  Medical Center
Shreveport, Louisiana 71130

BALARAMAN KALYANARAMAN
Biophysics Research Institute
Medical College of Wisconsin
Milwaukee, Wisconsin 53228

RICHARD A. LARSON
Institute For Environmental Studies
University of Illinois
Urbana, Illinois 61801

ALLEN M. MILES
Department of Physiology and
  Biophysics
Louisiana State Medical Center
Shreveport, Louisiana 71130

SIDNEY J. STOHS
School of Pharmacy and Allied
  Health Professions
Creighton University Health
  Sciences Center
Omaha, Nebraska 68178

Preface

The ground-state of molecular oxygen, O_2, is essential to many indispensible metabolic processes of all aerobic life forms ranging from prokaryotes, protists, plants, and fungi to animals. Research by mammalian toxicologists and clinicians has unravelled persuasive evidence that O_2 dependence imposes universal toxicity to all aerobic life processes. The basis of this paradox is that one-electron reduction of O_2 generates the superoxide anion free radical, $O_2^{\cdot-}$, from numerous biological sources; for example, redox-active autoxidizable molecules such as catecholamines, oxidoreductases, and subcellular organelles such as mitochondria, endoplasmic reticulum (microsomes), nuclei, and chloroplasts. Oxygen is also activated in biologically relevant photosensitizing reactions to highly reactive singlet oxygen, $^1O_2$.

In all biological systems, $O_2^{\cdot-}$ undergoes further reduction to H_2O_2 via Fenton reaction to the hydroxyl radical, $\cdot OH$. These, and some other forms of activated O_2, constitute reactive oxygen species (ROS) and/or metabolites (ROM). Both $\cdot OH$ and $^1O_2$ are the most reactive forms of ROS known and among their deleterious reactions are oxidation of proteins, DNA, steroidal compounds, and peroxidation of the cell membrane's unsaturated lipids to form unstable hydroperoxides. Their many breakdown products include malondialdehyde and hydroxynonenals that are themselves highly reactive and threaten cellular integrity and function. More importantly, they decompose to free radicals that can continue to propagate the vicious lipid peroxidation chain reaction. This is the so-called endogenous oxidative stress with which all aerobic organisms must cope.

Thus the evolutionary process that in the first instance harnessed oxygen for the physiological support of aerobic organisms resulted in toxicity as a side effect of ROS attack, rendering them targets for peroxidative attack, cell and tissue degeneration and, ultimately, organismal death. Earlier texts have emphasized the nature of pathological processes mediated by this stress which include the leakage of cell membranes, dysfunction of mitochondria, depletion of glutathione and disturbed redox states of cells, and the depletion of ATP. These processes which affect cells and DNAs lead to aging, tumor promotion and cancer, inflammatory diseases, post-ischemic injury and numerous serious ailments.

Faced with the inevitable consequences of O_2 toxicity, evolution began in the earliest aerobic cells to acquire appropriate defensive strategies. As evolution proceeded towards more complex aerobic life forms, it also favored the appropriate elaboration of antioxidant defenses. The first line of this defensive strategy was the deployment of antioxidant substances such as vitamin C (ascorbic acid), vitamin E (α-tocopherol), urate, glutathione, and carotenoids. In addition, a battery of antioxidant enzymatic defenses prevent the O_2 radical cascade and terminate the lipid peroxidation cycle. The enzymatic defenses are regarded as crucial in that only they can ameliorate O_2 toxicity when the antioxidant molecule supply is meager or exhausted.

With this antioxidant machinery, aerobic cells seem competent to cope with endogenous oxidative stress. Unfortunately aerobic organisms are subject to oxidative injury from prooxidants arising from natural products in dietary sources and from environmental pollutants released by humans as mining and industrial wastes, therapeutic drugs, and agrochemicals of all sorts. These interactions of prooxidants lead to the O_2 radical cascade and lipid peroxidation, thereby exacerbating the endogenous oxidative stress. In turn, the organisms seem to mobilize the very same defenses that in the first place were designed to cope with endogenous oxidative stress. Often, however, despite the induction of chain-terminating antioxidant enzymes, the toll of prooxidative insult is far more severe, resulting in early initiation of the aging process and numerous pathologies including cancer.

The implication of oxidative stress in far more pathologies and disorders than previously imagined has provided the impetus for numerous national and international meetings, symposia, and congresses. In addition, specialized journals have appeared in recent years, which are now major outlets for reports on this subject matter. The field of free-radical based biological and pathological interface is virtually exploding with new discoveries; for example, nitric oxide, ·NO, is now known to be an endothelial releasing factor with an essential physiological role as a vasodilator. Most volumes on oxyradicals published recently either as symposia proceedings or edited books are devoted to highly specialized topics that are clinically oriented. The effort extended to the fundamental studies of oxidative stress in organisms such as plants, fungi, and invertebrates was never neglected, but it was subordinated to the research effort devoted to mammalian models. That this relative neglect occurred in a field still being primarily driven by medically oriented scientists is hardly surprising. The enormous difficulty attendant upon accessing and synthesizing the vastly scattered literature on non-mammalian organisms is amply illustrated in this volume and in fact represents its unique character.

Oxidative stress is an area of contemporary research that is rapidly gaining momentum, as scientists from diverse disciplines seek more details about the deleterious effects ROS have on cells. It is therefore salutary that an attempt has been made to address in this volume oxidant-induced stress and antioxidant defenses in a broad biological context. Its broad coverage should be highly appealing to biologists as diverse as ecologists, entomologists, microbiologists, mycologists, molecular biologists, animal and plant physiologists, and biochemists. In addition, the specialists will find it a valuable reference source not only because the information is current, but is relevant to the progress made on other organisms, which previously had not received as much attention because of continued emphasis on mammalian species.

For the book to be as thorough in coverage as possible, it comprises of eleven chapters. The progression of chapters is logical, and their combined bibliographies provide a comprehensive view of our current knowledge.

E. Cadenas in chapter 1 describes the chemistry of oxygen activation in a manner that is easily comprehended by non-specialists and specialists alike. The chapter is comprehensive and, aside from the mechanisms of O_2 activation to ROS per se, it also deals with important reactions between ROS with carbon-centered free radicals and antioxidant molecules. In addition, the chemistry of the oxo-ferryl complex, a matter of contemporary interest, is addressed. The origin of these complexes and their connection with the oxidation of cell constituents (distinct from that by ˙OH radical) and drugs is described. Another novel aspect addressed is the biochemistry of ˙NO and its reactions with oxyradicals.

In chapter 2, Y. Chen et al. cover the numerous pathologies that seem to be ROS-mediated, which are primarily the outcome of vascular and tissue injury. ROS mediated ontogeny of cancer is also treated. Despite the extensive evidence for these pathologies, the authors are correct in pointing out that many lacunae still exist in defining the extent and particular forms of ROS that have an effect on particular pathologies. Lastly, both E. Cadenas in chapter 1 and Y. Chen et al. in chapter 2 describe the interaction of two radicals, ˙NO and O_2^-, which may react under certain conditions (e.g., pH and the flux of O_2^-) to generate the two most reactive free radical species, ˙OH and nitrogen dioxide ($NO_2^˙$) radicals, which might be the specific ROS species involved in the microvascular injury produced during ischemia and reperfusion of various organ systems. Since `NO research is new, the involvement of `NO in ischemic injury and other pathologies warrants experimental validation.

In chapter 3, B. Kalyanaraman addresses the pathological consequences of oxidation of the low-density lipoprotein (LDL, the major cholesterol and other lipid transport protein). Of several lipids associated with this protein, linoleate is most sensitive to peroxidative damage. A consequence of this damage is the degenerative disease atherosclerosis. LDL interactions involving oxyradicals, peroxynitrite (ONOO⁻) and metals such as Cu, and phenolic antioxidants, has been discussed at length. The author reaches the conclusion that the oxidation of LDL and food fat can be prevented by vitamins C and E.

Chapter 4 by S.J. Stohs is very comprehensive on oxidative stress exerted by environmental contaminants, including drugs and pesticides. The author emphasizes that not all contaminants act as prooxidants in a similar fashion. Some of these materials are able to exert oxidative stress directly, e.g., therapeutic drugs for malaria, antibiotics such as adriamycin for cancer therapy, and tetracyclines as broad-spectrum antibiotics, may exert O_2 toxicity including photosensitizing reactions. Others such as mercury raise oxidative stress through a pleiotropic response that generates H_2O_2. In addition, many halogenated compounds such as dioxins release iron from transport or storage proteins. Once iron is released, it leads to the production of the lipid peroxidizing `OH radical. The chapter also addresses the difficulty in pinpointing whether oxidative stress mediated pathology is an early, mid or late event in defining the sequence of events that lead to cell damage and death. Nevertheless, all compounds listed by Stohs exert oxidative stress, but he is judicious in arguing that more research and information will yield a better picture of the role of ROS in the action of prooxidative xenobiotics.

Chapter 5 by M.R. Berenbaum tackles a unique aspect of either behavioral escape or metabolic resistance to naturally occurring prooxidants. Some specialized feeders are capable of rapidly metabolizing photodynamic prooxidants with cytochrome P-450, and an assortment of other detoxification enzymes. While many redox-active compounds are well tolerated by herbivores including man (e.g., the flavonol quercetin), others are not. The compounds not well tolerated are highly redox-active and react with O_2 to generate ROS. The metabolism of these compounds is not well studied. Clearly then, more research is needed in this interesting area despite the fact that as the author states, ". . . unifying themes in the disposition of prooxidants is not an easy task."

Chapter 6 by R.A. Larson serves as an overview of antioxidant molecules that are components of natural products. Larson provides a good account of the initiation of the lipid peroxidation chain reaction and of termination processes involving vitamins C and E, flavonols, β-carotene, tertiary amines, bilirubin and metal complexing compounds. He addresses this subject well with kinetics and rate constants for interactions of various ROS with different antioxidants, and processes associated with the antioxidant action. For example, these molecules are effective via their reducing property and radical and excited-state quenching, which may be via physical or chemical interactions. He cites the novel aspect of naturally-occurring compounds such as phytic acid which complex with metals, such as iron, that are known to generate ROS.

Chapter 7 by S. Ahmad reviews the antioxidant defenses of enzymes and proteins. Substantive coverage is given to a group of antioxidant enzymes which act in a concerted or sequential manner to prevent the O_2 radical cascade and to terminate the lipid peroxidation chain reaction. Where the available information is considered authentic, Ahmad has provided the reaction schemes and kinetics of these enzymes, the preference for and range of substrates attacked, inhibitors, molecular weights of native proteins and subunits, and their isoenzymes. The coverage of ancillary antioxidant enzymes and proteins is brief, since D.A. Dalton and G.W. Felton, in their respective chapters, have provided an in-depth account of these processes.

Chapter 8 by R.P. Cunningham and H. Ahern provides an account of the antioxidant defenses of prokaryotes using the bacterial species *Escherichia coli* and *Salmonella typhimurium* as prime examples. It is evident that prokaryotes share many if not most of the same antioxidant defenses as eukaryotes which leads to the conclusion that countermeasures against O_2 toxicity are of ancient evolutionary origin. The chapter delves further into other repair mechanisms of oxidized lipids, proteins, and especially of DNA. The prokaryotic model provides insight into the mechanisms whereby oxidative lesions are indirectly repaired by an arsenal of enzymes that degrade oxidized molecules. The degraded products are either released for excretion or are conserved for reutilization. Recent evidence suggests that these indirect mechanisms of repair are also operative in eukaryotic systems.

Chapter 9 by D.A. Dalton on alleviation of oxidative stress in plants
and fungi is notable for its pioneering coverage. He identifies cellular locations in plants such as chloroplasts and nodules where the
generation of ROS is higher. He discusses oxidative stress as it arises
normally and in conditions of nitrogen fixation, and also points to
the benefits of ROS production. Plants' antioxidant machinery is essentially similar to that of other organisms with the exception of the
absence of selenium-dependent glutathione peroxidases typical of
vertebrates. He argues that an ascorbate-specific peroxidase is more
crucial for the destruction of H_2O_2 than catalase. He provides a thorough account of the enzymes, ascorbyl free radical reductase and
dehydroascorbate (DHA) reductase. Respectively, these enzymes
regenerate ascorbate from ascorbyl free radical and DHA, which result from ascorbate reactions with ROS. Based on his own research,
Dalton describes a fascinating interaction among ascorbate peroxidase, DHA reductase, and glutathione reductase. Dalton also pon-
ders the functional roles of many peroxidases present in plants. Peroxidase (horseradish peroxidase) is an enzyme of ubiquitous
occurrence, and is abundant in plants. Dalton traces the origin and
homologies of the various peroxidases, and concludes that in plants
the crucial form is ascorbate peroxidase. The account of antioxidant
defenses of fungi is relatively meager because of the paucity of data.
Nonetheless, according to Dalton, a cytochrome-*c* peroxidase is crucial in fungi for the destruction of H_2O_2. This enzyme is associated
with the fungal wall, where the generation of ROS is apparently
highest and the necessary co-substrate cytochrome *c* is in abundant
supply.

Chapter 10 by G.W. Felton on antioxidant defenses of vertebrates
and invertebrates is contemporary and comprehensive in its coverage. The fundamental processes of oxidant-induced disease states
are the same in all aerobic organisms, but many invertebrates do
not live long enough to exhibit the typical diseases of vertebrates.
Using insects as examples, he shows how ROS affect nutrition with
severe reduction in growth rate and lifespan resulting ultimately in
death. Fortunately, invertebrates are also endowed with an elaborate system of antioxidant defenses to avoid oxidative stress. Invertebrates resemble in this respect both vertebrates and plants except the selenium-dependent glutathione peroxidases are absent, but
in these animals the peroxidase activity of glutathione transferase
has been elaborated for the removal of lipid peroxides. H_2O_2 destruction follows the same path as described for plants. Felton also
claims the existence of an ascorbyl free radical reductase. There is

evidence to support this contention, although the animal enzyme may not be homologous to the plant enzyme. Another strength of this chapter are details on antioxidant proteins and proteases that degrade oxidized proteins. The latter aspect has been receiving considerable scrutiny, but as of this writing the specifics were unfortunately not worked out.

In the last chapter, chapter 11, H. Ahern and R.P. Cunningham describe the genetic regulation of the antioxidant defenses based on extensive work with *E. coli* and *S. typhimurium*. Genetic responses to ROS-mediated stress are complex and multi-layered, involving stimulons, regulons, regulatory circuits, and specific genes for transcriptional processes. H_2O_2 and O_2^- are the two ROS to which the genetic machinery responds, especially the H_2O_2 stimulon and oxyR regulon, and the O_2^- stimulon and soxRS regulon. The authors account for the induction of antioxidant enzymes and for the repair of proteins and enzymes targeted as oxidatively damaged macromolecules including DNA. Many other proteins are also transcribed in this process, but their role remains to be defined. The genetic regulation area is bound to be an area of research thrust. Since the antioxidant defenses of both prokaryotic species are remarkably similar (chapter 8) to those of eukaryotes, it seems logical to conclude that eukaryotes possess similar genetic machinery for the regulation of antioxidant defenses.

I am very grateful to the authors of this book for their fine contributions and to the staff of Chapman & Hall for their invaluable help. I would also like to thank my wife Farhana, daughter Didi, and sons Omar and Amer for their patience, understanding and support, especially when the demand on my time for this volume far exceeded my expectation. I thank my senior colleague, Dr. R. S. Pardini, for valuable discussions about this field. This publication was made possible through the support of ParaProfessional Services, Inc., and for successive grant supports from USDA, NSF, NIEHS and NIH.

April 1994   Sami Ahmad

List of Abbreviations

A; dehydroascorbate
$A^{.-}$; ascorbyl radical
AH^-; ascorbate mono-anion
AH_2; ascorbic acid
ACON; acetone
ACT; acetaldehyde
ADP; adenosine diphosphate
AFR; ascorbate free radical reductase
AHP/AHP reductase; alkyl hydroperoxidase
AP; ascorbate peroxidase
5-ASA; 5-aminosalicylic acid
ATP; adenosine triphosphate
ATPase; adenosine triphosphatase
BCNU; bischloronitrosourea
CAT; catalase
CCP; cytochrome-c peroxidase
CD18; cluster of differentiation
CoQ; Co-enzyme Q (UQ_{10})
$CoQH_2$; reduced CoQ ($UQ_{10}H_2$)
CP; ceruloplasmin
CuDIPS; copper di-isopropyl salicylate
CuZnSOD; copper and zinc SOD
D; oxidized DH_2
DCFH; 2′,7′-dichlorofluorscen
DDT; 1,1,1-dichloro-2,2-bis(p-chlorophenyl) ethane
DFX/desferal; desferrioxamine
DH_2; hydrogen donor
DHA; dehydroascorbate (A)
DHA reductase; dehydroascorbate reductase
DMPO; 5,5-dimethylpyrrolidone-(2)-oxyl-(1)
DMSO; dimethylsulfoxide
DT; di- or triphosphopyridine nucleotide
DT-diaphorase; quinone reductase
e^-; electron
EC-SOD; extracellular SOD
EDRF; endothelium derived relaxing factor
ESR; electron spin resonance

FA; formaldehyde

$Fe^{IV}=0$; oxoferryl complex in ferrylmyoglobin or ferrylhemoglobin

FeSOD; iron SOD

GPOX/Se-GPOX; selenium-dependent glutathione peroxidase

GPOX-GI; gastrointestinal GPOX

GR; glutathione (GSSG) reductase

GSH; glutathione (reduced form)

GS$\cdot$; glutathionyl radical

GSSG; glutathione disulfide (oxidized form of GSH)

GSSG$\cdot^-$; glutathione disulfide (anion) radical

GSOH; sulfenic acid of glutathione

GST; glutathione transferase

GST_{px}; GST's peroxidase activity

Hb; hemoglobin

$HOO\cdot/HO_2\cdot$; hydroperoxyl radical

HP; hydroperoxidase

HPETE; ((S)-5-hydroperoxy-6-trans-8,11,14-*cis*-eicosatetranoic acid)

HETE; metabolite of HPETE where hydroperoxy group is reduced to a hydroxyl group

HRP; horseradish peroxidase

HSP; heat shock protein

$HX-Fe^{IV}=O$; ferrylmyoglobin or ferrylhemoglobin

$HX-Fe^{III}$; metmyoglobin or methemoglobin

$HX-Fe^{II}O_2$; oxymyoglobin

IBD; inflammatory bowel disease

IOM; inside out membrane

I/R; ischemia/reperfusion

LH/RH; unsaturated lipid/organic compound

L$\cdot$/R$\cdot$; lipid/organic radical

LOH/ROH; lipid/organic alcohol

LOOH/ROOH; lipid/organic hydroperoxide

$LOO\cdot/ROO\cdot$ ($LO_2\cdot/RO_2\cdot$); lipid or peroxyl radical

$LO\cdot/RO\cdot$; alkoxyl radical

LDL; low-density lipoprotein

LT; leukotreine

LTB_4; leukotreine B_4

MDA; melondialdehyde

MnSOD; manganese SOD

NMR; nuclear magnetic resonance

MPO; myeloperoxidase

NASA; N-acetyl-5-ASA

˙NO (or NO); nitric oxide
NO₂˙; nitric dioxide
NPh-O˙; 1-naphthoxyl radical
NPhOH; 1-naphthol
$^1O_2$; singlet oxygen
O_2; molecular oxygen
$O_2^{\cdot-}$; superoxide anion radical
˙OH/HO˙; hydroxyl radical
PA; D-penicillamine
PAF; platelet activating factor
PBB's; polybromobiphenyls
PCB's; polychlorobiphenyls
PG; prostaglandin
PGD_2; prostaglandin D_2
PGE_2; prostaglandin E_2
PGG_2; prostaglandin G
PGH_2; prostaglandin H_2
PG/PS; peptidoglycan/polysaccharide
PH-GPOX; phospholipid hydroperoxide GPOX
PL-GPOX; plasma GPOX
PMNs; polymorphonuclear leukocytes
POD; peroxidase (also HRP)
PUFA; polyunsaturated fatty acid
Q; quinone
$Q^{\cdot-}$; semiquinone (radical)
QH_2; hydroquinone (dihydroquinone)
RBC; red blood cell
ROMs; reactive oxygen metabolites (=ROS)
ROS; reactive oxygen species
RSH; thiol compound
RS⁻; thiolate anion
RS˙⁻; thiyl radical
RS_2^-; aliphatic dithiol (e.g. α-dihydrolipoic acid)
RSSR˙⁻; disulfide anion radical
SAZ; sulfasalazine
SOD; superoxide dismutase
SP; sulphapyridine
TBHQ; 2 (3)-tert-bytyl-4-hydroquinone
TCDD; 2,3,7,8-tetrachlorodibenzo-p-dioxin
TCP; tetrachloro-1,4-benzoquinone
TEMPO; 4-hydroxy-2,2,6,6-tetramethylpiperidine-N-oxyl

α-T-OH/α-TH; α-tocopherol
α-T-O˙/α-T˙; α-tocopheroxyl radical
γ-TH; γ-tocopherol
γ-T˙; γ-tocopheroxyl radical
U; uric acid
U˙; uric acid radical
UH^-; urate
UQ_{10}; ubiquinone-10
$UQ_{10}^{\cdot-}$/$UQH^{\cdot}$; ubi-semiquinone radical
$UQ_{10}H_2$; ubiquinol-10
Vitamin C; ascorbic acid
Vitamin (vit) E; α-tocopherol
Vit E˙; α-tocopheroxyl radical
XD; xanthine dehydrogenase
XO; xanthine oxidase

CHAPTER 1

Mechanisms of Oxygen Activation and Reactive Oxygen Species Detoxification

Enrique Cadenas

1.1 INTRODUCTION

The biochemistry of 'reactive oxygen species' (ROS) is an important field with practical implications, because whereas oxygen is an essential component for living organisms, the formation of reactive oxygen intermediates seems to be commonplace in aerobically metabolizing cells. In addition to aerobic metabolism-encompassing electron transfer chains and certain enzyme activities, environmental sources, such as air pollutants, photochemical smog, industrial chemicals, and ionizing radiation, as well as the metabolism of xenobiotics, contribute to the cellular steady-state concentration of ROS. Further, reactive species are formed as a response to diverse stimuli by specialized physiological reactions: the formation of oxyradicals during the respiratory burst and the release of the endothelium-derived releasing factor, identified as nitric oxide, are such examples.

The stability of a variety of reactive species, whether of free radical character or not, varies substantially. Regardless of their source, it could be stated that in an appropriate setting virtually all cell components—lipids, nucleic acids, proteins, and carbohydrates-are sensitive to damage by ROS.[1]

Cells convene substantial resources to protect themselves from the potentially damaging effects of reactive oxygen species. Several vitamins and micronutrients, which are active at quenching these free radical species or required for their enzymic detoxification, as well as enzymes, such as superoxide dismutases (SODs), glutathione peroxidases (GPOXs), and catalases (CATs), constitute a first line of defense against ROS, and are generally referred to as primary antioxidants.

That the cellular generation of oxidants overwhelms or bypasses these defenses is attested by several lines of evidence. First, the growing evidence for the participation of free radicals in several pathologies indicate that these defenses are not entirely effective. Second, the occurrence of 'fingerprints' for oxidative damage has been found in DNA, proteins, carbohydrates, and lipids. This appears to stimulate the systematic activity of 'repair' enzymes regarded as secondary antioxidant defenses and involved in the removal of non-functional cell components or in the actual repair of damaged biomolecules. Third, oxidants (e.g., hydrogen peroxide H_2O_2) or processes leading to their formation (e.g., redox cycling) induce the synthesis of several proteins, as shown by work carried out with *E. coli* and *Salmonella typhimurium* (Christman et al., 1985; Greenberg and Demple, 1989; Demple and Levin, 1991) where two groups of oxidative stress-inducible proteins can be distinguished controlled by the *oxyR* and *soxR* loci. The former is activated by H_2O_2 and of the thirty proteins encoded by the *oxyR* regulon some are antioxidant enzymes, such as CAT and alkylhydroperoxide reductase. The latter is suggested to be linked to increased levels of superoxide anion O_2^- and appears to regulate oxidative tolerance in *E. coli*, probably due to the expression of enzymes such as Mn-superoxide dismutase (MnSOD). Tumor necrosis factor and x-rays also activate the expression of the human MnSOD (Wong and Goeddel, 1988). 'Repair' enzymes (secondary antioxidants) appear also to be induced by oxidative challenges (Chan and Weiss, 1987; Fornace et

---

[1] The term 'reactive oxygen species' (or similar terms often used) was adopted at some stage (in favor of oxyradicals or oxygen radicals) in order to include nonradical oxidants, such as hydrogen peroxide and singlet oxygen. However, this term still fails to reflect the rich variety of reactive species in free radical biology, for it does not include some oxidants or free radicals which are significant in a biological context, such as carbon-, nitrogen-, and sulfur-centered radicals.

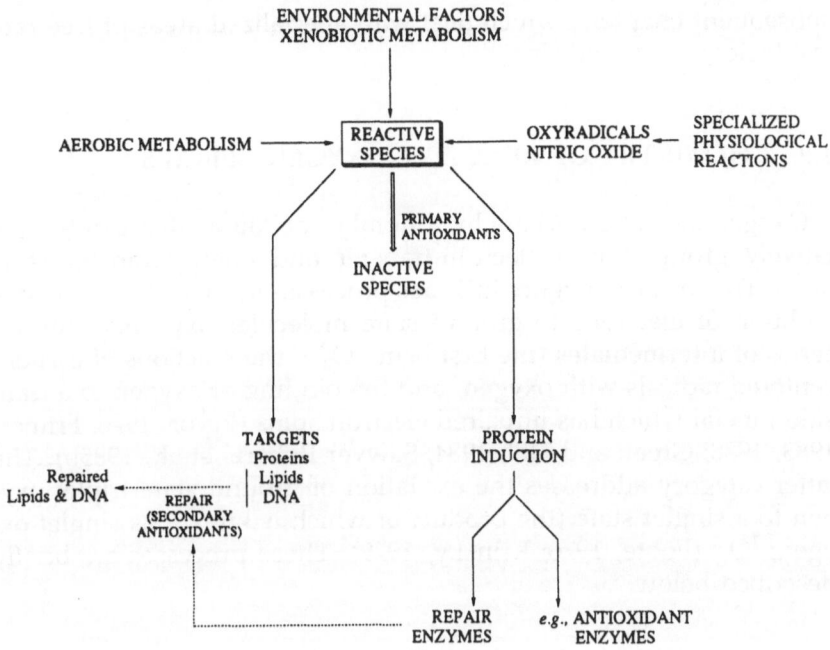

Fig. 1.1 Schematic overview of the main sources of reactive oxygen species and their biological effects: oxidation of bioconstituents, stimulation of 'repair' mechanisms, and induction of protein synthesis.

al., 1988), as suggested by the increase of DNA repair enzyme levels, such as endonuclease IV, in *E. coli* after treatment with redox active drugs.

These relationships, including the diverse sources of ROS and their reactivity with biological targets along with the antioxidant coordinated responses to oxidative stress, are outlined in the scheme in Fig. 1.1. It should be observed, however, that it is not possible to provide a complete and unambiguous perspective on the diversity of research areas contributing to many specific aspects of free radical biology and medicine within one scheme.

This chapter provides a survey on some relevant physico-chemical properties of free radicals and related oxidants and discusses the mechanisms of their formation and chemical reactivities. Another section examines the reactions of a variety of free radicals with a number of small molecular antioxidants, with emphasis on the decay pathways of the antioxidant radical species ensuing from these interactions. This overview is intended to serve as a background for subsequent chapters, which deal with specialized areas of free radical biology and medicine.

1.2 CHEMISTRY OF REACTIVE OXYGEN SPECIES

Oxygen activation occurs by a number of routes that can be tentatively grouped into electron-transfer and energy-transfer reactions. The former category includes processes such as the successive addition of electrons to ground state molecular oxygen to form a series of intermediates (the first being $O_2^{.-}$), the reactions of carbon-centered radicals with oxygen, and the binding of oxygen to a transition metal which has unpaired electron spins (Pryor, 1976; Frimer, 1983, 1988; Green and Hill, 1984; Sawyer Roberts et al., 1985b). The latter category addresses the excitation of ground state triplet oxygen to a singlet state (the product of which is known as singlet oxygen; $^1O_2$) (Foote, 1976; Krinsky, 1979). These processes are briefly described below.

1.2.1   *The Generation of Reactive Species by Electron Transfer Reactions*

i. Superoxide radical

The one-electron transfer to molecular O_2 yields the $O_2^{.-}$ radical, a species that has received considerable attention among the oxyradicals generated in biological systems. The chemistry of $O_2^{.-}$ in aqueous and aprotic media has been extensively characterized. In aqueous, neutral pH solutions, the chemical reactivity of $O_2^{.-}$ encompasses mainly three types of reactions:

[a] $O_2^{.-}$ behaves as a weak base undergoing protonation to the hydroperoxyl radical ($HO_2^{.}$) (reaction 1.1; $pK_a = 4.8$) followed by disproportionation to H_2O_2. The reaction between $O_2^{.-}$ and its conjugate acid $[O_2^{.-} + HO_2^{.}]$ is faster ($k = 8 \times 10^7$ $M^{-1}s^{-1}$) than that calculated for physiological pH and originating from the respective concentrations of $O_2^{.-}$ and $HO_2^{.}$ (reaction 1.2; $k_2 = 4.5 \times 10^5$ $M^{-1}s^{-1}$) (Bielski, 1978). When assessing the reactivity of $O_2^{.-}$ in biological

$$O_2^{.-} + H^+ \rightarrow HO_2^{.} \tag{1.1}$$

$$O_2^{.-} + O_2^{.-} + 2H^+ \rightarrow H_2O_2 + O_2 \tag{1.2}$$

systems, the participation of its conjugated acid ($HO_2^{.}$) should be critically considered, for the above pK_a value is only two units below biological pH. Moreover, the pH value could be considerably lower in the proximity of biological membranes; hence, the actual concentration of the conjugate acid of O_2^- would be expected to be much higher.

[b] O_2^- behaves as a one-electron reductant of adequate electron acceptors, such as oxidized transition metal complexes of Fe^{III} and Cu^{II}, ferricytochrome c, and quinones with suitable redox potentials, such as p-benzoquinone.

[c] O_2^- behaves as a one-electron oxidant of molecules bearing acidic protons, such as ascorbate, and it can oxidize Fe^{II}-EDTA complexes to form a ferric-peroxo complex (Bielski, and Cabelli, 1991; see also Sawyer et al., 1985b). Whereas O_2^- displays a moderate reactivity, HO_2^- is capable of, for example, abstracting a bis-allylic hydrogen atom from a polyunsaturated fatty acid (PUFA) thus initiating lipid autoxidation (Gebicki and Bielski, 1981); H abstraction by HO_2 proceeds more readily with lipid hydroperoxides (LOOHs), probably because the hydroperoxide moiety is a kinetically preferred reaction site (Aikens and Dix, 1991).

The chemistry of O_2^- in aprotic media has been critically surveyed (Frimer, 1983, 1988, and references therein; Sawyer et al., 1985b). It serves as a model for the chemical environment in biological membranes, and it encompasses four basic modes of action of O_2^- electron transfer (the most common mode in biosystems), nucleophilic substitution, deprotonation, and H-atom abstraction. It is clear that both types of chemical reactivity of O_2^--in aqueous and non-aqueous media—are relevant in biological systems.

ii. Hydrogen peroxide

The spontaneous or SOD-catalyzed disproportionation of O_2^- yields H_2O_2 (reaction 1.1), a species which is generated directly via two-electron transfers by several oxidases (see below). H_2O_2 is not a radical, for it has no unpaired electrons, and it displays a moderate chemical reactivity. However, this chemical reactivity is substantially enhanced by two features of H_2O_2: First, and at variance with its precursor O_2^-, H_2O_2 can cross freely biological membranes, a property apparently shared by the conjugate acid of O_2^-, HO_2 (Halliwell and Gutteridge, 1986). Second, H_2O_2 is required for the formation of more potent oxidants, such as the hydroxyl radical (HO·) and oxoferryl complexes ($Fe^{IV}=O$), upon its reaction with metal chelates and hemoproteins, respectively (see below).

iii. Hydroxyl radical

The metal-dependent decomposition of H_2O_2 is a source of $HO\cdot$ radical, a species with a diffusion-collision radius of 5-10 molecular diameters. The concept of a O_2^--driven Fenton-type reaction requires both O_2^- and H_2O_2 as precursors of $HO\cdot$; it proceeds via an intermediate catalyst, such as a transition metal chelate (e.g., Fe or Cu), which is reduced by O_2^- and reacts with H_2O_2 in a "Fenton-like" reaction to produce $HO\cdot$ (reactions 1.3 and 1.4). The requirements of a transition metal chelate for this reaction are two-fold:

$$Fe^{III} + O_2^- \rightarrow Fe^{II} + O_2 \qquad (1.3)$$

$$Fe^{II} + H_2O_2 \rightarrow Fe^{III} + HO^- + HO\cdot \qquad (1.4)$$

on the one hand, the chelators alters the redox potential of the transition metal, thereby facilitating electron transfer reactions with O_2^- and H_2O_2 (e.g., the $E°$ values for Fe^{3+}/Fe^{2+} and $Fe\text{-}EDTA^-/Fe\text{-}EDTA^{2-}$ are +770 and +120 mV, respectively; Koppenol and Butler, 1985) and, on the other hand, it maintains the transition metal in solution. These well-known reaction sequences give as net balance (Haber-Weiss reaction):

$$O_2^- + H_2O_2 \rightarrow O_2 + HO^- + HO\cdot \qquad (1.5)$$

O_2^- in reaction 1.3 can be replaced by a suitable electron donor, such as ascorbate (AH^-) (reaction 1.6) (Winterbourn, 1981). The prevalence of reaction 1.3 and 1.6 will depend, of course, on the actual steady-state concentrations of O_2^- or AH^-.

$$Fe^{III} + AH^- \rightarrow Fe^{II} + AH^{\cdot -} \qquad (1.6)$$

Reaction 1.7, analogous to reaction 1.6, serves to rationalize the

$$Fe^{III} + Q^{\cdot -} \rightarrow Fe^{II} + Q \qquad (1.7)$$

previous suggested $HO\cdot$ formation by antitumor antibiotic semiquinones ($Q^{\cdot -}$), paraquat radical, and other drugs, in a fashion not involving a direct organic radical-dependent H_2O_2 breakdown (Winterbourn et al., 1985).

The occurrence of reactions 1.3-1.4, or 1.3 and 1.6, will depend on the availability of transition metals in their complexed forms in the cell to catalyze the decomposition of H_2O_2. In a biological milieu, the exact identification and chemical nature of an iron pool to catalyze reactions 1.3-1.4 remains unclear. Halliwell and Gutteridge (1990) have extensively and critically evaluated the role of several proteins and hemoproteins and the inherent mechanisms leading to Fe release and its subsequent participation in HO⋅ production.

Regardless of the exact nature of this pool in in vivo situations, transition metal complexes may play two roles in biosystems (Goldstein & Czapski, 1986): on the one hand, as protectors against damage and, on the other hand, as sensitizers of toxic effects of $O_2^{\cdot-}$. The expression of one or another role could be accounted for by the kinetic properties of the metal ligands as well as the steady-state concentrations of the required O_2-derived species. The former role is supported by the disproportionation of $O_2^{\cdot-}$ catalyzed by several copper ligands through a ping-pong mechanism analogous to a SOD reaction. The latter role is explained in terms of the production of HO⋅ within a $O_2^{\cdot-}$-driven Fenton cycle as illustrated in reactions 1.3-1.4 above, where HO⋅ is generated [a] in the bulk solution (non site-specific mechanism of damage), thus accounting for a random attack of target(s), or [b] close to the biological target when the metal ion or its complex is bound to the target, which might serve as an effective ligand for, for example, Cu (site-specific mechanism of damage) (Goldstein and Czapski, 1986).

This distinction is biologically relevant, for the production of HO⋅ in the vicinity of DNA may control the type of chemistry observed. HO⋅ is endowed with unique properties: due to a combination of high electrophilicity, high thermochemical reactivity, and a mode of production that can occur near DNA, it can both add to DNA bases and abstract H atoms from the DNA helix (Pryor, 1988). Residues such as histidine in proteins are important binding sites for metals which can localize damage initiated by peroxides on particular functional groups close to the metal-binding amino acid (Stadman, 1990a,b; Davies, 1987), as in the case of inactivation of glutamine synthetase (Rivett and Levine, 1990). Alternatively, other action mechanisms are possible: protein-bound metals can participate in reactions which generate bulk-phase radicals capable of exerting oxidative damage outside the protein (Simpson and Dean, 1990) and the binding of metals to proteins can restrict their availability for free radical generation (Dean, 1991). The occurrence of long-lived, potentially reactive moieties in proteins, such as protein hydroperoxides, might provide a *locus* for Fenton chemistry (Dean, 1991). The overall process of free radical-mediated oxidation of proteins is complex, and it depends on the relative localization of radical generation, the occurrence and localization of antioxidants, and the type of target proteins (Dean et al., 1991).

An alternative route for HO˙ production in biological systems would be from the interaction of nitric oxide with O_2^- (Beckman et al., 1990) (see below).

iv. Peroxyl radicals

In addition to the electron-transfer reactions involving O_2 described above, another type of one-electron transfer that contributes to the formation of oxyradicals involves the quenching of carbon-centered radicals by molecular O_2. This reaction leads to the formation of peroxyl radicals (reaction 1.9; $k_9 \approx 10^9 \ M^{-1}s^{-1}$) and it constitutes, along with the reactivity of peroxyl radicals towards unsaturated fatty acids (reaction 1.10), the *propagation* steps of lipid peroxidation, following the *initiation* step (reaction 1.8).

$$RH \rightarrow R˙ \tag{1.8}$$

$$R˙ + O_2 \rightarrow ROO˙ \tag{1.9}$$

$$ROO˙ + RH \rightarrow ROOH + R˙ \tag{1.10}$$

The termination steps of lipid peroxidation encompass biradical reactions leading to the formation of stable, nonradical products.

$$R˙ + R˙ \rightarrow R{-}R \tag{1.11}$$

$$ROO˙ + R˙ \rightarrow ROOR \tag{1.12}$$

$$ROO˙ + ROO˙ \rightarrow ROOR + O_2 \tag{1.13}$$

The relative contribution of homotermination (reactions 1.11 and 1.13) and cross-termination (reaction 1.12) reactions to the termination process is strongly influenced by the concentration of O_2: at low O_2 pressures ([R˙] > [ROO˙]), equation 1.11 will be the most important termination step, whereas at high O_2 pressure ([ROO˙] > [R˙]), equation 1.13 will prevail (see Pryor, 1976). The half-life of peroxyl radicals at 37°C (with linoleate as substrate) has been estimated to be about 7 s, whereas that of the far more reactive alkoxyl radical (RO˙) is about 10^{-6} s (Pryor, 1986).

The generation of electrophilic peroxyl radicals could be accomplished by reactions other than the propagation step in equation 1.9, as it occurs in a variety of experimental models such as the lipoxygenase reaction, the conversion of linoleate hydroperoxide to different products catalyzed by hematin, and organic hydroperoxide (ROOH) hematin mixtures. The metal-ion and lipoxygenase-catalyzed breakdown of peroxidized fatty acids produces initially alkoxyl radicals (RO˙) by one-electron reductive cleavage of the hydroperoxide (reaction 1.14), and the subsequent reaction of $RO^{\cdot}$ with peroxides yields peroxyl radicals (reaction 1.15) as well as a further species tentatively identified as an acyl-$[RC(O^{\cdot})]$-radical adduct (Davies and Slater, 1987). Peroxyl radicals are also expected to be formed upon H abstraction of a hydroperoxide by an oxoferryl com-

$$ROOH + Fe^{II} \rightarrow RO^{\cdot} + Fe^{III} + HO^{-} \tag{1.14}$$

$$ROOH + RO^{\cdot} \rightarrow ROO^{\cdot} + ROH \tag{1.15}$$

plex ($Fe^{IV}{=}O$) (see below) (reaction 1.16) (Dix et al., 1985). HO_2 is capable of, for example, abstracting a *bis*-allylic hydrogen atom from a PUFA, thus initiating lipid autoxidation (Gebicki and Bielski,

$$ROOH + Fe^{IV}{=}O \rightarrow ROO^{\cdot} + Fe^{III} + HO^{-} \tag{1.16}$$

1981); H abstraction by HO_2 proceeds more readily with LOOHs (reaction 1.17), probably because the hydroperoxide moiety is a kinetically preferred reaction site (Aikens and Dix, 1991).

$$ROOH + HO_2^{\cdot} \rightarrow ROO^{\cdot} + H_2O_2 \tag{1.17}$$

Regardless of the molecular mechanism supporting the formation of $ROO^{\cdot}$ radicals, the propagation step of lipid peroxidation (reaction 1.9) or peroxide breakdown (reactions 1.15–1.17), their participation in recombination reactions does not proceed in the oversimplified manner of reaction 13. Peroxyl radical recombination is known to proceed via a tetroxide intermediate with elimination of a ketone, an alcohol, and O_2 (reaction 1.18).

$$2ROO^{\cdot} \rightarrow [ROOOOR] \rightarrow RO + ROH + O_2 \tag{1.18}$$

Although $ROO^{\cdot}$ radical recombination is considered a termination reaction leading to the formation of nonradical products, these are not necessarily less reactive: the products can be generated in an electronically-excited state which possesses high chemical reactivity such as triplet carbonyl compounds and singlet oxygen. The recombination process (reaction 1.18) is concerted concentrating part of the energy in the ketone fragment resulting in an excited carbonyl of triplet multiplicity ($^3RO^*$), a ground state alcohol, and triplet state ground O_2 (reaction 1.19; (Russell, 1957; Kellogg, 1969). In addition, an efficient quenching of the triplet carbonyl by molecular O_2 eliminated in reaction 19 and retained in the solvent cage could sup-

$$2ROO^{\cdot} \rightarrow [ROOOOR] \rightarrow {}^3RO^* + ROH + O_2 \tag{1.19}$$

port the formation of singlet molecular oxygen ($^1O_2$; see below) (reaction 1.20). The rate of triplet-triplet transfer quenching in the solvent cage has been estimated to be $\sim 10^{11}$ s^{-1}. The generation of $^1O_2$ originating from the combination of reactions 1.19 and 1.20 is usu-

$$[^3RO^* + {}^3O_2]_{cage} \rightarrow RO + {}^1O_2 \qquad (1.20)$$

ally written as in reaction 1.21. Alternatively, the tetroxide in reaction 1.19 can decompose to give RO˙ radicals and O_2 in the solvent

$$2ROO˙ \rightarrow [ROOOOR] \rightarrow RO + ROH + {}^1O_2 \qquad (1.21)$$

cage; the former can recombine to yield ROOR or initiate further oxidation (see Pryor, 1976).

v. The oxoferryl complex in myoglobin and hemoglobin

The inclusion of this section serves following purposes: first, to dispel ambiguities when addressing the different pathways by which the interaction between H_2O_2 and hemoglobin or myoglobin may cause cellular damage. In general, this issue is addressed without regard for the exact chemical nature and specific chemistry attainable by the electrophilic centers inherent in the high oxidation state of these hemoproteins resulting from the above interaction. Second, to evaluate the potential formation of HO˙, which might hypothetically originate from this interaction. Third, to provide an overview of the actual chemical reactivity of this high oxidation state of the hemoproteins and of its potential implication in particular pathophysiological processes.

The oxidation of metmyoglobin by H_2O_2 is currently explained in terms of a heterolytic cleavage of the O-O bond of the coordinated peroxide yielding a two-electron oxidation product of the hemoprotein, known as ferrylmyoglobin and described over 40 years ago (George and Irvine, 1952). The first oxidation equivalent provided by this reaction is retained in the form of the oxoferryl complex, $Fe^{IV}=O$, similar to compound II of peroxidases (George and Irvine, 1952), and the second oxidation equivalent is detected as a transient protein radical. The heterolytic cleavage of the peroxide encompassing the overall two-electron oxidation of myoglobin by H_2O_2 could be written as in reaction 1.22 (where HX— stands for an aromatic amino acid radical and —Fe for the heme iron).

$$HX-Fe^{III} + H_2O_2 \rightarrow {}˙X-Fe^{IV}=O + H_2O \qquad (1.22)$$

Although a common mechanism for the two-electron oxidation of hemoproteins was proposed to involve primarily an oxo-Fe^{IV} porphyrin π cation radical (Dolphin, 1985), this species was not confirmed experimentally during the oxidation of myoglobin or hemoglobin. A recent study with native and recombinant sperm whale myoglobins indicated that this process involved twice as much heterolytic as homolytic scission of the peroxide bond (Allentoff et al., 1992).

Reaction 1.22 indicates that there are two electrophilic centers in ferrylmyoglobin, the oxoferryl moiety and the aromatic amino acid radical. The specific chemistry attainable by these centers is different, and perhaps the most distinctive feature is their lifetime: [a] the oxoferryl complex is a long-lived strong oxidant (with a redox potential of about +0.99 V; Koppenol and Liebman, 1984), which decays slowly to metmyoglobin; this process is termed autoreduction (reaction 1.23; Uyeda and Peisach, 1981). The stability of the oxo-

$$Fe^{IV}=O \rightarrow Fe^{III} \qquad (1.23)$$

ferryl complex decreases in the presence of an excess of H_2O_2, and is dependent on the specific, species-dependent amino acid composition of the protein moiety (Uyeda and Peisach, 1981). The latter aspect suggests a critical role for specific tyrosyl residues in intra- or intermolecular reduction of the hypervalent heme iron. [b] The lifetime of the amino acid radical ranges between 50 and 280 ms (Miki et al., 1989). Another salient feature of the amino acid radical is that its yield ranges between 8 and 16%, and that the pathways for its dissipation remain largely unknown.

Although the oxoferryl complex and the amino acid radical occur simultaneously in the high oxidation state of the hemoprotein for a short period, they are endowed with different chemical reactivities. That of the oxoferryl complex in ferrylmyoglobin and ferrylhemoglobin has been amply described in terms of oxidation of various cell components and oxidation and cooxidation of drugs (Table 1.1). It could be speculated that this high oxidation state is a requirement for the formation of spectrin-hemoglobin complexes (Snyder et al., 1988). A more detailed reference to the wide spectrum of biological molecules (with emphasis on antioxidants) reacting with the high oxidation state of these hemoproteins is described in another section.

The ability of the oxoferryl complex to initiate lipid peroxidation is well documented (Table 1.1). Of biological interest is the potential formation of ferrylmyoglobin upon oxidation of metmyoglobin by LOOHs instead of H_2O_2. That this reaction can occur is supported by the oxidation of virtually all myoglobin in isolated myocytes by 15-hydroperoxy-5,8,11,13,-eicosatetraenoic acid (Walters et al., 1983)

Table 1.1
Reactivity of the Oxoferryl Complex Towards Cell Constituents and Drugs

CELL CONSTITUENTS	
Fatty acids	Kanner and Harel, 1985a
	Grisham, 1985
	Kanner et al. 1987
	Galaris et al. 1990
	Yamada et al. 1991
Cholesterol	Galaris et al. 1988
Lipoproteins	Bruckdorfer et al. 1990a,b
	Paganga et al. 1992
	Rice-Evans et al. 1993
Nitric Oxide	Dee et al. 1991
	Kanner et al. 1991
3-Hydroxykynurenine	Ishii et al. 1992
DRUGS AND MISCELLANEOUS COMPOUNDS	
Quinones and quinone thioethers	Buffinton and Cadenas, 1988
	Buffinton et al. 1988
N-Methylcarbazole	Kedderis et al. 1986
Sulfasalazine	Yamada et al. 1991
Dihydroriboflavin	Xu and Hultquist, 1991
Chlorpromazine	Kelder et al. 1989, 1991a,b
Mercaptopropionylglycine	Puppo et al. 1990
Ethanol	Harada et al. 1986
Desferrioxamine	Kanner and Harel, 1987 and Rice-Evans et al. 1989
Pyrroloquinoline quinol	Xu et al. 1993

as well as its oxidation by low density lipoprotein (LDL) containing LOOHs (Rice-Evans, 1993), a reaction which is also associated with iron release from the activated myoglobin (Rice-Evans et al., 1993). However, the oxidation of myoglobin by LOOHs cannot be generalized because it seems to depend on the physico-chemical properties of the lipids: although metmyoglobin rapidly peroxidizes palmitoyl-linoleyl-phosphatidyl choline large unilamellar vesicles containing 3% lipid peroxide and decomposes efficiently palmitoyl-linoleylhydroperoxide-phosphatidyl choline to an array of molecular products (mainly carbonyl compounds with the conjugated diene moiety), these reactions are not associated with heme iron valence changes of the hemoprotein (Maiorino et al., 1994). It is clear that a mechanism other than the activation of myoglobin to ferrylmyoglobin by lipid peroxides need be summoned to account for these latter observations. Heme iron valence changes encompassing the $Fe^{III} \leftrightarrow Fe^{II}$ transition (reactions 24–25; here ROOH = LOOH) (Davies and Slater, 1987) seem also unlikely because of the lack of absorption spectral evidence; it could be hypothesized, however, that these changes cannot be evidenced because they are so small, i.e., an oxidation state of the hemoprotein acting as an initiator of peroxidation, rather than in a catalytic fashion.

$$ROOH + Mb—Fe^{III} \rightarrow ROO^{\cdot} + H^+ + Mb—Fe^{II} \qquad (1.24)$$

$$Mb—Fe^{II} + ROOH \rightarrow Mb^{III}—Fe^{II} + HO^- + RO^{\cdot} \qquad (1.25)$$

There is ample evidence concerning the formation of a protein radical in myoglobin observed by electron spin resonance (ESR) spectroscopy (Gibson et al., 1958; King and Winfield, 1963; Harada and Yamazaki, 1987; Miki et al., 1989; Davies, 1991; Davies and Puppo, 1992) or in conjunction with the spin trap 5,5-dimethylpyrrolidone-(2)-oxyl-(1) (DMPO) (Davies, 1990; Kelman & Mason, 1992)—which decays after the addition of H_2O_2 by a poorly understood mechanism. Phenylalanine and tyrosine residues have been considered the primary loci of the protein free radical density. Three tyrosine residues (Tyr_{103}, Tyr_{146}, and Tyr_{151}) are present in sperm whale myoglobin, two residues (Tyr_{103} and Tyr_{146}) in horse heart myoglobin, and one (Tyr_{146}) in kangaroo myoglobin. Studies carried out with recombinant sperm whale myoglobins showed that all the proteins, including those devoid of tyrosine residues, reacted with H_2O_2 to yield an oxoferryl complex and a protein radical. This suggests that the radical character—probably centered on an imidazole or tryptophan residue in tyrosine-lacking myoglobins—was rapidly transferred from one amino acid to another (Wilks and Ortiz de Montellano, 1992).

Most of our understanding on the specific chemistry attainable by the amino acid radical in ferrylmyoglobin or ferrylhemoglobin originates from the work of Ortiz de Montellano and his collaborators. They have characterized three different pathways involving dissipation of the apoprotein radical character: [a] diradical cross-linking of Tyr_{103} of one chain of sperm whale myoglobin to Tyr_{151} of another leading to myoglobin dimer formation (Tew and Ortiz de Montellano, 1988). However, it was recently shown that Tyr_{103} was not essential for cross-linking (Wilks and Ortiz de Montellano, 1992); [b] covalent binding of the heme group to the protein (Rice et al., 1983; Catalano et al., 1989), and [c] protein radical mediated epoxidation of styrene (Ortiz de Montellano and Catalano, 1985) in a manner that may involve tyrosyl residues, although His_{64} appears to be essential for co-oxidative epoxidation (Rao et al., 1993).

The formation of HO˙ during the interaction of myoglobin (or, by inference, methemoglobin) with H_2O_2 merits further comment. First, Allentoff et al. (1992) compared the contribution of homolytic versus heterolytic cleavage of peroxides by several hemoproteins and concluded, for the particular case of myoglobin, that the interaction of the hemoprotein with H_2O_2 entails twice as much heterolytic as homolytic breakdown of the peroxide. On this basis, it could be assumed that a fraction of the peroxide might be decomposed to HO˙. However, there is no ESR evidence for the occurrence of HO˙ during the interaction of myoglobin with H_2O_2; of course, it could be argued that several situations prevented the trapping of HO˙ in this system: the HO˙ formed reacts at a diffusion controlled rate with a neighboring amino acid yielding the characterized aromatic amino acid radical and that the spin trap used, generally DMPO, has no access to a HO˙ formed in the vicinity of the heme crevice. Identification of a DMPO-HO˙ adduct in such a system is further complicated by the observation that ferrylmyoglobin reacts with the above adduct to give non-paramagnetic products (Yang et al., 1993). Conversely, when HO˙ and/or RO˙ radicals are detected by ESR in conjunction with spin traps other than DMPO, the signal is attributed to iron released from metmyoglobin (Mehlhorn and Gomez, 1993).

vi. Nitric oxide

Endothelial cells produce a short lived relaxing factor (endothelium-derived relaxing factor; EDRF), a vasodilator originating from the enzymic conversion of L-arginine to L-citrulline, and which has been identified as nitric oxide (NO). The importance of this subject has produced an explosion of research supporting the numerous functions of NO in physiological and pathological situations (see Moncada et al., 1988, 1991, 1992 and references therein). NO (˙N—O) is paramagnetic, has a half-life of 1–10s, and may decay by several pathways, the biochemical significance of which needs to be evaluated in light of the complexity of the experimental model. The rapid exothermic reaction of NO with molecular O_2 yields nitrogen dioxide (NO_2, a radical itself) (reaction 1.26) (Moncada et al., 1988), whereas that with O_2^- yields peroxynitrite (reaction 1.27; $k_{27} = 5.6 \times 10^7 \, M^{-1}s^{-1}$ at 37°C) (Blough and Zafiriou, 1985; Saran et al., 1990).

Reaction 1.27 is expected to compete efficiently with the disproportionation of O_2^- (reaction 2; $k_2 = 5 \times 10^5$ $M^{-1}s^{-1}$).

$$NO + NO + O_2 \rightarrow 2\,NO_2 \tag{1.26}$$

$$NO + O_2^- \rightarrow ONOO^- \tag{1.27}$$

The significance of reaction 1.27 purporting the reactivity of O_2^-
towards NO and, hence, its inactivation, is strengthened by the fact
that SOD has been consistently reported to stabilize EDRF. Conversely, processes which generate O_2^-, such as hydroquinone, pyrogallol oxidation, and metal complex autoxidation, are linked to NO
inactivation (Griffith et al., 1984; Hutchinson et al., 1987) probably
via reaction 27 and in a SOD-sensitive manner. A mechanism for
NO stabilization by SOD other than the scavenging of O_2^- by the
enzyme entails a reversible reduction of NO to NO^- (Murphy and
Sies, 1991) as in the electron transfer depicted in reaction 1.28.

$$SOD\text{-}Cu^I + NO \leftrightarrow SOD\text{---}Cu^{II} + NO^- \tag{1.28}$$

The products of reactions 1.26 and 1.27 above are further decomposed: NO_2 disproportionates to nitrate (NO_3^-) and nitrite (NO_2^-) in
aqueous solutions (reaction 1.29), whereas $ONOO^-$ is converted to
NO_3^- (reaction 1.30) or, as recently proposed, can be homolytically
cleaved to NO_2 and $HO^.$ (reaction 1.31) (Beckman et al., 1990). The
radical yield during the cleavage of ONOOH (reaction 1.31), however, requires consideration of the extent to which $HO^.$ and NO_2
escape the solvent cage rather than undergoing isomerization to
NO_3^- within the cage; an estimate of 40% free radical yield for this
reaction has been recently reported (Yang et al., 1992).

$$2\,NO_2 + H_2O \rightarrow NO_3^- + NO_2^- + 2H^+ \tag{1.29}$$

$$ONOO^- \rightarrow NO_3^- \tag{1.30}$$

$$ONOO^- + H^+ \leftrightarrow ONOOH \rightarrow [NO_2 + HO^.]_{cage} \rightarrow NO_2 + HO^. \tag{1.31}$$

The chemistry implied in reactions 1.27 and 1.31 has further implications: on the one hand, NO could be viewed as an antioxidant,
inasmuch as its ability to react with O_2^- may be associated with a
protection against the toxic effects of this species. On the other hand,
the decay of $ONOO^-$ via reaction 1.31 leads to $HO^.$ (Beckman et al.,
1990), the formation of which does not require transition metals (Hogg
et al., 1992). Based on thermodynamic considerations, however, it
was concluded that $HO^.$ was not involved in a process such as that
illustrated in reaction 1.31, and that an intermediate related to the
transition state for isomerization of ONOOH might be the oxidizing

intermediate responsible for hydroxyl-radical like oxidations mediated by ONOOH (Koppenol et al., 1992). Conversely, an ESR spin trapping study demonstrated the formation of HO˙ during the proton-catalyzed decomposition of $ONOO^-$ and, also, that the reaction of the latter with either glutathione or cysteine resulted in the formation of thiyl radicals (Augusto et al., 1994). The chemistry entailed in reaction 1.31 has obvious cytotoxic implications: the simultaneous production of NO and O_2^- by the sydnonimine, SIN-1, initiates peroxidation of LDL and also converts the lipoprotein to a more negatively charged form (Darley-Usmar et al., 1992). A mechanism which is likely to involve HO˙ could account for the cytotoxic features of NO in terms of sulfhydryl group oxidation (Radi et al., 1991a) and membrane lipid peroxidation (Radi et al., 1991b). It was also proposed that this reaction might play a central role in microvascular injury during ischemia-reperfusion (Beckman et al., 1990).

An antioxidant function for nitric oxide has been described, inasmuch as [a] NO facilitated the ferrylmyoglobin → metmyoglobin transition (the extent of this reaction being dependent on the relative concentration of both H_2O_2 and NO) (Dee et al., 1991), and [b] the treatment of the NO-myoglobin complex with H_2O_2 did not yield the expected oxoferryl complex, but metmyoglobin (Kanner et al., 1991).

It is not possible, within the scope of this overview, to survey the chemistry of NO, its countless pathophysiological implications, as well as the numerous lines of evidence which support its identification with EDRF. The reader is referred to a collection of manuscripts Moncada et al., (1992) and some recent review articles (Moncada et al. 1991; Nathan, 1992).

1.2.2   Energy Transfer to O_2: Formation of Singlet Oxygen

In addition to electron transfer, an alternative way of increasing the reactivity of O_2 is to move one of the unpaired electrons in a way that alleviates the spin restriction. Transfer of excitation energy from a triplet molecule to molecular O_2 produces an electronically excited singlet state of oxygen ($^1\Delta g$; $^1O_2$). Singlet oxygen has no unpaired electrons, and its lower energy state ($^1\Delta g = 22$ kcal·mol^{-1}) is relatively long lived (about 10^{-6}s in H_2O) and responsible for most of the $^1O_2$ chemistry in solution: at variance with the triplet ground state of molecular O_2, it can react both as a nucleophile and as an electrophile (Symons, 1985).

The five types of reactions of $^1O_2$ include: [1] ene reactions with olefins, [2] 1,2 addition to electron-rich olefins, [3] 1,4 addition to dienes and heterocycles, [4] oxidation of sulfides to sulfoxides, and [5] photooxidation of phenols (Foote, 1976). These different reaction types account for the reactivity of $^1O_2$ towards biological targets, a process viewed in terms of photooxidation of lipids, amino acids, nucleotides, and other bioconstituents.

The first type of reaction is biologically important, for $^1O_2$ reacts with unsaturated fatty acids shifting the bond to the allylic position and producing the corresponding fatty acid hydroperoxide (reaction 1.32).

$$R-CH=CH-CH_2-R' + {}^1O_2 \rightarrow R-\underset{\underset{OOH}{|}}{CH}-CH=CH-R' \qquad (1.32)$$

The second type of $^1O_2$ reaction, its 1,2 addition to electron-rich olefins lacking α-hydrogens or rigid olefins to yield dioxetanes (reaction 1.33), has been extensively quoted in connection with its subsequent breakdown to triplet carbonyls (reaction 1.34) and the contribution of the latter to the ultraweak chemiluminescence observed during lipid peroxidation. However, a recent study showed that

$$\diagdown\!\!=\!\!\diagup + {}^1O_2 \rightarrow \underset{O-O}{\square} \qquad (1.33)$$

$$\underset{O-O}{\square} \rightarrow {}^3\left[\diagdown\!\!=\!\!O\right]^* + \diagdown\!\!=\!\!O \qquad (1.34)$$

this mechanism is not associated with lipid peroxidation-supported formation of electronically excited states (Di Mascio et al., 1992).

The fourth and fifth type of $^1O_2$ reactions could be exemplified with the oxidation of methionine to the corresponding sulfoxide (reaction 1.35), and the oxidation of tyrosine could be considered as an example of the reaction of $^1O_2$ with phenols, although this reaction proceeds very slowly (see Foote, 1976).

$$CH_3-S-R + {}^1O_2 \rightarrow CH_3-\underset{\underset{O^-}{|}}{S^+}-R \qquad (1.35)$$

One of the characteristics of $^1O_2$, along with the triplet excited carbonyls referred to above (reactions 19–20), is its radiative decay to the ground state. This process is termed ultraweak or low-level chemiluminescence, and it implies solely the relaxation of an electronically excited state (P*) to the ground state with emission of a photon [A + B → P* → P + $h\nu$], without reference to a particular excited species in a system unless it is aided by spectral analysis and some other chemical approaches. Low-level chemiluminescence has been observed in connection with certain oxidative experimental models, thereby mimicking biological situations, and has been applied to perfused and exposed in situ organs as a non-invasive technique monitoring continuously oxidative reactions (Cadenas, 1984; Cadenas et al., 1984).

The contribution of excited triplet carbonyls to chemiluminescence during lipid peroxidation has been documented as a weak direct emission (reaction 1.36) or enhanced in the presence of specific fluorescers (Cadenas, 1984; Durán and Cadenas, 1987), although the chemical mechanisms underlying their formation are still an open question (Di Mascio et al., 1992). Alternatively, the contribution of

$$^3RO^* \rightarrow RO + h\nu_{450-550 \text{ nm}} \qquad (1.36)$$

$^1O_2$ dimol emission (reaction 1.37) to ultraweak chemiluminescence during lipid peroxidation is more controversial due to the requirement of high $^1O_2$ steady-state concentrations for this bimolecular collision to occur and more difficult to assess due to the low efficiency of this reaction and possible overlap of other uncharacterized emissions. However, unambiguous evidence for $^1O_2$ formation has been furnished by monitoring its monomol emission (reaction

$$2\ ^1O_2 \rightarrow 2\ ^3O_2 + h\nu_{633 \text{ and } 703 \text{ nm}} \qquad (1.37)$$

1.38) during halide-dependent, H_2O_2 decomposition by myelo-, chloro-, and lactoperoxidase/halide mixtures (Khan, 1984; Kanofsky, 1984). A mechanism similar to that expressed in reaction 1.21

$$^1O_2 \rightarrow\ ^3O_2 + h\nu_{1268 \text{ nm}} \qquad (1.38)$$

above entailing peroxyl radical recombination could be summoned to account for the monomol emission observed during the lipoxygenase-catalyzed oxidation of linoleate (Kanofsky and Axelrod, 1986).

1.3   BIOLOGICAL SOURCES OF FREE RADICALS

In principle, two species need be considered when addressing the biological sources of free oxyradicals: O_2^- and H_2O_2. This seems a conclusion derived from the brief chemistry of O_2 radicals outlined above. Although $O_2^{\cdot-}$ and H_2O_2 display a modest chemical reactivity, they provide the ingredients—via different redox reactions—for the formation of more reactive species.

$O_2^{\cdot-}$ seems to be produced in all aerobically metabolizing cells. $O_2^{\cdot-}$ is not, though, a major product of O_2 reduction, as judged by the tetravalent reduction of O_2 to H_2O by cytochrome oxidase and other H_2O-producing oxidases. The production of $O_2^{\cdot-}$ in biosystems does constitute a minor pathway, and an upper limit of about 5% could be set for the fraction of total O_2 reduced to $O_2^{\cdot-}$ (Chance et al., 1979; Boveris and Cadenas, 1982; Naqui et al., 1986). The steady-state concentration of H_2O_2 in liver—originating from various sources—was estimated about $10^{-7}-10^{-9}$ M and that of $O_2^{\cdot-}$ about 10^{-11} M (Chance et al., 1979). This figure for H_2O_2 steady-state concentration is expected to be higher in organs with less effective mechanisms than liver for disposal of H_2O_2. Thus, in most normal cells the ratio of the steady-state concentrations of H_2O_2 and $O_2^{\cdot-}$ is represented by $[H_2O_2]/[O_2^{\cdot-}] \approx 10^3$ (Boveris and Cadenas, 1982).

Several reactions in biological systems contribute to maintain the steady-state concentrations of H_2O_2 and $O_2^{\cdot-}$ referred to above. The electron-transfer chain of mitochondria is a well-documented source of H_2O_2 from disproportionating $O_2^{\cdot-}$, and several components of complex I, II, and III exhibit thermodynamic properties appropriate for the reduction of O_2 to $O_2^{\cdot-}$ (Cadenas et al., 1977; Boveris and Cadenas, 1982; Forman and Boveris, 1982). A $O_2^{\cdot-}$ generator in heart mitochondria, which is not involved in energy-linked respiration, has been described (Nohl, 1987): it requires NADH to initiate its generation and releases $O_2^{\cdot-}$ into the extramitochondrial space. This mitochondrial generator of $O_2^{\cdot-}$ seems to be organ-specific, for it is absent in liver mitochondria (Nohl, 1987). This activity in heart mitochondria might be important for the one-electron activation of quinones, such as adriamycin, thus explaining the selective cardiotoxicity of this antitumor drug.

In addition to the mitochondrial electron transfer chain, other cellular sources of $O_2^{\cdot-}$ and/or H_2O_2 in mammalian cells include the microsomal transfer chain (probably entailing a slow electron-transfer to O_2 via NADPH-cytochrome P_{450} and NADH-cytochrome b_5 reductase activities) and the NADP(H) oxidases in phagocytes (neutrophils and macrophages): one oxidase (in phagosomal membranes) prefers NADPH and produces $O_2^{\cdot-}$ and the other is highly specific for NADH and produces both $O_2^{\cdot-}$ and H_2O_2 (Forman and Thomas, 1986). Oxygen radicals formed during the respiratory burst appear

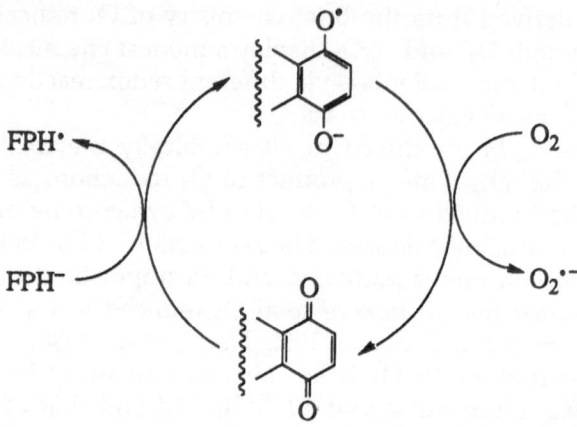

Fig. 2.2   One-electron redox cycling of quinonoid compounds

to originate mainly from the NADPH-dependent enzyme. Xanthine oxidase, aldehyde oxidase, and tryptophan dioxygenase are additional sources of O_2^-, the first enzyme being implicated in the extensively addressed ischemia-reperfusion model proposed by McCord (see for a recent review Omar et al., 1991). Two-electron reduction of O_2 to H_2O_2 occurs during the course of reactions catalyzed by a number of oxidases, such as monoamine oxidase, urate oxidase, acyl-CoA oxidase, and L-gulonolactone oxidase.

It is difficult to assess the extent of the contribution of environmental factors and autoxidation reactions (involving cellular components) to the cellular steady-state concentration of O_2^- and H_2O_2 referred to above. Certainly, the continuous exposure to toxic chemicals and radiation should contribute to this steady state and, hence, the above values estimated from well-characterized biological sources and reaction rate constants (Chance et al., 1979; Boveris and Cadenas, 1982)-might well represent an underestimation. Autoxidation reactions are clearly important and it is well established that O_2^- is formed when electronegative compounds intercept electrons from normal cellular electron transport and then reduce O_2, a process termed redox cycling (Borg and Schaich, 1984) encompassing a two-step redox cycle, which is usually described in the activation of certain anticancer and antiparasitic drugs (Fig. 1.2).

Hydroquinone formation, catalyzed by a unique two-electron transfer flavoprotein, DT-diaphorase (Cadenas et al., 1992), can also support a redox cycling process, contrary to the established dogma that hydroquinones are redox stable compounds in comparison with

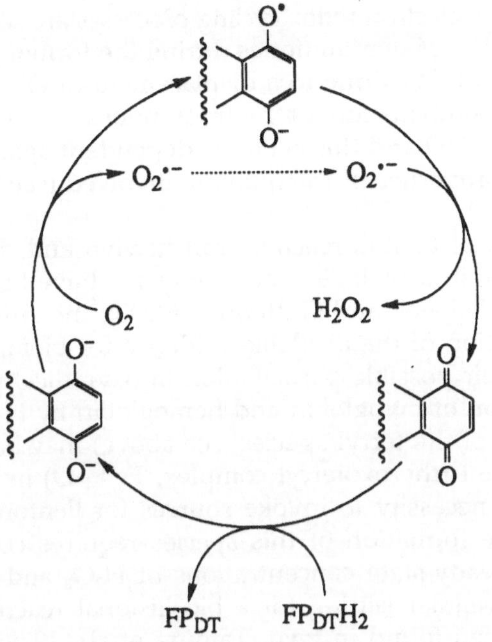

Fig. 3.3 Redox cycling of quinonoid compounds during their two-electron activation by DT-diaphorase

semiquinones. The redox stability of hydroquinones is an expression of their functional group chemistry. Redox cycling following the initial two-electron reduction by DT-diaphorase can be envisaged as a process where O_2^- is the propagating species in the hydroquinone autoxidation chain. The hydroquinone form of unsubstituted and methyl-substituted naphthoquinones is redox stable; however, slight modifications in their substitution pattern, such as the presence of a glutathionyl substituent (quinone-thioether derivatives are expected products in biological systems) or a methoxyl substituent, substantially enhance hydroquinone oxidation by a mechanism as that illustrated in Fig. 1.3 (Buffinton et al., 1989). In addition, semiquinone formation is not a feature only inherent in one-electron transfer activation of quinones: the semiquinone form of the anticancer quinone diaziquone is observed during its activation by DT-diaphorase, observation explained in terms a prevalence of comproportionation ($QH_2 + Q \rightarrow {}^2Q^{--}$) reactions and reduction of the quinone by O_2^- ($Q + O_2^- \rightarrow Q^{--} + O_2$) (Ordoñez and Cadenas, 1992).

One- and two-electron redox cycling processes are sensitive to SOD. The autoxidation of semiquinones during the former process is usually increased by SOD due to a displacement of $O_2^{\cdot-}$ from the equilibrium. The autoxidation of hydroquinones can be inhibited or stimulated by SOD and this action is dependent again on the physico-chemical properties of the quinone involved (see Öllinger et al., 1989).

The sources of Fenton-reactive iron in vivo and, hence, the formation of HO˙ in a biological environment, have been extensively reviewed by Halliwell and Gutteridge (1990); the authors provide a critical evaluation of the available evidence for biological iron complexes and their possible participation in oxyradical production.

The oxidation of myoglobin and hemoglobin by H_2O_2 (leading to the formation of the ferryl species; see above) may be the source of a strong oxidant (the oxoferryl complex, $Fe^{IV}=O$) in biological systems without necessity to invoke sources for Fenton-reactive iron. Of course, the formation of this species requires consideration of the relative steady-state concentrations of H_2O_2 and the hemoprotein and the cellular setting for a bicollisional reaction. Ferrylmyoglobin has been found in vivo (Tamura et al., 1978; Walter et al., 1983) and visualized by reflectance spectroscopy in rat diaphragm (Eddy et al., 1990) and the isolated ischemic rat heart (Arduini et al., 1990a) after its derivatization with Na_2S to form sulfmyoglobin. The latter approach permitted the observation of this species in myocytes supplemented with 15-hydroperoxy-5,8,11,13-eicosatetraenoic acid (Walters et al., 1983) and of ferrylhemoglobin in erythrocytes (Giulivi & Davies, 1990). Also, ferrylmyoglobin is readily reduced to metmyoglobin by ascorbate in vitro (see Table II below); this, along with the demonstration of the ferryl species in vivo and the reported protective role for ascorbate in induced ischemic arrest associated with cardiopulmonary bypass (Eddy et al., 1990), suggests a new model for ischemia reperfusion, whereby ferrylmyoglobin would play a key role (Galaris et al., 1989a; Turner et al., 1989, 1990). The biological implications for the chemical reactivity of ferrylmyoglobin (partly discussed above) are greater when considering that this species possesses a pseudoperoxidatic activity, whereby a process of cyclic characteristics ensues. This is centered on the $Fe^{IV}=O$ → Fe^{III} transition and involves the net reduction of H_2O_2 and oxidation of bioconstituents (RH in the Fig. 1.4A below). Furthermore, the ability of ferrylmyoglobin to initiate peroxidation of lipids -likely involving H abstraction from the bis-allylic methylene bond in PUFAs along with the reactivity of LOOHs towards metmyoglobin

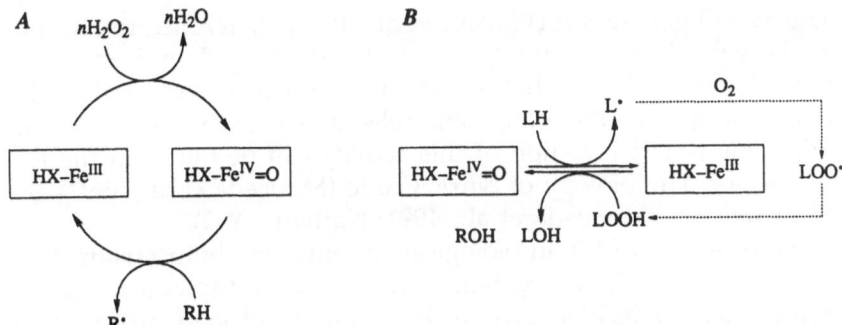

Fig. 4.4 (A) Pseudoperoxidatic activity of myoglobin. RH, parent compound (electron donor); R·, its radical form. (B) Hypothetical role of ferrylmyoglobin in lipid peroxidation. LH, unsaturated fatty acid. L·, fatty acid alkyl radical. LOO·, secondary lipid peroxyl radical. LOOH, lipid hydroperoxide. LOH, lipid hydroxide.

(discussed above) provides another model of cycling characteristics which might have deep biological implications in connection with lipid peroxidation (Fig. 1.4B). This is complicated by the fact that not all LOOHs oxidize the hemoprotein to the ferryl species, but despite the apparent absence of heme iron valence changes, lipid peroxides are decomposed efficiently to a set of products, among them conjugated diene carbonyls (Maiorino et al., 1994).

Conversely, an antioxidant role has been proposed for oxyhemoglobin in red blood cells: this process is dependent on the comproportionation reaction of this species with ferrylhemoglobin to yield the met derivative (reaction 1.39) (Giulivi and Davies, 1990), a reaction also described for myoglobin (Harada et al., 1986).

$$Hb-Fe^{II}-O_2 + Hb-Fe^{IV}=O + 2H^+ \rightarrow 2\,Hb-Fe^{III} + O_2 + H_2O \quad (1.39)$$

Finally, the enzyme activities—usually referred as the NO synthetase enzyme—which catalyze the formation of NO (endothelium-derived relaxing factor) from arginine encompass a complex set of reactions. There are relatively few reports on the mechanism of its biosynthesis and the mechanism of arginine oxidation is incompletely understood. NO synthesis is the result of a five-electron oxidation of one of the guanidium nitrogens on the amino acid arginine yielding as coproduct citrulline. NO synthetase appears to be a cytochrome P_{450} type hemoprotein (White and Marletta, 1992) and a N-hydroxylated derivative of arginine is a biosynthetic intermediate in NO generation (Wallace et al., 1991). Reference to NO as a secretory product of mammalian cells (Nathan, 1992, and references therein) and to NO synthetase activity in macrophages, endothelial cells, neutrophils, adenocarcinoma cells, and neuronal tissues, along with some characterization of this activity can be found in the two volumes on The Biology of Nitric Oxide (Moncada et al., 1992) and recent reviews (Moncada et al., 1991; Nathan, 1992).

The formation of $^1O_2$ in biological systems via photosensitized reactions, i.e., transfer of excitation energy from a triplet sensitizer to ground state molecular oxygen, is a well established process with obvious biological and pathological implications. This area has been extensively covered by Foote (1976). Conversely, the biological generation of $^1O_2$, i.e., involving chemiexcitation processes, such as free radical interactions (for example, those described in reactions 1.18-1.21) and enzymatic reactions, is a more controversial issue. The visible-range and infrared photoemission arising from several types of biological oxidations has been ascribed in some instances to the radiative decay of $^1O_2$ (for a discussion see: Cadenas, 1984; Durán and Cadenas, 1987; Murphy and Sies, 1990; Cadenas et al., 1994). There are two examples of these processes which seem relevant in a biological context:

First, the recombination of peroxyl radicals, as it occurs in the termination reaction of lipid peroxidation, appears to be a source of $^1O_2$. This has often been suggested on the basis of a weak $^1O_2$ dimol emission (reaction 37) and later confirmed by the observation of $^1O_2$ monomol emission (reaction 1.38) (Kanofsky, 1986). A similar mechanism would account for the generation of $^1O_2$ during the soybean lipoxygenase-catalyzed oxidation of linoleate (Kanofsky and Axelrod, 1986).

Second, the disproportionation of peroxides catalyzed by specialized peroxidases appears to be a source of $^1O_2$. $^1O_2$ ensuing from these peroxidase-catalyzed reactions requires an halide (undergoing the Cl^- ↔ HOCl transition) or heme iron (undergoing the $Fe^{IV}=O$ ↔ Fe^{III} transition). Examples of the former are given by $^1O_2$ production (observed as monomol emission) during the halide-dependent, peroxidase-catalyzed decomposition of H_2O_2 (Kanofsky, 1984; Khan, 1984) (the $^1O_2$ yield by myeloperoxidase—present in polymorphonuclear leukocytes—is about 15–20%). The prostaglandin synthase-catalyzed conversion of prostaglandin G_2 to H_2 is an example of the latter, a process associated with $^1O_2$ production (dimol emission) via a reaction involving an intermediate oxoferryl complex (Cadenas et al., 1983). Similarly, the H_2O_2 decomposition during the pseudoperoxidatic activity of myoglobin (entailing the formation of an oxoferryl complex as indicated in Fig. 1.4) is accompanied by co-oxidation of cholesterol to its 5-α-hydroperoxide adduct, which is indicative of $^1O_2$ involvement (Galaris et al., 1988).

1.4 THE REACTIVITY OF FREE RADICALS WITH NON-ENZYMIC SMALL MOLECULAR ANTIOXIDANTS

The current view of cellular oxidant defenses can be categorized into primary and secondary defense systems (Davies, 1986). The primary defenses consist of the broadly studied antioxidant compounds, such as α-tocopherol, ascorbic acid, β-carotene, and uric acid, along with a variety of antioxidant enzymes, where SODs, CATs, and the GPOXs are notable examples. The kinetic properties of small antioxidant molecules are covered in chapter 6, whereas chapter 7 focuses on enzymic antioxidant defenses. Secondary defenses include proteolytic and lipolytic enzymes, as well as the DNA-repair systems and are very briefly described below.

1.5 PRIMARY ANTIOXIDANT DEFENSES

The antioxidant activity of several natural and synthetic compounds is encompassed by a redox transition involving the donation of a single electron (or H atom, equivalent to donation of an electron and a H^+) to a free radical species. During the course of this electron transfer (reaction 1.40), the radical character is transferred to the antioxidant, yielding the antioxidant-derived radical. Thermodynamic consideration of the efficiency of the transfer of the radical character to the antioxidant molecule furnishes information mainly on the equilibria of these reactions, but it is of limited use to assess the actual antioxidant efficiency in a multicomponent system. Wardman (1988) has provided unambiguous evidence that the production of thiyl radicals in biological systems depends as much on kinetic factors as it does on the thermodynamics of individual steps or equilibria. The kinetic control of electron transfers of the type illustrated in reaction 1.40, is central to understanding the basis of antioxidant activity and requires careful consideration of the decay pathways of the antioxidant-derived radical, $A^.$.

$$AH + X \rightarrow A^. + XH \tag{1.40}$$

1.5.1 Vitamin E

Vitamin E (α-tocopherol) is known to react with a variety of free radicals at fairly high rates, yielding the corresponding chromanoxyl radical (αT—$O^.$). The reaction is formally regarded as a H-atom transfer (reaction 1.41), although an electron transfer reaction is favored in nonpolar media, such as lipids and membranes, followed by deprotonation of the antioxidant radical cation in a reaction with H_2O immersed in the lipid phase (Simic, 1991). $HO_2^.$, but not $O_2^{.-}$, reacts with α-tocopherol (αT-OH) and its water-soluble analog, Trolox,

$$\alpha T{-}OH + ROO^. \rightarrow \alpha T{-}O^. + ROOH \tag{1.41}$$

at a rate of $2 \times 10^5 \, M^{-1}s^{-1}$ (see Bielski and Cabelli, 1991); however, the lack of evidence for a phenoxyl radical intermediate, suggests that this reaction involves likely the formation of a hydroperoxyl radical adduct rather than the electron transfer depicted in reaction 41. The tocopheroxyl radical is also produced by H-abstraction from α-T—OH by excited carbonyls, such as triplet state ketones, as shown by ESR spectroscopy in anoxic media (Jore et al., 1991).

Although not a free radical, the oxoferryl complex ($Fe^{IV}{=}O$) in myoglobin and hemoglobin is a strong oxidant: its reactivity towards the water-soluble analog of vitamin E and vitamin E itself encompasses the formation of a chromanoxyl radical and reduction of the high oxidation state of the hemoprotein to metmyoglobin (reaction 1.42), consistent with ESR and absorption spectroscopy evidence (Giulivi et al., 1992; Nakamura and Hayashi, 1992). Table 1.2 lists several antioxidants which have been characterized in their reaction with the oxoferryl species in myoglobin or hemoglobin.

$$\alpha T{-}OH + Fe^{IV}{=}O \rightarrow \alpha T{-}O^. + Fe^{III} + HO^- \tag{1.42}$$

Following decay pathways for the chromanoxyl radical-generated as in reaction 41 above have been described:

[1] Its recovery by ascorbic acid proceeds efficiently (reaction 1.43). Other antioxidants, such as urate, also accomplish this reaction as

$$\alpha T{-}O^. + AH^- \rightarrow \alpha T{-}OH + A^{.-} \tag{1.43}$$

well as high concentrations of serotonine and other hydroxyindole

Table 1.2
Reactivity of the Oxoferryl Complex Towards Antioxidants

Vitamin E	Giulivi et al. 1992
	Nakamura and Hayashi (1992)
Trolox C	Davies, 1990
	Giulivi et al. 1992
	Nakamura and Hayashi (1992)
	Giulivi and Cadenas, 1993a
Ascorbic acid	Kanner and Harel, 1985b
	Galaris et al. 1989a
	Rice-Evans et al. 1989
	Giulini and Cadenas, 1993b
Ubiquinol	Mordente et al. 1993
β-carotene	Kanner and Harel, 1985b
Urate	Kaur and Halliwell, 1990
	Arduini et al. 1992
Dietary Phenols	Laranjinha et al. 1993

derivatives present in the nerve cells. GSH cannot recover the vitamin E chromanoxyl radical and thiyl radicals do not react with Trolox C, the water-soluble analog of α-tocopherol. GSH, however, has been proposed to [a] establish a link between H-atom transfer and electron-transfer reactions in a synergistic fashion (Willson et al., 1985) and, [b] support a reaction mediated by a heat-sensitive factor (a free radical reductase) involved in vitamin E recovery (McCay et al., 1989). There are also indications that the reduction of α-T—O$^{\cdot}$ radicals is linked to mitochondrial and microsomal electron transports (Maguire et al., 1989; Packer et al., 1989).

[2] $O_2^{\cdot-}$ has been shown in a pulse radiolysis study to reduce effectively the Trolox radical (reaction 1.44; $k_{44} = 4.5 \times 10^8 \ M^{-1}s^{-1}$) (Cadenas et al., 1989). The significance of such a reaction in a bio-

$$\alpha T—O^{\cdot} + O_2^{\cdot-} \rightarrow \alpha T—O^- + O_2 \qquad (1.44)$$

logical milieu is not determined and, of course, it will be dependent on the build up of a high steady-state concentration of $O_2^{\cdot-}$ in the vicinity of αT—O$^{\cdot}$.

[3] Sustained antioxidant protection is apparently dependent on the efficiency of the one-electron redox cycle encompassed by reactions 1.41 and 1.43. In the absence of suitable electron donors, however, the αT—O$^{\cdot}$ radical can undergo a second oxidation to a 8a-hydroxytocopherone, a process which proceeds via the formation of

a carbocation intermediate (reaction 1.45) (Marcus and Hawley,

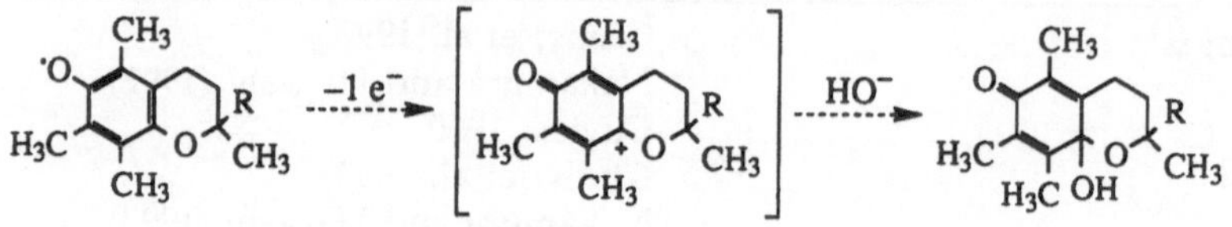

$$(1.45)$$

1970). The 8a-hydroxytocopherone, a two-electron oxidation product of vitamin E, can decay as follows:

[3a] the 8a-hydroxy derivative is unstable and, in the absence of reducing agents, spontaneously decays to tocopherylquinone (reaction 1.46). Of note, the two-electron transfer is already accomplished by reactions 1.41 and 1.45, whereas reaction 1.46 only entails a H^+ or HO^--assisted rearrangement and chromane ring

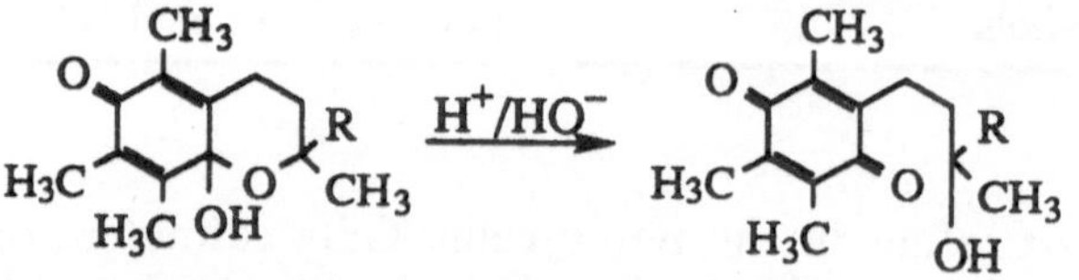

$$(1.46)$$

opening. The relevance of such a mechanism in biological systems is supported by the identification of stable oxidation products of vitamin E, such as tocopherylquinone, in vivo (Csallany et al., 1962) and an increase of the α-tocopherylquinone/α-tocopherol ratio during aortic crossclamping ischemia (Murphy et al., 1992). Tocopheroxyl quinone is also formed upon exposure of microsomal membranes to a flux of NO and O_2^- or upon incubation of tocopherol with NO (de Groot et al., 1993). Tocopherylquinones are encountered in rat liver at a concentration which is 10-12% of that of α-T—OH (Bieri and Tolliver, 1981); however, they are not necessarily end-products, for further oxidations to quinone epoxides and other products can occur.

[3b] Conversely, Liebler et al. (1989, 1990, 1991) have suggested a two-electron redox cycle which might be operative under certain cellular situations: in the presence of high concentrations of ascorbic acid and acidic conditions, 8a-hydroxytocopherone can be efficiently recovered to α-T—OH via the carbocation species. Support for this contention is obtained by the regeneration of vitamin E by ascorbic acid (via a mechanism other than that in reaction 43) during the oxidative metabolism of arachidonate in human platelets (Chan et al., 1991) and the hemoprotein-mediated oxidation of neutrophils

(Ho and Chan, 1992). 8a-Hydroxytocopherone is of critical importance for this reaction to occur, for its rearrangement product, tocopherylquinone (reaction 1.46), cannot be reduced back to α-tocopherol, although it can be reduced to tocopherylhydroquinone by a NADH or NADPH-dependent activity present in hepatocytes (Hayashi et al., 1992). Of note, α-tocopherylquinone is produced upon the reaction of α-tocopherol with O_3 via 8a-hydroxytocopherone, likely involving the rearrangement in reaction 1.46 (Giamalva et al., 1986).

[4] The reaction of α-T—O˙ radicals with themselves leads to the formation of spirodiene dimers and trimers. During the slow autoxidation of methyl-linoleate a dimer is formed transiently and an accumulation of the trimer is observed (Yamauchi et al., 1988).

[5] It was recently proposed (Bowry et al., 1992; Bowry and Stocker, 1993) that tocopheroxyl radicals can propagate peroxidation within lipoprotein particles upon reaction of this radical with PUFA moieties in the lipid, i.e., upon hydrogen abstraction from a bisallylic methylene group of PUFAs (reaction 1.47). The calculated rate constant value for this reaction is $0.1 \pm 0.05 \text{ M}^{-1}\text{s}^{-1}$ and that of reaction 1.48, which might be expected to occur in oxidized LDL is about 10-fold higher (Ingold et al., 1993).

$$\alpha T—O˙ + RH \rightarrow \alpha T\text{-OH} + R˙ \qquad (1.47)$$

$$\alpha T—O˙ + ROOH \rightarrow \alpha T—OH + ROO˙ \qquad (1.48)$$

The fact that tocopheroxyl radicals in LDL act as chain-transfer agents (in the absence of ascorbate and/or ubiquinol-10 ($UQ_{10}H_2$) and involving the sequence of reactions 1.41, 1.47, and 1.9), is apparently in contrast with the role of α-T—OH as chain-breaking, ROO radical-trapping antioxidant in bulk lipids (Burton and Ingold, 1986). This discrepancy can be bridged by considering the different behaviors of tocopheroxyl radicals in LDL and bulk solutions—as logical consequences of autoxidation and antioxidation in small lipid particles dispersed in water—as well as other features inherent in the antioxidant status of LDL particles (Ingold et al., 1993). In the bulk solution, is expected to react with another ROO˙ radical to yield non-radical end products, according to the sequence in reactions 1.45-1.46.

1.5.2   Ascorbic Acid

Perhaps the most salient feature of ascorbic acid (AH_2) in connection with its antioxidant activity is its involvement in the electron transfer-mediated recovery of vitamin E radical (reaction 1.43 above). Ascorbic acid also reacts with $ROO^{\cdot}$, triplet carbonyl compounds (Encinas et al., 1985), $^1O_2$ (Rougee and Bensasson, 1986), $O_2^{\cdot-}/HO_2^{\cdot}$ (see Bielski & Cabelli, 1991), and the oxoferryl complex in ferryl-myoglobin (Table 1.2). The reaction of $O_2^{\cdot-}/HO_2^{\cdot}$ with AH_2 proceeds with a rate constant of $1.2 \times 10^7\ M^{-1}s^{-1}$, which represents a composite of reactions 1.49 and 1.50. Recovery of ferrylmyoglobin by ascorbate (Kanner and Harel, 1985b; Galaris et al., 1989a; Rice-Evans

$$AH_2 + O_2^{\cdot-} \rightarrow A^{\cdot-} + H_2O_2 \qquad (1.49)$$

$$AH^- + HO_2^{\cdot} \rightarrow A^{\cdot-} + H_2O_2 \qquad (1.50)$$

et al., 1989) proceeds efficiently (reaction 1.51) and results in ascorbyl radical ($A^{\cdot-}$) formation (Giulivi & Cadenas, 1993b). Ascorbic acid (or the ascorbyl radical ensuing from reaction 1.51) further reduces metmyoglobin to oxymyoglobin, albeit at slower rates (reaction 1.52) (the reaction is thermodynamically possible: $E_{A/A^{\cdot-}} = -174$ mV; $E_{Mb \cdot Fe^{III},O2/Mb \cdot Fe^{II}O2} = +220$ mV; Koppenol and Butler, 1985).

$$AH^- + Fe^{IV}{=}O \rightarrow A^{\cdot-} + Fe^{III} + HO^- \qquad (1.51)$$
$$AH^- + Fe^{III} + O_2 \rightarrow A^{\cdot-} + Fe^{II}O_2 \qquad (1.52)$$

The reactivity of triplet carbonyls towards AH_2 is similar to that of $RO^{\cdot}$ radicals and they promote H abstraction from the enediol groups of ascorbate ($k_{53} = 1.2 \times 10^9\ M^{-1}s^{-1}$; Encinas et al., 1985). This is an important consideration, for—regardless of mechanistic aspects—triplet carbonyls contribute substantially to the electronically-excited species formed during the recombination of peroxyl radicals in lipid peroxidation.

$$AH^- + {}^3[RO]^* \rightarrow {\cdot}R{-}OH + A^{\cdot-} \qquad (1.53)$$

Finally, ascorbate is involved in thiyl radical repair (see below) (reaction 1.54; $k_{54} = 6 \times 10^8\ M^{-1}\ s^{-1}$) (Forni et al., 1983).

$$AH^- + RS^{\cdot} \rightarrow A^{\cdot-} + H^+ + RS^- \qquad (1.54)$$

1.5.3 Ubiquinol

In addition to its role as redox component of the mitochondrial electron-transport system, ubiquinone may function in its reduced form as an antioxidant (Beyer and Ernster, 1990) and membrane labilizer (Valtersson et al., 1985). Indeed, there is a wealth of information and circumstantial evidence (see Lenaz et al., 1990) regarding the antioxidant properties of ubiquinol, evidence derived from experiments with reconstituted membrane systems, mitochondrial membranes, and low density lipoproteins to observations on intact animals as well as in the clinical setting. Surprisingly, there is very little kinetic data on the mechanistic aspects by which this compound exerts its antioxidant activity and the precise molecular and cellular mechanism(s) underlying the antioxidant function of ubiquinol in various biological membranes remains to be elucidated. Also, ubiquinone has been implicated in the production of free radicals (Cadenas et al., 1977; Boveris and Cadenas, 1982).

It could be speculated that the hydrogen-donating activity by which ubiquinol presumably acts as an antioxidant (Cadenas et al., 1992) resembles the reaction of O_2^- with hydroquinones (reaction 1.55), a facile electron/hydride transfer (Sawyer, Calderwood, et al., 1985),

$$QH_2 + O_2^- \rightarrow Q^{\cdot-} + H_2O_2 \qquad (1.55)$$

which sustains the chain propagation in hydroquinone autoxidation (Öllinger et al., 1989). In addition to the peroxyl radicals suggested in reaction 1.56, it could be expected that $UQ_{10}H_2$ reacts readily with perferryl complexes of the type $ADP-Fe^{III}-O_2^-$, usually involved in lipid peroxidation experimental models (Forsmark et al., 1991).

$$(1.56)$$

The ubisemiquinone ($UQ_{10}^{\cdot-}$) formed in reaction 1.56 can decay by pathways involving disproportionation, autoxidation, as well as its reduction to the hydroquinone by enzymic systems. The latter aspect may be visualized to occur via the cytochrome b_{562} component of the mitochondrial "Q" cycle (reaction 1.57). At least for the case of mitochondrial membranes, it can be hypothesized that

$$UQ_{10}^{\cdot-} + b_{562}^{2+} + 2H^+ \rightarrow UQ_{10}H_2 + b_{562}^{3+} \qquad (1.57)$$

the cycle encompassed by reactions 1.56-1.57 (redox cycling in the classical sense) will provide an efficient antioxidant system. Alternatively, the ubisemiquinone species may autoxidize (reaction 1.58). Semiquinone autoxidation is a well documented process and there is an extensive body of work in the literature providing second

$$UQ_{10}^{\cdot-} + O_2 \rightarrow UQ_{10} + O_2^{\cdot-} \tag{1.58}$$

order rate constant values for the $Q^{\cdot-} + O_2 ;\leftrightarrow Q + O_2^{\cdot-}$ redox transition. For the case of $UQ_{10}^{\cdot-}$ the reaction is a thermodynamically feasible ($E_{UQ/UQ^{\cdot-}} = -230$ mV; $E_{O2/O2^{\cdot-}} = -155$ mV) (Koppenol and Butler, 1985) (the rate constant for $UQ_{10}^{\cdot-}$ autoxidation has been estimated as $1.32\ M^{-1}s^{-1}$; Cadenas et al., 1977).

These relationships are important when addressing the antioxidant status of LDLs, their oxidative modification, and their potential involvement in the early stages of atherosclerotic lesions (Steinberg et al., 1989; Gebicki et al., 1991; Stocker and Frei, 1991). Thus, the high effectivity of $UQ_{10}H_2$ as a chain-breaking antioxidant in LDL has been recently evaluated (Ingold et al., 1993): despite the lack of kinetic information on reaction 1.56, its occurrence in conjunction with reaction 1.58 was claimed to serve to export the radical character from the LDL particle to the aqueous phase. The $O_2^{\cdot-}$ generated during this process (reaction 1.58) was suggested to react with the vitamin E radical (Ingold et al., 1993), a reaction already described with the Trolox radical and which proceeds with a rate constant of $4.5 \times 10^8\ M^{-1}s^{-1}$ (reaction 1.44 above) (Cadenas et al., 1989). That the latter reaction has been studied in aqueous medium does not diminish its importance, for the proposed recovery of tocopheroxyl radical in LDL is expected to occur in the aqueous phase. Whether this reaction actually occurs in physiological conditions remains to be determined.

Viewed under this perspective, the antioxidant efficacy of $UQ_{10}H_2$ in LDL (reactions 1.56 and 1.58) is apparently higher than that of the $\alpha T-OH/AH^-$ couple (reactions 1.41 and 1.43); however, the latter seems to be the major antioxidant system in LDL under physiological conditions (Stocker et al., 1991), for only 50–60% of the LDL particles contain a molecule of $UQ_{10}H_2$ (Mohr et al., 1992).

Studies on the ability of ubiquinol homologs with different isoprenoid chain length to prevent lipid peroxidation in biomembranes indicated that those compounds with long isoprenoid chains were much less efficient in preventing membrane lipid peroxidation than the short-chain homologs (Kagan et al., 1990a). Conversely, in mitochondrial membranes with reconstituted electron-transport, ubiquinols show an opposite order of efficiency, long chain homologs being more active than the short-chain homologs. In addition, the question of a potential synergism between ubiquinol and vitamin E in terms of antioxidant activity has been addressed (Kagan et al.,

1990b) by considering that the effect of ubiquinol may not be due to its direct radical scavenging reactivity, but rather result from its ability to maintain a more efficient recycling of α-T-OH than AH_2 does (reaction 1.59). This, along with the evidence provided by ex-

$$\alpha T{-}O^{\cdot} + UQ_{10}H_2 \rightarrow \alpha T\text{-}OH + UQ_{10}^{\cdot-} \qquad (1.59)$$

periments on lipid peroxidation of mitochondrial membranes before and after extraction of ubiquinone and vitamin E and indicating that $UQ_{10}H_2$ can inhibit peroxidation in the absence of the tocopherol (Forsmark et al., 1992), suggests that the antioxidant activity of $UQ_{10}H_2$ cannot be ascribed to a single process.

1.5.4 Uric Acid

Pulse radiolysis studies provided information on the rate constants for the reactions of uric acid with $HO^{\cdot}$ (7.1×10^9 $M^{-1}s^{-1}$), $CCl_3OO^{\cdot}$ ($3.2 - 7 \times 10^8$ $M^{-1}s^{-1}$), $NO_2^{\cdot}$, and guanyl radicals (1.2×10^9 $M^{-1}s^{-1}$) (Willson et al., 1985; Simic and Jovanovic, 1989). Uric acid is also efficient in the repair of glutathionyl radicals (1.4×10^7 $M^{-1}s^{-1}$) and it was proposed to react with $O_2^{\cdot-}/HO_2^{\cdot}$ (Willson et al., 1985). The reaction of these radicals ($R^{\cdot}$) with the monoanion urate (UH^-) would follow the general electron transfer in reaction 1.60.

$$UH^- + R^{\cdot} \rightarrow U^{\cdot} + RH \qquad (1.60)$$

The loss of an electron from uric acid was assumed to take place at O^8 because this hydroxy group was the strongest acid (reaction 1.61) (Simic and Jovanovic, 1989); alternatively, Maples and Mason (1988) placed the radical character at N^9 or N^7 (reaction 1.62).

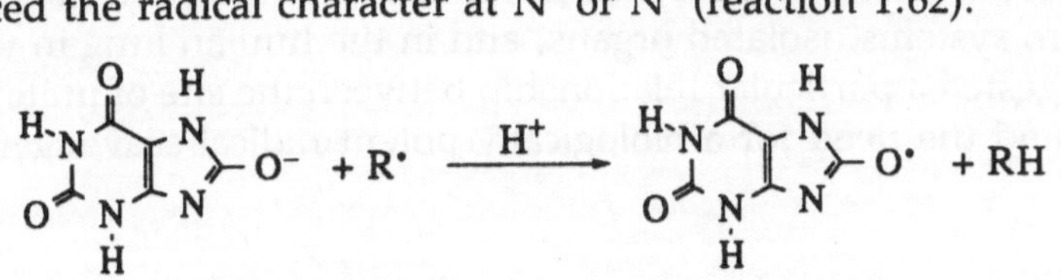

(1.61)

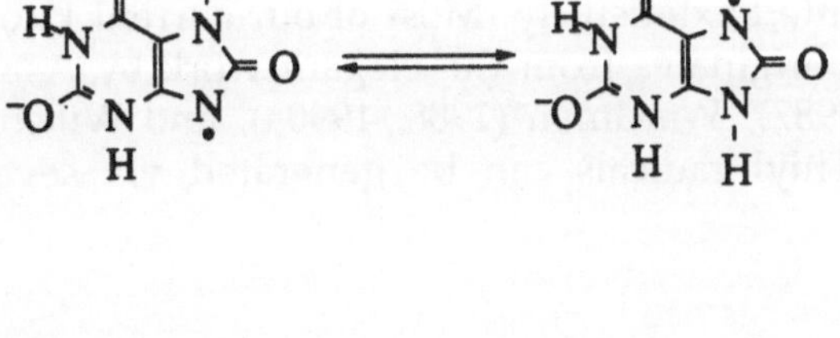

(1.62)

The radical form of uric acid decays to stable inert molecular products, such as allantoin, glyoxylate, oxalate. However, the free radical form(s) of uric acid seems to be quite reactive (although it does not react with O_2; Simic and Jovanovic, 1989) and with a longer life time than that of the radical initially scavenged. This is an issue to be considered when assessing the antioxidant properties of uric acid (Ames et al., 1981). The free radical form(s) of urate or those derived from further free radical processes as yet uncharacterized appear to be responsible for the inhibition of yeast alcohol dehydrogenase (Kittridge and Wilson, 1984) and human α_1-antiproteinase (Aruoma and Halliwell, 1989). The putative antioxidant properties of uric acid could be enhanced by ascorbic acid, for the uric acid radical can be repaired by ascorbate (reaction 1.63; $k_{63} = 10^6\ M^{-1}s^{-1}$) (Simic and Jovanovic, 1989). This reaction is an example of transfer of the radical character to a less reactive species, which is less likely to have toxicological implications in a biological milieu. For example, scavenging of $HO^{\cdot}$ by urate, followed by recovery of the urate radical by ascorbic acid involves the sequential formation of species with decreasing oxidation potential: $[E_{HO^{\cdot}/HO^-} = +2.31\ V] \gg [E_{U^{\cdot}/UH^-} = +0.51\ V] > [E_{A^{\cdot-}/AH^-} = +0.28\ V]$.

$$U^{\cdot} + AH^- \rightarrow UH^- + A^{\cdot-} \tag{1.63}$$

A role of uric acid as an antioxidant in vivo could be inferred from its consumption during exposure of plasma to O_3 (Cross et al., 1992) and $NO_2^{\cdot}$ (Halliwell et al., 1992). Although the subsequent redox reactions of the one-electron oxidation product of uric acid were not evaluated, uric acid reduces ferrylmyoglobin to metmyoglobin (Arduini et al., 1992) and allantoin was identified as the major product during this reaction (Kaur and Halliwell, 1990). Becker (1993) reviewed the evidence for uric acid as a selective antioxidant molecule in in vitro systems, isolated organs, and in the human lung in vivo, and suggested a particular relationship between the site of urate formation and the need for a biologically potent radical scavenger.

1.5.5   Thiols

The reactivity of sulfur-containing compounds towards free radicals is documented extensively. Most of our current knowledge on thiol chemistry originates from the elegant work by Asmus (1990), von Sonntag (1987), Wardman (1988, 1990a), and Willson (Willson et al., 1985). Thiyl radicals can be generated via several routes

(Wardman, 1988), which briefly include the following redox transitions involving either hydrogen or electron transfer: O_2^{-} reacts slowly with thiols (e.g., cysteine; $k_{64} = 15 \ M^{-1}s^{-1}$), whereas, expectedly, its conjugated acid, HO_2^{-} reacts somewhat more rapidly ($k_{65} = 6 \times 10^2 \ M^{-1}s^{-1}$) (Bielski and Shiue, 1979). $HO^{\cdot}$ reacts with GSH and other thiols at diffusion-controlled rates (reaction 1.66) (Buxton et al., 1988). The reaction of other oxygen-centered radicals, such as the 1-naphthoxyl radical ($NPh\text{-}O^{\cdot}$), with GSH proceeds at a slower rate than that of $HO^{\cdot}$; D'Arcy-Doherty et al., (1986) estimated k_{67} value of about $3 \times 10^5 \ M^{-1}s^{-1}$. Although no absolute rate constant values are available, the glutathionyl radical is formed during the oxidation of the thiol by the phenoxyl radical form of acetaminophen (Ross and Moldéus, 1986); likewise, nitrogen-centered radicals, such as arylamino radicals derived from the metabolism of phenacetin, do form $GS^{\cdot}$ upon their interaction with GSH as evidenced by ESR spectroscopy in conjunction with the spin trap

$$RSH + O_2^{-} + H^+ \rightarrow RS^{\cdot} + H_2O_2 \qquad (1.64)$$

$$RSH + HO_2^{\cdot} \rightarrow RS^{\cdot} + H_2O_2 \qquad (1.65)$$

$$RSH + HO^{\cdot} \rightarrow RS^{\cdot} + H_2O \qquad (1.66)$$

$$GSH + NPh\text{—}O^{\cdot} \rightarrow GS^{\cdot} + NPh\text{—}OH \qquad (1.67)$$

DMPO (Ross and Moldéus, 1986). The repair of carbon-centered radicals (*e.g.*, from glucose, $R^{\cdot}$) by GSH in biological systems is expected to proceed rapidly ($k_{68} = 7 \times 10^6 \ M^{-1}s^{-1}$) (Tamba & Quintiliani, 1984).

$$GSH + R^{\cdot} \rightarrow GS^{\cdot} + RH \qquad (1.68)$$

In addition to the radical species described above, the oxoferryl complex ($Fe^{IV}{=}O$) of myoglobin is efficiently reduced by thiols, among them glutathione, cysteine, N-acetylcysteine, to metmyoglobin (Fe^{III}) (Table 1.3). During this electron transfer thiyl radicals are formed as demonstrated via a characteristic ESR signal from the spin trap DMPO (reaction 1.69) (Romero et al., 1992). The second order rate constant values for these reactions vary between 0.1-and $2.2 \times 10^5 \ M^{-1}s^{-1}$, (calculated as a function of $-d[Fe^{IV}{=}O]/dt$) (Romero et al., 1992). The aromatic thiol, ergothioneine, also reduces efficiently ferrylmyoglobin (Arduini et al., 1990b; Akannu et al., 1991); however, evidence for thiyl radical formation could not be furnished by the above approach, for aromatic thiols are highly resonance-stabilized and do not react with DMPO (Romero et al. 1992).

$$Fe^{IV}{=}O + RSH \rightarrow Fe^{III} + HO^- + RS^{\cdot} \qquad (1.69)$$

Table 1.3
Reactivity of the Oxoferryl Complex Towards Thiols and Disulfides

	Redox transition	References
Thiols		
Glutathione	$Fe^{IV}=O \rightarrow Fe^{III}$	Mitsos et al. 1988
		Galaris et al. 1989
		Turner et al. 1991
		Romero et al. 1992
Cysteine	$Fe^{IV}=O \rightarrow Fe^{III}$	Romero et al. 1992
N-Acetylcysteine	$Fe^{IV}=O \rightarrow Fe^{III}$	Romero et al. 1992
		Mitsos et al. 1988
Ergothioneine	$Fe^{IV}=O \rightarrow Fe^{III}$	Arduini et al. 1990b
		Romero et al. 1992
		Akannu et al. 1991
α-Dihydrolipoate	$Fe^{IV}=O \rightarrow Fe^{II} O_2$	Romero et al. 1992
	$Fe^{III} \rightarrow Fe^{II} O_2$	
Disulfides		
α-Lipoate	$Fe^{IV}=O \rightarrow Fe^{III}$	Romero et al. 1992

Reactions 1.64–1.69 above indicate that thiols react with oxygen-, nitrogen-, and carbon-centered radicals as well as with the high oxidation state of myoglobin at different rates. Thiyl radicals (RS·) generated in the course of these hydrogen- or electron-transfer processes decay by several routes. Wardman (1988) considers two pathways which exert a kinetic control of- and are critical to the free radical biochemistry of thiols. These pathways entail conjugation reactions of the thiyl radical with thiolate (RS⁻) or O_2. A requisite condition for the former reaction is the ionization of the thiol (reaction 1.70; a small percent of GSH ($pK_a = 9$) will be present as GS⁻ at physiological pH).

$$RSH \rightarrow RS^- + H^+ \qquad (1.70)$$

Conjugation of thiyl radicals with RS⁻ yields the disulfide anion radical [RSSR·⁻] (reaction 1.71), whereas that with O_2 yields the GSOO· (reaction 1.72). The dimerization of thiyl radicals to GSSG (reaction 1.63) is expected to contribute very little to their removal, for a high steady-state concentration of these radicals is required for a bimolecular collision.

$$GS^· + GS^- \rightarrow [GSSG] \qquad (1.71)$$

$$GS^· + O_2 \rightarrow GSOO^· \qquad (1.72)$$

$$GS^· + GS^· \rightarrow GSSG \qquad (1.73)$$

Of note, the disulfide anion radical formed upon conjugation of glutathionyl radicals with glutathione itself (reaction 1.71) is a very strong reducing agent ($E_{RS^-, RS^-/RSSR^{--}}$ = -1.6 V) (Surdhar and Armstrong, 1986) and its removal may be coupled to the diffusion-controlled electron transfer to O_2 in reaction 1.74 (k_{74} = 0.9×10^9 M^{-1}s^{-1}) (Willson, 1970).

$$[RSSR]^{--} + O_2 \rightarrow RSSR + O_2^{--} \tag{1.74}$$

It is worth noting that the sequence of reactions encompassing, on the one hand, repair of free radicals by GSH (involving the GSH $\rightarrow$ GS$^{\cdot}$ transition) and, on the other hand, removal of the thiyl radical upon its conjugation with thiolate (GS$^{\cdot}$ + GS$^-$ $\rightarrow$ GSSG^{--}) are endowed with different thermodynamic properties: whereas the former reaction yields a powerful oxidant (GS$^{\cdot}$), the latter yields a strong reductant (GSSG^{--}). This concept is central to the theme developed

$$\underset{E \sim +0.7\,V}{GS^- \rightarrow GS^{\cdot}} \xrightarrow[\text{WITH GS}-]{\text{CONJUGATION}} \underset{E \sim -1.6\,V}{GSSG^{--} \rightarrow GSSG} \tag{1.75}$$

by Wardman (1988, 1990a,b), which establishes that the thermodynamic restrictions of reactions involving thiyl free radicals in chemical and biological systems are overcome by kinetic factors. An elegant example is the thermodynamically unfavorable repair of aminopyrine (Wilson et al., 1986) and 1-naphthoxyl radicals (D'Arcy-Doherty et al., 1986) by GSH. In the latter instances, at pH 7, the one-electron redox potential of the naphthoxyl radical/naphthol couple is +560 mV and that of the GS$^{\cdot}$/GS$^-$ couple is +840 mV; thus, despite that the electron transfer reaction 1.67 above is unfavorable, the removal of GS$^{\cdot}$ from the equilibrium via its conjugation with the thiolate (reaction 1.71) and the further removal of GSSG^{--} via reaction 1.74, drives the equilibrium of reaction 1.67 towards the right (D'Arcy-Doherty et al., 1986). The importance of this concept in thiyl radical biochemistry has been subsequently exemplified in the reaction of GSH with alloxan radicals and products of dihydropyrimidines autoxidation (Winterbourn and Munday, 1989, 1990; Winterbourn, 1989) and during the GSH-mediated reductive decay of the semiquinone form of the anticancer compound diaziquone (Ordoñez and Cadenas, 1992). The oxidizing and reducing power of GS$^{\cdot}$ and GSSG^{--} (according to reaction 1.75 above), is illustrated by the efficient oxidation of p-hydrobenzoquinones by the former and reduction of p-benzoquinones by the latter. The k_{76} values ranged between 0.04$-$ and 1.4×10^7 M^{-1}s^{-1} increasing with methyl substitution, whereas k_{77} values ($0.06-1.7 \times 10^9$ $M^{-1}s^{-1}$) decreased with methyl substitution of the p-benzoquinone (Butler and Hoey, 1992). The oxidizing character of thiyl radicals is further substantiated by

$$GS^{\cdot} + QH_2 \rightarrow GSH + Q^{\cdot-} + H^+ \tag{1.76}$$

$$GSSG^{\cdot-} + Q \leftrightarrow GSSG + Q^{\cdot-} \tag{1.77}$$

their ability to react with polyunsaturated fatty acids via two reactions: abstraction of a bis-allylic hydrogen forming pentadienyl radicals (with rate constants of about $10^6- 10^7$ $M^{-1}s^{-1}$ (Schöneich et al., 1989), and addition to the double bonds (Schöneich et al., 1992). The former reaction suggests a role for thiyl radicals as initiators of lipid peroxidation. A protection of membrane lipids against attack by thiyl radicals was proposed for vitamin A (retinol), which reacts rapidly with glutathionyl radicals (k = 1.4×10^9 $M^{-1}s^{-1}$) (D'Aquino et al., 1989).

Dithiols, such as α-dihydrolipoic acid, merit further comment: first, because of their particular redox chemistry and, second, because of the recent numerous reports attesting the strong antioxidant activity of α-dihydrolipoic acid as well as that of its oxidized counterpart, α-lipoic acid. A preliminary distinction can be made concerning the oxidative reactions involving aliphatic monothiols and dithiols and leading to the accumulation of aliphatic and cyclic disulfides, respectively. The reactions above indicate that following free radical repair, the resulting thiyl radical is subjected to distinct decay routes, their importance determined by physico-chemical and environmental factors. On the other hand, free radical reactions leading to the formation of cyclic disulfides, such as α-lipoate, are likely to proceed with intermediate formation of a disulfide anion radical formed upon rapid intramolecular complexation of the thiyl radical with the adjacent SH group (reaction 1.78). Similar to what was discussed for reaction 1.71, this intermediate is a powerful reducing agent with a redox potential of about −1.6 V (Surdhar and Armstrong, 1986). The

$$\tag{1.78}$$

formation of such an intermediate may be a feature inherent in the one-electron oxidation of a α-dihydrolipoate, whereas the occurrence of such a species during the redox transitions of other aliphatic monothiols would depend on the significance of the conjugation of the thiyl radical with thiolate (reaction 1.71) relative to that with O_2 (reaction 1.72) or even its dimerization to the aliphatic disulfide (reaction 1.73).

Among the aliphatic or aromatic monothiols listed in Table III, α-dihydrolipoic acid reveals unique features (Romero et al., 1992): first, it reduces the hypervalent iron of the oxoferryl complex ($Fe^{IV}=O$) to the ferrous state (Fe^{II}) whereas all other thiols facilitate only the $Fe^{IV}=O \rightarrow Fe^{III}$ transition (reaction 1.69 above). Second, α-dihydrolipoate reduces metmyoglobin to oxymyoglobin (i.e., facilitating the $Fe^{III} \rightarrow Fe^{II}$ transition), a reaction which is not accomplished by the other sulfur-containing compounds. Third, at variance with GSH and cysteine, oxidation of α-dihydrolipoate was not associated with the formation of a thiyl radical which could be identified by ESR in conjunction with the spin trap DMPO; this could be explained by the close proximity of the SH groups in the dithiol, and the fact that the thiyl radical formed rapidly complexes with the adjacent RS^- group to yield the intramolecular radical anion complex, making the trapping by DMPO of the initial thiyl radical virtually impossible.

The formation of an intramolecular radical anion complex (reaction 1.78) with strong reducing power seems of utmost importance in the evaluation of these results, for this species is likely to be responsible for the unique feature of α-dihydrolipoate in its reactions with ferryl- and metmyoglobin. The lipoic acid radical has been reported to reduce FAD to $FADH^.$ at a rate of $2.4 \times 10^8\ M^{-1}s^{-1}$ (Chan et al., 1974). Also, the reduction of metmyoglobin by α-dihydrolipoate is a thermodynamically unfavorable process as indicated by the reduction potential values of the redox couples involved (i.e., $E_{RS^./RS^-}) = +650$ mV (Surdhar and Armstrong, 1986) and $E_{Fe^{III}/Fe^{II}O_2}$ $= +220$ mV (Koppenol and Butler, 1985). This might constitute another example in which thermodynamic constrains are overcome by kinetic factors, for the disulfide anion radical—an intermediate inherent in the redox transitions of dihydrolipoate—could readily reduce metmyoglobin. Although the decay of this intermediate could be coupled to the diffusion-controlled electron transfer to O_2 to form O_2^- (Willson, 1970), it is unlikely that the latter contributes to hemoprotein reduction, for the process is insensitive to superoxide dismutase.

The disulfide, α-lipoate, is also unique in that it reduces the hypervalent iron in $Fe^{IV}=O$ to Fe^{III}, a reaction that other disulfides, such as GSSG and cystine, cannot accomplish. The higher reactivity towards electrophiles of the disulfide bond of the dithiolane ring in lipoic acid over that of the disulfide bond in open chain seems to be a function of the low activation energy caused by the ring strain in the former (Schmidt *et al.*, 1969). This observation is analogous to the reported scavenging of hypochlorous acid (HOCl) by α-lipoate which would likely involve the formation of an S-oxide product (Haennen and Bast, 1991).

The antioxidant properties of α-dihydrolipoate have been examined in a great range of experimental models encompassing the reduction of peroxyl, ascorbyl, and chromanoxyl radicals (Kagan et al., 1992; Suzuki et al., 1991), protection against microsomal lipid peroxidation (Bast and Haennen, 1988; Scholich et al., 1989), as a cofactor for the peroxidase activity of ebselen (Haennen et al., 1990), and the scavenging of hypochlorous acid (Haennen and Bast, 1991). Furthermore, the relationships between dihydrolipoate and ferrylmyoglobin described above might be of relevance to the interpretation of the protective effect displayed by the dithiol against ischemia-reperfusion injury in the isolated rate heart (Serbinova et al., 1992).

Although the intracellular concentration of α-dihydrolipoate is in the low μM range and it does not occur free but bound to protein complexes, hepatic tissue can uptake lipoate by a carrier-mediated process at low concentrations of the disulfide and by diffusion at high concentrations (Peinado et al., 1989). This, along with the antioxidant properties of dihydrolipoate itself or in connection with α-tocopherol and a deeper knowledge of its pharmacokinetic properties, strengthens the view that the dithiol may be used as a drug to lessen or prevent oxidative stress conditions.

1.6 SECONDARY ANTIOXIDANT DEFENSES

Secondary defenses may be considered as a varied host of enzymes (cf. chapter 7), many of which participate in metabolic systems in response to oxidant challenge and/or injury and they serve as a repair system to eliminate molecules or cell components that were damaged by oxidants or free radical reactions which escaped the primary antioxidant defenses. Thus, the enzymes considered as secondary defenses contribute to the repair of membrane phospholipids, proteins, and DNA, processes which are briefly referred to below and in detail in subsequent chapters by R. P. Cunningham and H. Ahern, (chapter 8) and G. W. Felton, (chapter 10). Digestion of critically damaged RNA and proteins represents their actual 'repair' process; repair of phospholipids and DNA involves removal of oxidized portions and followed by lysophospholipid reacylation or proper base replacement, respectively.

An early event resulting from oxidative injury to membrane phospholipids (encompassing the chemical reactions discussed above) is the formation of fatty acyl hydroperoxides. The disruptive effects of these hydroperoxides—or derived products—on membrane structures apparently triggers a lipid hydrolytic activity, which could be considered the onset of the repair sequence consisting on removal of the LOOH (Van Kuijk et al., 1987). This sequence entails following steps: first, hydrolysis of the phospholipid hydroperoxide by phospholipase A_2 to yield a free fatty acid hydroperoxide and lysophospholipid. Second, the fatty acid hydroperoxide is reduced in the cytosol by a 'primary' antioxidant enzyme; GPOX. Third, the lysophospholipid in the membrane is reacylated through the normal acyl-CoA/acyltransferase reactions which participate in phospholipid synthesis and turnover, a process which is stimulated during membrane oxidant stress and, as such, can be viewed as a repair reaction (van Kuijk et al., 1987; Lubin and Kuypers, 1991; Sevanian, 1991; Pacifici et al., 1993). Thus, it appears that two enzymes are central to membrane phospholipid repair, phospholipase A_2 and fatty acyl tranferases.

Oxidative damage to nucleic acids occurs in vivo and it is thought to be quite extensive (Ames, 1990) with estimates as high as one base modification per 130,000 bases in nuclear DNA (Richter et al., 1988) and per 8,000 bases in mitochondrial DNA; furthermore, the fragments of oxidized mitochondrial DNA have been implicated in cancer and aging (Richter, 1988). This damage disrupts transcription, translation, and DNA replication and also gives rise to mutations. There is evidence that DNA repair enzyme activity—requiring excision of the damaged strand followed by a endogenous-polymerases-mediated insertion of the proper nucleotides—is induced by oxidative damage (Howard-Flanders, 1981; Lindhal, 1987; Teebor et al., 1988). In mammalian systems, N-glycosylase activities eliminate the damaged base generating apurinic/apyrimidinic sites; mammalian redoxyendonuclease has a broader substrate specificity and removes pyrimidine base damage products via an N-glycosylase activity followed by an AP lyase activity catalyzing a β,δ-elimination reaction which results in removal of the deoxyribose and generation of a strand scission product (Doetsch, 1991; Demple and Levin, 1991).

Likewise, inactivation of protein function is an early expression of free radical exposure and oxidatively damaged proteins—involving amino acid alterations, increased hydrophobic interactions, partial unfolding, and/or covalent cross-linking. At variance with membrane phospholipids and DNA, there appears to be no mechanism for the repair of oxidatively damaged proteins, if repair entails a selective removal and replacement of the damaged amino acid (except for oxidized cysteinyl and methionyl residues which can be repaired via a enzyme-catalyzed disulfide exchange (Stadtman, 1991)). However, oxidatively damaged proteins possess an increased proteolytic susceptibility and, it follows, that its proteolytic degradation will increase the pool of free amino acids for the de novo synthesis of a new protein. The significance of this process is highlighted by the observation that the age-related loss of protease activity is accompanied with an enhancement of catalytically inactive enzymes and of total oxidized proteins, which may account for a large percent of the total cellular protein (Starke-Reed and Oliver, 1989). The oxidized proteins are selectively degraded by a 600-700 kDa multicatalytic/multifunctional proteinase complex (isolated from a variety of tissues and with many different names; see Rivett, 1989). Although there is agreement about the function of this multicatalytic proteinase complex, the exact physico-chemical modification in oxidized proteins which provides the signal for selective proteolysis is less clear: [a] partial protein unfolding and exposure of previously buried hydrophobic moieties (Davies et al., 1987; Pacifici and Davies, 1991); [b] covalent modification of critical amino acid residues has been proposed as reactions which represent the marking steps in protein turnover (Rivett and Hare, 1987; Rivett and Levine, 1990; Stadtman, 1990b; Levine et al., 1991).

1.7 OXIDANTS AND ANTIOXIDANTS

Table IV lists the reduction potentials of several species which have been considered above as oxidant or antioxidant molecules. The basic concept in antioxidant activity is summarized in the electron- or hydrogen transfer illustrated in reaction 40: by means of this reaction, the radical character (initially centered on a strong oxidant, e.g., $HO^{\cdot}$) is transferred to an 'antioxidant' molecule. The formation of an antioxidant-derived radical is a process inherent in their oxidant scavenging activity and understanding the decay pathways of the antioxidant-derived radical has important biological implications.

A distinction between oxidants and antioxidants under the terms of Table IV is not necessarily correct and the species therein could be referred to as strong oxidants and weak oxidants. The species

Table 1.4.
Reduction Potentials of Some Relevant Oxidants and Antioxidants

	$E°/V$
HO$\cdot$, H$^+$/H$_2$O	+2.18
RO$\cdot$, H$^+$/ROH	+1.60
HO$_2\cdot$, H$^+$/H$_2$O$_2$	+1.06
ROO$\cdot$, H$^+$/ROOH	+1.00
Fe^{IV}=O/Fe^{III}	+0.99
NO$_2$/NO$_2^-$	+0.87
GS$\cdot$/GS$^-$	+0.85
RSSR$\cdot^-$/RS$_2^-$	+0.65
U$\cdot$/U$^-$	+0.52
α-TO$\cdot$/α-TOH	+0.48
UQ$_{10}$H$\cdot$/UQ$_{10}$H$_2$	+0.35
A$\cdot^-$/AH$^-$	+0.28

Reduction potential values taken from Koppenol and Butler (1985); Surdhar and Armstrong (1986), and Buxton et al. (1988).

listed at the top of Table IV (HO$\cdot$, RO$\cdot$, the oxoferryl complex (myoglobin/hemoglobin), HO$_2\cdot$, and ROO$\cdot$) possess a very positive reduction potential and their chemical reactivity with different biological targets has been briefly described above. The species listed at the bottom of Table 1.4 (uric acid, α-tocopherol, coenzyme Q or ubiquinol$_{10}$, and ascorbic acid) are clearly weak oxidants and their interaction with the above reactive oxygen species yields the corresponding antioxidant-derived radicals. This has been illustrated numerous times by a redox cycling involving the reaction of various oxidants with α-tocopherol, and recovery of the α-TO$\cdot$ radical by ascorbic acid.

The antioxidant-derived radical, with a reduction potential and a chemical reactivity lower—and a life time usually longer than that of the oxidant initially scavenged, is not inert and can have toxic implications provided the proper cellular settings are given. Examples of such a behavior can be encountered in the inhibition of alcohol dehydrogenase and α_1-antiproteinase (Kittridge and Wilson, 1984; Aruoma and Halliwell, 1989) by the radical form(s) of uric acid. α-Tocopheroxyl radicals can apparently propagate peroxidation within lipoprotein particles upon reaction of this radical with polyunsaturated fatty acid moieties in the lipid (Bowry et al., 1992; Ingold et al., 1993). It is likely that these unwanted effects may be overcome by the presence of ascorbic acid, which efficiently repairs urate radical and α-tocopheroxyl radical, thus potentiating the antioxidant efficiency of these molecules.

The reduction potentials listed in Table 1.4 suggest that the concerted activity of ascorbic acid with other 'antioxidant' molecules would result in ultimate transfer of the radical character to a weak oxidant, the ascorbyl radical. It may be inferred that this radical is nontoxic and it decays readily by a second-order disproportionation process (reaction 1.79; $k_{79} = 2 \times 10^5 \ M^{-1}s^{-1}$) yielding the ascorbate mono-anion and dehydroascorbate (Bielski, 1982). These issues were recently addressed by Buettner and Jurkiewicz (1993), who also suggested that the intensity of the ascorbyl radical ESR signal could serve as a marker of oxidative stress.

$$A^{\cdot -} + A^{\cdot -} + H^+ \rightarrow AH^- \tag{1.79}$$

The situation with thiols is far more complex: their position in Table 1.4 indicates that thiyl radicals are strong oxidants and, indeed, they react with a variety of biomolecules, including unsaturated fatty acids (Schöneich et al., 1989, 1992). Understanding of the decay pathways of thiyl radicals is critical to the evaluation of their reactivity in a biological environment (Wardman, 1988): two reactions seem relevant in this context, the conjugation of thiyl radicals with thiolate anions and their reaction with ascorbic acid. The former reaction leads to the production of a strong reductant (reaction 1.75), whereas the latter yields a mild oxidant, the ascorbyl radical (reaction 1.54), which can decay by disproportionation to nonradical products.

The practical implications of these thermodynamic aspects (as intended with the reduction potentials listed in Table 1.4) in a biological environment are obviously difficult to assess. A detailed analysis of the current condition of antioxidant therapy (Rice-Evans and Diplock, 1993) has focused on the antioxidant status of individuals and its role in protection against amplification of certain disease processes which are known to be associated with oxidative stress. Although this concept is gaining significance in coronary heart disease, inflammation, and atherosclerosis, and there is some evidence for the beneficial effects of free radical scavenging drugs, the authors conclude that the implementation of antioxidant therapies requires a better understanding of the involvement of free radicals and the molecular mechanisms by which they exert cytotoxicity in disease states. It is also self-evident that knowledge is required on the concerted activity of antioxidant molecules in a cellular or extracellular setting as well as identification of potential specific sites for antioxidant action.

1.8 SUMMARY

This chapter contains three main sections dealing with the chemistry of reactive oxygen species, their sources in biological systems, and the electron- or hydrogen transfer reactions implied in their interaction with antioxidant molecules.

The first section covers the electron-transfer and energy transfer routes for activation of dioxygen: the former route is described in terms of the successive addition of electrons to ground state molecular oxygen to form a series of intermediates (such as superoxide radical [O_2^-], hydrogen peroxide [H_2O_2], and hydroxyl radical [$HO^.$]) and the reaction of carbon-centered radicals with oxygen (yielding peroxyl radicals). The latter group, activation of oxygen by energy transfer, deals with singlet oxygen ($^1O_2$) formation during photosensitization processes as well as during chemiexcitation. The chemical reactivity of these reactive species is described in terms of reactions which are relevant in a biological context. In addition, the chemistry attainable by the oxoferryl complex, a hypervalent heme iron state resulting from the oxidation of myoglobin or hemoglobin by H_2O_2 and distinct from the $HO^.$ radical, is described in connection with oxidation of cell constituents and some drugs. Some novel aspects of nitric oxide biochemistry, especially concerned with its reaction with oxyradicals, is addressed in this section.

The second section surveys briefly the biological sources of O_2^- and H_2O_2, emphasizing those which contribute greatly to their cellular steady-state concentration. Cellular activation of electronegative compounds is addressed in terms of the classical redox cycling process and distinguishing those supported by initial one-electron activation from those initiated by two-electron activation, i.e., during catalysis by the unique flavoenzyme, DT-diaphorase. The biological and pathological implications of the high oxidation state of myoglobin, based on the chemical reactivity of the oxoferryl complex described in the first section, are discussed in terms of the reaction of this species with membrane phospholipids and oxidized lipoproteins as well as the potential significance of this species in coronary heart disease. Finally, two biological sources of $^1O_2$ are described: the disproportionation of secondary lipid peroxyl radicals (of importance in lipid peroxidation) and the disproportionation of hydroperoxides catalyzed by specific peroxidases.

The final section in this chapter describes the electron- or hydrogen-transfer processes involved in the interaction of reactive oxygen species with small antioxidant molecules, such as α-tocopherol, ascorbic acid, ubiquinol-10, uric acid, and thiols. The thermodynamic and kinetic aspects that control the decay routes of the antioxidant-derived radical originating from these interactions are emphasized as being critical to the antioxidant functions of these molecules.

References

Aikens, J. and Dix, T.A. (1991) Peroxyl radical (HOO˙)-initiated lipid peroxidation. The role of fatty acid hydroperoxides. *J. Biol. Chem.* **266**, 15091–15098.

Akanmu, D., Cecchini, R., Aruoma, O.I. and Halliwell, B. (1991) The antioxidant action of ergothioneine. *Arch. Biochem. Biophys.* **288**, 10–16.

Allentoff, A.J., Bolton, J.L., Wilks, A., Thompson, J.A. and Ortiz de Montellano, P.R. (1992) Heterolytic versus homolytic peroxide bond cleavage by sperm whale myoglobin and myoglobin mutants. *J. Am. Chem. Soc.* **114**, 9744–9749.

Ames, B.N. (1990) Endogenous oxidative DNA damage, aging, and cancer. *Free Rad. Res. Commun.* **7**, 121–128.

Ames, B.N., Cathcart, R., Schwiers, E. and Hochstein, P. (1981) Uric acid provides an antioxidant defence in humans against oxidant- and radical-caused aging and cancer: a hypothesis. *Proc. Natl. Acad. Sci. USA* **78**, 6858–6862.

Arduini, A., Eddy, L. and Hochstein, P. (1990a) Detection of ferrylmyoglobin in the isolated ischemic rat heart. *Free Rad. Biol. Med.* **9**, 511–513.

Arduini, A., Eddy, L. and Hochstein, P. (1990b) The reaction of ferrylmyoglobin with ergothioneine: a novel function for ergothioneine. *Arch. Biochem. Biophys.* **281**, 41–43.

Arduini, A., Mancinelli, G., Radatti, G.L., Hochstein, P. and Cadenas, E. (1992) Possible mechanism of inhibition of nitrite-induced oxidation of oxyhemoglobin by ergothioneine and uric acid. *Arch. Biochem. Biophys.* **294**, 398–402.

Aruoma, O.I. and Halliwell, B. (1989) Inactivation of α₁-antiproteinase by hydroxyl radicals. The effect of uric acid. *FEBS Lett.* **244**, 76–80.

Asmus, K.-D. (1990) Sulfur-centered free radicals. *Methods Enzymol.* **186**, 168–180.

Augusto, O., Gatti, R.M. and Radi, R. (1994) Spin-trapping studies of peroxynitrite decomposition and of 3-morpholinosydnonimine N-ethylcarbamide autoxidation: direct evidence for metal-independent formation of free radical intermediates. *Arch. Biochem. Biophys.* **310**, 118–125.

Bast, A. and Haennen, G.R.M.M. (1988) Interplay between lipoic acid and glutathione in the protection against microsomal lipid peroxidation. *Biochim. Biophys. Acta* **963**, 558–561.

Becker, B.H. (1993) Towards the physiological function of uric acid. *Free Rad. Biol. Med.* **14**, 615–631.

Beckman, J.S., Beckman, T.W., Chen, J., Marshall, P.A. and Freeman, B.A. (1990) Apparent hydroxyl radical production by peroxynitrite: Implication for endothelial injury from nitric oxide and superoxide. *Proc. Natl. Acad. Sci. USA* **87**, 1620–1624.

Beyer, R.E. and Ernster, L. (1990) The antioxidant role of coenzyme Q. In *Highlights in Ubiquinone Research* (Lenaz, G., Barnabei, O., Rabbi, A. and Battino, M., eds.), Taylor & Francis, London, pp. 191–213.

Bielski, B.H.J. (1978) Reevaluation of the spectral and kinetic properties of HO_2 and O_2-free radicals. *Photochem. Photobiol.* **28**, 645–649.

Bielski, B.H.J. (1982) Chemistry of ascorbic acid radicals. In *Ascorbic Acid: Chemistry, metabolism, and Uses* (Seib, P.A. and Tolbert, B.M., eds.), American Chemical Society, Washington, pp. 81–100.

Bielski, B.H.J. and Cabelli, D.E. (1991) Highlights of current research involving superoxide and perhydroxyl radicals in aqueous solutions. *Int. J. Radiat. Biol.* **59**, 291–319.

Bielski, B.H.J. and Shiue, G.G. (1979) Reaction rates of superoxide radicals with the essential amino acids. In *Oxygen Free Radicals and Tissue Damage* (Ciba Foundation Symposium 65 (new series)), Excerpta Medica, Amsterdam, pp. 43–48.

Bieri, J.G. and Tolliver, T.J. (1981) On the occurrence of α-tocopherylquinone in rat tissue. *Lipids* **16**, 777–779.

Blough, N.V. and Zafiriou, O.C. (1985) Reaction of superoxide with nitric oxide to form peroxonitrite in alkaline aqueous solution. *Inorg. Chem.* **24**, 3502–3504.

Boiteux, S., O'Connor, T.R., Lederer, F., Gouyette, A. and Laval, J. (1990) Homogeneous *Escherichia coli* FPG protein. A DNA glycosylase which excises imidazole ring-opened purines and nicks DNA at apurinic/apyrimsidinic sites. *J. Biol. Chem.* **265**, 3916–3922.

Borg, D.C. and Schaich, K. (1984) Cytotoxicity from coupled redox cycling of autoxidizing xenobiotics and metals. *Israel J. Chem.* **24**, 38–53.

Boveris, A. and Cadenas, E. (1982) Production of superoxide radicals and hydrogen peroxide in mitochondria. In *Superoxide Dismutase* (Oberley, L.W., ed.), Vol. II, CRC Press, Boca Raton, FL, pp. 15–30.

Bowry, V.W., Ingold, K.U. and Stocker, R. (1992) Vitamin E in human low-density lipoprotein. When and how this antioxidant becomes a pro-oxidant. *Biochem. J.* **288**, 341–344.

Bowry, V.W. and Stocker, R. (1993) Tocopherol-mediated peroxidation. The prooxidant effect of vitamin E on the radical-initiated oxidation of human low-density lipoprotein. *J. Am. Chem. Soc.* **115**, 6029–6044.

Bruckdorfer, K.R., Dee, G., Jacobs, M. and Rice-Evans, C. (1990*a*) The protective action of nitric oxide against membrane damage induced by myoglobin radicals. *Biochem. Soc. Trans.* **18**, 285–286.

Bruckdorfer, K.R., Jacobs, M. and Rice-Evans, C. (1990*b*) Endothelium-derived relaxing factor (nitric oxide), lipoprotein oxidation and atherosclerosis. *Biochem. Soc. Trans.* **18**, 1061–1063.

Buettner, G.R. and Jurkiewicz, B.A. (1993) Ascorbate free radical as a marker of oxidative stress: an EPR study. *Free Rad. Biol. Med.* **14**, 49–55.

Buffinton, G. and Cadenas, E. (1988) Reduction of ferrylmyoglobin by quinonoid compounds. Chem.-Biol. Interact. **66**, 233–250.

Buffinton, G., Mira, D., Galaris, D., Hochstein, P. and Cadenas, E. (1988) Reduction of ferryl- and met-myoglobin to ferrous myoglobin by menadione-glutathione conjugate. Spectrophotometric studies under aerobic and anaerobic conditions. Chem.-Biol. Interact. **66**, 205–222.

Buffinton, G., Öllinger, K., Brunmark, A. and Cadenas, E. (1989) DT-diaphorase-catalyzed reduction of 1,4-naphthoquinone derivatives and glutathionyl-quinone conjugates. Effect of substituents on autoxidation rates. Biochem. J. **257**, 561–571.

Burton, G.W. and Ingold, K.U. (1986) Vitamin E: application of the principles of physical organic chemistry to the exploration of its structure and function. Acc. Chem. Res. **19**, 194–201.

Butler, J. and Hoey, B.M. (1992) Reactions of glutathione and glutathione radicals with benzoquinones. Free Rad. Biol. Med. **12**, 337–345.

Buxton, G.V., Greenstock, C.L., Helman, W.P. and Ross, A.B. (1988) Critical review of rate constants for reactions of hydrated electrons, hydrogen atoms and hydroxyl radicals ($\cdot OH/\cdot O^-$) in aqueous solutions. J. Phys. Chem. Ref. Data **17**, 513–886.

Cadenas, E. (1984) Biological chemiluminescence. Photochem. Photobiol. **40**, 823–830.

Cadenas, E. (1989) Biochemistry of oxygen toxicity. Ann. Rev. Biochem. **58**, 79–110.

Cadenas, E., Boveris, A. and Chance, B. (1984) Low-level chemiluminescence of biological systems. In Free Radical Biology and Medicine (Pryor, W.A., ed.), Vol. VI, Academic Press, New York, pp. 212–242.

Cadenas, E., Boveris, A., Ragan, C.I. and Stoppani, A.O.M. (1977) Production of superoxide radicals and hydrogen peroxide by NADH-ubiquinone reductase and ubiquinol-cytochrome c reductase from beef heart mitochondria. Arch. Biochem. Biophys. **180**, 248–257.

Cadenas, E., Giulivi, C., Ursini, F. and Boveris, A. (1994) Electronically-excited state formation during lipid peroxidation. Methods Toxicol. **1B**, 384–397.

Cadenas, E., Hochstein, P. and Ernster, L. (1992) Pro- and antioxidant functions of quinones and quinone reductases in mammalian cells. Adv. Enzymol. **65**, 97–146.

Cadenas, E., Merényi, G. and Lind, J. (1989) Pulse radiolysis study on the reactivity of Trolox C phenoxyl radical with superoxide anion. FEBS Lett. **253**, 235–238.

Cadenas, E., Sies, H., Nastaincyzk, W. and Ullrich, V. (1983) Singlet oxygen formation detected by low-level chemiluminescence during the enzymatic reduction of prostaglandin G_2 to H_2. Hoppe- Seyler's Z. Physiol. Chem. **364**, 519–528.

Catalano, C.E., Choe, Y.S. and Ortiz de Montellano, P.R. (1989) Reactions of the protein radical in peroxide-treated myoglobin. Formation of a heme-protein cross-link. J. Biol. Chem. **264**, 10534–10541.

Chan, A.C., Tran, K., Raynor, T., Ganz, P.R. and Chow, C.K. (1991) Regeneration of vitamin E in human platelets. J. Biol. Chem. **266**, 17290–17295.

Chan, E. and Weiss, B. (1986) Endonuclease IV of Escherichia coli is induced by paraquat. Proc. Natl. Acad. Sci. USA **84**, 3189–3193.

Chance, B., Sies, H. and Boveris, A. (1979) Hydroperoxide metabolism in mammalian organs. Physiol. Rev. **59**, 527–605.

Christman, M.F., Morgan, R.W., Jacobson, F.S. and Ames, B.N. (1985) Positive control of a regulon for defenses against oxidative stress and some heat shock proteins in *Salmonella typhimurium. Cell* 41, 753–762.

Cross, C.E., Motchnik, P.A., Bruener, B.A., Jones, D.A., Kaur, H., Ames, B.N. and Halliwell, B. (1992) Oxidative damage to plasma constituents by ozone. *FEBS Lett.* 298, 269–272.

Csallany, A.S., Draper, H.H. and Shah, S.N. (1962) Conversion of d-α-tocopherol-C^{14} to tocopherol quinone in vivo. *Arch. Biochem. Biophys.* 98, 142–145.

D'Aquino, M., Dunster, C. and Willson, R.L. (1989) Vitamin A and glutathione-mediated free radical damage: competing reactions with polyunsaturated fatty acids and vitamin C. *Biochem. Biophys. Res. Commun.* 161, 1199–1203.

D'Arcy-Doherty, M., Wilson, I., Wardman, P., Basra, J., Patterson, L.H. and Cohen, G.M. (1986) Peroxidase activation of 1-naphthol to naphthoxy or naphthoxy-derived radicals and their reactions with glutathione. *Chem.-Biol. Interact.* 58, 199–215.

Darley-Usmar, V., Hogg, N., O'Leary, V.J., Wilson, M.T. and Moncada, S. (1993) The simultaneous generation of superoxide and nitric oxide can initiate lipid peroxidation in human low density lipoprotein. *Free Rad. Res. Commun.* 17, 9–20.

Davies, K.J.A. (1986) Intracellular proteolytic systems may function as secondary antioxidant defenses: a hypothesis. *J. Free Rad. Biol. Med.* 2, 155–173.

Davies, K.J.A. (1987) Protein damage and degradation by oxygen radicals. II. Modification of amino acids. *J. Biol. Chem.* 262, 9902–9907.

Davies, K.J.A., Lin, S.W. and Pacifici, R.E. (1987) Protein damage and degradation by oxygen radicals. IV. Degradation of denatured protein. *J. Biol. Chem.* 262, 9914–9920.

Davies, M.J. (1990) Detection of myoglobin-derived radicals on reaction of metmyoglobin with hydrogen peroxide and other peroxidic compounds. *Free Radical Res. Commun.* 10, 361–370.

Davies, M.J. (1991) Identification of a globin free radical in equine myoglobin treated with peroxides. *Biochim. Biophys. Acta* 1077, 86–90.

Davies, M.J. and Puppo, A. (1992) Direct detection of a globin-derived radical in leghaemoglobin treated with peroxides *Biochem. J.* 281, 197–201.

Davies, M.J. and Slater, T.F. (1987) Studies on the metal-ion and lipoxygenase-catalysed breakdown of hydroperoxides using electron-spin-resonance spectroscopy. *Biochem. J.* 245, 167–173.

Dean, R.T. (1991) Protein damage and repair: an overview. in *Oxidative Damage and Repair: Chemical, Biological and Medical Aspects* (Davies, K.J.A., ed.), Pergamon Press, New York, pp. 341–347.

Dean, R.T., Hunt, J.V., Grant, A.J., Yamamoto, Y. and Niki, E. (1991) Free radical damage to proteins: the influence of the relative localization of radical generation, antioxidants, and target proteins. *Free Rad. Biol. Med.* 11, 161–168.

Dee, G., Rice-Evans, C., Obeyesekera, S., Meraji, S., Jacobs, M. and Bruckdorfer, K.R. (1991) The modulation of ferryl myoglobin formation and its oxidative effects on low density lipoproteins by nitric oxide. *FEBS Lett.* 294, 38–42.

deGroot, H., Hegi, U. and Sies, H. (1993) Loss of α-tocopherol upon exposure to nitric oxide or the sydnonimine SIN-1. *FEBS Lett.* 315, 139–142.

Demple, B. and Levin, J.D. (1991) Repair system for radical-damaged DNA. In *Oxidative Stress: Oxidants and Antioxidants* (Sies, H., ed.), Academic Press, London, pp. 119–154.

DiMascio, P., Catalani, L.H. and Bechara, E.J.H. (1992) Are dioxetanes chemiluminescent intermediates in lipoperoxidation? *Free Rad. Biol. Med.* 12, 471–478.

Dix, T.A., Fontana, R., Panthani, A. and Marnett, L.J. (1985) Hematin-catalyzed epoxidation of 7,8-dihydroxy-7,8-dihydrobenzo[a]pyrene by polyunsaturated fatty acid hydroperoxides. *J. Biol. Chem.* 260, 5358–5365.

Doetsch, P.W. (1991) Repair of oxidative DNA damage in mammalian cells. In *Oxidative Damage and Repair: Chemical, Biological and Medical Aspects* (Davies, K.J.A., ed.), Pergamon Press, New York, pp. 192–196.

Dolphin, D. (1985) Cytochrome P_{450}: substrate and prosthetic-group free radicals generated during the enzymatic cycle. *Phil. Trans. R. Soc. Lond. B.* 311, 579–591.

Durán, N. and Cadenas, E. (1987) The role of singlet oxygen and triplet carbonyls in biological systems. *Rev. Chem. Intermediates* 8, 147–187.

Eddy, L., Arduini, A. and Hochstein, P. (1990) Reduction of ferrylmyoglobin in rat diaphragm. *Am. J. Physiol.* 259, C995–C997.

Encinas, M.V., Lissi, E.A. and Olea, A.F. (1985) Quenching of triplet benzophenone by vitamins E and C and by sulfur containing amino acids and peptides. *Photochem. Photobiol.* 42, 347–352.

Foote, C.S. (1976) Photosensitized oxidation and singlet oxygen: consequences in biological systems. In *Free Radicals in Biology* (Pryor, W.A., ed.), Vol. II, pp. 85–133, Academic Press, New York.

Forman, H.J. and Boveris, A. (1982) Superoxide radical and hydrogen peroxide in mitochondria. In *Free Radicals in Biology* (Pryor, W.A., ed.), vol. V, pp. 65–90, Academic Press, New York.

Forman, H.J. and Thomas, M.J. (1986) Oxidant production and bactericidal activity of phatocytes. *Ann. Rev. Physiol.* 48, 669–680.

Fornace Jr., A.J., Alamo Jr., I. and Hollander, M.C. (1988) DNA damage-inducible transcripts in mammalian cells. *Proc. Natl. Acad. Sci. USA* 85, 8800–8804.

Forni, L.G., Monig, J., Mora-Arellano, V.O. and Willson, R.L. (1983) Thiyl free radicals: direct observation of electron transfer reactions with phenothiazines and ascorbate. *J. Chem. Soc. Perkin Trans. II*, pp. 961–965.

Forni, L.G. and Willson, R.B. (1986a) Thiyl and phenoxyl free radicals and NADH. Direct observation of one-electron oxidation. *Biochem. J.* 240, 897–903.

Forni, L.G. and Willson, R.B. (1986b) Thiyl free radicals and the oxidation of ferrocytochrome *c*. Direct observation of coupled hydrogen-atom- and electron-transfer reactions. *Biochem. J.* 240, 905–907.

Forsmark, P. A°berg, F., Norling, B., Nordenbrand, K., Dallner, G. and Ernster, L. (1991) Vitamin E and ubiquinol as inhibitors of lipid peroxidation in biological membranes. *FEBS Lett.* 285, 39–43.

Frimer, A.A. (1983) The organic chemistry of superoxide anion radical. In *The Chemistry of Functional Groups: Peroxides* (Patai, S., ed), John Wiley, New York, pp. 429–461.

Frimer, A.A. (1988) Superoxide chemistry in non-aqueous media. In *Oxygen Radicals in Biology and Medicine* (Simic, M.G., Taylor, K.A., Ward, J.F. and von Sonntag, C., eds.), Plenum Press, New York, pp. 29–38.

Galaris, D., Cadenas, E. and Hochstein, P. (1989a) Redox cycling of myoglobin and ascorbate: a potential protective mechanism against oxidative reperfusion injury in muscle. *Arch. Biochem. Biophys.* 273, 497–504.

Galaris, D., Cadenas, E. and Hochstein, P. (1989b) Glutathione-dependent reduction of peroxides during ferryl- and metmyoglobin interconversion: a potential protective mechanism in muscle. *Free Rad. Biol. Med.*, 6, 473–478.

Galaris, D., Mira, D., Sevanian, A., Cadenas, E. and Hochstein, P. (1988) Co-oxidation of salicylate and cholesterol and the generation of electronically-excited states during the oxidation of metmyoglobin by H_2O_2. *Arch. Biochem. Biophys.* 262, 221–231.

Galaris, D., Sevanian, A., Cadenas, E. and Hochstein, P. (1990) Ferrylmyoglobin-catalyzed linoleic acid peroxidation. *Arch. Biochem. Biophys.* 281, 163–169.

Gebicki, J.M. and Bielski, B.H.J. (1981) Comparison of the capacities of the perhydroxyl and the superoxide radicals to initiate chain oxidation of linoleic acid. *J. Am. Chem. Soc.* 103, 7020–7022.

Gebicki, J.M., Jürgens, G. and Esterbauer, H. (1991) Oxidation of low-density lipoprotein *in vitro*. In *Oxidative Stress: Oxidants and Antioxidants* (Sies, H., ed.), Academic Press, London, pp. 371–397.

George, P. and Irvine, D.H. (1952) The reaction between metmyoglobin and hydrogen peroxide. *Biochem. J.* 52, 511–517.

Giamalva, D.H., Church, D.F. and Pryor, W.A. (1986) Kinetics of ozonization. IV. Reactions of ozone with alpha-tocopherol and oleate and linoleate esters in carbon tetrachloride and in aqueous micellar solves. *J. Am. Chem. Soc.* 108, 6646–6651.

Gibson, J.F., Ingram, D.J.E. and Nichols, P. (1958) Free radical produced in the reaction of metmyoglobin with hydrogen peroxide. *Nature* 181, 1398–1399.

Giulivi, C. and Cadenas, E. (1993a) Inhibition of protein radical reactions of ferrylmyoglobin by the water-soluble analog of vitamin E, Trolox C. *Arch. Biochem. Biophys.* 303, 152–158.

Giulivi, C. and Cadenas, E. (1993b) The reaction of ascorbic acid with different heme iron redox states of myoglobin. Antioxidant and Pro-oxidant aspects. *FEBS Lett.* 332, 287–290.

Giulivi, C. and Davies, K.J.A. (1990) A novel antioxidant role for hemoglobin. The comproportionation of ferrylhemoglobin with oxyhemoglobin. *J. Biol. Chem.* 265, 19453–19460.

Giulivi, C., Romero, F.J. and Cadenas, E. (1992) The interaction of trolox C, a water-soluble vitamin E analog, with ferrylmyoglobin: reduction of the oxoferryl moiety. *Arch. Biochem. Biophys.* 299, 302–312.

Goldstein, S. and Czapski, G. (1986) The role and mechanism of metal ions and their complexes in enhancing damage in biological systems or in protecting these systems from the toxicity of O_2^-. *J. Free Rad. Biol. Med.* 2, 3–11.

Green, M.J. and Hill, H.A.O. (1984) Chemistry of dioxygen. *Methods Enzymol.* 105, 3–22.

Greenberg, J.T. and Demple, B. (1989) A global response induced in *Escherichia coli* by redox-cycling agents overlaps with that induced by peroxide stress. *J. Bacteriol.* **171**, 3933–3939.

Griffith, T.M., Edwards, D.H., Lewis, M.J., Newby, A.C. and Henderson, A.H. (1984) The nature of endothelium-derived vascular relaxant factor. *Nature* **308**, 645–647.

Grisham, M.B. (1985) Myoglobin-catalyzed, hydrogen peroxide-dependent arachidonic acid peroxidation. *Free Rad. Biol. Med.* **1**, 227–232.

Haennen, G.R.M.M. and Bast, A. (1991) Scavenging of hypochlorous acid by lipoic acid. *Biochem. Pharmacol.* **42**, 2244–2246.

Haennen, G.R.M.M., de Rooij, B.M., Vermeulen, N.P.E. and Bast, A. (1990) Mechanism of the reaction of ebselen with endogenous thiols: dihydrolipoate is a better cofactor than glutathione in the peroxidase activity of ebselen. *Mol. Pharmacol.* **37**, 412–422.

Halliwell, B. and Gutteridge, J.M.C. (1986) Oxygen free radicals and iron in relation to biology and medicine: some problems and concepts. *Arch. Biochem. Biophys.* **246**, 501–514.

Halliwell, B. and Gutteridge, J.M.C. (1990) Role of free radicals and catalytic metal ions in human disease: an overview. *Methods Enzymol.* **186**, 1–85.

Halliwell, B., Hu, M.-L., Louie, S., Duvall, T.R., Tarkington, B.K., Motchnik, P. and Cross, C.E. (1992) Interaction of nitrogen dioxide with human plasma. Antioxidant depletion and oxidative changes. *FEBS Lett.* **313**, 62–66.

Harada, K. Tamura, M. and Yamazaki, I. (1986) The two-electron reduction of sperm whale ferrylmyoglobin by ethanol. *J. Biochem.* **100**, 499–504.

Harada, K. and Yamazaki, I. (1987) Electron spin resonance spectra of free radicals formed in the reaction of metmyoglobins with ethylhydroperoxide. *J. Biochem.* **101**, 283–286.

Hayashi, T., Kanetoshi, A., Nakamura, M., Tamura, M. and Shirahama, H. (1992) Reduction of α-tocopherylquinone to α-tocopherylhydroquinone in rat hepatocytes. *Biochem. Pharmacol.* **44**, 489–493.

Ho, C.T. and Chan, A.C. (1992) Regeneration of vitamin E in rat polymorphonuclear leukocytes. *FEBS Lett.* **306**, 269–272.

Hogg, N., Darley-Usmar, V.M., Moncada, S.M. and Wilson, M.T. (1992) Production of hydroxyl radicals from the simultaneous generation of superoxide and nitric oxide. *Biochem. J.* **281**, 419–424.

Howard-Flanders, P. (1981) Inducible repair of DNA. *Sci. Am.* **245**, 72–80.

Hutchinson, P.J.A., Palmer, R.M.J. and Moncada, S. (1987) Comparative pharmacology of EDRF and nitric oxide on vascular strips. *Eur. J. Pharmacol.* **141**, 445–451.

Ingold, K.U., Bowry, V.W., Stocker, R. and Walling, C. (1993) Autoxidation of lipids and antioxidation by α-tocopherol and ubiquinol in homogeneous solution and in aqueous dispersions of lipids: unrecognized consequences of lipid particle size as exemplified by oxidation of human low density lipoprotein. *Proc. Natl. Acad. Sci. USA* **90**, 45–49.

Ishii, T., Iwahashi, H., Sugata, R. and Kido, R. (1992) Oxidation of 3-hydroxykyn-urenine catalyzed by methemoglobin with hydrogen peroxide. *Free Rad. Biol. Med.* 13, 17–20.

Jore, D., Ferradini, C., Madden, K.P. and Patterson, L.K. (1991) Spectra and structure of α-tocopherol radicals produced in anoxic media. *Free Radical Biol. Med.* 11, 349–352.

Kagan, V.E., Serbinova, E.A., Koynova, G.M., Kitanova, S.A., Tyurin, V.A., Stoychev, T.S., Quinn, P.J. and Packer, L. (1990a) Antioxidant action of ubiquinol homologues with different isoprenoid chain length in biomembranes. *Free Rad. Biol. Med.* 9, 117–126.

Kagan, V., Serbinova, E. and Packer, L. (1990b) Antioxidant effects of ubiquinones in microsomes and mitochondria are mediated by tocopherol recycling. *Biochem. Biophys. Res. Commun.* 169, 851–857.

Kagan, V.E., Shvedova, A., Serbinova, E., Khan, S., Swanson, C., Powel, R. and Packer, L. (1992) Dihydrolipoic acid—A universal antioxidant both in the membrane and in the aqueous phase. Reduction of peroxyl, ascorbyl, and chromanoxyl radicals. *Biochem. Pharmacol.* 44, 1637–1649.

Kanner, J., German, J.B. and Kinsella, J.E. (1987) Initiation of lipid peroxidation in biological systems. *Crit. Rev. Food Sci. Nutr.* 25, 317–364.

Kanner, J. and Harel, S. (1985a) Initiation of membranal lipid peroxidation by activated metmyoglobin and methemoglobin. *Arch. Biochem. Biophys.* 237, 314–321.

Kanner, J. and Harel, S. (1985b) Lipid peroxidation and oxidation of several compounds by H_2O_2 activated metmyoglobin. *Lipids* 20, 625–628.

Kanner, J. and Harel, S. (1987) Desferrioxamine as an electron donor. Inhibition of membranal lipid peroxidation initiated by H_2O_2-activated metmyoglobin and other peroxidizing systems. *Free Rad. Res. Commun.* 3, 309–317.

Kanner, J., Harel, S. and Granit, R. (1991) Nitric oxide as an antioxidant. *Arch. Biochem. Biophys.* 289, 130–136.

Kanofsky, J.R. (1984) Singlet oxygen production by lactoperoxidase: halide dependence and quantitation of yield. *J. Photochem.* 25, 105–113.

Kanofsky, J.R. (1986) Singlet oxygen production from the reactions of alkylperoxy radicals. Evidence from 1268 nm chemiluminescence. *J. Org. Chem.* 51, 3386–3388.

Kanofsky, J.R. and Axelrod, B. (1986) Singlet oxygen production by soybean lipoxygenase isozymes. *J. Biol. Chem.* 261, 1099–1104.

Kaur, H. and Halliwell, B. (1990) Action of biologically-relevant oxidizing species upon uric acid. Identification of uric acid oxidation products. *Chem.-Biol. Interact.* 73, 235–247.

Kedderis, G.L., Rickert, D.E., Pandey, R.N. and Hollenberg, P.F. (1986) [18]O studies of the peroxidase-catalyzed oxidation of N-methylcarbazole. *J. Biol. Chem.* 261, 15910–15914.

Kelder, P.P., deMol, N.J. and Janssen, L.H. (1989) Is hemoglobin a catalyst for sulfoxidation of chlorpromazine? An investigation with isolated purified hemoglobin and hemoglobin in monooxygenase and peroxidase mimicking systems. *Biochem. Pharmacol.* 38, 3593–3599.

Kelder, P.P., deMol, N.J. and Janssen, L.H. (1991) Mechanistic aspects of the oxidation of phenothiazine derivatives by methemoglobin in the presence of hydrogen peroxide. *Biochem. Pharmacol.* **42**, 1551–1559.

Kelder, P.P., Fischer, M.J., deMol, N.J. and Janssen, L.H. (1991) Oxidation of chlorpromazine by methemoglobin in the presence of hydrogen peroxide. Formation of chlorpromazine radical cation and its covalent binding to methemoglobin. *Arch. Biochem. Biophys.* **284**, 313–319.

Kellogg, R.E. (1969) Mechanism of chemiluminescence from peroxy radicals. *J. Am. Chem. Soc.* **91**, 5433–5436.

Kelman, D.J. and Mason, R.P. (1992) The myoglobin-derived radical formed on reaction of metmyoglobin with hydrogen peroxide is not a tyrosine peroxyl radical. *Free Radical Res. Commun.* **16**, 27–33.

Khan, A.U. (1984) Discovery of enzyme generation of $^1\Delta g$ molecular oxygen: spectra of $(0,0)^1\Delta g \rightarrow {}^3\Sigma g^-$ IR emission. *J. Photochem.* **25**, 327–334.

King, K.N. and Winfield, M.E. (1963) The mechanism of metmyoglobin oxidation. *J. Biol. Chem.* **238**, 1520–1528.

Kittridge, K. and Willson, R.L. (1984) Uric acid substantially enhances the free radical inactivation of alcohol dehydrogenase. *FEBS Lett.* **170**, 162–164.

Koppenol, W.H. and Butler, J. (1985) Energetics of interconversion reactions of oxyradicals. *Adv. Free Radical Biol. Med.* **1**, 91–132.

Koppenol, W.H., Moreno, J.J., Pryor, W.A., Ischiropoulos, H. and Beckman, J.S. (1992) Peroxynitrite, a cloaked oxidant formed by nitric oxide and superoxide. *Chem. Res. Toxicol.* **5**, 834–842.

Koppenol, W.H. and Liebman, J.F. (1984) The oxidizing nature of the hydroxyl radical. A comparison with the ferryl ion (FeO^{2+}). *J. Phys. Chem.* **88**, 99–101.

Krinsky, N.I. (1979) Biological roles of singlet oxygen. In *Singlet Oxygen* (Wasserman, H.H. and Murray, W.A., eds.), Academic Press, New York, pp. 597–641.

Laranjinha, J.A.N., Almeida, L.M. and Madeira, M.C. (1993) Ferrylmyoglobin (Fe^{IV}) reduction by phenolic microcomponents of diet. Inhibition of lipid peroxidation in human low density liproteins. In *International Conference on Critical Aspects of Free Radicals in Chemistry, Biochemistry, and Medicine* (H. Nohl and H. Esterbauer, Organizers), Book of Abstracts, p. 239.

Lenaz, G., Barnabaei, O., Rabbi, A. and Battino, M. (eds.) (1990) *Highlights in Ubiquinone Research*, Taylor & Francis, London.

Levine, R.L., Climent, I., Farber, J.M., Shames, B.D., Sahakian, J.A. and Rivett, A.J. (1991) Metal-catalyzed oxidation of glutamine synthetase: determinants of proteolytic susceptibility. in *Oxidative Damage and Repair: Chemical, Biological and Medical Aspects* (Davies, K.J.A., ed.), Pergamon Press, New York, pp. 373–379.

Liebler, D.C., Baker, P.F. and Kaysen, K.L. (1990) Oxidation of vitamin E: evidence for competing autoxidation and peroxyl radical trapping reactions of the tocopheroxyl radical. *J. Am. Chem. Soc.* **112**, 6995–7000.

Liebler, D.C., Kaysen, K.L. and Burr, J.A. (1991) Peroxyl radical trapping and autoxidation reactions of α-tocopherol in lipid bilayers. *Chem. Res. Toxicol.* **4**, 89–93.

Liebler, D.C., Kaysen, K.L. and Kennedy, T.A. (1989) Redox cycles of vitamin E: hydrolysis and ascorbic acid dependent reduction of 8a-(alkyldioxy)tocopherones. *Biochemistry* **28**, 9772–9777.

Lindhal, T. (1987) Regulation and deficiencies in DNA repair. *Brit. J. Cancer* **56**, 91–95.

Lubin, B.H. and Kuypers, F.A. (1991) Phospholipid repair in human erythrocytes. in *Oxidative Damage and Repair: Chemical, Biological, and Medical Aspects* (Davies, K.J.A., ed.), Pergamon Press, New York, pp. 557–563.

Maguire, J.J., Wilson, D.S. and Packer, L. (1989) Mitochondrial electron transport-linked tocopheroxyl radical reduction. *J. Biol. Chem.* **264**, 851–857.

Maiorino, M., Ursini, F. and Cadenas, E. (1994) The reactivity of myoglobin towards lipid hydroperoxides. *Free Rad. Biol. Med.* **16**, 661–667.

Maples, K.R. and Mason, R.P. (1988) Free radical metabolite of uric acid. *J. Biol. Chem.* **263**, 1709–1712.

Marcus, M.F. and Hawley, M.D. (1970) Electrochemical studies of the redox behaviour of α-tocopherol. *Biochim. Biophys. Acta* **201**, 1–8.

McCay, P.B., Brueggemann, G., Lai, E.K. and Powell, S.R. (1989) Evidence that α-tocopherol functions cyclically to quench free radicals in hepatic microsomes. Requirement for glutathione and a heat labile factor. *Ann. N.Y. Acad. Sci.* **570**, 32–45.

Melhorn, R.J. and Gomez, J. (1993) Hydroxyl and alkoxyl radical production by oxidation products of metmyoglobin. *Free Rad. Res. Commun.* **18**, 29–41.

Miki, H., Harada, K., Yamazaki, I., Tamura, M. and Watanabe, H. (1989) Electron spin resonance spectrum of Tyr-151 free radical formed in reactions of sperm whale myoglobin with ethyl hydroperoxide and potassium irridate. *Arch. Biochem. Biophys.* **275**, 354–362.

Mitsos, S.E., Kim, D., Lucceshi, B.R. and Fantone, J.C. (1988) Modulation of myoglobin-H_2O_2-mediated peroxidation reactions by sulfhydryl compounds. *Lab. Invest.* **59**, 824–830.

Mohr, D., Bowry, V.W. and Stocker, R. (1992) Dietary supplementation with coenzyme Q_{10} results in increased levels of ubiquinol-10 within circulating lipoproteins and increased resistance of human low-density lipoprotein to the initiation of lipid peroxidation. *Biochim. Biophys. Acta* **1126**, 247–254.

Moncada, S., Marletta, M.A., Hibbs, Jr. J.B. and Higgs, E.A. (Eds) (1992) *The Biology of Nitric Oxide.* Vol. I and II, Porland Press, London.

Moncada, S., Palmer, R.M.J. and Higgs, E.A. (1991) NO: Physiology, pathophysiology and pharmacology. *Pharmacol. Rev.* **43**, 109–142.

Moncada, S., Radomski, M.W. and Palmer, R.M.J. (1988) Endothelium-derived relaxing factor. Identification as nitric oxide and role in the control of vascular tone and platelet function. *Biochem. Pharmacol.* **37**, 2495–2501.

Mordente, A., Martorana, G.E., Santini, S.A., Miggiano, G.A.D., Petitti, T., Giardina, B. and Littarru, G.P. (1993) Effect of coenzyme Q on ferrylmyoglobin. In *International Conference on Critical Aspects of Free Radicals in Chemistry, Biochemistry, and Medicine* (H. Nohl and H. Esterbauer, Organizers), Book of Abstracts, p. 94.

Murphy, M.E., Kolvenbach, R., Aleksis, M., Hansen, R. and Sies, H. (1992) Antioxidant depletion in aortic crossclamping ischemia: increase of the plasma α-tocopherylquinone/α-tocopherol ratio. *Free Rad. Biol. Med.* **13**, 95–100.

Murphy, M.E. and Sies, H. (1990) Visible-range low-level chemiluminescence in biological systems. *Methods Enzymol.* **186**, 595–610.

Murphy, M.E. and Sies, H. (1992) Reversible conversion of nitroxyl anion to nitric oxide by superoxide dismutase. *Proc. Natl. Acad. Sci. USA* **88**, 10860–10864.

Nakamura, M. and Hayashi, T. (1992) Oxidation mechanism of vitamin E analogue (Trolox C, 6-hydroxy-2,2,5,7,8-pentamethylchroman) and vitamin E by horseradish peroxidase and myoglobin. *Arch. Biochem. Biophys.* **299**, 313–319.

Naqui, A., Cadenas, E. and Chance, B. (1986) Reactive oxygen intermediates in biochemistry. *Ann. Rev. Biochem.* **55**, 137–166.

Nathan, C. (1992) Nitric oxide as a secretory product of mammalian cells. *FASEB J.* **6**, 3051–3064.

Nohl, H. (1987) Demonstration of the existence of an organo-specific NADH dehydrogenase in heart mitochondria. *Eur. J. Biochem.* **169**, 585–591.

Öllinger, K., Buffinton, G.D., Ernster, L. and Cadenas, E. (1990) Effect of superoxide dismutase on the autoxidation of substituted hydro-and semi-naphthoquinones. *Chem.-Biol. Interact.* **73**, 53–76.

Omar, B., McCord, J. and Downey, J. (1991) Ischaemia-reperfusion. In *Oxidative Stress: Oxidants and Antioxidants* (Sies, H., ed.), Academic Press, London, (pp. 493–527.)

Ordoñez, I. and Cadenas, E. (1992) Thiol oxidation coupled to DT-diaphorase-catalysed reduction of diaziquone. Reductive and oxidative pathways of diaziquone semiquinone modulated by glutathione and superoxide dismutase. *Biochem. J.* **286**, 481–490.

Ortiz de Montellano, P.R. and Catalano, C.E. (1985) Epoxidation of styrene by hemoglobin and myoglobin. Transfer of oxidizing equivalents to the protein surface. *J. Biol. Chem.* **260**, 9265–9271.

Pacifici, E.H.K., McLeod, L.L. and Sevanian, A. (1993) Lipid hydroperoxide-induced peroxidation and turnover of endothelial cell phospholipids. *Free Rad. Biol. Med.*, in press.

Pacifici, R.E. and Davies, K.J.A. (1991) Selective proteolysis of oxidatively modified proteins by macroxyproteinase (M.O.P.). In *Oxidative Damage and Repair. Chemical, Biological, and Medical Aspects* (Davies, K.J.A., ed.), Pergamon Press, New York, pp. 364–372.

Packer, L., Maguire, J., Mehlhorn, R., Serbinova, E. and Kagan, V. (1989) Mitochondria and microsomal membranes have a free radical reductase that prevents chromanoxyl radical accumulation. *Biochem. Biophys. Res. Commun.* **159**, 229–235.

Paganga, G., Rice-Evans, C., Rule, R. and Leake, D. (1992) The interaction between ruptured erythrocytes and low-density lipoproteins. *FEBS Lett.* **303**, 154–158.

Peinado, J., Sies, H. and Akerboom, T.P.M. (1989) Hepatic lipoate uptake. *Arch. Biochem. Biophys.* **273**, 389–395.

Pryor, W.A. (1976) The role of free radication reactions in biological systems. In *Free Radicals in Biology* (Pryor, W.A., ed.), Vol. I, Academic Press, New York, pp. 1–49.

Pryor, W.A. (1986) Oxy-radicals and related species: their formation, lifetimes, and reactions. *Ann. Rev. Physiol.* **48**, 657–667.

Pryor, W.A. (1988) Why is the hydroxyl radical the only radical that commonly adds to DNA? Hypothesis: It has a rare combination of high electrophilicity, high thermochemical reactivity, and a mode of production that can occur near DNA. *Free Rad. Biol. Med.* **4**, 219–223.

Puppo, A., Cicchini, R., Aruoma, O.I., Bolli, R. and Halliwell, B. (1990) Scavenging of hypochlorous acid and of myoglobin-derived oxidants by the cardioprotective agent mercaptopropionylglycine. *Free Rad. Res. Commun.* **10**, 371–381.

Radi, R., Beckman, J.S., Bush, K.M. and Freeman, B.A. (1991a) Peroxynitrite oxidation of sulfhydryls. *J. Biol. Chem.* **266**, 4244–4250.

Radi, R., Beckman, J.W., Bush, K.M. and Freeman, B.A. (1991b) Peroxynitrite induced membrane lipid peroxidation: the cytotoxic potential of superoxide and nitric oxide. *Arch. Biochem. Biophys.* **288**, 481–487.

Rao, S.I., Wilks, A. and Ortiz de Montellano, P.R. (1993) The roles of His-64, Tyr-103, Tyr-146, and Tyr-151 in the epoxidation of styrene and β-methylstyrene by recombinant sperm whale myoglobin. *J. Biol. Chem.* **268**, 803–809.

Rice, R.H., Lee, Y.M. and Brown, W.D. (1983) Interaction of heme proteins with hydrogen peroxide: protein cross-linking and covalent binding of benzo[a]pyrene and 17β-estradiol. *Arch. Biochem. Biophys.* **221**, 417–427.

Rice-Evans, C. (1993) Personal communication.

Rice-Evans, C.A. and Diplock, A.T. (1993) Current status of antioxidant therapy. *Free Rad. Biol. Med.* **15**, 77–96.

Rice-Evans, C., Green, E., Paganga, G., Cooper, C. and Wrigglesworth, J. (1993) Oxidised low density lipoproteins induce iron release from activated myoglobin. *FEBS Lett.* **326**, 177–182.

Rice-Evans, C., Okunade, G. and Khan, R. (1989) The suppression of iron release from activated myoglobin by physiological electron donors and by desferrioxamine. *Free Rad. Res. Commun.* **7**, 45–54.

Richter, C., Park, J.W. and Ames, B.N. (1988) Normal oxidative damage to mitochondrial and nuclear DNA is extensive. *Proc. Natl. Acad. Sci. USA* **85**, 6455–6467.

Richter, C. (1988) Do mitochondrial DNA fragments promote cancer and aging? *FEBS Lett.* **241**, 1–5.

Rivett, A.J. (1985) Preferential degradation of the oxidatively modified form of glutamine synthetase by intracellular proteases. *J. Biol. Chem.* **260**, 300–305.

Rivett, A.J. (1989) The multicatalytic proteinase of mammalian cells. *Arch. Biochem. Biophys.* **268**, 1–8.

Rivett, A.J. and Hare, J.F. (1987) Enhanced degradation of oxidized glutamine synthetase *in vitro* and after microinjection into hepatoma cells. *Arch. Biochem. Biophys.* **259**, 423–430.

Rivett, A.J. and Levine, R.L. (1990) Metal catalysed oxidation of *E. coli* glutamine synthetase: correlation of structural and functional changes. *Arch. Biochem. Biophys.* **278**, 26–34.

Romero, F.J., Ordoñez, D., Arduini, A. and Cadenas, E. (1992) The reactivity of thiols and disulfides with different redox states of myoglobin. Redox and addition reactions and formation of thiyl radical intermediates. *J. Biol. Chem.* **267**, 1680–1688.

Ross, D. and Moldéus, P. (1986) Thiyl radicals—Their generation and further reactions. In *Biological Reactive Intermediates III. Mechanisms of Action in Animal Models and Human Disease* (Kocsis, J.J., Jollow, D.J., Witmer, C.M., Nelson, J.P. and Snyder, R., eds.), Plenum Press, New York, pp. 329–335.

Rougée, M. and Bensasson, R. (1986) Détermination des constantes de vitesse de désactivation de l'oxygène singulet ($^1\Delta g$) en presence de biomolécules. *C. R. Acad. Sci. Paris* **20**, 1223–1226.

Russell, G.A. (1957) Deuterio-isotope effects in the autoxidation of aralkyl hydrocarbons. Mechanism of the interaction of peroxy radicals. *J. Am. Chem. Soc.* **79**, 3871–3877.

Saran, M., Michel, C. and Bors, W. (1990) Reaction of NO with O_2^-. Implications for the action of endothelium-derived relaxing factor (EDRF). *Free Rad. Res. Commun.* **10**, 221–226.

Sawyer, D.T., Calderwood, T.S., Johlman, C.L. and Wilkins, C.L. (1985a) Oxidation by superoxide anion of catechols, ascorbic acid, hydrophenazine, and reduced flavins to their respective anion radicals. A common mechanism with a sequential proton-hydrogen atom transfer. *J. Org. Chem.* **50**, 1409–1412.

Sawyer, D.T., Roberts, J.L. Jr., Calderwood, T.S., Sugimoto, H. and McDowell, M.S. (1985b) Reactivity and activation of dioxygen-derived species in aprotic media (a model matrix for biomembranes). *Phil. Trans. R. Soc. London* **B311**, 483–503.

Schmidt, U., Grafen, P., Altland, K. and Goedde, H.W. (1969) Biochemistry and chemistry of lipoic acids. *Adv. Enzymol.* **32**, 423–467.

Schöneich, C., Asmus, K.-D., Dillinger, U. and von Bruchhausen, F. (1989) Thiyl radical attack on polyunsaturated fatty acids: a possible route to lipid peroxidation. *Biochem. Biophys. Res. Commun.* **161**, 113–120.

Scholich, H., Murphy, M.E. and Sies, H. (1989) Antioxidant activity of dihydrolipoate against microsomal lipid peroxidation and its dependence on α-tocopherol. *Biochim. Biophys. Acta* **1001**, 256–261.

Schöneich, C., Dillinger, U., von-Bruchhausen, F. and Asmus, K.-D. (1992) Oxidation of polyunsaturated fatty acids and lipids through thiyl and sulfonyl radicals: reaction kinetics, influence of oxygen and structure of thiyl radicals. *Arch. Biochem. Biophys.* **292**, 456–467.

Serbinova, E., Reznick, S.K.A.Z. and Packer, L. (1992) Thioctic acid protects against ischemia-reperfusion injury in the isolated perfused Langendorf heart. *Free Rad. Res. Commun.* **17**, 49–58.

Sevanian, A. (1991) Lipid damage and repair. In *Oxidative Damage and Repair: Chemical, Biological, and Medical Aspects* (Davies, K.J.A., ed.), Pergamon Press, New York, pp. 543–549.

Simic, M.G. (1991) Antioxidant compounds: an overview. In *Oxidative Damage and Repair. Chemical, Biological, and Medical Aspects* (Davies, K.J.A., ed.), Pergamon Press, New York, pp. 47–56.

Simic, M.G. and Jovanovic, S.V. (1989) Antioxidation mechanisms of uric acid. *J. Am. Chem. Soc.* **111**, 5778–5782.

Simpson, K.E. and Dean, R.T. (1990) Stimulatory and inhibitory actions of proteins and amino acids on copper-catalysed free radical generation in the bulk phase. *Free Rad. Res. Commun.* **10**, 303–312.

Snyder, L.M., Fortier, N.L., Leb, L., McKenney, J., Trainor, J., Sheerin, H. and Mohands, N. (1988) The role of membrane protein sulfhydryl groups in hydrogen peroxide-mediated membrane damage in human erythrocytes. *Biochim. Biophys. Acta* **937**, 229–240.

Stadtman, E.R. (1990*a*) Metal ion-catalyzed oxidation of proteins: biochemical mechanism and biological consequences. *Free Rad. Biol. Med.* **9**, 315–325.

Stadtman, E.R. (1990*b*) Covalent modification reactions are marking steps in protein turnover. *Biochemistry* **29**, 6323–6331.

Stadtman, E.R. (1991) Protein damage and repair. In *Oxidative Damage and Repair: Chemical, Biological, and Medical Aspects* (Davies, K.J.A., ed.), Pergamon Press, New York, pp. 348–354.

Starke-Reed, P.E. and Oliver, C.N. (1989) Protein oxidation and proteolysis during aging and oxidative stress. *Arch. Biochem. Biophys.* **275**, 559–567.

Steinberg, D., Parthasarathy, S., Carew, T.E., Khoo, J.C. and Witztum, J.L. (1989) Beyond cholesterol. Modifications of low-density liprotein that increase atherogenicity. *New Engl. J. Med.* **320**, 915–924.

Stocker, R., Bowry, V.W., and Frei, B. (1991) Ubiquinol-10 protects human low density lipoprotein more efficiently against lipid peroxidation than does α-tocopherol. *Proc. Natl. Acad. Sci. USA* **88**, 1646–1650.

Stocker, R. and Frei, B. (1991) Endogenous antioxidant defences in human blood plasma. In *Oxidative Stress: Oxidants and Antioxidants* (Sies, H., ed.), Academic Press, London, pp. 213–244.

Surdhar, P.S. and Armstrong, D.A. (1986) Redox potentials of some sulfur-containing radicals. *J. Phys. Chem.* **90**, 5915–5917.

Suzuki, Y., Tsuchiya, M. and Packer, L. (1991) Thioctic acid and dihydrolipoic acid are novel antioxidants which interact with reactive oxygen species. *Free Rad. Res. Commun.* **15**, 255–263.

Tamba, M. and Quintiliani, M. (1984) Kinetic studies of reactions involved hydrogen transfer from glutathione to carbohydrate radicals. *Radiat. Phys. Chem.* **23**, 259–263.

Tamura, M., Oshino, N., Chance, B. and Silver, I. (1978) Optical measurements of intracellular oxygen concentration of rat heart *in vitro*. *Arch. Biochem. Biophys.* **191**, 8–22.

Teebor, G.W., Boorstein, R.J. and Cadet, J. (1988) The repairability of oxidative free radical mediated damage to DNA: a review. *Int. J. Radiat. Biol.* **43**, 131–150.

Tew, D. and Ortiz de Montellano, P.R. (1988) The myoglobin protein radical. Coupling of Tyr-103 in the H_2O_2-mediated cross-linking of sperm whale myoglobin. *J. Biol. Chem.* **263**, 17880–17886.

Turner, J.J.O., Rice-Evans, C.A., Davies, M.J. and Newman, E.S. (1990) Free radicals, myocytes and reperfusion injury. *Biochem. Soc. Trans.* **18**, 1056–1059.

Turner, J.J.O., Rice-Evans, C.A., Davies, M.J. and Newman, E.S.R. (1991) The formation of free radicals by cardiac myocytes under oxidative stress and the effects of electron-donating drugs. *Biochem. J.* **277**, 833–837.

Uyeda, M. and Peisach, J. (1981) Ultraviolet difference spectroscopy of myoglobin: assignment of pK values of tyrosyl phenolic groups and the stability of the ferryl derivatives. *Biochemistry* **20**, 2028–2035.

Valtersson, C., van Duÿn, G., Verkleij, A.J., Chojnacki, T., de Kruijff, B. and Dallner, G. (1985) The influence of dolichol, dolichol esters, and dolichyl phosphate on phospholipid polymorphism and fluidity in model membranes. *J. Biol. Chem.* **260**, 2742–2751.

van Kuijk, F.J.G.M., Sevanian, A., Handelman, G. and Dratz, E.A. (1987) A new role for phospholipase A₂. *Trends Biochem. Sci.* **12**, 31–34.

von Sonntag, C. (1987) *The Chemical Basis of Radiation Biology*, Taylor & Francis, London.

Wallace, G.C., Gulate, P. and Fukuto, J.M. (1991) N-Hydroxy-L-arginine: A novel arginine analog capable of causing vasodilation in bovine intrapulmonary artery. *Biochem. Biophys. Res. Commun.* **176**, 528–534.

Walters, F.P., Kennedy, F.G. and Jones, D.P. (1983) Oxidation of myoglobin in isolated adult rat cardiac myocytes by 15-hydroperoxy-5,8,11,13-eicosatetraenoic acid. *FEBS Lett.* **163**, 292–296.

Wardman, P. (1988) Conjugation and oxidation of glutathione via thiyl free radicals. In *Glutathione Conjugation. Mechanisms and Biological Significance* (Sies, H. and Ketterer, B., eds.), Academic Press, London, pp. 44–72.

Wardman, P. (1990a) Thiol reactivity towards drugs and radicals: some implications in the radiotherapy and chemotherapy of cancer. In *Sulfur-centred Reactive Intermediates in Chemistry and Biology* (Chatgilialoglu, C. and Asmus, K.-D., eds.), Plenum Press, New York, pp. 415–427.

Wardman, P. (1990b) Bioreductive activation of quinones: redox properties and thiol reactivity. *Free Rad. Res. Commun.* **8**, 219–229.

White, K.A. and Marletta, M.A. (1992) Nitric oxide synthase is a cytochrome P₄₅₀ type hemoprotein. *Biochemistry* **31**, 6627–6631.

Wilks, A. and Ortiz de Montellano, P.R. (1992) Intramolecular translocation of the protein radical formed in the reaction of recombinant sperm whale myoglobin with H₂O₂. *J. Biol. Chem.* **267**, 8827–8833.

Willson, R.L. (1970) Pulse radiolysis studies of electron transfer in aqueous disulphide solutions. *Chem. Commun.*, p. 1425–1426.

Willson, R.L., Dunster, C.A., Forni, L.G., Gee, C.A. and Kittridge, K.J. (1985) Organic free radicals and proteins in biochemical injury: electron- or hydrogen-transfer reactions? *Phil. Trans. R. Soc. Lond. B* **311**, 545–563.

Wilson, I., Wardman, P., Cohen, G.M. and d'Arcy-Doherty, M. (1986) Reductive role of glutathione in the redox cycling of oxidizable drugs. *Biochem. Pharmacol.* **35**, 21–22.

Winterbourn, C.C. (1981) Hydroxyl radical production in body fluids. Roles of metal ions, ascorbate and superoxide. *Biochem. J.* **198**, 125–131.

Winterbourn, C.C. (1989) Inhibition of autoxidation of divicine and isouramil by the combination of superoxide dismutase and reduced glutathione. *Arch. Biochem. Biophys.* **271**, 447–455.

Winterbourn, C.C., Gutteridge, J.M.C. and Halliwell, B. (1985) Doxorubicin-dependent lipid peroxidation at low partial pressures of O₂. *J. Free Rad. Biol. Med.* **1**, 43–49.

Winterbourn, C.C. and Munday, R. (1989) Glutathione-mediated redox cycling of alloxan: mechanisms of superoxide dismutase inhibition and of metal-catalysed ·OH formation. *Biochem. Pharmacol.* **38**, 271–277.

Winterbourn, C.C. and Munday, R. (1990) Concerted action of reduced glutathione and superoxide dismutase in preventing redox cycling of dihydropyrimidines, and their role in antioxidant defence. *Free Rad. Res. Commun.* **8**, 287–293.

Wong, G.H.W. and Goeddel, D.V. (1988) Induction of manganous superoxide dismutase by tumor necrosis factor: possible protective mechanism. *Science* **242**, 941–944.

Xu, F. and Hultquist, D.E. (1991) Coupling of dihydroriboflavin oxidation to the formation of the higher valence states of hemeproteins. *Biochem. Biophys. Res. Commun.* **181**, 197–203.

Xu, F., Mack, C.P., Quandt, K.S., Shlafer, M., Massey, V. and Hultquist, D.E. (1993) Pyrroloquinoline quinone acts with flavin reductase to reduce ferrylmyoglobin *in vitro* and protects isolated heart from re-oxygenation injury. *Biochem. Biophys. Res. Commun.* **193**, 434–439.

Yamada, T., Volkmer, C. and Grisham, M.B. (1991) The effects of sulfasalazine metabolites on hemoglobin-catalyzed lipid peroxidation. *Free Rad. Biol. Med.* **10**, 41–49.

Yamauchi, R., Kato, K. and Ueno, Y. (1988) Formation of trimers of α-tocopherol and its model compound, 2,2,5,7,8-pentamethylchroman-6-ol, in autoxidizing methyl linoleate. *Lipids* **23**, 779–783.

Yang, G., Candy, T.E.G., Boaro, M., Wilkin, H.E., Jones, P., Nazhat, N.B., Saadalla-Nazhat, R.A. and Blake, D.R. (1992) Free radical yields from the homolysis of peroxynitrous acid. *Free Rad. Biol. Med.* **12**, 327–330.

Yang, W.D., De Bono, D. and Symons, M.C.R. (1993) The effects of myoglobin and apomyoglobin on the formation and stability of the hydroxyl radical adduct of 5,5'-dimethyl-1-pyrroline-N-oxide. *Free Rad. Res. Commun.* **18**, 99–106.

CHAPTER 2

Pathophysiology and Reactive Oxygen Metabolites

Yan Chen, Allen M. Miles and Matthew B. Grisham

2.1 INTRODUCTION

Aerobic metabolism provides an organism with a distinct metabolic advantage in that it allows for the complete combustion of glucose to carbon dioxide and water, and in the process produces 36 moles of ATP. However, all aerobic organisms pay a price for this metabolic advantage. It is known that during normal metabolism a small but significant flux of reactive oxygen metabolites (ROMs) are produced by the mitochondrial electron transport chain and by a variety of different oxidases. More specialized cells such as erythrocytes may generate even greater amounts of ROMs via the spontaneous autooxidation of important biological substrates such as hemoglobin. Thus, virtually all aerobic organisms possess several different enzymatic and nonenzymatic antioxidants including superoxide dismutase (SOD) or SOD-like chelates, catalase and/or peroxidases, as well as a variety of metal binding proteins and low molecular weight antioxidants.

It is becoming increasingly apparent that the inadvertent overproduction of ROMs may overwhelm the protective oxidant defenses resulting in oxidative tissue injury. ROMs have been implicated in several different disease states including cardiovascular disease, chronic gut inflammation, arthritis, shock, sickle cell anemia, cancer and AIDS (Halliwell et al., 1992a, 1992b). In addition, it has been suggested that continuous oxidative stress over many years may in part be responsible for the aging process (Halliwell et al., 1992a, 1992b). Although numerous pathophysiological situations have been suggested to involve ROMs, this discussion will focus on just a few conditions in which there is abundant experimental evidence that implicates these reactive species as mediators of the cell injury and dysfunction.

2.2 ISCHEMIA AND REPERFUSION INJURY

The concept that ROMs play an important role in mediating the microvascular dysfunction associated with reperfusion of the ischemic tissue was first proposed in 1981 (Granger et al., 1981). Information derived from numerous studies performed over the past several years has led to a biochemical hypothesis to explain oxygen-dependent reperfusion injury. This hypothesis states that xanthine oxidase-derived oxidants produced following reoxygenation of ischemic tissue promotes adherence of circulating neutrophilic polymorphonuclear leukocytes (PMNs) to venular endothelium where these cells ultimately mediate reperfusion-induced microvascular injury (Granger, 1988). This discussion will summarize the evidence supporting this hypothesis and present newer data that may eventually lead to additional modifications of this hypothesis.

An important assumption in the oxygen radical hypothesis of ischemia/reperfusion (I/R) injury is that the tissue injury observed following reperfusion is due to the reintroduction of oxygen rather that a delayed manifestation of injury incurred during the ischemic period. Although the validity of this assumption has not been definitively resolved, there are several lines of evidence that tend to support it. Parks and Granger (1986c) have demonstrated mucosal injury produced by 4 hours of ischemia without reperfusion. They also found that reperfusion of intestine with deoxygenated perfusate after 3 hours of ischemia produced significantly less injury than that observed following reperfusion with oxygenated whole blood. Microvascular permeability to plasma protein has proven to be a useful and sensitive index for evaluating the influence of ischemia/reperfusion on microvascular integrity. For example, if the cat small bowel is subjected to one hour of ischemia (blood flow reduced to about 20% of control) without reperfusion, microvascular permeability increases by approximately 2-fold. However, the same period of ischemia followed by reperfusion increases microvascular permeability by approximately 5-fold. The assumption that reoxygenation accounts for the greater rise in permeability after reperfusion is supported by the observation that antioxidants and inhibition of oxy-radical formation (e.g., allopurinol) attenuate only the increased permeability induced by reperfusion. Another argument that is used to defend the position that reperfusion per se is largely responsible for the injury observed in intestinal models of ischemia-reperfusion is that inhibitors of oxy-radical production (e.g., allopurinol, oxypurinol) offer protection when administered at the time of reperfusion.

The concept that cellular ATP levels decrease while hypoxanthine accumulates in ischemic tissues is well documented for many organs including the intestine (Blum et al., 1986; Schoenberg et al. 1985). Schoenberg et al. (1985) have demonstrated that 2 hours of ischemia reduces ATP concentration by approximately 40% of the preischemic value (intestinal arterial pressure decreased to 25-30 mmHg). The depletion of ATP is associated with increases in tissue levels of AMP (8-fold), hypoxanthine (10-fold) and uric acid (4-fold). Ten minutes following reperfusion, the levels of AMP and hypoxanthine remain significantly elevated. The relationship between duration of ischemia and the extent of ATP depletion has not been defined for the cat intestine. However, ^{31}P nuclear magnetic resonance (NMR) spectroscopy has been used to follow the ATP depletion that occurs in the ischemic rat intestine (Blum et al. 1986). The results of this analysis reveal that depletion of mucosal ATP is rapid and complete within 20 minutes of total ischemia, i.e., longer periods of ischemia do not produce a further decline in ATP levels. This observation is consistent with reports that only 30 minutes of ischemia are needed to produce prolonged functional and structural changes in the rat intestine (Robinson et al. 1981). It has been demonstrated that the tissue hypoxanthine levels in the normally perfused intestinal mucosa are approximately 20 μM, increasing to more than 200 μM during ischemia (Schoenberg et al., 1985). Mousson et al. (1983) have determined that the K_m for hypoxanthine is 11μM for xanthine oxidase isolated from human jejunum. These observations suggest that the tissue concentration of hypoxanthine in both normal and ischemic bowel is not rate-limiting in the production of uric acid by xanthine dehydrogenase and/or xanthine oxidase.

There is considerable evidence to suggest that oxidants play a role in the increased microvascular permeability produced by ischemia followed by reperfusion. SOD and copper di-isopropyl salicylate (CuDIPS), a lipophilic SOD mimetic, both attenuate I/R-induced increases in microvascular permeability (Granger, 1988). Although the beneficial effects of SOD and CuDIPS implicate a role for the superoxide anion radical O_2^- in I/R-induced microvascular injury, more recent evidence suggests that other oxidants derived from O_2^- may play an equally important role in the injury process. Decomposition of hydrogen peroxide (H_2O_2) by the intravenous administration of catalase (CAT) significantly attenuates the I/R-induced increase in microvascular permeability. Similar results are obtained with dimethylsulfoxide (DMSO), a hydroxyl radical ($^.OH$) scavenger (Granger, 1988). Originally, it was proposed that the protective effects of SOD, CAT and dimethylsulfoxide (DMSO) were consistent with the concept that secondarily-derived radicals such as $^.OH$ are formed during reperfusion by the iron-catalyzed interaction between xanthine oxidase-generated O_2^- and its dismutation product H_2O_2. However, it should be noted that DMSO has other biological effects including the ability to inhibit neutrophil adherence (Sekizuka et al. 1989). Furthermore, SOD and CAT are also known to inhibit the adherence of PMNs to post capillary venules in vivo (Suzuki et al., 1989, 1991).

The low to moderate reactivity of O_2^- and H_2O_2, coupled to the fact that DMSO was shown to be protective in inhibiting I/R-induced microvascular injury, suggested that the microvascular injury produced by O_2^- and/or H_2O_2 may actually result from the generation of secondary, highly reactive free radicals. One mechanism by which these ROMs may interact to generate damaging oxygen species is via their reaction with certain transition metals, such as iron (Fe) or copper (Cu) and/or their low molecular chelates, to yield the highly reactive hydroxyl radical ($^.OH$) (see Chapter 1):

$$O_2^- + Fe^{3+} \longrightarrow O_2 + Fe^{2+} \qquad (2.1)$$

$$H_2O_2 + Fe^{2+} \longrightarrow {}^.OH + OH^- + Fe^{+3} \qquad (2.2)$$

$$\text{Sum: } O_2^- + H_2O_2 \longrightarrow {}^.OH + OH^- + O_2 \qquad (2.3)$$

In this type of reaction Fe acts as catalyst, being cyclically reduced by O_2^- (eq. 2.1), and then oxidized by H_2O_2 (eq. 2.2). Reactions 2.1 and 2.2 have been termed the metal-catalyzed Haber-Weiss reaction or, more appropriately, the O_2^- driven Fenton reaction (eq. 2.3). Metal-containing compounds that have been implicated as biological catalysts for this reaction are the low molecular weight chelates of Fe such as with adenosine diphosphate, $ADP\text{-}Fe^{3+}$, citrate-Fe^{3+}, and certain amino acid chelates of Fe^{3+}. It should be noted that although iron-loaded ferritin, transferrin or lactoferrin may not function directly as catalysts for this reaction, $O_2^{\cdot-}$ is known to reductively release Fe from these proteins, thereby liberating redox active iron that may participate in the $O_2^{\cdot-}$-driven Fenton reaction. Potent iron chelators such as desferrioxamine (desferal) bind iron and inhibit this reaction. Because of the potential to generate noxious free radicals such as ˙OH, the body has taken great steps to sequester low molecular transition metals such as iron or copper into macromolecules to prevent their participation in redox reactions. Thus, it is very difficult to demonstrate mobilization of redox active iron, except in extreme cases such as iron overload (hemochromatosis) or hemmorhagic shock. As discussed below, this presents an interesting dilemma in explaining the mechanism by which small amounts of an iron chelator desferal protect the microcirculation from the injurious effects of ischemia and reperfusion.

Ramos et al. (1992) have also proposed an alternative mechanism by which ˙OH may be generated in an iron-independent manner. They propose that $O_2^{\cdot-}$ and hypochlorous acid (HOCl) generated by stimulated neutrophils may interact to yield ˙OH in the following reaction (eq. 2.4):

$$O_2^{\cdot-} + HOCl \longrightarrow {\cdot}OH + O_2 + Cl^- \qquad (2.4)$$

Evidence that this reaction may occur in vivo remains tenuous, and has not been directly demonstrated using reagent HOCl and $O_2^{\cdot-}$ (Dix and Aikens, 1992).

Hydroxyl radical is an extremely reactive species, reacting immediately with virtually all known biomolecules at diffusion limited rates ($\sim 10^7 - 10^{10}$ M-sec). *Thus, it is very short lived and will react at the site where it is formed, i.e., in a site-specific fashion.* This reactivity dictates that the radical will be injurious to cells only if the metal catalyst is localized on biomolecules essential for survival. For example, DNA, membrane phospholipids, or proteins required for essential metabolic processes could be potential targets for such a mechanism. For example, ˙OH radicals are able to abstract methylene hydrogen atoms from membrane-associated polyunsaturated fatty acids (PUFA). This reaction initiates the process of lipid peroxidation and leads to the formation of lipid-derived free radicals such as conjugated dienes, lipid hydroperoxide radicals, and hydroperoxides (LOOH). Measurements of conjugated dienes are commonly used as an index of lipid peroxide formation. Schoenberg and co-workers (1985; Younes et al., 1987) have reported increased levels of conjugated dienes in the reperfused cat intestine. The reperfusion-induced increase in mucosal conjugated dienes was not observed in animals treated with either SOD or allopurinol. These observations suggest that powerful oxidants (e.g., ˙OH) mediate the lipid peroxidation associated with reperfusion of the small intestine. If, on the other hand, ˙OH is generated on biomolecules present in high concentrations (e.g., albumin or glucose), the physiological consequences of such degradative reactions may be minimal. The role of iron in mediating I/R induced intestinal injury has been tested using desferrioxamine (an iron chelator), or apotransferrin (an iron-binding protein), and the results showed a significant protection provided by both of the two regents in I/R induced intestinal microvascular permeability enhancement (Hernandez et al., 1987a). The observation that iron-loaded desferrioxamine or transferrin did not offer protection against I/R injury argues against a nonspecific protective influence of these iron binding substances. Although these observations are consistent with the formation of ˙OH or ˙OH-like species via the superoxide-driven Fenton reaction, more recent data suggests an equally attractive alternative mechanism for the generation of ˙OH in an iron-independent reaction (see below).

The source of the reactive oxygen species produced during reperfusion was initially assumed to be xanthine oxidase (XO). Compared to other tissues, the intestinal mucosa has a tremendous capacity to oxidize hypoxanthine via XO. Xanthine dehydrogenase (XD) plus XO activities in the intestinal mucosa is approximately 100 mU per gram wet weight of tissue in most species (Parks and Granger, 1986a). The cytotoxic potential of this high enzyme activity is exemplified by the observation that isolated cells are injured when exposed to XO levels as low as 2 mU/ml (Simon et al., 1981). Intestinal XO activity is found primarily in the mucosal layer, with an increasing gradient of activity from villus base to tip (Granger et al., 1986). This is consistent with the observation that the villus tip is more sensitive to ischemic injury than the base. Immunolocalization studies suggest that mucosal XO is found exclusively in endothelial cells lining the microvasculature in the villus core (Jarasch et al., 1986). However, histochemical studies demonstrate preferential localization of the enzyme in epithelial cells (Granger et al., 1986). We have found that fresh isolated villus epithelium accounts for the majority of mucosal XO in cat ileal mucosa (Hernandez et al., 1987b). Nonetheless, the significant enzyme activity remaining in the lamina propria may be concentrated in microvascular endothelium.

XO exists in normal healthy cells predominantly as a NAD^+-reducing XD. Conversion of XD to XO can be initiated either by limited proteolysis, oxidation of sulfhydryl groups, or both (Parks and Granger, 1986b). The XO formed by limited proteolysis is irreversible, while sulfhydryl-reducing agents (e.g., dithiothreitol) can reverse the XO formed by sulfhydryl oxidation of XD. Relatively little is known about the mechanisms and kinetics of XD to XO conversion in the ischemic small intestine. Even after extensive efforts to limit artifactual conversion of XD to XO during tissue preparation, 10-20% of the enzyme exists as the oxidase form in normal bowel. Using extensive controls and careful maintenance of temperature during ischemia, Parks et al. (1988) found that only 50% of XD is converted to XO in rat bowel after 2 hours of ischemia. This represents a 2-3-fold increase in activity of the oxidant producing form of the enzyme. A slower rate of conversion was noted in cat ileal mucosa.

The assumption that XO plays a critical role in I/R injury is supported by studies using the inhibitor allopurinol. Parks et al. (1982) and Schoenberg et al. (1985) have reported a marked reduction in the severity of mucosal lesions induced by I/R in allopurinol-treated animals. Similar protection is noted in the following treatment with oxypurinol, the long-lived metabolite of allopurinol. Allopurinol has also been shown to attenuate the increased intestinal microvascular permeability to plasma proteins induced by I/R (Granger, 1988). Enteral administration of either pterin aldehyde or folic acid also attenuates the I/R-induced increase in microvascular permeability. Both compounds are effective inhibitors of XO, with the potency of pterin aldehyde comparable to that of oxypurinol (Hernandez et al., 1987c).

Another approach that has been used to assess the role of XO in I/R-injury is to place animals on a molybdenum-deficient, tungsten (W) supplemented diet. This regimen leads to incorporation of tungsten, rather than molybdenum, into XO and thereby inactivates it. Using the W-supplemented diet, Parks and et al. (1986a) observed a 75% reduction in mucosal XD+XO activity and a corresponding attenuation of the I/R-induced increase in intestinal microvascular permeability. A limitation of this approach is that other molybdenum containing enzymes (e.g., aldehyde oxidase) are also inactivated by tungsten supplementation. Although the protection afforded by allopurinol has generally been attributed to decreased production of XO-derived oxidants, preservation of the nucleotide pool and free radical scavenging are frequently invoked as alternate explanations. Allopurinol does appear to attenuate the reduction in cellular ATP produced by intestinal ischemia; however, maintenance of ATP levels by administration of exogenous purines (e.g., inosine) does not afford protection against I/R injury in cat intestine (Schoenberg et al., 1985). The observation that administration of either allopurinol or oxypurinol at the onset of reperfusion is equally as effective as pretreatment (administration before ischemia) in attenuating reperfusion injury also argues against a major role for purine salvage (Morris et al., 1987). However, the latter observation does not negate the possibility that allopurinol affords protection by acting as a free radical scavenger.

There are several reports which suggest that the beneficial effects of allopurinol in models of I/R may be due to direct free radical scavenging properties of the drug rather than its ability to inhibit XO. Moorhouse et al. (1987) have reported that allopurinol and oxypurinol are powerful scavengers of ˙OH radicals in vitro. They also reported that oxypurinol is a scavenger of the neutrophil-derived oxidant HOCl, which may be involved in the microvascular injury associated with I/R. We found that the regimen of allopurinol administration used in most I/R studies leads to an extracellular allopurinol/oxypurinol concentration of 10-20 μM, which effectively inhibits XO activity but does not significantly enhance the scavenging properties of extracellular fluid (Zimmerman et al., 1988). In vitro studies suggest that the allopurinol/oxypurinol concentration must exceed 500 μM to significantly enhance the radical scavenging potential of extracellular fluid (Moorhouse et al., 1987).

Another potential source of oxidants in the small intestine are the neutrophilic PMNs. Measurements of mucosal myeloperoxidase (MPO) activity in cat intestine suggest that there are approximately 10 million granulocytes per gram tissue (Grisham et al., 1986; Granger, 1988). When maximally activated, these cells can generate a superoxide flux of approximately 35 nmol/min/g tissue, a rate which may be directly or indirectly cytotoxic. Activated granulocytes also secrete a variety of enzymes (myeloperoxidase, collagenase, elastase) that can injure parenchymal cells and the microvasculature (Weiss, 1986). We have examined the influence of ischemia and reperfusion on granulocyte fluxes in the cat intestinal mucosa using tissue associated MPO activity (Grisham et al., 1986). We observed a significant increase in mucosal MPO during the ischemic period, while reperfusion produced an even more dramatic enhancement of MPO activity. In an attempt to determine whether I/R-induced tissue granulocyte infiltration is related to the formation of XO-derived oxidants, we examined the influence of treatment with either SOD or allopurinol on the I/R-induced increase in mucosal MPO activity (Grisham et al., 1986). The results of these studies indicate that both SOD and allopurinol significantly attenuate the increased mucosal MPO activity observed after reperfusion. This attenuation of MPO activity reflected a decrease in the amount of enzyme rather than inhibition of MPO catalytic activity by SOD and allopurinol. It has also been demonstrated that the reperfusion-induced granulocyte infiltration in intestine is largely prevented by pretreatment with either catalase, desferrioxamine or dimethylthiourea, a ·OH radical scavenger (Zimmerman and Granger, 1988).

In an attempt to directly observe the response of granulocytes to ischemia and reperfusion, we applied intravital microscopic techniques to study adherence, rolling velocity and extravasation of leukocytes in cat mesenteric venules during low flow ischemia and reperfusion (Granger et al., 1989). The cat mesentery contains 50-80 mU/g tissue XO activity. Leukocyte adherence increased during ischemia at 10 min and 50 min after reperfusion. The number of extravasated leukocytes increased during the same periods. The responses of venular blood flow, wall shear rate, and leukocyte rolling velocity to ischemia and reperfusion do not differ between control (untreated) animals and animals treated with either allopurinol or SOD. However, the numbers of adherent and extravasated leukocytes following reperfusion are significantly lower in both allopurinol and SOD treated animals. These preliminary observations support the view that XO-derived oxidants play a critical role in the attraction and activation of granulocytes following reperfusion of ischemic tissues.

The ability of XO inhibitors, oxy-radical scavengers, and an iron chelator to interfere with I/R-induced granulocyte infiltration suggests that XO-derived oxidants play a role in the recruitment of granulocytes in the post-ischemic intestine (Granger, 1988). Originally, the working hypothesis was that XO-derived oxidants, produced by epithelial and endothelial cells, initiate the production and release of proinflammatory agents which subsequently attract and activate PMNs within the microcirculation. This scheme would explain why agents such as SOD, catalase, desferrioxamine, and allopurinol attenuate both the granulocyte infiltration and microvascular injury induced by I/R.

There are several naturally occurring substances that could mediate the granulocyte infiltration associated with reperfusion of the ischemic intestine. These include bacterial products that gain access to the mucosal interstitium via the gut lumen (N-formylated peptides, endotoxin), leukotrienes, activated complement components (C5a), LOOHs, and O_2^--dependent chemoattractants. We have recently examined the chemotactic potential obtained from these studies, which indicate that the chemotactic potential of feline extracellular fluid is not enhanced by exposure to O_2^- (Zimmerman et al., 1987). Extracellular fluid levels of leukotrienes B_4 (LTB_4), and activated complement components have not been determined in intestines subjected to I/R. We are currently attempting to determine which proinflammatory agent links XO-derived oxidants to reperfusion-induced granulocyte infiltration.

An important question that arises from the data relating granulocyte infiltration to XO-derived oxidants is whether neutrophils are a cause or an effect of ischemia/reperfusion injury. Hernandez et al. (1987c) have recently assessed the importance of granulocytes in mediating the microvascular injury associated with ischemia/reperfusion in the small intestine using two approaches; one approach used neutrophil depletion, whereas the other approach used prevention of neutrophil adherence with a monoclonal antibody directed against a specific membrane-associated glycoprotein that modulates prevention of granulocyte adherence to attenuate the I/R-induced increase in microvascular permeability (Granger, 1988). The observation that granulocyte depletion and prevention of granulocyte adherence are equally effective in attenuating the injury suggests that adherence of granulocytes to endothelium is a limiting factor in I/R-induced microvascular injury. In some tissues, the protective effect of granulocyte depletion could be attributed to an increased perfusion during the ischemic phase (Schmid-Schoenbein and Engler, 1987); however, this potential influence was eliminated in our studies by precise control of total intestinal blood flow during the ischemic period in all experimental groups. Additional studies using intravital microscopy confirmed that ischemia and reperfusion results in dramatic increases in leukocyte adherence and emigration, which was attenuated by pretreating with allopurinol or SOD (Suzuki et al., 1989). Furthermore, lipoxygenase inhibitors such as nordihydroguaiaretic acid and L663, 536, and the LTB_4 receptor antagonist SC-41930, significantly reduced the neutrophil infiltration and microvascular injury following reperfusion of the ischemic bowel, suggesting that LTB_4 may be important in mediating neutrophil adherence and activation (Zimmerman et al., 1990). It has also been demonstrated that a platelet activating factor (PAF) receptor antagonist (e.g., WEB 2086) was able to attenuate leukocyte adherence and extravasation during ischemia reperfusion (Kubes et al., 1990). Indeed, intravenous administration of the pro-inflammatory mediator PAF produces a model of acute intestinal inflammation resulting in large increases in microvascular permeability which is attenuated by blocking CD18-mediated adhesion of neutrophils or by scavenging O_2^- with SOD, suggesting that neutrophils and oxidants are important mediators of PAF-induced microvascular injury (Kubes et al., 1990). CD18 is the β-subunit of CD (cluster of differentiation) CD receptors found on surface of leukocytes and mediates adhesions. These receptors are also known as integrins.

The relative importance of XO-derived oxidants and PMNs in mediating the increased intestinal microvascular permeability produced by ischemia per se remains undefined. Although there is a significant increase in mucosal PMNs during the ischemic period (Grisham et al., 1986), the altered chemical composition of extracellular fluid induced by ischemia may lead to a suppression of granulocyte function. Neutrophilic O_2^- production can be inhibited by hypoxia, acidosis, and adenosine (Cronstein et al., 1986; Gabig et al., 1979). Neutrophilic O_2^- production is half maximal at oxygen tensions of 3-10 mm Hg (Gabig et al., 1979; Jones, 1985). Inasmuch as resting tissue pO_2 at the apex of rat intestinal villi (the region that is most vulnerable to ischemia-reperfusion injury) is as low as 5 mm Hg, it is likely that tissue pO_2 falls well below the K_m for neutrophilic O_2^- production when intestinal blood flow is reduced to 20-30% of normal (Bohlen, 1980). Estimates of intestinal mucosa pH during ischemia generally fall between 6.0 and 6.7 (Blum et al., 1987). In vitro studies indicate that acidosis dramatically suppresses O_2^- production by human neutrophils, such that O_2 production is reduced by 60% and 80% at a pH of 6.5 and 6.0, respectively (Gabig et al., 1979). Finally, adenosine, which accumulates in the extracellular fluid of ischemic tissues, has been shown to suppress O_2^- production and inhibit the adherence of neutrophils to microvascular endothelium (Cronstein et al., 1986). We have observed that intra-arterial infusion of adenosine (to achieve a blood level of 2 nmol/ml) significantly attenuates the intestinal microvascular injury induced by I/R (Grisham et al., 1989). Whether adenosine affords this protection by suppressing neutrophilic superoxide production or prevention of adherence remains unclear.

Although the results of our studies support a role for granulocytes in I/R-induced microvascular injury, the chemical mediators of this injury remain undefined. It is tempting to attribute the injury process entirely to oxidants since oxy-radical scavengers also protect against ischemia/reperfusion injury. However, this observation alone does not constitute strong support for oxidants as final mediators of injury since SOD and CAT also prevent neutrophil infiltration, indicating that oxidants may function primarily to recruit granulocytes into postischemic tissue. In addition to oxidants, activated granulocytes release a variety of proteins that are capable of damaging the microvasculature. These include lactoferrin, elastase, collagenase, and cationic proteins. There is also evidence indicating that oxidant production is required in order for some of the neutrophilic proteases to produce tissue injury, e.g., collagenase activation requires HOCl (Weiss et al., 1985).

Although it has been well accepted that oxidants appear to directly or indirectly promote the microvascular dysfunction associated with reperfusion of the ischemic gut, there remain several puzzling questions. For example, how does the extracellular administration of SOD mediate a supposed intracellular event? Some studies have attempted to determine whether SOD may gain access into the intracellular space, thereby providing a more rational explanation for the consistent observations that intravenous administration of SOD is protective. Whether SOD protects by interfering with O_2^- production by endothelial cells, or PMNs, or by some other mechanism remains speculative. Unfortunately, these studies have been by and large inconclusive. Another very perplexing question that arises from the work on ischemia/reperfusion injury of the small intestine is the apparent dependence upon low molecular weight iron complexes as a catalyst for the apparent formation of $\cdot$OH. This conclusion is based primarily upon the observations that intravenous infusion of the iron chelator desferrioxamine or iron binding protein transferrin attenuates the microvascular dysfunction mediated by reperfusion of the ischemic gut. Normal blood plasma contains large amounts of the iron binding protein transferrin. This large molecular weight glycoprotein very avidly binds low molecular weight ferric iron, and is normally only 30% saturated with respect to iron. Since the small amount of iron liberated during reperfusion would be bound by the extra binding sites on transferrin, it is not apparent how the addition of relatively small amounts of another iron chelator would prove beneficial in this situation. Finally, as outlined above, the $\cdot$OH is extremely reactive and will react instantaneously at the site where it is formed. Thus, it would be virtually impossible to achieve high enough concentrations of any "hydroxyl radical scavenger" in vivo to effectively compete with $\cdot$OH for reaction with surrounding plasma components. Taken together, these types of considerations lead one to speculate that the precise biochemical mechanisms involved in reperfusion injury have not been entirely defined.

Recent observations by Beckman and coworkers suggest a potential answer to some of these questions (Beckman et al., 1990). They suggest that the free radical nitric oxide (NO; usual abbreviation, but see chapter 1 wherein actual radical form = $\cdot N{=}O$), an endothelial-derived relaxing factor, may react rapidly with $O_2^{\cdot-}$ in a radical-radical coupled interaction to generate the peroxynitrite anion ($ONOO^-$) (eq. 2.5):

$$O_2^{\cdot-} + \cdot NO \longrightarrow ONOO^- \tag{2.5}$$

Although $ONOO^-$ is relatively stable at alkaline pH (pH = 14), it has a pKa of 6.6 which dictates that it will be protonated at physiological pH to yield peroxynitrous acid (ONOOH) (eq. 2.6). This compound is very unstable and will rapidly decompose to yield a $\cdot OH$ like species and the potent oxidizing agent nitrogen dioxide (NO_2):

$$ONOO^- + H^+ \longleftrightarrow ONOOH \longrightarrow \text{''}OH + NO_2\text{''} \tag{2.6}$$

The simultaneous formation of $O_2^{\cdot-}$ by reoxygenated endothelial and/or epithelial cells and endothelial cell-derived NO may represent an important pathway by which two relatively nontoxic radicals may interact to form very noxious oxidants in vivo. Beckman et al. (1990) have also demonstrated that desferrioxamine, but not the iron-loaded form, inhibits peroxynitrite-mediated oxidative damage to a variety of biomolecules. They propose that this mechanism may account for protective effects of SOD and desferrioxamine during the $O_2^{\cdot-}$ dependent microvascular injury produced by ischemia and reperfusion of various organ systems. Obviously, more experimental documentation is needed in this area.

2.3 INFLAMMATORY BOWEL DISEASE

Idiopathic inflammatory bowel disease (IBD) is a general description of two major disease processes called ulcerative colitis and Crohn's disease. Ulcerative colitis is a diffuse, recurrent inflammation of the colon and rectum that affects predominately the colonic mucosa. Crohn's disease, on the other hand, is a patchy transmural inflammation that may affect any part of the alimentary tract from the mouth to the rectum. Patients who suffer from IBD exhibit rectal bleeding, diarrhea, fever, pain, anorexia and weight loss. Active episodes of IBD are characterized by extravasation and infiltration of large numbers of phagocytic leukocytes such as neutrophilic PMNs, eosinophils, monocytes, and macrophages into the mucosal interstitium (lamina propria) (Riddell, 1988). This enhanced inflammatory cell infiltrate is accompanied by extensive mucosal injury, including disruption of the interstitial matrix, edema, epithelial cell necrosis and, ultimately, erosions and ulcerations (Riddell, 1988). This apparent association between phagocyte infiltration and mucosal injury suggests that these cells may play an important role in the pathogenesis of interstitial and epithelial cell damage. Mononuclear and polymorphonuclear leukocytes have the potential to synthesize and release a variety of potentially toxic agents into the extracellular environment where they may injure cells and tissue. One mechanism by which these cells may mediate tissue injury is by the release of ROMs. There are several lines of indirect evidence suggesting that the chronically inflamed intestine and/or colon may be subjected to considerable oxidative stress and thus susceptible to oxidative injury. First, it is well known that inflammatory phagocytes are activated by certain pro-inflammatory mediators such as leukotriene B_4 (LTB_4) and PAF to release large amounts of potentially cytotoxic oxidants into the extracellular space (Samuelsson, 1983; Ingram et al., 1982). Enhanced synthesis of LTB_4 and PAF has been demonstrated in mucosal samples obtained from patients with active IBD (Sharon and Stenson, 1984; Wengrower et al., 1987). Second, several studies have demonstrated that phagocytic leukocytes (neutrophils, monocytes, macrophages) obtained from patients with active IBD respond to various proinflammatory stimuli with enhanced reactive oxygen metabolism when compared to cells obtained from healthy volunteers (Kitahora et al., 1988; Anton et al., 1989; Shirabata et al., 1989). In addition, Keshavarzian et al. (1992) and Simmonds et al. (1992) have demonstrated the overproduction of oxyradicals in mucosa obtained from experimental animals with ulcerative colitis and patients with active IBD. Furthermore, a recent study by Emerit and coworkers reports that intramuscular injections of bovine CuZnSOD proved beneficial in attenuating the inflammation and mucosal injury observed in 26 patients with severe Crohn's disease (Emerit et al., 1981, 1989). Bovine CuZnSOD was evaluated in a phase II trial of patients who had failed corticosteroid therapy over an eight year period. During the early period of the study (1981–1982) patients received SOD that had been encapsulated by liposomes via subcutaneous injections. From 1982 patients were treated with the purified SOD by intramuscular injections of the protein which had been dissolved in isotonic saline. Following intramuscular or subcutaneous administration, the concentration of the SOD reached peak circulating concentrations within 2 hours, with a circulating half life of approximately 3-4 hours. It should be noted that the CuZnSOD was not used concomitantly with corticosteroids or sulfasalazine (SAZ). Using nonspecific criteria, the authors reported that 81% of the patients (20 of 26) experienced "good results" with an average follow-up of 4.2 years. Endoscopic and/or radiographic examination revealed complete healing or significant improvement in 15 patients (Emerit et al., 1981). Interestingly, a subgroup of patients received concomitant desferrioxamine therapy from 1985–1987 and showed signs of clinical improvement. The authors proposed that chelation of iron by desferrioxamine may prove beneficial in the treatment of inflammatory bowel disease via its ability to bind iron and prevent the formation of ·OH. The authors concluded that SOD may be a useful anti-inflammatory drug in the treatment of IBD, especially in combination with desferrioxamine. In another uncontrolled clinical study, bovine CuZnSOD encapsulated in liposomes was given to 3 patients with radiation-induced enteritis and to 3 patients with Crohn's disease (Emerit et al., 1989). Two of the patients with Crohn's disease showed a regression of pyoderma gangrenosa lesions and vulvar ulcerations following local application of the liposomal SOD. Although these uncontrolled case reports should be viewed with caution, they do provide intriguing clinical data to suggest that antioxidant therapy may prove useful in the treatment of IBD.

A third piece of indirect data suggesting that the chronically inflamed bowel is subjected to significant oxidative stress is the fact that the well characterized antioxidant and free radical scavenger 5-aminosalicylic acid (5-ASA) is used clinically to attenuate the mucosal injury and inflammation associated with IBD (Miyachi et al., 1987; Carlin et al., 1985, 1989; Betts et al., 1985; Aruoma et al., 1987; Craven et al., 1987). It is well known that oral administration of SAZ is effective in attenuating the mucosal injury and inflammation associated with ulcerative colitis and Crohn's disease (Dick et al., 1964; Anthonisen et al., 1974; Peppercorn, 1990; Schroder and Campbell, 1972; Azad-Khan et al., 1977; van Hess et al., 1980). SAZ reaches the distal ileum and colon unmodified, where it is metabolized by endogenous bacteria to yield 5-ASA and sulphapyridine (SP) (Dick et al., 1964). It is now well accepted that the pharmacologically active moiety of SAZ is 5-ASA (Dick et al., 1964; Anthonisen et al., 1974; Peppercom, 1990, Schroder and Campbell, 1972; Azad-Khan et al., 1977; van Hess et al., 1980). Although SAZ (i.e., 5-ASA) has been used for over 40 years, the mechanism by which 5-ASA exerts its anti-inflammatory activity in vivo remains only speculative. It has been suggested that 5-ASA protects the gut mucosa by inhibiting cyclooxygenase and/or lipoxygenase activities as well as by inhibiting mitogen-stimulated secretion of immunoglobulins from mononuclear leukocytes (MacDermott et al., 1989). However, the concentrations of 5-ASA required to inhibit these reactions range from 1–10 mM which is considerably higher than the 0.1–0.2 mM range that has been determined experimentally for the normal colonic mucosal interstitium (Grisham and Granger, 1989). An alternative mechanism for the protective effective of 5-ASA is its potent antioxidant and free radical scavenger properties (Aruoma et al., 1987; Dull et al., 1987; Craven et al., 1987; Grisham, 1988a,b; Williams and Hallett, 1989; Ahnfelt-Ronne and Nielsen, 1987; Hoult and Page, 1981; Fedorak et al., 1990; Yamada et al., 1990).

It has been estimated that the inflamed colon contains approximately 5–10 million neutrophils, which will produce approximately 40–80 μM O_2^- when fully activated. Although O_2^- per se is not very reactive or cytotoxic, it will interact with iron to generate the toxic $^.OH$ as discussed previously. Thus any compound that decomposes O_2^- would attenuate O_2^- dependent formation of $^.OH$. A recent report by Craven et al. (1987) demonstrated that 5-ASA has potent SOD-like activity as measured by its ability to inhibit the reduction of cytochrome c by XO generated O_2^-. Using a more direct assay for the decomposition of O_2^-, i.e., the direct spectrophotometric determination of the O_2^- radical at 250 nm, we find that 5-ASA directly interacts with O_2^-, causing the rapid decomposition of this radical (Yamada et al., 1990). The disappearance of O_2^- may occur by two possible pathways: One pathway would require the one electron reduction of O_2^- by 5-ASA in the presence of protons to yield H_2O_2; the other way in which 5-ASA may decompose O_2^- is by the one electron reduction of 5-ASA by O_2^- to yield O_2. Because 5-ASA would acquire an odd electron by either pathway it would become, by definition, a free radical itself. The significance of these observations is currently under investigation in our laboratory. Phagocytic leukocytes also produce large amounts of H_2O_2 by the spontaneous or enzymatic (SOD) dismutation of O_2^-. We have found that 5-ASA does not react with H_2O_2 to any significant extent, and thus is unlikely to participate in the decomposition of this oxidant in vivo (Yamada et al., 1990). As already mentioned O_2^- and H_2O_2 will interact in the presence of iron to yield the highly reactive oxidant ˙OH. This oxidant is capable of mediating the oxidative degradation of a variety of biomolecules including lipids and carbohydrates. Work from several laboratories, including our own, have demonstrated that 5-ASA is very effective in scavenging the ˙OH; however, the parent compound SAZ and the metabolically inactive metabolites N-acetyl-5-ASA (NASA) and SP are equally effective (Reviewed in Yamada et al., 1990). These data suggest that ˙OH scavenging does not account for the therapeutic action of 5-ASA. In contrast with these studies we have found that 5-ASA, but not SAZ, NASA, nor SP, is effective in inhibiting lipid peroxidation (I_{50} concentration = $8\mu M$) initiated by peroxyl radicals generated from the thermal decomposition of 2,2'-azobis (amidinopropane) dihydrochloride (A—N=N—A)(eq. 2.7–2.11):

$$A-N=N-A \longrightarrow A^{\cdot} + N_2 + A^{\cdot} \qquad (2.7)$$

$$A^{\cdot} + O_2 \longrightarrow AOO^{\cdot} \qquad (2.8)$$

$$AOO^{\cdot} + LH \longrightarrow AOOH + L^{\cdot} \qquad (2.9)$$

$$L^{\cdot} + O_2 \longrightarrow LOO^{\cdot} \qquad (2.10)$$

$$LOO^{\cdot} + LH \longrightarrow LOOH + L^{\cdot} \qquad (2.11)$$

where ˙A, ˙AOO, and AOOH represent the alkyl radical, peroxyl radical and lipid hydroperoxide, respectively. Because lipid peroxidation is not initiated by the O_2^- dependent, iron catalyzed formation of ˙OH in this system, the inhibitory effect of 5-ASA is due solely to its ability to scavenge the peroxyl free radicals. These data agree with and extend the findings of Ahnfelt-Ronne and Nielsen, (1987) who have demonstrated potent scavenging of a nitrogen-centered free radical by 5-ASA (I_{50} concentration = 5 μM) but not by SAZ or SP. Another property of 5-ASA that may contribute to its antioxidant activity is its ability to chelate iron. Obviously, any compound capable of binding iron and rendering it poorly redox active would be very effective in inhibiting the formation of ˙OH. We have recently demonstrated that 5-ASA inhibits the iron-catalyzed, ˙OH-mediated degradation of deoxyribose by chelating iron and preventing its interaction with O_2^- and H_2O_2 (I_{50} = 300 μM). NASA and SAZ were only modestly effective, whereas SP was inactive, suggesting a relatively selective effect by the therapeutically active metabolite.

MPO-catalyzed oxidation of Cl^- by H_2O_2 to yield HOCl represents another significant pathway of oxidant production in inflamed tissue. Recent work by von Ritter et al. (1989) has demonstrated that 5-ASA as well as SP, NASA, and SAZ are all very effective at scavenging HOCl in vitro (I_{50} concentration = 25–30 μM). However, Aruoma and coworkers (1987) have shown that 5-ASA was selective in its ability to protect α-1-protease inhibitor against inactivation by HOCl (I_{50} = 400 μM) suggesting that some of the beneficial effects of 5-ASA may be due to its ability to selectively interact with and decompose HOCl in the presence of other biological compounds. The MPO-catalyzed formation of HOCl requires the interaction between H_2O_2 and the enzyme to form a potent hemoprotein-associated oxidant termed compound I. This porphyrin cation radical is a potent oxidizing agent capable of oxidizing a wide variety of biological compounds in addition to Cl^-:

$$P-Fe^{3+} + H_2O_2 \longrightarrow P^{\cdot+}-Fe^{4+}=O + H_2O \qquad (2.12)$$

$$P^{\cdot+}-Fe^{4+}=O + AH \longrightarrow P-Fe^{3+} + A^+ + {}^-OH \qquad (2.13)$$

where $P-Fe^{+3}$, $P^{\cdot+}Fe^{+4}=O$, AH and A^+ represent the hemoprotein, porphyrin cation radical, substrate, and oxidized substrate, respectively.

We have found that 5-ASA and 4-ASA were much more effective in decomposing Compound I of MPO, and thus much more potent at inhibiting the activity of MPO, than were SP or NASA. An I_{50} concentration of 20 μM and 25 μM were determined for 5-ASA and 4-ASA, respectively. Apparently 5-ASA and 4-ASA act as alternative substrates for Compound I, preferentially becoming oxidized instead of the substrate. It is quite possible that this is the reason why Ahnfelt-Ronne et al. detected significant levels of oxidation products of 5-ASA in SAZ-treated patients with active inflammatory bowel disease (Ahnfelt-Ronne et al., 1990). It has also been suggested that the interstitial hemoglobin (Hb) released during intestinal bleeding may mediate some of the mucosal injury by interacting with phagocyte-derived H_2O_2 to generate ferryl (Fe^{+4}) Hb, a hemoprotein-associated free radical similar to MPO Compound I. Ferryl Hb is a potent oxidant capable of initiating lipid peroxidation (Yamada et al., 1991). We have found that 5-ASA, and to a lesser extent SAZ or SP, selectively inhibit Hb-catalyzed lipid peroxidation by acting as an alternative substrate for ferryl Hb (I_{50} concentration = 50 μM). These types of reactions may help explain observations made by Hoult and Page (1981) who demonstrated enhanced production of certain prostaglandins by colonic mucosa in the presence of relatively small amounts of 5-ASA (0.5 mM). It is known that prostaglandin synthetase contains two enzymatic activities: cyclooxygenase and hemoprotein hydroperoxide peroxidase. During the enzymatic reaction there is a progressive inhibition of the enzyme due to the oxidative inactivation of the hemoprotein peroxidase. It is well known that certain antioxidants (e.g., phenolic compounds) inhibit this inactivation process and prolong prostaglandin production. Apparently the LOOH generated by cyclooxygenase (PGG_2) combines with the hemoprotein peroxidase to generate a Compound I-like oxidant. In the absence of an exogenous electron donating substrate (antioxidant) the hemoprotein-localized free radical oxidizes certain amino acid residues proximal to the active site, ultimately resulting in inactivation of the enzyme. It is intriguing to speculate that 5-ASA, by virtue of its antioxidant activity, may protect the mucosa by enhancing the formation of protective prostaglandins (PG) such as prostacyclin and/or prostaglandin E (PGE) derivatives. Indeed, recent reports suggest that certain PGE analogs are potent antiulcer compounds for the colon (Fedorak et al., 1990; Yamada et al., 1991).

2.4 ARTHRITIS

Another pathophysiological condition in which ROMs have been implicated is rheumatoid arthritis. Much of this interest was originally based upon observations made by McCord (1974) in which he demonstrated that exposure of synovial fluid to an oxygen radical generating system (xanthine/xanthine oxidase) resulted in a significant reduction in its viscosity such that it approached the viscosity obtained from patients with active disease. Furthermore, McCord showed that exposure of the glycosaminoglycan hyaluronic acid to ROMs dramatically reduced its viscosity as well, suggesting that this intra-articular lubricant may be degraded by free radicals in vivo. Oxy radical-induced depolymerization of this important biopolymer has been shown to be mediated by ˙OH generated by the superoxide-driven Fenton reaction, as discussed previously. Neither O_2^- nor H_2O_2 are capable of interacting with or degrading hyaluronic acid. In addition to ˙OH it has also been shown that other oxidants such as HOCl are also capable of degrading hyaluronic acid (Green, 1990). There are at least two sources of ROMs proposed for the inflamed joint. One source is the phagocytic leukocytes such as neutrophils, monocytes and macrophages (Halliwell, 1992). As discussed previously, activated PMNs release large quantities of ROMs, including O_2^-, H_2O_2, and HOCl. In addition, the pannus overgrowing the cartilage contains monocyte or macrophage like cells which could produce ROMs and thus degrade the underlying cartilage.

Another source of ROM has been proposed to be from cyclic ischemia and reperfusion occurring in rheumatoid joint during exercise and resting. The intra-articular pressure measured in rheumatoid arthritis joint fluid is higher than normal joint, which may occlude the vessels supplying the synovium during exercise and during resting. In this case, the pressure drops and the pO_2 level rises, and reperfusion may lead to generation of oxidative species which may perpetuate the oxidative damage to the joint (John et al., 1987). Besides, the localization of XO activity in human synovial membrane also suggested a role of I/R induced ROM (Gutteridge, 1987a).

Recent data by Skaleric et al. (1991) demonstrate that ROM may be directly involved in mediating the joint injury and inflammation induced by the systemic administration of the arthritis producing bacterial polymer peptidoglycan/polysaccharide (PG/PS). They found that intra-articular injection of either CAT or SOD significantly reduced the PG/PS-induced joint inflammation and evolution of erosive arthritis in female Lewis rats. These data suggested that both H_2O_2 and O_2^- are directly or indirectly involved in the pathobiology of this model of arthritis.

The source of iron necessary to catalyze the formation of $^.OH$ in vivo remains only speculative. It is known that low molecular weight, redox active iron has been demonstrated in synovial fluid from many patients with rheumatoid arthritis (Gutteridge, 1987a). The source of this catalytic metal is not known, but it has been suggested that the degradation of hemoglobin present during microvascular bleeding into the intra-articular space would liberate significant amounts of iron (Gutteridge, 1987a). In addition, ferritin may represent another source of iron which could be liberated by O_2^- and/or protease-mediated events (Gutteridge, 1987a).

2.5 CENTRAL NERVOUS SYSTEM INJURY

A number of factors predispose brain tissue to oxidative damage. For example, the brain is highly enriched in easily oxidizable $22:6$ and $20:4$ unsaturated fatty acids. In addition, certain regions of the brain possess relatively high concentrations of iron capable of catalyzing Fenton-type reactions leading to formation of reactive oxygen metabolites and oxidative tissue damage. Furthermore, the brain utilizes a large amount of oxygen relative to its weight, and the integrity of the blood-brain barrier must be maintained for proper oxygenation and normal brain function (Floyd and Carney, 1992).

Docosohexanoic acid may comprise up to 20% of the total polyunsaturated fatty acids (PUFAs) in the brain and 35% of the synaptic vesicles (Breckenridge et al., 1973). Rehncrona et al. (1980) showed that incubates of iron ascorbate and brain homogenates exhibited loss of arachidonic and docosohexanoic acids by lipid peroxidation. Brain tissue contains non-heme iron at levels estimated at about 0.074 μg/mg protein (Youdim and Green, 1978) and it has been suggested that regional differences in the lipoperoxidative capacity of brain areas are governed by local iron content (Zaleska and Floyd, 1980). Additionally, whole blood contains 150 mg of hemoglobin per mL which amounts to about 0.5 mg of Fe^{+2} per mL. (Triggs and Willmore, 1984). Disruption of the structural integrity of the brain, as in head and spinal trauma, correspondingly results in the decompartmentalization of PUFA, ascorbate, oxygen, and heme-iron, which are normally kept separate. These same components, when added to aqueous solutions of tissue homogenates or PUFA, resulted in formation of highly reactive oxygen moieties competent to initiate and propagate lipid peroxidation (Aust et al., 1985). In vivo, this may lead to central nervous system (CNS) dysfunction.

Intracortical injection of rats with purified bovine hemoglobin has been observed to be epileptogenic. Interestingly, the appearance of epileptiform activity corresponded with the expected rate of hemoglobin breakdown and release of iron (Rosen and Frumin, 1979). Earlier, Willmore et al. (1978a, 1978b) induced recurrent seizures in rat and cat by cortical injection of iron salts resulting in acute epileptiform discharges. Pretreatment of rats with antiperoxidants has been shown to effectively attenuate the development of seizures and related histopathologic effects of Fe^{+2} injection (Wilmore and Ruvin, 1984). Sadrzadeh et al. (1984) showed that both hemoglobin and free iron salts markedly inhibited Na/K ATPase activity in vitro in CNS homogenates and in vivo in spinal cords of living cats. Desferrioxamine completely abrogates the Hb-induced inhibition of Na/K ATPase activity; however, since desferrioxamine is thought not to interact with Hb, the mechanism of inhibition may involve prior release of iron. In accordance with this, iron-free hematoprotoporphyrin injected into rat neurophil failed to initiate brain injury responses associated with free radical reactions (Wilmore and Triggs, 1991). Since the extracellular fluids do not contain substantial amounts of antioxidants, any free hemoglobin released due to spontaneous hemolysis or trauma must be quickly removed or these may contribute to CNS dysfunction via iron-driven oxidative reactions. Haptoglobins, glycoproteins found in the alpha-globin fraction of serum, constitute a specialized system for removal of active metal complexes before these can react to produce reduced oxygen metabolites (Putman, 1975). In support of this, artificially induced hypohaptoglobinemia in mice caused attenuated clearance of free hemoglobin and thus lead to increased peroxidation of brain lipids (Panter et al., 1985). These results also suggested that impaired clearance of hemoglobin from the CNS may be a determining factor in the development of certain forms of familial idiopathic epilepsy (Panter et al. 1985). Gutteridge (1987 b) demonstrated in vitro that inhibition of hemoglobin-dependent peroxidation by albumin, transferrin, and desferrioxamine was somewhat less effective than that exhibited by haptoglobin.

The damaging effects of hemoglobin on the CNS probably involves iron-catalyzed oxidative reactions at submicromolar concentrations of iron (Panter et al., 1985; Sadrzadeh and Eaton, 1988; Gutteridge 1992). In in vitro tests, Hb-derived iron at submicromolar levels initiated the generation of large amounts of thiobarbituric acid reactive substances. This necessitates redox cycling of iron between the ferric and ferrous redox states. Sadrzadeh and Eaton (1988) demonstrated that both desferrioxamine and ferene, Fe^{+3} and Fe^{+2} specific chelators, respectively, inhibit iron mediated lipid peroxidation. They further proposed that endogenous CNS ascorbic acid (present in large amounts) may react as a reducing agent to facilitate repetition of Fe^{+3} and Fe^{+2} redox cycling. Although the exact reaction mechanism by which Hb-derived iron/ascorbate-mediated CNS damage occurs has not been fully eludicated; several reaction steps may be involved (eq. 2.14-2.20).

$$Hb\text{—}Fe^{+3} \longrightarrow Fe^{+3} + apoprotein \tag{2.14}$$

$$Fe^{+3} + ascorbate \longrightarrow Fe^{+2} + ascorbyl\ radical \tag{2.15}$$

$$Fe^{+2} + O_2 \longrightarrow Fe^{+3} + O_2^- \tag{2.16}$$

$$2O_2^- + 2H^+ \longrightarrow H_2O_2 + O_2 \tag{2.17}$$

$$Fe^{+2} + H_2O_2 \longrightarrow Fe^{+3} + \,^{\cdot}OH + \,^-OH \tag{2.18}$$

$$LH + \,^{\cdot}OH \longrightarrow L^{\cdot} + HOH \tag{2.19}$$

$$L^{\cdot} + O_2 + LH \longrightarrow LOOH + L^{\cdot} \tag{2.20}$$

where LH and LOOH are membrane PUFA and PUFA hydroperoxide, respectively. It follows therefore that ascorbic acid, which normally reacts to maintain CNS components in a reduced state, may, under conditions of head and spinal cord trauma, participate in reactions which lead to peroxidative dysfunction of the CNS.

2.6   ACUTE RENAL FAILURE

Acute tubule necrosis, or hemoglobinuric nephropathy, is associated with injury and the resultant ischemia due to decreased blood flow. The condition is characterized by the presence of myoglobin and/or hemoglobin in the circulation and urine (Braun et al., 1970). While it is known that intramuscular injection of glycerol in rats is associated with severe renal failure, glycerol itself is not directly responsible (Thiel et al., 1967). Rather, intramuscular injections of hypotonic glycerol solution cause muscle necrosis and intravascular hemolysis with release of myoglobin and hemoglobin, respectively (Paller, 1988). Coupled with ischemia, the presence of heme-protein results in severe tubule dysfunction (Braun et al., 1970). In vitro studies have shown that both myoglobin and hemoglobin can participate in Fenton type reactions to promote production of ˙OH (Sadrzadeh and Eaton, 1988; Grisham, 1985). In vitro assays performed in the presence of XO-generated H_2O_2 substantiated the ability of myoglobin to initiate peroxidation of arachidonic acid (Grisham, 1985). Paller (1988) demonstrated in vivo that heme protein-derived iron-catalyzed ˙OH free radical formation initiated lipid-peroxidation, thereby causing acute renal failure in rats. Desferrioxamine (a Fe^{+3} specific chelator) markedly attenuated hemoglobin-induced injury. Recently, in vitro and in vivo investigations (Gamelin and Zager, 1988; Borkan and Schwartz, 1989) have questioned the role of reactive oxygen intermediates in ischemia-reperfusion renal injury, as well as the beneficial effects of administering antioxidants which detoxify them. More recently, however, Nath and Paller (1990) successfully demonstrated the deleterious effects of dietary deficiency of vitamin E and selenium on rat kidney function and structure. Deprivation of oxidant scavenging ability markedly exacerbated ischemic injury, suggesting a pathogenetic role for reactive oxygen species.

2.7   SICKLE CELL ANEMIA

Sickle cell hemoglobin (HbS) results from an aberrant mutant gene product in which glutamine in the sixth position of the beta globin chains has been replaced by valine (Hebbel, 1990). Although the ability to transport oxygen is retained, HbS is unstable and exhibits the tendency to polymerize at low oxygen tension. The results of an investigation conducted by Asakura et al. (1973) showed that oxyHbS denatures almost ten-fold faster than oxyHb A. A very perturbing consequence of HbS instability is an accelerated rate of autoxidation and decompartmentalization of iron (Hebbel et al., 1988). HbS has been shown to undergo autoxidation at a rate nearly twice that of normal Hb. Usually, under physiological conditions, oxygenation of Hb is a reversible process resulting in the release of oxygen, leaving heme-iron in the ferrous oxidation state. Occasionally, however, autoxidation occurs in which oxygen is released as the O_2^- radical with the heme Fe^{+2} ion oxidized to Fe^{+3} (Hebbel et al., 1988). This process is accelerated in the case of HbS and potentially exposes the sickle red blood cell (RBC) to pathologic levels of ROS (Misra and Irwin, 1972). In accordance with this, Hebbel et al. (1982) showed earlier that sickle RBC spontaneously produced about twice the normal amount of O_2^- compared with normal erythrocytes.

Iron chelates are known to stimulate peroxidation of unsaturated lipid, and therefore separation of iron and the cell membrane components must be maintained to prevent initiation of damaging oxidative reactions. The instability of HbS results in failure of iron compartmentation and leads to abnormally high iron deposits intimately associated with membrane structures (Hebbel, 1990). Inside out sickle (IOM) cell membranes have a 2- to 3-fold increase in membrane associated heme over that found associated with IOM for normal RBC (Kuross and Hebbel, 1988; Kuross et al., 1988). Nonheme iron has been reported to average as high as 4.0 nmol/mg membrane protein in sickle IOM (Kuross and Hebbel, 1988). In addition, abnormally high deposits of ferritin-like or hemosiderin-like iron have been observed in sickle RBC cytosol (Jacobs et al., 1981). The combination of heme and nonheme iron in sickle IOM averages about nine times the amount associated with normal IOM membranes (Kuross and Hebbel, 1988). Therefore, accelerated autoxidation and/ or release of iron are sufficient chemical events which potentially provide for oxidative damage of HbS RBC. An increase in formation of H_2O_2 via spontaneous and enzymatic dismutation of superoxide in conjunction with decompartmentalized iron may result in generation of the more damaging ˙OH radical via the Fenton reaction. Of course, propagation of this process is limited by the turnover of ferrous iron (i.e., via redox cycling) by cytosolic reducing agents as well as the presence of antioxidants. Generally, SOD, CAT, and GSH levels are significantly lower in sickle RBCs than in normal erythrocytes (Schacter et al., 1985). Yet, reduced GSH has been reported to play an important role in the recovery of Ca^{++} ATPase from oxidant injury (Hebbel et al., 1986).

2.8   CANCER

During normal metabolism, the rate of oxidative damage to DNA is very high. Ames and coworkers have estimated that the total number of oxidative "hits" to DNA per cell per day approaches 10^4 in humans and 10^5 in rats (Ames and Shipgenaga, 1992). Furthermore, investigators have demonstrated that mammals with a high metabolic rate, short life span, and high incidence of age-related cancer have much higher rates of oxidative damage to their DNA when compared to those rates determined for long-lived species with lower metabolic rates and low age-specific cancer rates (Ames and Shigenaga, 1992). Indeed, Cutler et al. (1984) have demonstrated that a lower basal metabolic rate such as occurs in man vs small animals such as mice or rats contributes significantly to longevity. Since cancer rates increase with approximately the fourth power of age in both humans and rodents, cancer has been suggested to be a degenerative disease of old age. There are four endogenous processes that may lead to DNA damage and possibly increased mutation frequency in vivo, and these include oxidation, methylation, deamination, and depurination reactions (Ames and Shigenaga, 1992). The presence of specific repair systems for each of these alterations emphasizes the importance of these reactions as possible mediators of mutagenesis in vivo.

Although reactive metabolites of oxygen may contribute to deamination and depurination reactions, we will consider only their ability to produce oxidation products from DNA. The very high incidence of mutations in bacteria that genetically lack functional SOD lends support to the concept that oxidants and free radicals are potent mutagens in vivo, and antioxidant enzymes are important defenses against the genotoxic effects of these reactive metabolites (Ames and Shigenaga, 1992). It is well known that neither O_2^- nor H_2O_2 per se are capable of interacting with either the deoxyribose or base portions of DNA, suggesting that the secondarily derived ·OH radical may be the primary reactive species. As mentioned previously, ·OH or an oxidant with ·OH-like reactivity may be produced by the O_2^- driven Fenton reaction, or possibly by the interaction between NO· and O_2^-. Since ·OH will react only at the site of its production (i.e., at the site of the metal catalyst), one must assume that OH is generated on or very near the DNA (Halliwell and Aruoma, 1992). Furthermore, one must assume that normal DNA contains some sort of transition metal (Fe or Cu) associated with the "naked" regions of DNA, i.e., those regions which are not covered by histone protein (Halliwell and Aruoma, 1992). Once generated, OH may interact with DNA to produce at least 20 different oxidation products of DNA (Ames and Shigenaga, 1992). Several investigators have concentrated on thymine glycol, thymidine glycol, hydroxymethyluracil, hydroxylmethlyuridine and 8-hydroxydeoxyguanosine as probes to measure DNA oxidation in vivo. Hydroxylated products of guanine and thymine have been demonstrated to be mutagenic, which is apparently due to misreading of the DNA during replication (Wood, 1990). Some of these oxidation products have been detected in the urine of healthy individuals, indicating the efficient removal of "damaged" bases and/or nucleosides from oxidized DNA (Ames and Shigenaga, 1992). However, it should be remembered that endogenous bacteria and diet may contribute to the presence of some of these metabolites in vivo.

An interesting concept proposed by Ames and Gold (1990) as well as Preston-Martin et al. (1990) is the idea that a major risk factor for cancer is mitogenesis. It has long been appreciated by geneticists that dividing cells are at an increased risk for developing mutations compared to quiescent cells because: a) the rate of conversion of DNA adducts to mutations may occur before they are repaired is much greater in dividing cells; b) the chances of mutations due to DNA replication (i.e., mitotic recombination, gene conversion, and nondisjunction) are increased in dividing cells; and c) replicating DNA is more vulnerable to the actions of genotoxic agents. Indeed, chronic mitogenesis as would occur in chronic inflammatory conditions (e.g., inflammatory bowel disease) represents a significant risk for cancer development. Oxidants produced by inflammatory leukocytes may induce mutagenesis and possibly carcinogenesis by enhancing mitogenesis as well as by modifying the DNA bases.

2.9 SUMMARY

Reactive oxygen metabolites have been implicated in a variety of different disease states. One may conclude that oxidants and free radicals are most probably produced in virtually all of these pathophysiological conditions. Whether ROMs play a critical role as mediators of the pathobiology, or are simply produced in response to some other form of injury, is an essential question that needs to be addressed. This discussion highlights several examples of conditions in which ROMs appear to play an active role in mediating vascular and tissue injury. However, the precise mechanisms and identity of the specific oxidants involved in many of these pathophysiological conditions remain to defined.

ACKNOWLEDGMENTS

Some of the work reported in this manuscript was funded by grants from the NIH (DK 43785; Project 6 and DK 47663).

References

Ahnfelt-Ronne, I. and Nielsen, O.H. (1987) The anti-inflammatory moiety of sulfasalazine, 5-aminosalicylic acid is a radical scavenger. *Agents Actions* 21, 191–194.

Ahnfelt-Ronne, I., Nielsen, O.H. Christensen, A., Langholz, E., and Riis, P. (1990) Clinical evidence supporting the radical scavenger mechanism of 5-aminosalicylic acid. *Gastroenterology* 98, 1162–1169.

Ames B.N. and Shigenaga MK. (1993) DNA damage by endogenous oxidants and mitogenesis as causes of aging and cancer. In *Molecular Biology of Free Radical Scavengers System* (J.G. Scandalios, ed.), Cold Spring Harbor Laboratory Press, Plainview, NY., pp. 1–22.

Ames, B.N. and Gold, L.S. (1990) Chemical carcinogenesis: too many rodent carcinogens. *Proc. Natl. Acad. Sci. USA* 87, 7772–7776.

Anthonisen, P. (1974) The clinical effect of salazosulphapyridine (SalazopyrinxR) in Crohn's disease: a controlled double-blinded study. *Scand. J. Gastro.* 9, 549–554.

Anton, P.A., Targan, S.R. and Shanahan, F. (1989) Increased neutrophil receptors for and response to the proinflammatory bacterial peptide formyl-methionyl-leucyl-phenylalanine in Crohn's disease. *Gastroenterology* 97, 20–28.

Aruoma, O.I., Wasil, M., Halliwell, B., Hoey, B.M. and Butler, J. (1981) The scavenging of oxidants by sulphasalazine and its metabolites: A possible contribution to their anti-inflammatory effects? *Biochem. Pharm.* 36, 3739–3742.

Asakura, T., Agarwal, P.L., Relman, D.A., McCray, J.A., Chance, B., Schwartz, E. Friedman, S. and Lubin, B. (1973) Mechanical instability of the oxy-form of sickle haemoglobin. *Nature* 244, 437–438.

Aust, S.D., Morehouse, L.A. and Thomas, C.E. (1985) Role of metals in oxygen radical reactions. *J. Free Rad. Biol. Med.* 1, 3–35.

Azad-Khan, A.H., Piris, J. and Truelove, S.C. (1977) An experiment to determine the active moiety of sulfasalazine. *Lancet* 2, 892–895.

Beckman, J.S., Beckman, T.W., Chen, J., Marshall, P.A. and Freeman, B.A. (1990) Apparent hydroxyl radical production by peroxynitrite: Implication for endothelial injury from nitric oxide and superoxide. *Proc. Natl. Acad. Sci. USA* 87, 1620–1624.

Betts, W.H., Whitehouse, M.W., Cleland, L.G. and Vernon-Roberts, B. (1985) In vitro antioxidant properties of potential biotransformation products of salicylate, sulphasalazine and amidopyrine. *J. Free Rad. Biol. Med.* **1**, 273–280.

Blum, H., Chance, B. and Buzby, G.P. (1987) In vivo noninvasive observation of acute mesenteric ischemia in rats. *Surg. Gyn. Obstet.* **164**, 409–414.

Blum, H., Summers, J.J., Schnall, M.D., Barlow, C., Leigh, J.S., Chance, B. and Buzby, G.P. (1986) Acute intestinal ischemia studies by phosphorous nuclear magnetic resonance spectroscopy. *Ann. Surg.* **204**, 83–88.

Bohlen, H.G. (1980) Intestinal tissue pO_2 and microvascular responses during glucose exposure. *Am. J. Physiol.* **238**, H164–H171.

Borkan, S.C. and Schwartz, J.H. (1989) Role of oxygen free radical species in in vitro models of proximal tubular ischemia. *Am. J. Physiol.* **257**, F298–F305.

Braun, S.R., Weiss, F.R., Keller, A.I., Ciccone, J.R. and Preuss, M.G. (1970) Evaluation of the renal toxicity of heme proteins and their derivatives: A role in the genesis of acute tubule necrosis. *J. Exp. Med.* **131**, 443–460.

Breckenridge, W.C., Morgan, I.G., Zanetta, J.P. and Vincendon, G. (1973) Adult rat synaptic vesicles. II. Lipid Composition. *Biochim. Biophys. Acta* **320**, 681–686.

Carlin, G., Djursater, R. and Smedegard, G. (1989) Inhibitory effects of sulfasalazine and related compounds on superoxide production by human polymorphonuclear leukocytes. *Pharmacol. Toxicol.* **65**, 121–127.

Carlin, G. Djursater, R. Smedegard, G. and Gerdin, B. (1985) Effect of anti-inflammatory drugs on xanthine oxidase induced depolymerization of hyaluronic acid. *Agents Actions* **16**, 377–384.

Craven, P.A., Pfanstiel, J., Saito, R. and DeRubertis, F.R. (1987) Actions of sulfasalazine and 5-aminosalicylic acid as reactive oxygen scavengers in the suppression of bile acid-induced increases in colonic epithelial cell loss and proliferative activity. *Gastroenterology* **92**, 1998–2008.

Cronstein, B.N., Levin, R.I., Belanoff, J., Weissmann, G. and Hirschhorn, R. (1986) Adenosine: an endogenous inhibitor of neutrophil mediated injury to endotheilial cells. *J. Clin. Invest.* **78**, 760–770.

Cutler, R.G. (1984) Antioxidants, aging and longevity. In *Free Radicals in Biology* (W.A. Pryor, ed.), Academic Press, New York, Vol. 6, pp. 371–423.

Dick, A.P., Grayson, M.J., Carpenter, R.G. and Petrie, A. (1964) Controlled trial of sulphasalazine in the treatment of ulcerative colitis. *Gut* **5**, 437–442.

Dix, T.A. and Aikens, J. (1992) Mechanisms and biological relevance of lipid peroxidation initiation. *Chem. Res. Toxicol.* **6**, 2–18.

Dull, B.J., Salata, K., Van Langenhove, A.V. and Goldman, P. (1987) 5-aminosalicylate: oxidation by activated leukocytes and protection of cultured cells from oxidative damage. *Biochem. Pharmacol.* **36**, 2467–2472.

Emerit, J. Loeler, J., Chomett, G. (1981) Superoxide dismutase and the treatment of post-radiotherapeutic necrosis and of Crohn's disease. *Bull. Eur. Physiopathol. Resp.* **17**, 287.

Emerit, J., Pelletier, S., Tosoni-Verilgnue, D. and Mollet, M. (1989) Phase II trial of copper zinc superoxide dismutase (CuZnSOD) in treatment of Crohn's disease. *Free Rad. Biol. Med.* **7**, 145–149.

Fedorak, R.N., Empey, L.R., MacArthur, C. and Jewell, L.D. (1990) Misoprostol provides a colonic mucosal protective effect during acetic acid-induced colitis in rats. *Gastroenterology* **98**, 615–25.

Floyd, R.A. and Carney, J.M. (1992) Free radical damage to protein and DNA: Mechanisms involved and relevant observations on brain undergoing oxidative stress. *Ann. Neurol.* **32**, S22–S27.

Gabig, T.G., Bearman, S.I. and Babior, B.M. (1979) Effects of oxygen tension and pH on the respiratory burst of human neutrophils. *Blood* **53**, 1133–1139.

Gamelin, L.M. and Zager, R.A. (1988) Evidence against oxidant injury as a critical mediator of postischemic acute renal failure. *Am. J. Physiol.* **225**, F450–F460.

Granger, D.N. (1988) Role of xanthine oxidase and granulocytes in ischemia-reperfusion injury. *Am. J. Physiol.* **255**, H1269–H1275.

Granger, D.N., Benoit, J.N. Suzuki, M. and Grisham, M.B. (1989) Leukocyte adherence to venular endothelium during ischemia-reperfusion. *Am. J. Physiol.* **257**, G683–G688.

Granger, D.N., Hollwarth, M.E. and Parks, D.A. (1986). Ischemia-reperfusion injury: role of oxygen-derived free radicals. *Acta Physiol. Scand. Suppl.* **548**, 47–63.

Granger, D.N. Rutili, G. and McCord, J.M. (1981) Superoxide radicals in feline intestinal ischemia. *Gastroenterology* **81**, 22–29.

Green, S.P., Baker, M.S. and Lowther, D.A. (1990) Depolymerization of synovial fluid hyaluronic acid by the complete myeloperoxidase system may involve the formation of a HA-MPO complex. *J. Rheum.* **17**, 1670–1675.

Grisham, M.B. (1985) Myoglobin-catalyzed hydrogen peroxide dependent arachidonic acid peroxidation. *J. Free Rad. Biol. Med.* **1**, 227–232.

Grisham, M.B. (1988a) Antioxidant properties of 5-aminosalicylic toward neutrophil-derived oxidants. In *Inflammatory Bowel Disease: Current Status and Future Approach* (R.P. MacDermott, ed.), Elsevier Science Publishers B.V. New York, pp. 261–266.

Grisham, M.B. (1988b) Effect of 5-aminosalicylic acid on ferrous sulfate-mediated damage to deoxyribose. *Biochem. Pharmacol.* **39**, 2060–63.

Grisham, M.B. and Granger, D.N. (1989) 5-aminosalicylic acid concentration in mucosal interstitium of cat small and large intestine. *Dig. Dis. Sci.* **34**, 575–578.

Grisham, M.B., Hernandez, L.A. and Granger, D.N. (1986) Xanthine oxidase and neutrophil infiltration in intestinal ischemia. *Am. J. Physiol.* **251**, G567–G574.

Grisham, M.B., Hernandez, L.A. and Granger, D.N. (1989) Adenosine inhibits ischemia-reperfusion-induced leukocyte adherence and extravasation. *Am. J. Physiol.* **257**, H1334–H1339.

Gutteridge, J.M.C. (1987a) Bleomyin-detectable iron in knee joint synovial fluid from arthritic patients and its relationship to the extracellular antioxidant activities of ceruloplasmin, transferrin and lactoferrin. *Biochem. J.* **245**, 415–421.

Gutteridge, J.M.C. (1987b) The antioxidant activity of haptoglobin towards hemoglobin-stimulated lipid peroxidation. *Biochim. Biophys. Acta* **917**, 219–223.

Gutteridge, J.M.C. (1992) Iron and oxygen radicals in brain. *Ann. Neurol.* **32**, S16–S21.

Halliwell, B. and Aruoma, O.I. (1992) DNA damage by oxygen-derived species: its mechanism and measurement using chromatographic methods. In *Molecular Bi-*

ology of Free Radicals Scavenger System, Vol. 9 (J.G. Scandalios, ed.), Cold Spring Harbor Laboratory Press, Plainview, NY, pp. 47–67.

Halliwell, B., Gutteridge, J.M.C. and Cross, E.E. (1992) Free radicals, antioxidants and human disease: Where are we now? *J. Lab. Clin. Med.* **119**, 598–620.

Hebbel, R.P. (1990) The sickle erythrocyte in double jeopardy: Autooxidation and iron decompartmentalization. *Sem. Hematol.* **27(1)**, 51–69.

Hebbel, R.P., Eaton, J.W., Balasingam, M. and Steinberg, M.H. (1982) Spontaneous oxygen radical generation by sickle cell erythrocytes. *J. Clin. Invest.* **70**, 1253–1259.

Hebbel, R.P., Morgan, W.T., Eaton, J.W. and Hedlund, S.E. (1988) Accelerated autoxidation and heme loss due to instability of sickle hemoglobin. *Proc. Natl. Acad. Sci. USA* **85**, 237–241.

Hebbel, R.P., Shaley, O., Foker, W. and Rank, B.H. (1986) Inhibition of erythrocyte Ca^{+2}-ATPase by activated oxygen through thiol- and lipid-dependent mechanisms. *Biochim. Biophys. Acta* **862**, 8–16.

Hernandez, L.A., Grisham, M.B. and Granger, D.N. (1987a) A role for iron in oxidant-mediated ischemic injury to intestinal microvasculature. *Am. J. Physiol.* **253**, G49.

Hernandez, L.A., Grisham, M.B., von Ritter, C. and Granger, D.N. (1987) Biochemical localization of xanthine oxidase in the cat small intestine. *Gastroenterology* **92**, 1433.

Hernandez, L.A., Grisham, M.B., Twohig, B., Arfors, K.E., Harlan, J.M. and Granger, D.N. (1987) Role of neutrophils in ischemia-reperfusion induced microvascular injury. *Am. J. Physiol.* **253**, H699–H703.

Hoult, J.R.S. and Page, H. (1981) 5-aminosalicylic acid, a co-factor for colonic prostacylin synthesis? *Lancet* **2**, 255.

Ingram, L.T., Coates, T., Aden, J., Higgens, C., Baebner, R. and Boxer, L. (1982) Metabolic, membrane and functional responses of human polymorphonuclear leukocytes to platelet activating factor. *Blood* **59**, 1259–1266.

Jacobs, A., Peters, S.W., Bauminger, E.R., Elkelboom, J., Ofer, S. and Rachmilewitz, E.A. (1981) Ferritin concentration in normal and abnormal erythrocytes measured by immunoradiometric assay with antibodies to heart and sperm ferritin and Mössbauer spectroscopy. *Brit. J. Hematol.* **49**, 201–207.

Jarasch, E.D., Bruder, G. and Heil, H.W. (1986) Significance of xanthine oxidase in capillary endothelial cells. *Acta Physiol. Scand. Suppl.* **548**, 39–46.

John, R., Outhwaite, J., Morris, C.J. and Blake, D.R. (1987) Xanthine oxidase-reductase is present in human synovium. *Ann. Rheum. Dis.* **46**, 843–845.

Jones, D.P. (1985) The role of oxygen concentration in oxidative stress: hypoxic and hyperoxic models. In *Oxidative Stress* (H. Sies, ed.), Academic Press, San Diego, pp. 152–195.

Keshavarzian, A., Sedghi, S., Kanofsky, J., List, T., Robinson, C., Ibrahim, C. and Winship, D. (1992) Excessive production of reactive oxygen metabolites by inflamed colon: Analysis by chemiluminescence probe. *Gastroenterology* **103**, 177–185.

Kitahora, T., Suzuki, K., Asakura, H., Yoshida, T., Suematsu, M., Watanabe, M., Aiso, S. and Tsuchiya, M. (1988) Active oxygen species generated by monocytes and polymorphonuclear leukocytes in Crohn's disease. *Dig. Dis. Sci.* **33**, 951–955.

Kubes, P., Arfors, K.E. and Granger, D.N. (1990) Platelet-activating factor-induced musocal dysfunction: role of oxidants and granulocytes. *Am. J. Phys.* **260**, G965–G971.

Kubes, P., Ibbotson, G., Russel, J., Wallace, J. and Granger, D.N. (1990) Role of platelet activating factor in ischemia/reperfusion-induced leukocyte adhesion. *Am. J. Physiol.* **259**, G300–G305.

Kuross, S.A. and Hebbel, R.P. (1988) Nonheme iron in sickle erythrocyte membranes: Association with phospholipids and potential role in lipid peroxidation. *Blood* **72**, 1278–1285.

Kuross, S.A., Rank, B.H. and Hebbel R.P. (1988) Excess heme in sickle erythrocyte inside-out membranes: Possible role of thiol oxidation. *Blood* **71(4)**, 876–882.

MacDermott, R.P., Scholoemann, S.R., Bertovich, M.J., Nash, G.S., Peters, M. and Stenson, W.F. (1989) Inhibition of antibody secretion by 5-aminosalicylic acid. *Gastroenterology* **96**, 442–448.

McCord, J.M. (1974) Free radicals and inflammation protection of synovial fluid by superoxide dismutase. *Science* **185**, 529–531.

Misra, H.P. and Irwin, F. (1972) The generation of superoxide radical during the autooxidation of hemoglobin. *J. Biol. Chem.* **247**, 6960–6962.

Miyachi, Y., Yoshioka, A., Imamura, S. and Niwa, Y. (1987) Effect of sulphasalazine and its metabolites on the generation of reactive oxygen species. *Gut* **28**, 190–195.

Moorhouse, P.C., Grootveld, M., Halliwell, B., Quinlan, J.G. and Gutteridge, J.M.C. (1987) Allopurinol and oxypurinol are hydroxyl radical scavengers. *FEBS Lett.* **213**, 23–28.

Morris, J.B., Haglund, U. and Bulkley, G.B. (1987) The protection from postischemic injury by xanthine oxidase inhibition: blockade of free radical generation or purine salvage. *Gastroenterology* **92**, 1542.

Mousson, B., Desjacques, P. and Baltasatt, P. (1983) Measurement of xanthine oxidase activity in some human tissues. *Enzyme* **29**, 32–43.

Nath, K.A. and Paller, M.S. (1990) Dietary deficiency of antioxidants exacerbates ischemic injury in the kidney. *Kidney Inter.* **38**, 1109–1117.

Paller, M.S. (1988) Hemoglobin- and myoglobin-induced acute renal failure in rats: Role of iron in nephrotoxicity. *Am. J. Physiol.* **255**, F539–F544.

Panter, S.S., Sadrzadeh, S.M.H., Hallaway, P.E., Haines, J.L., Anderson, V.E. and Eaton, J.W. (1985) Hypohaptoglobinemia associated with familial epilepsy. *J. Exp. Med.* **161**, 748–754.

Parks, D.A., Bulkley, G.B., Granger, D.N., Hamilton, S.R. and McCord, J.M. (1982) Ischemic injury in the cat small intestine: role of superoxide radicals. *Gastroenterology* **82**, 9–15.

Parks, D.A. and Granger, D.N. (1986) Contributions of ischemia and reperfusion to mucosal lesion formation. *Am. J. Physiol.* **250**, G749–G753.

Parks, D.A. and Granger, D.N. (1986a) Role of oxygen radicals in gastrointestinal ischemia. in *Superoxide and Superoxide Dismutase in Chemistry, Biology and Medicine.* (G. Rotilio, ed.), Elsevier, Amsterdam, pp. 614–617.

Parks, D.A. and Granger, D.N. (1986b) Xanthine oxidase: Biochemistry, distribution and physiology. *Acta Physiol. Scand. Suppl.* **548**, 47–63.

Parks, D.A., Williams, T.K. and Beckman, J.S. (1988) Conversion of xanthine dehydrogenase to oxidase in ischemic rat intestine: A re-evaluation. *Am. J. Physiol.* **254**, G768–G774.

Peppercorn, M.A. (1990) Advances in drug therapy for inflammatory bowel disease. *Ann. Modern Med.* **112**, 50–60.

Preston-Martin, S., Pike, M.C., Ross, R.K. and Jones, P.A. (1990) Increased cell division as a cause of human cancer. *Cancer Res.* **50**, 7415–7421.

Putman, F.W. (1975) In *The Plasma Proteins* (F.W. Putman ed.), Academic Press, New York, pp. 1–50.

Ramos, C.L., Par, S., Britigan, B.E., Cohen, M.E. and Rosen, G.M. (1992) Spin trapping evidence for myeloperoxidase-dependent hydroxyl radical formation by human neutrophils and monocytes. *J. Biol. Chem.* **267**, 8307–8312.

Rehncrona, S., Smith, D.S., Akesson, B., Westerberg, E. and Siesjo, B.K. (1980) Peroxidative changes in brain cortical fatty acids and phospholipids, as characterized during Fe^{+2} and ascorbic acid-stimulated lipid peroxidation in vitro. *J. Neurochem.* **34**, 1630–1638.

Riddell, R.H. (1988) Pathology of idiopathic inflammatory bowel disease. In *Inflammatory Bowel Disease* (J.B. Kirsner and R.G. Shorter, eds.), Lea and Febiger, Philadelphia, pp. 329–350.

Robinson, J.W.L., Mirkovitch, V., Winistorfer, B. and Saegesser, F. (1981) Response of the intestinal mucosa to ischemia. *Gut* **22**, 512–527.

Rosen, A.D. and Frumin, N.V. (1979) Focal epileptogenesis after intracortical hemoglobin injection. *Exp. Neurol.* **66**, 277–284.

Sadrzadeh, S.M.H., Anderson, D.K., Panter, S.S., Hallaway, P.E. and Eaton, J.W. (1984) Hemoglobin potentiates central nervous system damage. *J. Clin. Invest.* **79**, 662–664.

Sadrzadeh, S.M.H. and Eaton, J.W. (1988) Hemoglobin-mediated oxidant damage to the central nervous system requires endogenous ascorbate. *J. Clin. Invest.* **82**, 1510–1515.

Samuelsson, B. (1983) Leukotrienes: Mediators of immediate hypersensitivity reactions and inflammation. *Science* **220**, 568–575.

Schacter, J.R. Delvillano, B.C., Gordon, E.M., et al. (1985) Red cell superoxide dismutase and sickle cell anemia symptom severity. *Am. J. Hematol.* **19**, 137–144.

Schmid-Schoenbein, G.W. and Engler, R.L. (1987) Granulocytes as active participants in acute myocardial ischemia and infarction. *Am. J. Cardiovasc. Pathol.* **1**, 15–30.

Schoenberg, M.H., Fredholdm, B.B., Haglund, U., Jung, H., Sellin, D., Younes, M. and Schildberg, F.W. (1985) Studies on the oxygen radicals mechanism involved in small intestinal reperfusion damage. *Acta Physiol. Scand.* **124**, 581–589.

Schroder, H. and Campbell, D.E.S. (1972) Absorption, metabolism, and excretion of salicylazosulfapyridine in man. *Clin. Pharmacol. Ther.* **13**, 539–551.

Sekizuka, E., Benoit, J.N., Grisham, M.B. and Granger, D.N. (1989) Dimethylsulfoxide prevents chemoattractant-induced leukocyte adherence. *Am. J. Physiol.* **256**, pp. H594-H597.

Sharon, P. and Stenson, W.F. (1984) Enhanced synthesis of leukotriene B_4 by colonic mucosa in inflammatory bowel disease. *Gastroenterology* **86**, 453–460.

Shirabata, Y., Aoki, S., Takada, S., Kiriyama, H., Ohta, K., Hai, H., Teraoka, S., Matano, S., Matsumoto, K. and Kami, K. (1989) Oxygen derived free radical generating capacity of PMN in patients with ulcerative colitis. *Digestion* **44**, 163–171.

Simmonds, N.J., Allen, R.E., Stevens, T.R.J., Niall, R., van Someren, M., Blake, D.R. and Rampton, D.S. (1992) Chemiluminescence assay of mucosal reactive oxygen metabolites in inflammatory bowel disease. *Gastroenterology* **103**, 186–196.

Simon, R.H., Scoggin, C.H. and Patterson, D. (1981) Hydrogen peroxide caused the fatal injury to human fibroblasts exposed to oxygen radicals. *J. Biol. Chem.* **256**, 7181–7186.

Skaleric, U., Allen, J.B., Smith, P.D., Mergenhagen, S.E. and Wahl, S.M. (1991) Inhibitors of reactive oxygen intermediates suppress bacterial cell wall-induced arthritis. *J. Immunol.* **147**, 2559–2564.

Suzuki, M., Grisham, M.B. and Granger, D.N. (1991) Leukocyte-endothelial cell adhesive interactions: role of xanthine oxidase-derived oxidants. *J. Leukocyte Biol.* **50**, 488–494.

Suzuki, M., Asako, H., Kubes, P., Jennings, S., Grisham, M.B. and Granger, D.N. (1991) Neutrophil-derived oxidants promote leukocyte adherence in postcapillary venules. *Microvas. Res.* **42**, 125–138.

Suzuki, M., Inauen, W., Kvietys, P.R., Grisham, M.B., Meininger, C., Schelling, M.E., Granger, H.J. and Granger, D.N. (1989) Superoxide mediates reperfusion-induced leukocyte-endothelial cell interactions. *Am. J. Physiol.* **257**, H1740-H1745.

Thiel, G., Wilson, D.R., Arce, M.L. and Oken, D.E. (1967) Glycerol induced hemoglobinuric acute renal failure in the rat: II The Experimental model, predisposing factors, and pathophysiologic features. *Nephron.* **4**, 276–297.

Touati, D. (1989) The molecular genetics of superoxide dismutase in *E. coli*. *Free Rad. Res. Commun.* **8**, 1–8.

Triggs, W.J. and Willmore, L.J. (1984) In vivo lipid peroxidation in rat brain following intracortical Fe^{+2} injection. *J. Neurochem.* **42**, 976–980.

van Hess, P.A.M., Balsker, J.H. and Togeren, J.H.M. (1980) Effect of sulfapyridine, 5-aminosalicylic acid, and placebo in patients with idiopathic proctitis: a study to determine the active therapeutic moiety of sulfasalazine. *Gut* **21**, 632–635.

von Ritter, C., Grisham, M.B. and Granger, D.N. (1989) Sulfasalazine metabolites and dapsone attenuate formyl-methionyl-leucyl-phenylalanine-induced mucosal injury in rat ileum. *Gastroenterology* **96**, 811–816.

Weiss, S.J. (1986) Oxygen, ischemia and inflammation. *Acta Physiol. Scand. Suppl.* **548**, 9–37.

Weiss, S.J., Peppin, G., Oritz, X., Ragsdale, C. and Test, S.T. (1985) Oxidative autoactivation of latent collagenase by human neutrophils. *Science* **227**, 747–749.

Wengrower, D., Liakim, R., Karmeli, F., Razin, E. and Rachmilewitz, D. (1987) Pathogenesis of ulcerative colitis (UC): Enhanced colonic formation of inositol phosphates (IP) and platelet activating factor (PAF). *Gastroenterology* **92**, 1691.

Williams, J.G. and Hallett, M.B. (1989) The reaction of 5-aminosalicylic acid with hypochlorite. Implications for its actions by its mode of action in inflammatory bowel disease. *Biochem. Pharmacol.* **38**, 149–154.

Willmore, L.J. and Ruvin, J.J. (1984) Effects of antioxidants on $FeCl_2$-induced lipid peroxidation and focal edema in rat brain. *Exp. Neurol.* **83**, 62–70.

Willmore, L.J., Sypert, G.W. and Munson, J.B. (1978) Recurrent seizures induced cortical iron injection: A model of posttraumatic epilepsy. *Ann. Neurol.* **4**, 329–336.

Willmore, L.J., Sypert, G.W., Munson, J.B. and Hurd, R.W. (1978) Chronic focal epileptiform discharges induced by injection of iron into rat and cat cortex. *Science* **200**, 1501–1503.

Willmore, L.J. and Triggs, W.J. (1991) Iron-induced lipid peroxidation and brain injury responses. *J. Devel. Neurosci.* **9**, 175–180.

Wood, M.L., Dizdaroglu, M., Gajewski, E. and Essigmann, J.M. (1990) Mechanistic studies of ionizing radiation and oxidative mutagenesis: genetic effects of a single 8-hydroxyguanine (7-hydro-8-oxoguanine) residue inserted at a unique site in a viral genome. *Biochemistry* **29**, 7024–7032.

Yamada, T., Fujimoto, K., Tso, P., Fujimuto, T., Gaginella, T.S. and Grisham, M.B. (1992) Misoprostol accelerates colonic mucosal repair in acetic acid-induced colitis. *J. Pharmacol. Exp. Therap.* **260**, 313–318.

Yamada, T., Specian, R.D., Granger, D.N., Gaginella, T.S. and Grisham, M.B. (1991) Misoprostol attenuates acetic acid-induced increases in mucosal permeability and inflammation: role of blood flow. *Am. J. Physiol.* **261**, G332-G339.

Yamada, T., Volkmer, C. and Grisham, M.B. (1990) Antioxidant properties of 5-ASA: Potential mechanisms for its anti-inflammatory activity. *Can. J. Gastroenterology* **4**, 295–302.

Yamada, T., Volkmer, C. and Grisham, M.B. (1991) The effect of sulfasalazine and metabolites on hemoglobin-catalyzed lipid peroxidation. *J. Free Rad. Biol. Med.* **10**, 41–49.

Youdim, M.B.H. and Green, A.R. (1978) Iron deficiency and neurotransmitter synthesis and function. *Proc. Nutr. Soc.* **37**, 173–179.

Younes, M., Mohr, A., Schoenberg, M.H. and Schildberg, F.W. (1987) Inhibition of lipid peroxidation by superoxide dismutase following regional intestinal ischemia and reperfusion. *Res. Exp. Med.* **187**, 9–17.

Zaleska, M.M. and Floyd, R.A. (1980) Regional lipid peroxidation in rat brain in vitro: Possible role of endogenous iron. *Neurochem. Res.* **34**, 1630–1638.

Zimmerman, B.J. and Granger, D.N. Role of hydrogen peroxide, iron, and hydroxyl radicals in ischemia/reperfusion-induced neutrophil infiltration. *The Physiologist* **31**, A229.

Zimmerman, B.J., Grisham, M.B. and Granger, D.N. (1987) Role of superoxide-dependent chemoattractants in ischemia-reperfusion induced neutrophil infiltration. *Fed. Proc.* **46**, 1124.

Zimmerman, B.J., Guillory, D.J., Grisham, M.B., Gaginella, T.S. and Granger, D.N. (1990) Role of leukotriene B_4 in granulocyte infiltration into the postischemic feline intestine. *Gastroenterology* **99**, 1358–1363.

Zimmerman, B.J., Parks, D.A., Grisham, M.B. and Granger, D.N. (1988) Allopurinol does not enhance the antioxidant properties of extracellular fluid. *Am. J. Physiol.* **255**, H202-H206.

CHAPTER 3

Free Radical Mechanism of Oxidative Modification of Low Density Lipoprotein (or the Rancidity of Body Fat)

Balaraman Kalyanaraman

3.1 INTRODUCTION

Life is a constant battle to avoid becoming rancid.
Gary G. Duthie

Nearly five decades ago, Golumbic and Mattill (1941) reported that ascorbic acid (vitamin C), despite its low solubility in lipid, could enhance the antioxidant action of α-tocopherol (vitamin E) in a synergistic fashion. The summary of that seminal paper reads as follows:

> Ascorbic acid is an effective antioxidant for certain vegetable oils, their hydrogenated products and esters. It enhances the antioxygenic activity of tocopherols, hydroxy chromans, hydroquinones, and related compounds.

Subsequently, a number of reports elucidating the mechanism of synergistic reaction between vitamin C and vitamin E have appeared (Bascetta et al., 1983; Packer et al., 1979; Doba et al., 1985).

More recently, the participants of an NHLBI workshop on "Antioxidants in the Prevention of Human Atherosclerosis" (organized by Dr. Daniel Steinberg) concluded that since vitamin C may regenerate and "spare" vitamin E, the protection afforded by the two given together should be additive. The panel concluded, "The first clinical trials should probably not be undertaken with a single agent, but with an antioxidant 'cocktail' . . ." The consensus was that a trial of vitamin C, vitamin E, and β carotene in a 2 × 2 × 2 factorial design would be the best approach (Steinberg, 1992). Clearly, much more needs to be done with respect to clinical trials and so forth; however, it is ironic that the rancidity of body fat may be prevented in much the same way as rancidness of food fat. The purpose of this chapter is to provide a molecular mechanism of action of phenolic antioxidants in inhibiting the oxidative modification of low-density lipoprotein and its relevance to atherogenesis. In this chapter, the applications of electron spin resonance (ESR) technique to investigate the mechanism of free-radical reactions are also reviewed.

3.2 OXIDATIVE MODIFICATION OF LOW DENSITY LIPOPROTEIN AND ITS RELEVANCE TO ATHEROGENESIS

The low-density lipoprotein (LDL) is the major cholesterol-carrying lipoprotein in plasma. Human LDL consists of ~25% apo B-100 protein and approximately 75% lipid consisting almost entirely of cholesterol esters and some triglycerides. The major unsaturated fatty acyl group in LDL is linoleate associated with cholesteryl ester. The rich unsaturated fatty acid content of LDL is very susceptible to peroxidative degeneration under oxidizing conditions in the circulation. It is currently believed that the cholesterol carried in LDL is the most atherogenic form of serum cholesterol (Brown and Goldstein, 1983; Steinberg et al., 1989).

Several clinical and experimental studies have firmly established a correlation between elevated levels of LDL and the onset of atherosclerosis (Steinberg et al., 1989). There are few receptors for native LDL in macrophages. Consequently, LDL undergoes little degradation by these cells (Henriksen et al., 1981; Morel et al., 1984; Stenbrecher et al., 1984; Heinecke et al., 1984; Heinecke, 1987; Hennig and Chow, 1988; Sparrow et al., 1989). However, the LDL that has been modified by trace metals (Esterbauer et al., 1988, 1990), lipoxygenase (Kalyanaraman et al., 1990), or endothelial cells is rap-

Table 3.1.
Inhibitors of LDL Oxidation

Endogenous antioxidant	Plausible mechanisms
α-Tocopherol	LOO· + α-TH → LOOH + α-T·
	LOO· + α-T· → nonradical products
γ-Tocopherol	LOO· + γ-TH → LOOH + γ-T·
β-Carotene	LOO· + BC → LOO-BC·
	LOO-BC· + LOO → nonradical products
Ubiquinol-10	LOO· + UQH_2 → LOOH + UQH·
	α-T· + UQH_2 → α-TH + UQH·
Lycopene	LOO· + lycopene → products

α TH = α-tocopherol; α-T· = α-tocopheroxyl radical; γ-TH = γ-tocopherol; γ-T· = γ-tocopheroxyl radical; BC = β-carotene; LOO· = lipid peroxyl radical; UQH_2 = reduced ubiquinone$_{10}$; UQH· = ubisemiquinone radical.

idly degraded in macrophages by an alternate scavenger receptor, which ultimately leads to the formation of "foam cells" (Steinberg et al., 1989). Results from studies using antioxidants have shown a definite involvement of free radicals during the conversion of the "native" LDL to the more atherogenic "oxidatively modified" LDL (Steinberg et al. 1989).

3.3   EFFECT OF SUPPLEMENTATION WITH ANTIOXIDANTS

LDL itself contains several antioxidants such as α-tocopherol, γ-tocopherol, β-carotene, lycopene, and ubiquinol (Esterbauer et al. 1990). Only when these antioxidants are exhausted will oxidative modification occur, leading to the generation of macrophage-derived foam cells. Table 3.1 lists some of the plausible radical reactions involved in retarding LDL oxidation. It follows that increasing the antioxidant capacity should retard the development of atherogenesis. A number of structurally different compounds had been used to inhibit oxidative modification of LDL. Table 3.2 lists those compounds and their plausible mechanism of inhibition of LDL oxidation.

3.4   INHIBITION OF LDL OXIDATION BY PHENOLIC ANTIOXIDANTS

Phenoxyl radicals are formed by a one-electron oxidation of phenols, and in the absence of any reductants, they undergo polymer-

Table 3.2.
Inhibitors of LDL Oxidation

Exogenus antioxidants	Plausible mechanism
Probucol butylatedhydroxytoluene	Peroxyl radical scavenger
Ascorbic acid	Regeneration of vitamin E, probucol, and BHT
Probucol diglutarate	Water-soluble, intracellular peroxyl radical scavenger
Flavonoids	Peroxyl radical scavenger Metal-ion chelator
Ebselen (±GSH)	Lipid hydroperoxide and phospholipid hydroperoxide scavenger
Aminoguanidine	Aldehyde scavenger
NDGA	Peroxyl radical scavenger Metal-ion chelator 15-lipoxygenase inhibitor
Nifedipine	Radical scavenger
Hydroxamates	Peroxyl radical scavenge Metal-ion chelator Lipoxygenase inhibitor
α-Phenyl-*tert*-butyl N-nitrone	Lipid-derived radical scavenger

ization forming dimers, trimers, etc. In the presence of a reductant (AH⁻) the phenoxyl radicals are reduced to form the parent compound, with the concomitant formation of the reductant-derived radical (Fig. 3.1). Of significance is the reaction between phenoxyl radicals and ascorbic acid. This reaction occurs at a fairly rapid rate ($k = 10^5$ $M^{-1}s^{-1}$) (Schuler, 1977). Formation of dityrosine is often diagnostic of tyrosyl radical intermediacy (Heinecke et al., 1993; Huggins et al., 1993).

Some well-known phenolic antioxidants capable of inhibiting LDL oxidation are BHT, BHA, vitamin E (Vit-E), and Probucol (Fig. 3.2).

Vit-E inhibits LDL oxidation by acting as a chain-breaking oxidant (Esterbauer, 1990).

$$Vit\text{-}E + LOO^{\cdot} \rightarrow Vit\text{-}E^{\cdot} + LOOH$$

$$Vit\text{-}E^{\cdot} + LOO^{\cdot} \rightarrow Vit\text{-}E{-}OOL$$

Vit-E radical does not react with oxygen (Doba et al., 1984), but can react with nitric oxide ($^{\cdot}NO$) (Wilcox and Janzen, 1993)

$$Vit\text{-}E^{\cdot} + {}^{\cdot}NO \rightarrow Vit\text{-}E{-}NO$$

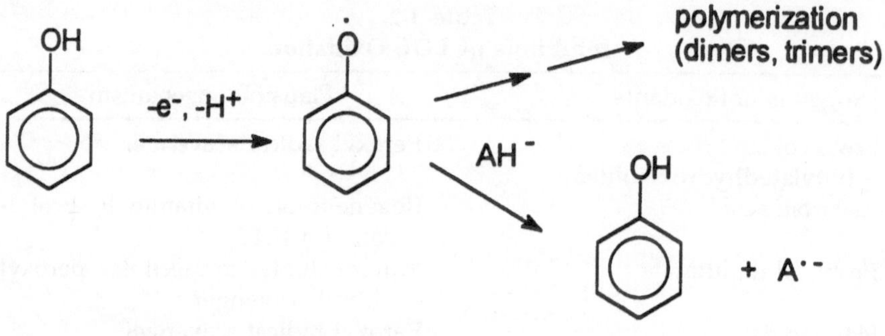

Fig. 3.1 Reactions of phenoxyl radical

BHT also inhibits LDL oxidation by reacting with the lipid peroxyl radical and forming the corresponding BHT phenoxyl radical (Janzen et al., 1993). In contrast to Vit-E radical, BHT reacts with molecular oxygen forming the corresponding hydroperoxide (BHT-OOH). BHT has also been shown to react with NO forming the nitroso product (BHT-NO) (Janzen et al., 1993). In this review, only reactions of Vit-E and Probucol in LDL oxidation will be discussed.

3.4.1 *Vitamin E Radical Formation in LDL*

Vitamin E radical is resonance-stabilized and shares the characteristics of a phenoxyl radical. The electron density on the oxygen

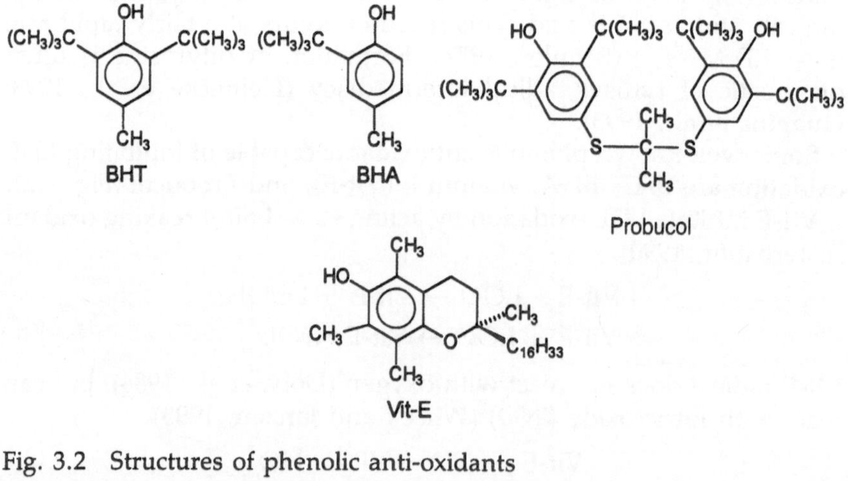

Fig. 3.2 Structures of phenolic anti-oxidants

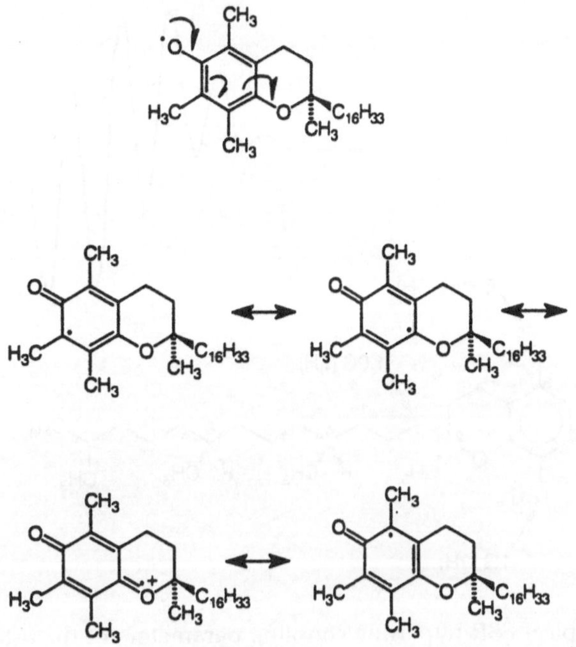

Fig. 3.3   Resonance-stabilized structures of vitamin E radical.

atom is distributed throughout the molecule as a result of resonance stabilization. Several canonical structures (Fig. 3.3) can be drawn as follows:

In accord, the ESR spectrum of Vit-E radical is a composite of electron interacting with several different nuclei as shown (Fig. 3.4). The numbers in parentheses denote hyperfine coupling in gauss. Figure 3.4 (inset) shows the seven-line ESR spectrum characteristic of the vitamin-E radical. (For ESR conditions please refer to Kalyaranaman et al., 1992)

LDL also contains γ-tocopherol. The question then is: how can one distinguish between the α-tocopheroxyl radical and the γ-tocopheroxyl radical. The ESR spectrum of the γ-tocopheroxyl radical is characterized by electron interacting with several nuclei as shown in Fig. 3.5. The difference between the α-tocopheroxyl radical and the γ-tocopheroxyl radical is the absence of a methyl group. Figure 3.5 (inset) shows a five-line spectrum characteristic of the γ-tocopheroxyl radical. In most in vitro oxidation of LDL, only the α-tocopheroxyl radical has been detected by ESR. Addition of Cu^{2+} or

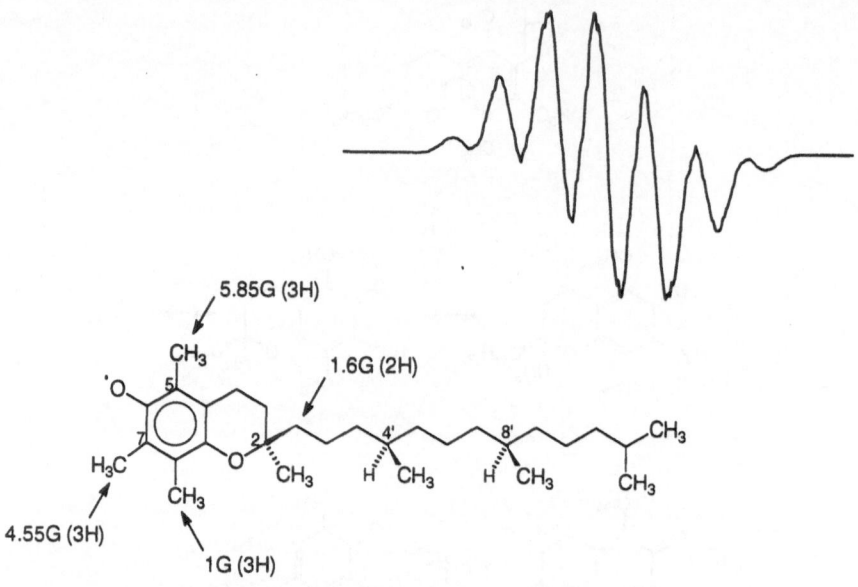

Fig. 3.4 Typical ESR hyperfine coupling parameters of the α-tocopheroxyl radical. (Inset) A computer simulated ESR spectrum based on the ESR hyperfine coupling parameters shown.

lipoxygenase to LDL results in the formation of Vit-E radical associated with LDL. In the presence of ascorbic acid, the ESR spectrum of Vit-E radical is replaced by the ascorbate radical spectrum (Fig. 3.6) (Kalyanaraman et al., 1992). After the depletion of ascorbate, Vit-E radical reappeared. Since ascorbate is water soluble, it cannot trap the peroxyl radical formed in the LDL lipid. However, it can still inhibit LDL peroxidation by regenerating Vit-E in LDL.

$$(\text{Vit-E}^{\cdot})_{\text{LDL}} + (\text{ascorbate})_{\text{aq}} \rightarrow (\text{Vit-E})_{\text{LDL}} + (\text{ascorbate radical})_{\text{aq}}$$

In the absence of ascorbic acid, Vit-E generally decays under peroxidizing conditions to form Vit-E quinone and other epoxides (Liebler et al., 1991). Such products have not been detected in LDL oxidation in the presence of Cu^{2+} or lipoxygenase or other peroxyl-radical initiators (such as ABAP). However, peroxynitrite (a species formed from the interaction between ˙NO and O_2^{-}) has been shown to oxidize LDL forming Vit-E quinone (Graham et al., 1993).

More recently, the prooxidant role of Vit-E in Cu^{2+}-induced LDL oxidation has been described (Bowry et al., 1992; Bowry and Stocker,

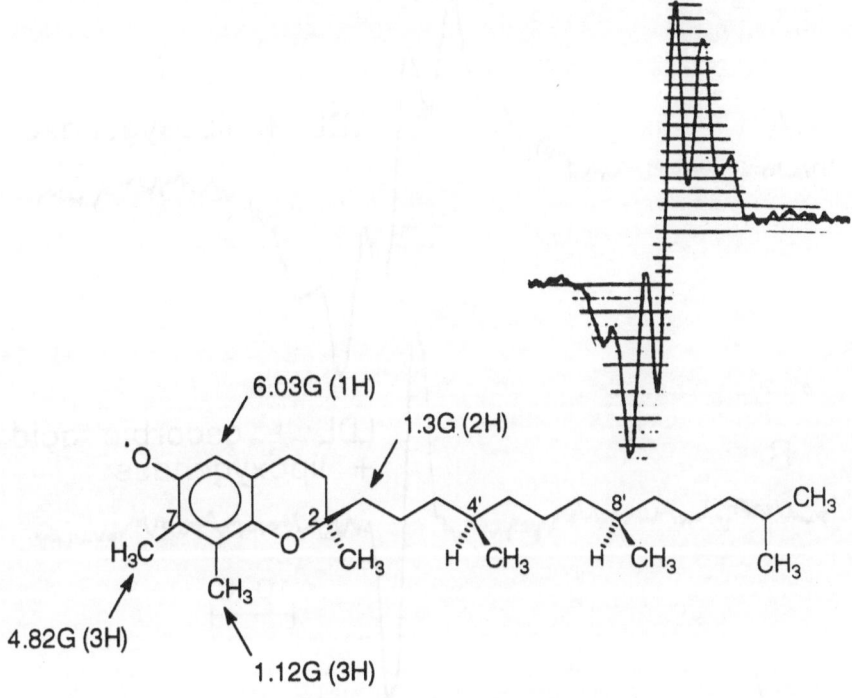

6.03G (1H)

1.3G (2H)

4.82G (3H)

1.12G (3H)

γ-Tocopheroxyl Radical

Fig. 3.5 Typical ESR hyperfine coupling parameters of the γ-tocopheroxyl radical. (Inset) An ESR spectrum of the γ-tocopheroxyl radical.

1993; Maiorino et al., 1993). The following reactions can account for such a prooxidant mechanism.

$$Cu^{2+} + LOOH \xrightarrow{-H+} Cu^{+} + LOO^{\cdot}$$

$$Cu^{+} + LOOH \rightarrow Cu^{2+} + LO^{\cdot} + {}^{-}OH$$

$$LO^{\cdot} + LH \rightarrow L^{\cdot} + LOH$$

$$L^{\cdot} + O_2 \rightarrow LOO^{\cdot}$$

$$LOO^{\cdot} + LH \rightarrow L^{\cdot} + LOOH$$

Since LO˙ formation is vital to the propagation reaction, Vit-E has been shown to facilitate its formation as follows:

$$Cu^{2+} + Vit\text{-}E \rightarrow Vit\text{-}E^{\cdot} + Cu^{+}$$

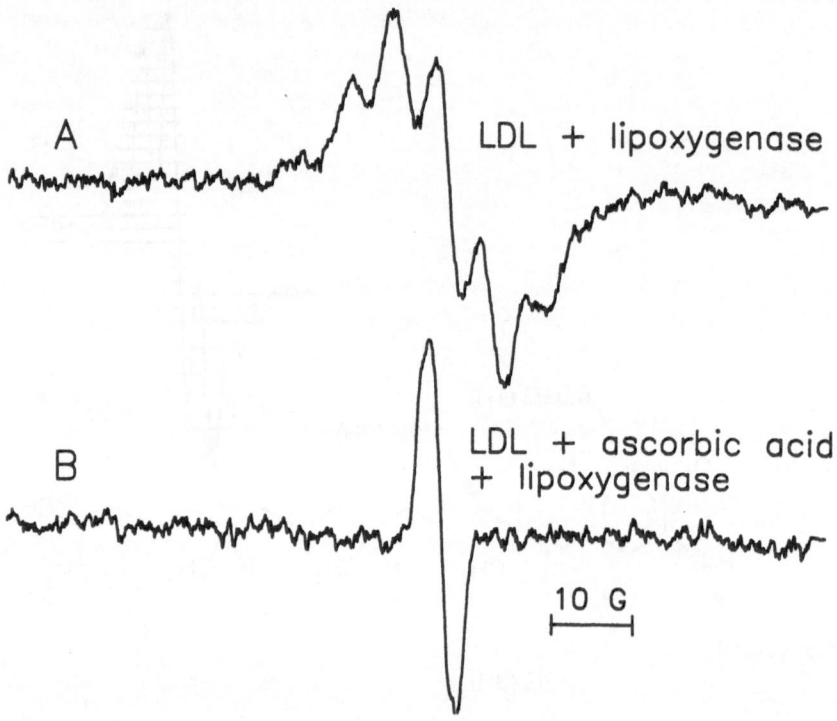

Fig. 3.6   A) ESR spectrum of the α-tocopheroxyl radical formed during oxidation of LDL by lipoxygenase. B) Same as above, but in the presence of ascorbic acid.

The Vit-E radical also has been suggested to initiate LDL oxidation (Bowry et al., 1992; Bowry and Stocker, 1993).

$$Vit\text{-}E^{\cdot} + LH \rightarrow Vit\text{-}E + L^{\cdot}$$

It is likely that reactions of this kind may simply represent a "test tube" phenomenon, and under physiological conditions, reactions between Vit-E radical, ascorbic acid, and Vit-E radical/glutathione may prove to be far more significant.

3.4.2   *Probucol Radical*

Evidence to support the oxidative modification of LDL hypothesis has been found in experimental models of atherosclerosis in which the development of atherosclerotic lesions was inhibited by Probucol, a lipid-soluble antioxidant (Carew et al., 1987; Kita et al., 1987; Jackson et al., 1993). Probucol, which lowers cholesterol, is transported in the lipoprotein fraction of plasma (Parthasarathy, 1992; Rankin et al., 1991; Steinberg, 1986; Parthasarathy et al., 1986 and 1992).

The ex vivo assessment of the antioxidant activity of LDL, isolated from experimental animals or patients undergoing treatment with Probucol, has shown that the compound increases the resistance of LDL to oxidation. In an attempt to investigate the type of free radical produced during oxidation of Probucol-enriched LDL, we had oxidized (using lipoxygenase) LDL isolated from plasma of patients treated with conventional dosages of Probucol. Aerobic incubation of LDL-containing Probucol and lipoxygenase produced a composite ESR spectrum arising from both the α-tocopheroxyl and Probucol phenoxyl radicals (Fig. 3.7) (Kalyanaraman et al., 1992). In the presence of ascorbate, the composite ESR spectrum was again replaced by the doublet spectrum of the ascorbate radical.

In order to determine whether the one-line spectrum (denoted by ● in Fig. 3.8) is due to the Probucol phenoxyl radical, we generated the authentic ESR spectrum of Probucol in a chemical system (Fig. 3.8) and resolved the hyperfine couplings of the two *meta*-protons when the scan range was reduced from 100 to 10 G (Kalyanaraman et al., 1992). The triplet spectrum, with $1:2:1$ intensity ratios, was assigned to an electron interacting with two *meta*-protons in one of the aromatic rings. Hyperfine couplings from the neighboring aromatic ring were negligible because of the lack of electron delocalization between the two aromatic rings.

It has been shown that the ultimate product of Cu^{2+}-catalyzed oxidation of Probucol was bis(2,6-di-*tert*-butylphenol) as shown below:

Since the mechanism of conversion of Probucol phenoxyl radical to the bis(2,6-di-*tert*-butylphenol) was not immediately apparent, we monitored with time the Ag_2O-catalyzed oxidation of Probucol, under alkaline conditions. We observed a time-dependent transformation of the Probucol phenoxyl radical to the phenoxyl radical from bis(2,6-di-*tert*-butylphenol). As shown in Fig. 3.9, the ESR spectrum of the phenoxyl radical formed form bis(2,6-di-*tert*-butylphenol) exhibits a distinctly different hyperfine coupling pattern (five-line) from the ESR spectrum of Probucol phenoxyl radical.

Based on this, we have shown a radical-mediated mechanism of transformation of Probucol to bis(2,6-di-*tert*-butylphenol). The following scheme is different from the previously proposed mecha-

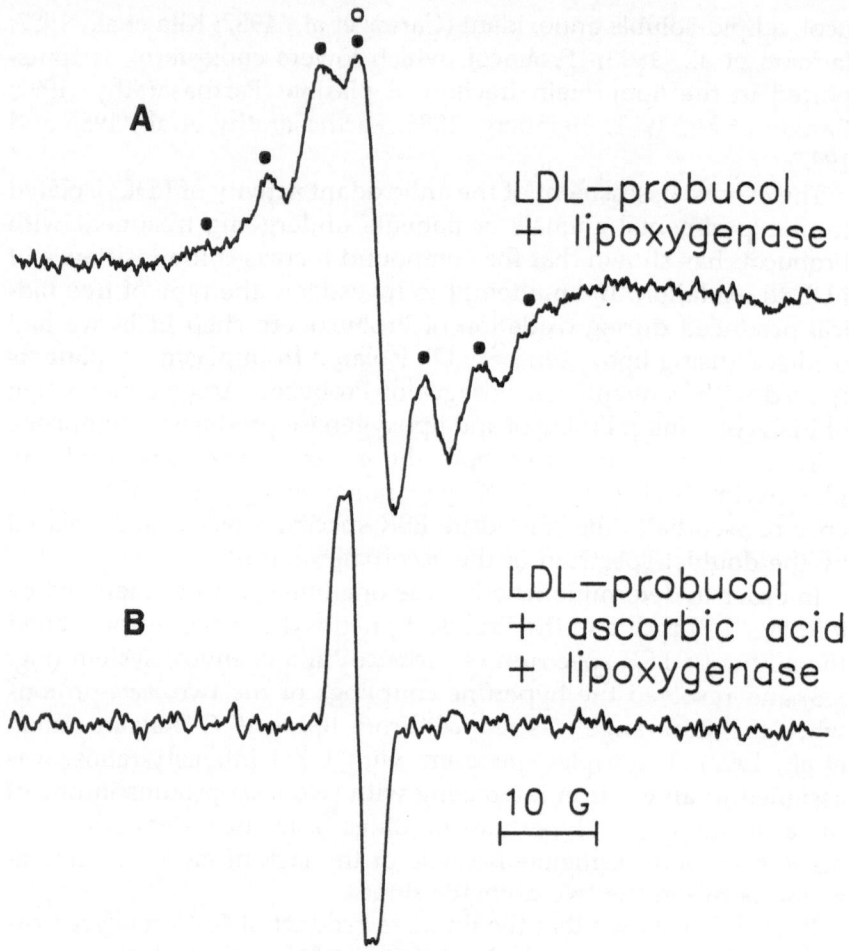

Fig. 3.7 A) ESR spectrum of a mixture of α-tocopheroxyl and Probucol phenoxyl radicals formed during oxidation of LDL enriched with Probucol by lipoxygenase, and B) same as above, but in the presence of ascorbic acid.

nism in which a bi-radical from Probucol was implicated (Barnhart et al., 1989).

Previously, we had shown that the combination of ascorbate and Probucol inhibits the oxidation of LDL, as measured by an increase in lag-time for the formation of conjugated dienes (Kalyanaraman et al., 1992). We then examined the possibility that this might also

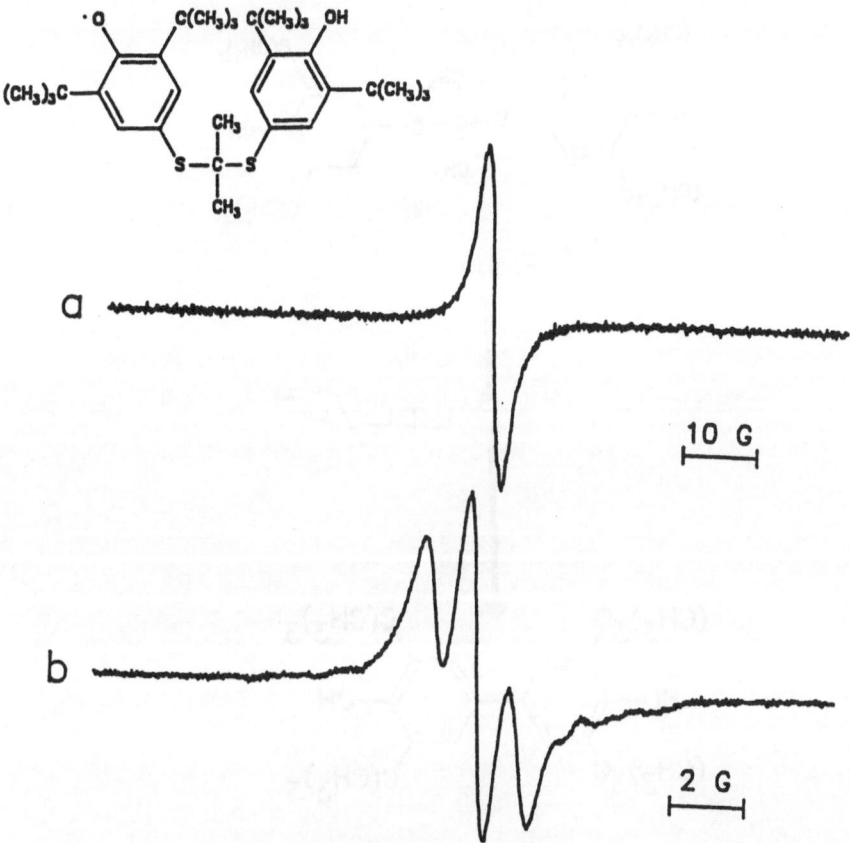

Fig. 3.8 ESR spectrum of the Probucol phenoxyl radical generated during Ag$_2$O-catalyzed oxidation of Probucol in methanol.

inhibit the uptake of LDL by macrophages. Figure 3.10 shows that LDL incubation in the presence of Cu^{2+} resulted in enhanced formation of thiobarbituric acid reactive substances (TBARS) and increased degradation of oxidized LDL by macrophages. Addition of Probucol inhibited both TBARS formation and macrophage degradation in a dose-dependent manner. These efforts were further enhanced in the presence of ascorbic acid.

It is likely that the potential of compounds such as Probucol to inhibit the oxidation of LDL in vivo could be significantly enhanced by the synergistic interaction with ascorbate in plasma and the artery wall (Aust et al., 1993).

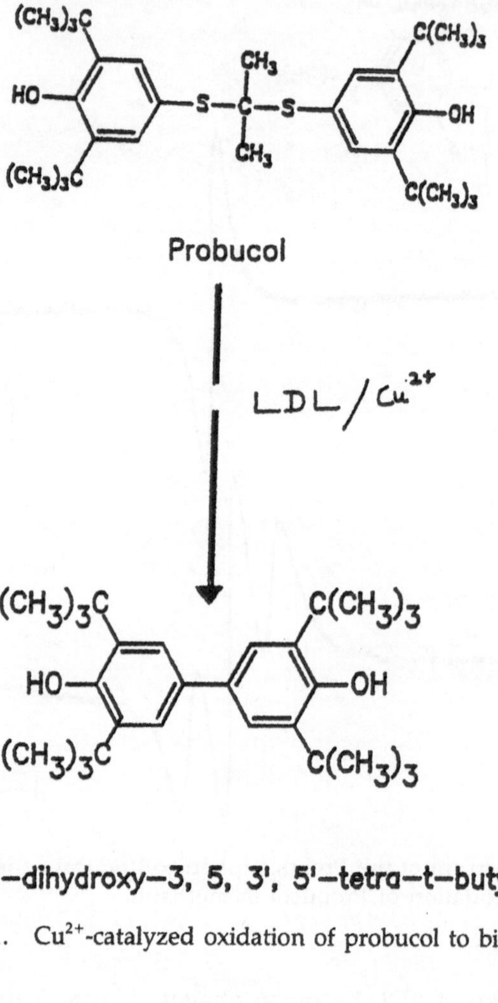

4, 4'–dihydroxy–3, 5, 3', 5'–tetra–t–butyl biphenyl

Scheme 3.1.   Cu^{2+}-catalyzed oxidation of probucol to bis(2,6-di-*tert*-butyl-phenol)

3.4.3   *Effect of Quinone and Hydroquinone on LDL Oxidation*

It is well known that quinones and hydroquinones undergo one-electron reduction and oxidation to form the corresponding *p*-semiquinones (Kalyanaraman et al. 1985). In the presence of oxygen, p-semiquinones undergo redox cycling to form O_2^- and the parent quinone. Para-semiquinones also exist in equilibrium with quinone and hydroquinone. These reactions are shown in Scheme 3.3.

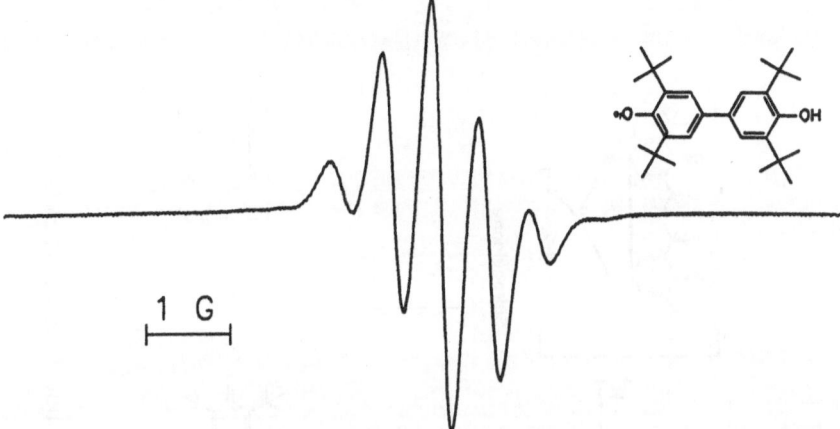

Fig. 3.9   ESR spectrum of the phenoxyl radical derived from autoxidation of 4,4'-dihydroxy-3,5,3',5'-tetra-*tert*-butyl-biphenyl.

Semiquinones are also produced during peroxidation, i.e., through the hydrogen abstraction reactions from hydroquinones by peroxyl and alkoxyl radicals.

$$QH_2 + LOO^{\cdot} \rightarrow QH^{\cdot} + LOOH.$$

$$QH_2 + LO^{\cdot} \rightarrow QH^{\cdot} + LOH$$

In this way, a hydroquinones also can act as a chain-breaking antioxidant. Quinones are also potent inhibitors of lipid peroxidation, although the mechanism of inhibition has not been fully established.

LDL contains ubiquinol-10($CoQH_2$). Oral supplementation of humans with its oxidized form, ubiquinol-10 (CoQ), has been shown to increase the levels of ubiquinol-10 in the plasma and in LDL (Mohr et al., 1992). Structures of $CoQH_2$ and CoQ are shown in Fig. 3.11. Ubiquinol-10 has been suggested to recycle the α-tocopheroxyl radical to α-tocopherol.

$$UQH_2 + α\text{-}T^{\cdot} \rightarrow α\text{-}TH + UQH^{\cdot}$$

where $UQH^{\cdot}$ is the ubi-semiquinone radical (Mukai et al., 1990). The presence of ubisemiquinone radical in LDL oxidation has not been detected by ESR. In the presence of oxygen, ubi-semiquinone forms ubiquinone and $O_2^{\cdot-}$. Although results indicate that dietary supplementation with CoQ increased the resistance of human LDL to the

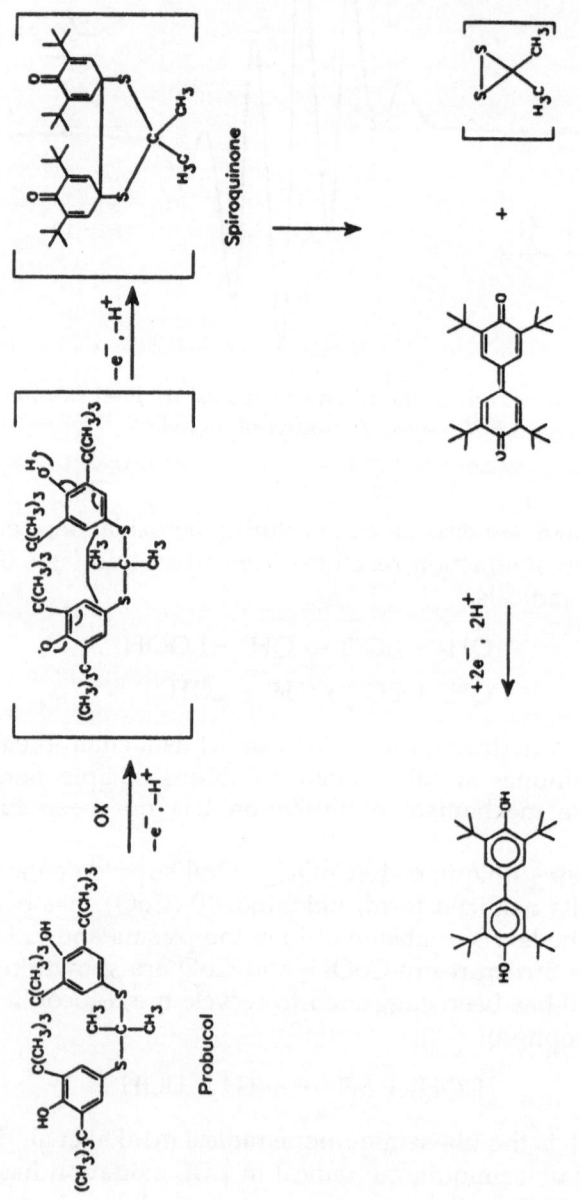

Scheme 3.2.   Mechanism of formation of bis-phenol from the Probucol phenoxyl radical

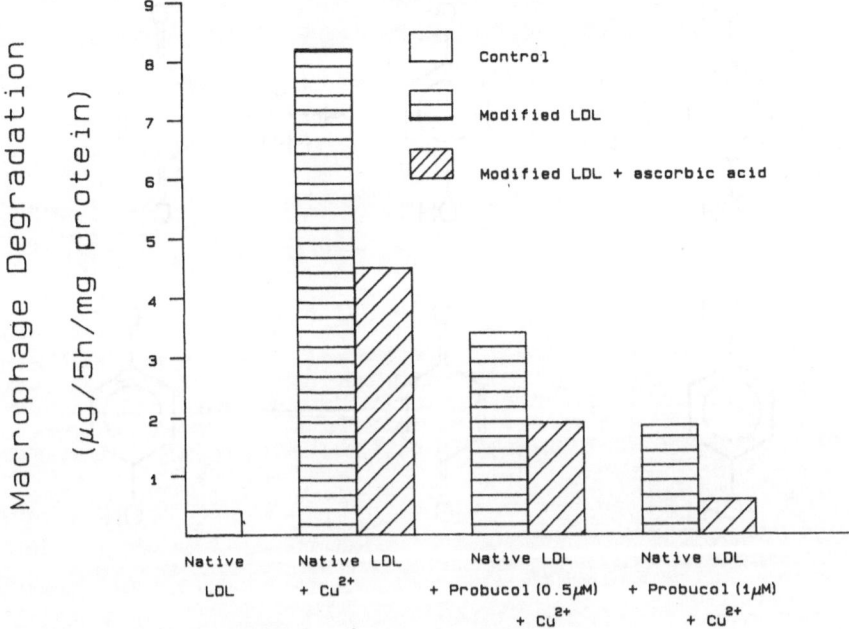

Fig. 3.10 The effect of ascorbic acid and Probucol on TBARS formation and macrophage degradation of Cu^{2+}-oxidized LDL.

initiation of lipid peroxidation (Mohr et al., 1992; Stocker et al., 1991), the mechanism of inhibition of LDL oxidation by $CoQH_2$ is far from clear.

3.4.4 *Effect of Catechols and Flavonoids on LDL Oxidation*

One-electron oxidation of catechols forms o-semiquinones, which undergo disproportionation to the corresponding o-quinones and the parent catechols. Ortho-semiquinones also exit in equilibrium with o-quinones and catechols (Scheme 3.4). Catechols and o-quinones are effective chelators of iron. Catechols also can reduce ferric iron to the ferrous form. One of the commonly used inhibitors of lipoxygenase is a catecholic antioxidant, nordihydroguaiaretic acid (NDGA). NDGA reduces the catalytically active ferric form of lipoxygenase to the inactive ferrous form (Kemal et al., 1987). Reactions between soybean lipoxygenase and NDGA have been shown to produce the corresponding o-semiquinone species, which was detected by ESR (Van der Zee et al., 1989). Recently, lipoxygenase

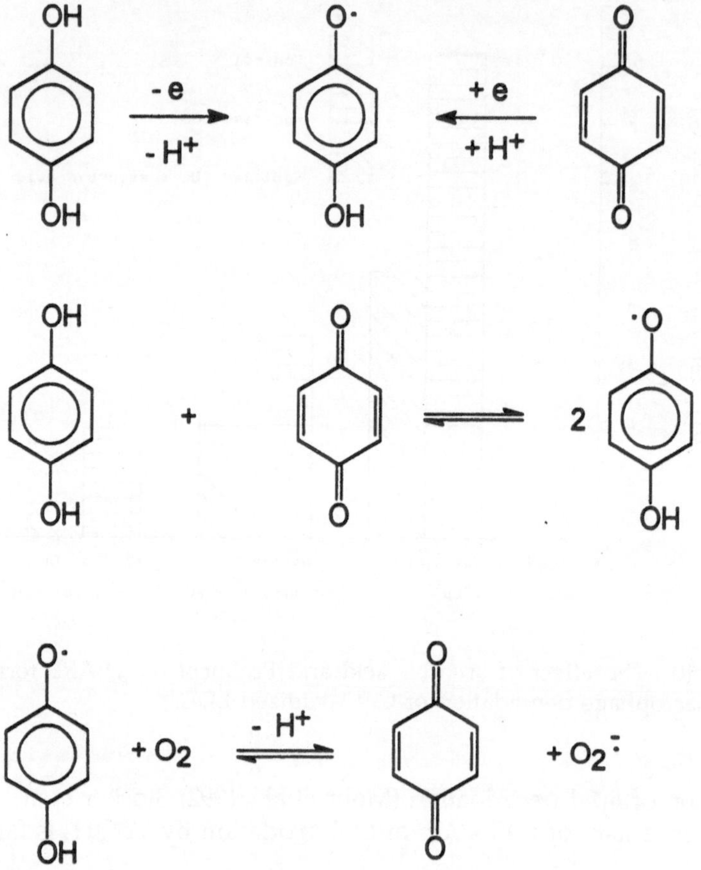

Scheme 3.3.   Reactions of quinones, hydroquinones, and semiquinones

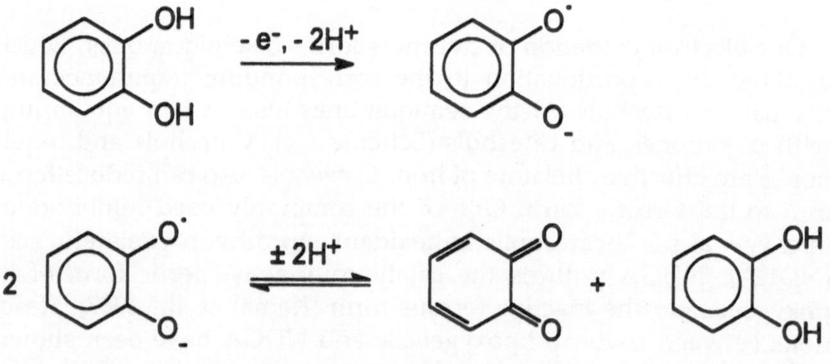

Scheme 3.4.   Reactions between o-quinones, catechols, and o-semiquinones.

Fig. 3.11 Structure of ubiquinol and ubiquinone.

inhibitors have been shown to block the cellular oxidative modification of LDL (Rankin et al., 1991), although the specificity of these compounds with respect to inhibition of lipoxygenases has recently been questioned (Sparrow and Olszewski, 1992). Irrespective of the mechanism involved, several catechol-type compounds inhibit the oxidative modification of LDL under in vitro conditions.

Recently, it has been shown that certain flavonoids, plant constituents found in the diet, are potent inhibitors of the oxidative modification of LDL (de Whalley et al. 1990). In a recent study, the lower incidence of heart disease in France has been attributed to the regular consumption of red wine. The French red wines, presumably, contain a flavonoid-type compound that decreases the oxidation of LDL.

3.5 SUMMARY

Many phenolic and catecholic compounds present in the diet are also antioxidants. Ascorbic acid synergistically enhanced the antioxidant potential of phenolic compounds in LDL oxidation. Antioxidants such as vitamins E and C have been shown to protect the oxidation of food fat. It is conceivable that oxidation of body fat could also be prevented in a similar fashion.

ACKNOWLEDGEMENTS

This research was funded by NIH grants HL47250 and RR01008.

References

Aust, S.D., Chignell, C.F. Bray, T.M. (1993) Kalyanaraman, B. and Mason, R.P. (1993) Contemporary issues in toxicology: free radicals in toxicology. *Toxicol. Appl. Pharmacol.* **120**, 168–178.

Barnhart, R.L., Busch, S.J. and Jackson, R.L. (1989) Concentration-dependent antioxidant activity of Probucol in low density lipoproteins in vitro: Probucol degradation precedes lipoprotein oxidation. *J. Lipid Res.* **30**, 1703–1710.

Bascetta, E., Gunstone, F.D. and Walton, J.C. (1983) Electron spin resonance study of the role of vitamin E and vitamin C in the inhibition of fatty acid oxidation in a model membrane. *Chem. Phys. Lipids* **33**, 207–210.

Bowry, V.W., Ingold, K.U. and Stocker, R. (1992) Vitamin E in human low-density lipoprotein—when and how this antioxidant becomes a pro-oxidant. *Biochem. J.* **288**, 341–344.

Bowry, V.W. and Stocker, R. (1993) Tocopherol-mediated peroxidation. The prooxidant effect of vitamin E on the radical-initiated oxidation of human low-density lipoprotein. *J. Am. Chem. Soc.* **115**, 6029–6044.

Brown, M.S. and Goldstein, J.L. (1983) Lipoprotein metabolism in the macrophage: implications for cholesterol deposition in atherosclerosis. *Ann. Rev. Biochem.* **52**, 223–261.

Carew, T.E., Schwenke, D.C. and Steinberg, D. (1987) Antiatherogenic effect of Probucol unrelated to its hypocholesterolemic effect: evidence that antioxidants *in vivo* can selectively inhibit low density lipoprotein degradation in macrophage-rich fatty streaks and slow the progression of atherosclerosis in the Watanabe heritable hyperlipidemic rabbit. *Proc. Natl. Acad. Sci. USA* **84**, 7725–7729.

de Whalley, C.V., Rankin, S.M., Hoult, J.R.S., Jessup, W. and Leake, D.S. (1990) Flavonoids inhibit the oxidative modification of low density lipoproteins by macrophages. *Biochem. Pharmacol.* **39**, 1743–1750.

Doba, T., Burton, G.W. and Ingold, K.U. (1985) Antioxidant and co-oxidant activity of vitamin C, either alone or in the presence of vitamin E or a water-soluble analogue upon the peroxidation of aqueous multilamellar phospholipid liposomes. *Biochim. Biophys. Acta* **835**, 298–303.

Doba, T., Burton, G.W., Ingold, K.U. and Matsuo, M. (1984) α-Tocopheroxyl decay: lack of effect of oxygen. *J. Chem. Soc. Chem. Commun.* 461–462.

Duthie, G.G. (1991) Measuring oxidation and antioxidant status in vivo. *Chem. Industry* **21**, 42–44.

Esterbauer, H., Dieber-Rothender, M., Waeg, G., Striegl, G. and Jurgens, G. (1990) Biochemical, structural, and functional properties of oxidized low-density lipoprotein. *Chem. Res. Toxicol.* **3**, 77–92.

Esterbauer, H., Quenhenberger, O. and Jurgens, G. (1990) The effect of vitamin C, either alone or in the presence of vitamin E or a water-soluble vitamin E analog, upon the peroxidation of aqueous multilamellar phospholipid liposomes. In *Free Radicals: Methodology and Concepts* (C. Rice-Evans and B. Halliwell ed.), Richelieu, London, pp. 243–268.

Golumbic, C. and Mattill, H.A. (1941) Antioxidants and the autoxidation of fats. XIII. The antioxygenic action of ascorbic acid in association with tocopherols, hydroquinones, and related compounds. *J. Am. Chem. Soc.* **63**, 1279–1280.

Graham, A., Hogg, N., Kalyanaraman, B., O'Leary, V., Darley-Usmara, V. and Moncada, S. (1993) Peroxynitric modification of low-density lipoprotein leads to recognition by the macrophage scavenger receptor. *FEBS Lett.* **330**, 181–185.

Heinecke, J.W. (1987) Free radical modification of low-density lipoprotein: mechanisms and biological consequences. *Free Rad. Biol. Med.* **3**, 65–77.

Heinecke, J.W., Li, W., Daehnke III, H.L. and Goldstein, J.A. (1993) Dityrosine, a specific marker of oxidation is synthesized by the myeloperoxidase-hydrogen peroxide system of human neutrophils and macrophages. *J. Biol. Chem.* **268**, 4069–73.

Heinecke, J.W., Rosen, H. and Chait, A. (1984) Iron and copper promote modification of low density lipoprotein by human arterial smooth muscle cells in culture. *J. Clin. Invest.* **74**, 1890–1894.

Hennig, B. and Chow, C.K. (1988) Lipid peroxidation and endothelial cell injury: implications in atherosclerosis. *Free Rad. Biol. Med.* **4**, 99–106.

Henriksen, T., Mahoney, E.M. and Steinberg, D. (1981) Enhanced macrophage degradation of low density lipoprotein incubated with cultured endothelial cells: recognition by receptor for acetylated low density lipoprotein. *Proc. Natl. Acad. Sci. USA* **78**, 6499–6503.

Huggins, T.G., Wells-Knecht, M.C., Detorie, N.A., Baynes, J.W. and Thorpe, S.R. (1993) Formation of o-tyrosine and dityrosine in proteins during radiolytic and metal-catalyzed oxidation. *J. Biol. Chem.* **268**, 12341–12347.

Jackson, R.L., Ku, G. and Thomas, C.E. (1993) Antioxidants: a biological defense mechanism for the prevention of atherosclerosis. *Med. Res. Rev.* **13**, 211–251.

Janzen, E.G., Wilcox, A.L. and Manoharan, V. (1993) Reactions of nitric oxide with phenolic antioxidants and phenoxyl radicals. *J. Org. Chem.* **58**, 3597–3599.

Kalyanaraman, B., Antholine, W.E. and Parthasarathy, S. (1990) Oxidation of low-density lipoprotein by Cu^{2+} and lipoxygenase: an electron spin resonance study. *Biochim. Biophys. Acta* **1035**, 286–292.

Kalyanaraman, B., Darley-Usmar, V.M., Wood, J. and Parthasarathy, S. (1992) Synergistic interaction between the Probucol phenoxyl radical and ascorbic acid in inhibiting the oxidation of low density lipoprotein. *J. Biol. Chem.* **267**, 6789–6795.

Kalyanaraman, B., Felix, C.C. and Sealy, R.C. (1985) Semiquinone anion radicals of cathecholamines, catechol estrogens, and their metal ion complexes. *Environ. Health Perspect.* **64**, 185–198.

Kemal, C., Louis-Flamberg, P., Krupinski-Olsen, R. and Shorter, A.L. (1987) Reductive inactivation of soybean lipoxygenase 1 by catechols: a possible mechanism for regulation of lipoxygenase activity. *Biochemistry* **26**, 7064–7072.

Kita, T., Nagano, Y., Yokode, M., Ishii, K., Kume, N., Ooshima, A., Yoshida, H. and Kawai, C. (1987) Probucol prevents the progression of atherosclerosis in Watanabe inheritable hyperlipidemic rabbit, an animal model for familial hypercholesterolemia. *Proc. Natl. Acad. Sci. USA* **84**, 5928–5931.

Liebler, D.C., Kaysen, K.L. and Burr, J.A. (1991) Peroxyl radical trapping and autoxidation reactions of α-tocopherol in lipid bilayers, *Chem. Res. Toxicol.* **4**, 89–93.

Maiorino, M., Zamburlini, A., Roveri, A. and Ursini, F. (1993) Prooxidant role of vitamin E in copper-induced lipid peroxidation. *FEBS Lett.* **330**, 174–176.

Mohr, D., Bowry, V.W. and Stocker, R. (1992) Dietary supplementation with coenzyme Q_{10} results in increased levels of ubiquinol −10 within circulatory lipoproteins and increased resistence of human low-density lipoprotein to the initiation of lipid peroxidation. *Biochim. Biophys. Acta* **1126**, 247–254.

Morel, D.W., Dicoreleto, P.E. and Chisolm, G.M. (1984) Endothelial and smooth muscle cells alter low density lipoprotein in vivo by free radical oxidation. *Atherosclerosis* **4**, 357–364.

Mukai, K., Kikuchi, S. and Urano, S. (1990) Stopped-flow kinetic study of the regeneration reaction of tocopheroxyl radical by reduced ubiquinone-10 in solution. *Biochim. Biophys. Acta* **1035**, 77–82.

Packer, J.E., Slater, T.F. and Wilson, R.A. (1979) Direct observation of a free radical interaction between vitamin E and vitamin C. *Nature* **278**, 737–738.

Parthasarathy, S. (1992) Evidence for an additional intracellular site of action of Probucol in the prevention of oxidative modification of low density lipoprotein. *J. Clin. Invest.* **89**, 1618–1621.

Parthasarathy, S. and Rankin, S.M. (1992) Role of oxidized low density lipoprotein in atherogenesis. *Prog. Lipid Res.* **31**, 127–143.

Parthasarathy, S., Steinberg, D. and Witztum, J.L. (1992) The role of oxidized low-density lipoproteins in the pathogenesis of atherosclerosis. *Ann. Rev. Med.* **43**, 219–225.

Parthasarathy, S., Young, S.G., Witztum, J.L., Pittman, R.C. and Steinberg, D. (1986) Probucol inhibits oxidative modification of low density lipoprotein. *J. Clin. Invest.* **77**, 641–644.

Rankin, S.M., Parthasarathy, S. and Steinberg, D. (1991) Evidence for a dominant role of lipoxygenase in the oxidation of LDL by mouse peritoneal macrophages. *J. Lipid Res.* **32**, 449–456.

Schuler, R.H. (1977) Oxidation of ascorbate anion by electron transfer to phenoxyl radicals. *Rad. Res.* **69**, 417–433.

Sparrow, C.P. and Olszewski, J. (1992) Cellular oxidative modification of low-density lipoprotein does not require lipoxygenases. *Proc. Natl. Acad. Sci. USA* **89**, 128–131.

Sparrow, C.P., Parthasarathy, S. and Steinberg, D. (1989) A macrophage receptor that recognizes low density lipoprotein but not acetylated low density lipoprotein. *J. Biol. Chem.* **264**, 2599–2605.

Steinberg, D. (1986) Studies on the mechanism of action of Probucol. *Am. J. Cardiol.* **57**, 16H–21H.

Steinberg, D. (1992) Antioxidants in the prevention of human atherosclerosis: summary of the proceedings of a National Heart, Lung, and Blood Institute workshop, September 5–6, 1991, Bethesda, MD. *Circulation* **85**, 2337–44.

Steinberg, D., Parthasarathy, S., Carew, T.E., Khoo, J.C. and Witztum, L.J. (1989) Beyond cholesterol, modification of low-density lipoprotein that increase its atherogenicity. *New Engl. J. Med.* **320**, 915–924.

Stenbrecher, U.P., Parthasarathy, S., Leake, D.S., Witztum, J.L. and Steinberg, D. (1984) Modification of low density lipoprotein by endothelial cells involves lipid peroxidation and degradation of low density lipoprotein phospholipids. *Proc. Natl. Acad. Sci. USA* **81**, 3883–3887.

Stocker, R., Bowry, V.W. and Frei, B. (1991) Ubiquinol-10 protects human low density lipoprotein more efficiently against lipid peroxidation than does α-tocopherol. *Proc. Natl. Acad. Sci. USA* **88**, 1646–1650.

Van der Zee, J., Eling, T.E. and Mason, R.P. (1989) Formation of free radical metabolites in the reaction between soybean lipoxygenase and its inhibitors. An ESR study. *Biochemistry* **28**, 8363–8367.

Wilcox, A.L. and Janzen, E.G. (1993) Nitric oxide reactions with antioxidants in model systems: sterically hindered phenols and α-tocopherol in sodium dodecyl sulfate (SDS) micelles. *J. Chem. Soc. Chem. Commun.* 1371–1379.

CHAPTER 4

Synthetic Pro-oxidants: Drugs, Pesticides and Other Environmental Pollutants

Sidney J. Stohs

4.1 INTRODUCTION

In recent years, a tremendous increase has occurred in our knowledge regarding the role of reactive oxygen species (ROS) in the toxicity of numerous xenobiotics and in the pathology of various disease states. The number of xenobiotics known to promote the formation of ROS has expanded greatly. Furthermore, oxidative stress is believed to be involved in the pathophysiology of aging, and various age-related diseases including cataracts, atherosclerosis, neoplastic diseases, diabetes, diabetic retinopathy, chronic inflammatory diseases of the gastrointestinal tract, aging of skin, diseases associated with cartilage, Alzheimer's disease, and other neurological disorders including Down's syndrome (see also chapter 2) (Emerit and Chance, 1992).

Frenkel (1992) has reviewed the experimental data that points to the formation of ROS and oxidative DNA base damage in cancer development. He has also reviewed possible sources of pathogenic reactive oxygen species including: phagocytic cells and non-phago-cytic cells (epidermal keratinocytes, HeLa, and 10 T 1/2) treated with various activators or tumor promoters; quinone-semiquinone redox cycling; xanthine oxidase; ischemia/reperfusion; induction of fatty acid CoA oxidases by peroxisome proliferators; ionizing radiation; cigarette smoke; ozone; inhibition of antioxidant enzymes; and various chemotherapeutic agents and pesticides. In addition to these sources of reactive oxygen, microsomal enzyme systems including cytochrome P-450 (Morehouse and Aust, 1988) and mitochondria (Paraidathathu et al., 1992) also constitute major sources of reactive oxygen species. Recent evidence indicates that multiple sites of pro-duction of ROS may be involved in response to xenobiotic exposure including mitochondria, microsomes and macrophage (Bagchi and Stohs, 1993).

Various forms of tissue damage occur in response to the formation of an oxidative stress. Several types of oxidative DNA damage may occur (Frenkel, 1992). Dargel (1992) has suggested that lipid per-oxidation may be a common pathogenic mechanism because it is considered a basic mechanism involved in reversible and irrevers-ible cell and tissue damage. Since ROS are implicated in toxic injury due to a broad range of xenobiotics as well as the pathogenesis of cancer, numerous disease states, ischemic/reperfusion damage and inflammatory processes, the production of ROS and the formation of an oxidative stress may represent common pathogenic mecha-nisms with lipid peroxidation and DNA and protein damage oc-curring as a consequence thereof.

In recent years, a wide range of structurally dissimilar xenobiotics has been shown to induce oxidative stress with the resultant pro-duction of lipid peroxidation and DNA damage. A variety of mech-anisms appears to be involved in the production of ROS by various xenobiotics, and therefore, a single, unified mechanism has not been developed, and may not be possible to establish involving toxicant-initiated oxidative stress. The following discussion will review the abilities of various chemically related groups of compounds to in-duce the formation of ROS, and produce an oxidative stress with resultant tissue damage. Examples of haloalkanes, polyhalogenated cyclic pesticides, phorbol esters, paraquat and diquat, quinones, quinolones, dioxin (TCDD) and its bioisosteres, transition metals and cation complexes, and miscellaneous inducers of oxidative stress will be reviewed. The rapid advances in our knowledge of xenobiotics which induce oxidative stress preclude the inclusion of all known classes of xenobiotics which enhance the formation of ROS.

4.2 HALOGENATED ALKANES AND ALKENES

Halogenated compounds are among the most common and most important xenobiotics with respect to human exposure and potential health hazards. Many compounds such as carbon tetrachloride (4.1), chloroform, dichloroethylene and ethylene bromide are widely used for industrial and agricultural purposes, and are commonly found in water supplies. These and other halogenated compounds represent major environmental toxicants. However, the occurrence of halogenated chemicals in our biosphere can not be blamed entirely on industry. Gribble (1992) reported that more than 1500 different halogenated chemicals are produced and discharged into the environment by plants, insects, bacteria, fungi, marine organisms, and other natural processes.

Many halogenated compounds have been shown to be potent inducers of lipid peroxidation, a common index of oxidative stress and free radical production. Fraga et al. (1987) screened 27 halogenated compounds as potential inducers of lipid peroxidation in rat liver, kidney, spleen and testes slices. Using the release of thiobarbituric acid reactive substances (TBARS) as an index of lipid peroxidation, these investigators observed that the amount of TBARS released from liver slices incubated with bromotrichloromethane, carbon tetrabromide, dichloromethane, bromobenzene, chloroform, bromoform, benzyl chloride and bromochloromethane correlated with the lethality of these compounds as evaluated by their oral LD_{50} in rats. Thus, a wide range of halogenated compounds are capable of inducing lipid peroxidation, and lipid peroxidation appears to play an important role in toxicity and cell death.

A role for lipid peroxidation in the toxicity of halogenated alkanes was initially demonstrated by Recknagel who has summarized the mechanisms of toxicity of carbon tetrachloride (4.1) (Recknagel et al., 1989). Over the past thirty years, studies have demonstrated that carbon tetrachloride hepatotoxicity depends on the reductive dehalogenation of carbon tetrachloride catalyzed by cytochrome P-450 in the liver endoplasmic reticulum. This metabolic activation results in the formation of trichloromethyl and trichloromethylperoxyl free radicals which are involved in a cascade of secondary

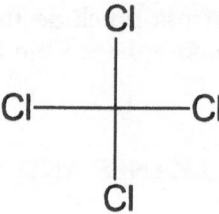

Fig. 4.1   Carbon Tetrachloride

mechanisms that are ultimately responsible for plasma membrane disruption and cell death.

The free radical metabolites of carbon tetrachloride initiate membrane lipid peroxidation which results in a sustained rise in the concentration of calcium ions in the cytoplasm. The elevated calcium levels activate lipases, proteases and endonucleases which contribute to cell necrosis (Orrenius et al., 1992; McConkey et al., 1989). Studies have shown that both bromotrichloromethane and carbon tetrachloride activate phospholipase A_2 which may play a critical role in mediating cell death. Phospholipase A_2 preferentially hydrolyzes peroxidized fatty acids present in membranes (van Kuijk et al., 1987). Vitamin E is the most abundant biologically occurring lipid-soluble antioxidant. Increasing the liver content of vitamin E affords a significant degree of protection against carbon tetrachloride-induced chronic liver damage and cirrhosis (Parola et al., 1992), providing evidence for the role of free radicals and ROS in the cytotoxicity of this xenobiotic.

Oxygen is one of the factors that markedly affects carbon tetrachloride hepatotoxicity. Low oxygen partial pressure enhances carbon tetrachloride hepatotoxicity in animals, perfused liver, and isolated hepatocytes (Masuda and Nakamura, 1990). These investigators have suggested that oxygen deficiency rather than lipid peroxidation itself, and the essential role of extracellular calcium may be important for carbon tetrachloride-induced hepatic cell necrosis. Recknagel et al. (1989) have also extensively summarized the possible role of calcium in the toxicity of carbon tetrachloride. Carbon tetrachloride not only can result in membrane damage with the ultimate influx of extracellular calcium, but also mobilizes calcium ions from microsomal and mitochondrial pools. It is the sustained rise in cytosolic calcium concentration that results in activation of the various hydrolytic enzymes (Orrenius et al., 1992).

Pretreatment of rats with large doses of vitamin A (retinol) potentiates the hepatotoxicity of carbon tetrachloride as well as other chemicals including acetaminophen, allyl alcohol and endotoxin which produce hepatic injury by diverse mechanisms (ElSisi, Hall, et al., 1993). The authors concluded that the vitamin A potentiated hepatic injury by altering a process involving the progression of cell injury. Further studies indicated that the potentiation of carbon tetrachloride hepatotoxicity by vitamin A pretreatment was associated with enhanced lipid peroxidation, and was independent of carbon tetrachloride by a transformation (ElSisi et al., 1993a). When animals were treated with the Kupffer cell inhibitor methylpalmitate (2 g/ kg), the vitamin A potentiated hepatotoxicity by carbon tetrachloride was blocked as was the production of ROS (ElSisi et al., 1993b). These results support the conclusion that vitamin A enhanced carbon tetrachloride liver injury is mediated at least in part by reactive oxygen species produced by Kupffer cells and possibly other macrophages which are activated by vitamin A.

Cox and Volp (1993) have tested perchloroalkanes and various chloropropanes to determine their ability to initiate lipid peroxidation under normoxic and hypoxic conditions following in vitro incubation with rat liver microsomes. Under normoxic conditions both carbon tetrachloride and octachloropropane induced a linear production of malondialdehyde (MDA) with incubation time, using the TBARS method. Hexachloroethane and other chloropropanes produced little or no MDA under these incubation conditions. Under hypoxic conditions, all chloropropanes produced MDA in a linear fashion over the incubation interval. The amount of MDA which was produced increased with successive chlorination of a particular carbon atom. These results demonstrate the importance of the bioactivation of chloropropanes, and confirm that low oxygen partial pressure enhances the potential cytotoxicity of these compounds.

Danni et al. (1991) have demonstrated that synergism can occur between carbon tetrachloride and other haloalkanes as 1,2-dibromoethane with respect to lipid peroxidation and irreversible cell damage in isolated rat hepatocytes. The use of hepatocytes from vitamin E pretreated rats prevented both the oxidant and cytotoxic effects of carbon tetrachloride while total nonprotein thiol content was not significantly modified by carbon tetrachloride alone, but was markedly decreased in the presence of 1,2-dibromoethane alone and in association with carbon tetrachloride. Thus, multiple mechanisms

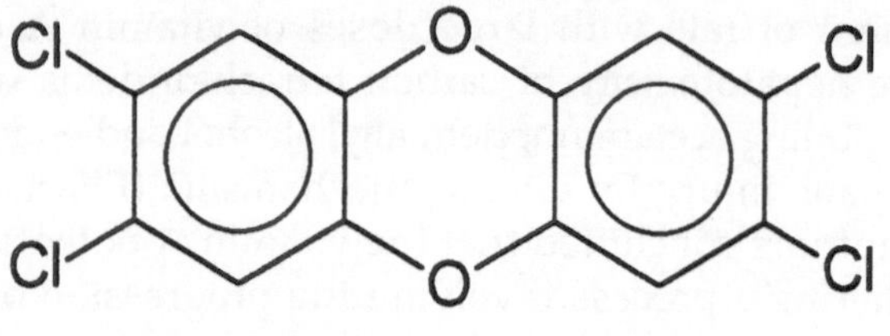

Fig. 4.2   2,3,7,8-Tetrachlorodibenzo-*p*-dioxin (TCDD)

appear to be involved in oxidative tissue damage associated with haloalkanes.

Carbon tetrachloride by its own reductive dehalogenation can also act as an uncoupler of oxidative phosphorylation in mitochondria, giving rise to the production of reactive oxygen species which in turn produce tissue damaging effects (Recknagel et al., 1989). Thus, a complex interplay exists between reactive metabolites and ROS as well as enzyme activation, lipid peroxidation, antioxidant depletion, membrane degradation and altered calcium homeostasis in the toxicity of carbon tetrachloride, and presumably other haloalkanes and haloalkenes.

4.3   DIOXIN AND ITS BIOISOSTERES

The dioxin, 2,3,7,8-tetrachlorodibenzo-*p*-dioxin (TCDD) (4.2) has been identified as one of the most potent toxins and tumor promoters that is known. It is prototypical of many halogenated polycyclic hydrocarbons including the polyhalogenated biphenyls (polychlorobiphenyl, PCB's; polybromobiphenyl's, PBB's and polyhalogenated dibenzofurans) (Poland and Knutson, 1982; Kociba and Schwetz, 1982). Substantial evidence has accumulated in recent years that TCDD and its bioisosteres induce an oxidative stress, and much of this literature has been previously reviewed (Stohs, 1990).

Administration of TCDD to rodents results in the accumulation of lipofuscin pigments in the heart, increases in hepatic lipid peroxidation as determined by formation of conjugated dienes, production of MDA by isolated microsomes, an increase in whole liver MDA content as determined by the formation of thiobarbituric acid reactive substances (TBARS), an increase in ethane exhalation, an increase in hepatic DNA single strand breaks, and decreases in hepatic microsomal, mitochondrial and plasma membrane fluidity

(Stohs, 1990; Stohs et al., 1990). Increases in the calcium content of various hepatic subcellular fractions parallel increases in lipid per-oxidation and DNA damage, and are inversely related to decreases in nonprotein sulfhydryl and NADPH content as well as membrane fluidity (Stohs et al., 1990).

Calcium may play an important role in chemically induced cyto-toxicity and the ability of calcium to activate endonucleases, pro-teases and phospholipases is well known (Orrenius et al., 1992). Studies by McConkey et al. (1989) have suggested that a sustained elevation of cytosolic calcium may be the rate limiting step in both the in vitro and in vivo ability of TCDD to kill thymocytes. Calcium activation of endonuclease was believed to be responsible for the fragmentation of DNA which was observed in response to TCDD. Thus, the calcium influx, which may occur as the result of plasma membrane damage, may contribute to the cascade of events asso-ciated with TCDD-induced cytotoxicity. TCDD also disturbs iron, copper and magnesium homeostasis in rat liver, and evidence in-dicates that altered calcium homeostasis may contribute to en-hanced production of ROS and lipid peroxidation (Wahba et al., 1990).

Evidence indicates that most toxicological responses to TCDD are mediated by the Ah (TCDD) receptor (Cook et al., 1987; Safe, 1988). The physicochemical properties of this receptor are well character-ized, and the receptor is analogous but not identical to glucocorti-coid and other steroid hormone receptors (Wilhelmsson et al., 1986; Cuthill et al., 1988; Cook and Greenlee, 1989). The Ah receptor pos-sesses distinct ligand and DNA-binding domains, and is able to reg-ulate gene elements other than the gene with which it is associated, thus being referred to a *trans*-acting effector of gene expression (Whitlock, 1989). The net effect is the production of a broad spec-trum of toxic effects which are mediated by the Ah receptor com-plex. Nebert et al. (1990) have shown that the Ah gene battery is composed of at least six genes. All six genes appear to be positively induced by TCDD and other ligands of the Ah receptor. This battery of mammalian genes is released from negative control in response to phytoalexin-induced oxidative stress, and DNA damage occurs. Studies by Bombick and Matsumura (1987) have shown that TCDD increases the activity of a protein kinase in TCDD-responsive mice but not in TCDD-resistant mice, suggesting that the increase in pro-tein kinase *c* activity and subsequent activation of macrophage NADPH oxidase are mediated through the Ah (TCDD) receptor.

Possible sources of ROS in response to TCDD have been reviewed by Stohs (1990). Based on this review, initial evidence indicates that macrophages, microsomes and mitochondria appear to be major sources of ROS, including superoxide anion ($O_2^{.-}$), hydrogen peroxide (H_2O_2) and hydroxyl ($^.OH$) radical. In addition, the xanthine oxidase (XO/XD) system, peroxisomes and membranes in general might serve as other possible sources of ROS. Furthermore, the enzyme selenium-dependent glutatione peroxidase (GPOX) is extensively inhibited following the administration of TCDD and PCBs to rodents and chickens. This enzyme eliminates H_2O_2 and hydroperoxides (ROOH), and this inhibition may contribute to the induction of an oxidative stress by these xenobiotics.

Several studies have provided evidence for the involvement of phagocytic cells in the production of ROS and the acute toxicity of TCDD (Clark et al., 1991). These investigators implicated tumor necrosis factor (α-TNF) in the acute toxicity of TCDD, and evidence indicated that the Ah receptor mediates this response. Since TNF sensitizes endotoxin-treated mice and activates phagocytic cells to agents that induce them to release ROS, TNF may act as an amplifying loop in oxidative stress and contribute to the debilitating effects observed following TCDD exposure. Further evidence supporting this hypothesis has been presented by Taylor et al. (1992) who demonstrated that dexamethasone and anti-TNF antibody treatment reversed the acute toxicity of TCDD. However, dexamethasone did not block the TCDD induction of cytochrome P-450 (Cyp 1A1), demonstrating a separation of this biochemical parameter from the acute toxic effects of TCDD. Dexamethasone is known to provide protection against endotoxin-induced toxicity and prevent the transcription of TNF mRNA (Taylor et al., 1992).

Studies by Alsharif et al. (1994) have shown that dexamethasone (2 mg/kg) and anti-TNF monoclonal antibody (40 μg/mouse) provided partial protection against TCDD-induced oxidative stress as measured by DNA single strand breaks in hepatic nuclei, lipid peroxidation in hepatic mitochondria and microsomes, and peritoneal lavage cell (primarily macrophages) activation in C57BL/6J mice. The combination of anti-TNF antibody and dexamethasone resulted in additive effects. These results suggest that TNF release may play a role in sensitizing and activating phagocytic cells following treatment with TCDD, contributing to the overall oxidative stress of animals exposed to this xenobiotic.

TCDD produces dose- and time-dependent increases in $O_2^{.-}$ production by peritoneal lavage cells from Sprague-Dawley rats (Alsharif et al., 1990). These investigators have also examined whether TCDD-induced production of $O_2^{.-}$ radical by peritoneal lavage cells is mediated through the Ah receptor by using congenic mice which differ at the Ah (TCDD) locus. The difference in sensitivity to TCDD toxicity is believed to be related to the presence of this specific TCDD binding complex. One day after the administration of 5, 25, 50 or 125 µg TCDD/kg p.o. as a single dose, 1.4-, 1.7-, 4.3-, and 4.5-fold increases, respectively, occurred in O_2^- production by peritoneal lavage cells from TCDD-responsive C57BL/6J [bb] mice which contain high levels of the Ah receptor. However, only 125 µg TCDD/kg produced a significant increase in O_2^- formation with peritoneal lavage cells from the non-responsive C57BL/6J [dd] congenic strain of mice. Similar results were obtained when using the TCDD-resistant DBA/2 strain of mice. These observations indicate that TCDD produces an increase in the production of O_2^- by peritoneal lavage cells which is mediated, at least in part, by the Ah receptor complex. The possibility exists that oxidative stress is a generalized mechanism of toxicity associated with many xenobiotics which may be produced by the activation of macrophages.

The ability of TCDD to induce the in vitro production of ROS when incubated with hepatic mitochondria and microsomes as well as peritoneal macrophages from rats has been examined (Bagchi and Stohs, 1993). At a concentration of 100 ng/ml, TCDD produced increases in chemiluminescence of 115, 160 and 400% in peritoneal macrophages, hepatic mitochondria and microsomes, respectively. TCDD also produced significant increases in the production of O_2^- as determined by the cytochrome c reduction assay with these three tissue fractions. When microsomal membranes were incubated with TCDD, a 2.3-fold increase in the apparent microviscosity occurred, indicating a significant decrease in membrane fluidity. Thus, the in vitro incubation of TCDD with membrane fractions is capable of inducing the production of ROS and altering membrane fluidity.

Methods for assessment of oxidative stress, more specifically lipid peroxidation, have included quantitation of lipofuscin pigments, TBARS content, rate of formation of TBARS using MDA as the standard, formation of conjugated dienes, and exhalation of ethane and pentane. With the exception of ethane and pentane exhalation, the above methods are relatively nonspecific and involve invasive procedures. Until recently, a highly specific, non-invasive method for assessing peroxidative damage has not been available. Shara et al. (1992) have developed a method for the simultaneous identification of the lipid metabolites formaldehyde (FA), MDA, acetaldehyde (ACT) and acetone (ACON) in the urine of experimental animals. Initial studies quantitated the urinary excretion of these lipid metabolites in response to TCDD, as well as endrin, carbon tetrachloride and paraquat. Urine samples of female Sprague-Dawley rats were collected over dry ice and derivatized with 2,4-dinitrophenylhydrazine. The hydrazones of the four lipid metabolic products were quantitated by high pressure liquid chromatography on a Waters 10-μm μ-Bondapak C_{18} column. The identities of FA, ACT, MDA and ACON in urine were confirmed by co-chromatography and gas chromatography-mass spectrometry (Shara et al., 1992). An oxidative stress was induced by orally administering 100 μg/kg TCDD, 75 mg/kg paraquat, 6 mg/kg endrin or 2.5 ml/kg carbon tetrachloride to the rats. These doses represent the approximate LD_{50}'s for these compounds. Urinary excretion of the four metabolites increased relative to control animals 24 hours after treatment with all xenobiotics.

A HPLC method has also been developed for the simultaneous determination of serum levels of FA, MDA, ACT and ACON (Bagchi, D., et al., 1992). A modification of the method employed for urine samples was used (Shara et al., 1992), involving a deproteination step. Acetonitrile-water (49:51) v/v was used as the mobile phase. Following the administration of 50 μg TCDD/kg to rats as a single oral dose, significant increases in the serum levels of the four lipid metabolites were observed.

Bagchi, D. and Stohs (unpublished) have compared the time-dependent changes in serum and urine levels of MDA, FA, ACT and ACON in response to a single, oral 50 μg/kg dose of TCDD in rats. The effects of TCDD were compared with ad libitum fed control animals and pair-fed animals. Serum and urine levels of the four metabolites were assayed on days 0, 3, 6, 9 and 12, and the percent changes for TCDD-treated animals and pair-fed control animals are presented as percent of ad libitum fed control animals in Figure 4.3a,b,c,d.

Following TCDD administration, significant increases in the four metabolites present in serum and urine were observed at all time points. For example, on day 6 post-treatment, MDA, FA, ACT and ACON increased approximately 2.6-, 2.5-, 2.4-, and 6.9-fold in serum respectively, and 3.0-, 2.3-, 3.8-, and 3.7-fold in urine, respectively. As can be seen in the figures, increases were also observed in the serum and urine levels of the four metabolites in pair-fed animals relative to the ad libitum fed control animals. However, the increases in the serum and urine levels of the four metabolites was significantly greater for TCDD animals as compared to the pair-fed control animals at most time points. When serum levels of MDA as determined by HPLC were compared with the results obtained by

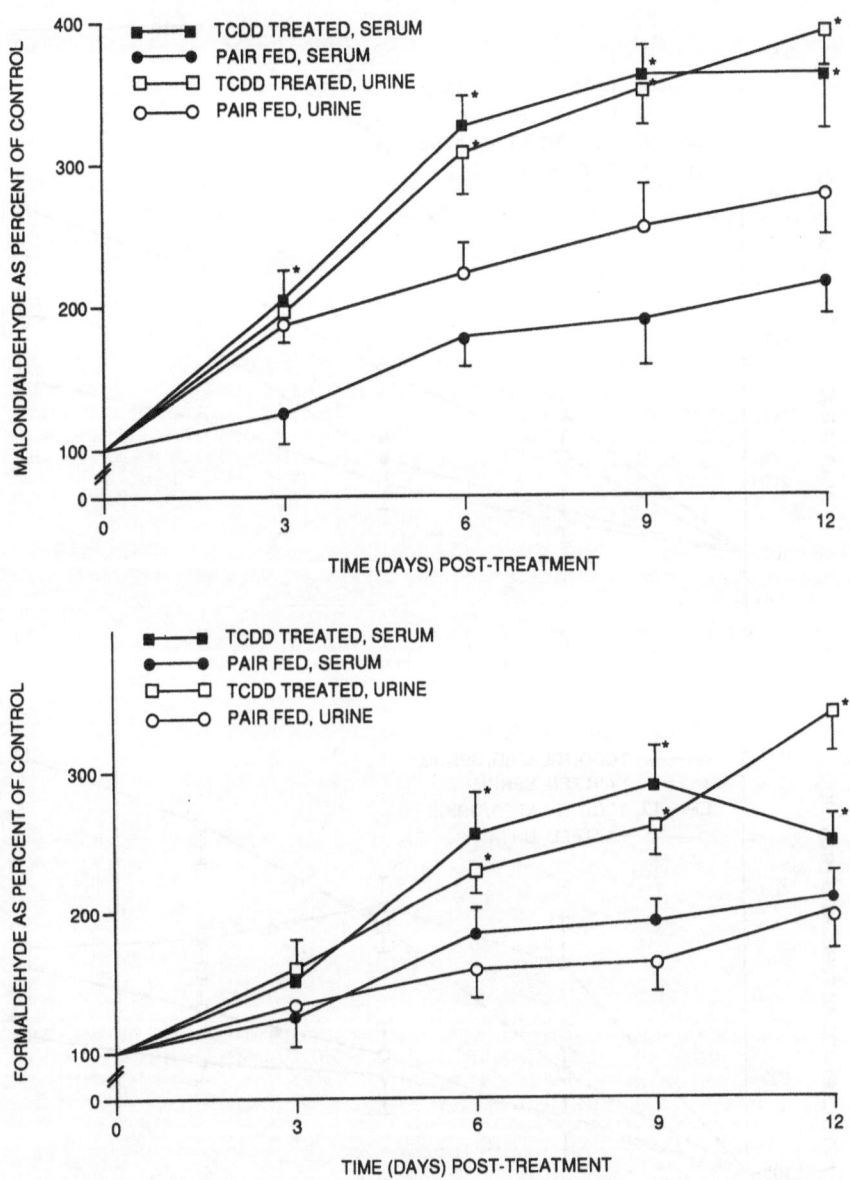

Fig. 4.3 A–D Time-dependent changes in: [A] malondialdehyde (MDA); [B] formaldehyde (FA); [C] acetone (ACON); and [D] acetaldehyde (ACT) in serum and urine of rats treated with 50 µg TCDD/kg, pair-fed rats, and *ad libitum* fed control animals. Serum and urine samples were prepared 3, 6, 9 and 12 days following initiation of the experiments. The four urinary lipid metabolites were quantitated by HPLC. Each value represents the mean and SD of four animals. *P < 0.05 with respect to the *ad libitum* fed control group. **P < 0.05 with respect to the pair-fed group.

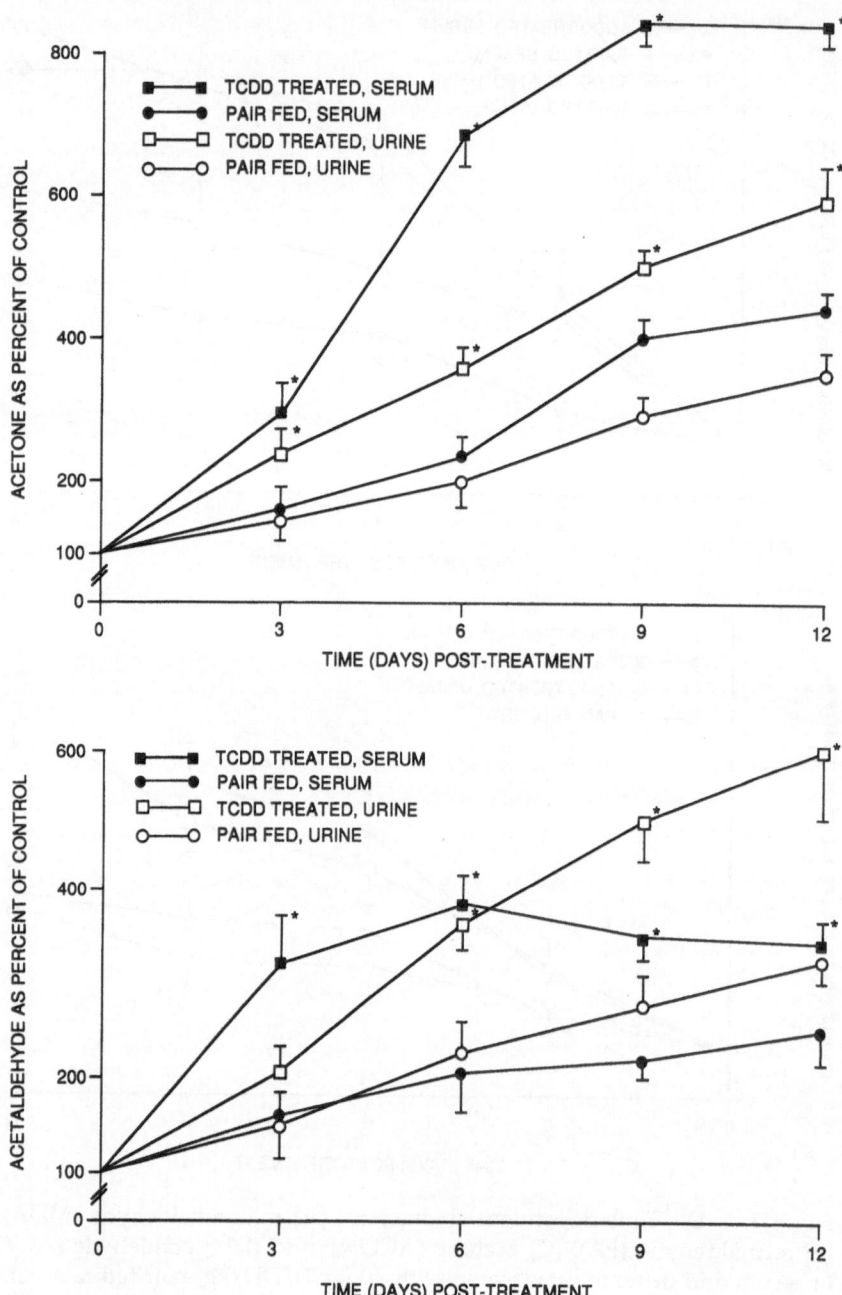

Fig. 4.3 C-D

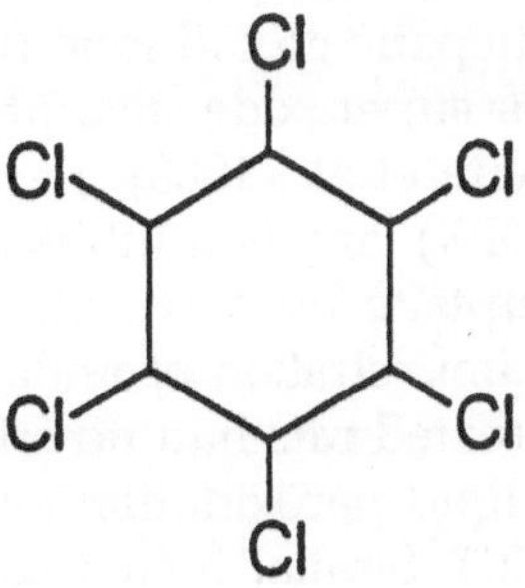

Fig.4.4 γ-Hexachlorocyclohexane (Lindane)

the TBARS colorimetric method, similar time courses were observed although higher values were obtained for the less specific TBARS method. The results clearly demonstrate that TCDD causes markedly elevated serum and urine levels of four specific products associated with lipid metabolism.

4.4 HALOGENATED CYCLIC PESTICIDES

A wide range of structurally dissimilar polyhalogenated cyclic hydrocarbons are used as insecticides and pesticides. An ever increasing number of these xenobiotics has been shown to induce an oxidative stress, and many of these compounds are environmental pollutants. Lindane (4.4) is an important organochlorine pesticide extensively used for agricultural and public health purposes. It is the gamma isomer of hexachlorocyclohexane. Lindane-induced liver oxidative stress has been extensively studied by Videla et al. (1990). Goel et al. (1988) demonstrated that lindane produces hepatic lipid peroxidation. Following the administration of an acute dose (50-100 mg/kg) of lindane to rats, a biphasic increase in hepatic lipid peroxidation is observed (Junqueira et al., 1998). The initial increase in lipid peroxidation which is seen as early as four hours post-treatment may occur as the result of lindane metabolism by microsomes involving a process of dehydrogenation and dehydrochlorination, giving rise to a variety of reactive metabolites which can undergo further metabolism. Some of these metabolites may initiate lipid peroxidation. Approximately 24 hours after lindane treatment, the insecticide elicits a dose-dependent increase in the rate of O_2^- production by liver microsomes. Additional studies have demonstrated that lindane decreases hepatic glutathione (GSH) content (Barros et al., 1988) and decreases superoxide dismutase (SOD) and catalase (CAT) activities (Junqueira et al., 1988). Furthermore, the iron chelator desferrioxamine (DFX) largely inhibits the lindane-induced enhancement of oxygen uptake by perfused rat livers (Videla et al., 1989, 1991), while the administration of lindane to phenobarbital and 3-methylcholanthrene-treated rats had no effect on the ability of lindane to induce hepatic lipid peroxidation and alter GSH homeostasis (Junqueira et al., 1991). Similar increases in the dose- and time-dependent production of lipid peroxidation have been demonstrated for the α and γ isomers of hexachlorocyclohexane (Barros et al., 1991).

Hydrogen peroxide formation occurs either by disproportionation of O_2^- by SOD, autooxidation of cytochrome P-450 or cytochrome P-450 reductase in response to lindane (Videla et al., 1990). Furthermore, lindane has been shown to be a potent initiator of O_2^- generation by macrophages and polymorphonuclear (PMN) leukocytes, which is associated with a marked alteration in calcium homeostasis. Thus, lindane intoxication involves enhanced generation of O_2^- and other reactive oxygen species and/or free radical metabolites of lindane.

The in vitro incubation of hepatic mitochondria and microsomes as well as peritoneal macrophages from Sprague-Dawley rats with up to 200 ng lindane/ml results in the production of ROS as determined by chemiluminescence and cytochrome c reduction (Bagchi and Stohs, 1993). Furthermore, a significant decrease in the fluidity of microsomal membranes was observed. When hepatic mitochondria and microsomes were incubated with lindane for 15 min at 100 ng/ml, 7.5- and 11.6-fold increases, respectively, in cytochrome c reduction were observed. Thus, the results indicate that lindane induces the production of ROS at multiple sites, and can produce direct effects on these membrane fractions. The lindane-induced decrease in microsomal membrane fluidity may reflect large changes in membrane function.

In summary, lindane induces the microsomal cytochrome P-450 system, enhances rates of O_2^- generation, increases lipid peroxidation, decreases activities of the enzymes SOD and CAT, and depletes GSH. In general, the changes in hepatic lipid peroxidation and antioxidant parameters are closely interrelated and coincide with the onset and progression of morphological lesions. In addition to macrophages and microsomes, mitochondria may also represent a potential source of ROS in response to lindane.

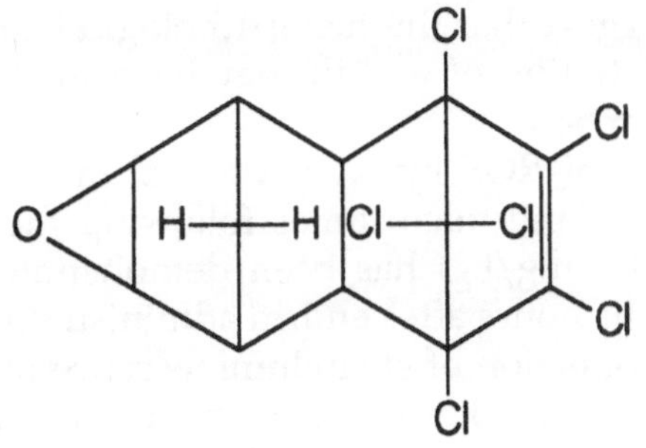

Fig. 4.5   Endrin

Endrin   (1,2,3,4,10,10-hexachloro-6,7-epoxyl-1,4,4α,5,6,7,8,8α-oc-tahydroendo,endo-1, 4:5,8-dimethanonaphthalene) (4.5) is a poly-halogenated cyclic pesticide which produces hepatic and neurologic toxicity. It is prototypical of a number of highly toxic cyclodiene in-secticides. Studies in rats have shown that endrin induces lipid per-oxidation in liver and kidneys with a concomitant decrease in the GSH content of these organs (Numan et al., 1990a, 1990b). In ad-dition, the enzyme GPOX is inhibited by endrin, facilitating the ac-cumulation of ROOHs (Numan et al., 1990b). Pretreatment of rats with butylated hydroxyanisole (BHA), vitamin E, vitamin C and cysteine significantly inhibited hepatic glutathione depletion and the lipid peroxidation induced by 4 mg endrin/kg body weight. These antioxidants also provided partial protection against lethality pro-duced by 8 mg endrin/kg body weight, suggesting that free radicals and ROS are involved in the toxic manifestations of endrin (Numan et al., 1990a).

Extensive interspecies variability exists in the sensitivity towards endrin. Histopathological changes and lipid peroxidation in the liv-ers and kidneys of rats, mice, hamsters, and guinea pigs have been examined after the administration of 4 mg endrin/kg body weight orally (Hassan et al., 1991). Degeneration and necrotic changes with inflammatory cell infiltration were observed in liver and kidneys, and interspecies variability occurred. Fatty changes in the form of foam cells with cytoplasmic vacuolation were present. Lipofuscin pigments, associated with lipid peroxidation, were observed in Kupffer cells and hepatocytes. The pretreatment of rats with BHA, vitamin E, vitamin C or cysteine prevented the histopathological ef-fects of endrin in rats. The extent of endrin-induced lipid peroxi-dation correlated well with the degree of histopathological changes.

These results suggest that the histopathological changes associated with the administration of endrin may be related to the induction of an oxidative stress.

The production of ROS by rat peritoneal macrophages, and hepatic mitochondria and microsomes following the oral administration of endrin (4.5 mg/kg) has been demonstrated (Bagchi et al., 1993). Twenty-four hours after endrin administration, significant increases in the production of chemiluminescence by the three tissue fractions were observed. Furthermore, peritoneal macrophages from endrin-treated animals resulted in 3.0- and 2.8-fold increases in cytochrome c and iodonitrotetrazolium (INT) reduction, respectively, indicating enhanced production of O_2^-

The effects of endrin on hepatic mitochondrial and microsomal lipid peroxidation as determined by the TBARS method and membrane fluidity as well as the incidence of hepatic nuclear DNA damage were also assessed following the oral administration of a single dose of endrin (Bagchi and Hassoun et al., 1993). Endrin administration resulted in significant increases in lipid peroxidation of mitochondrial and microsomal membranes. Twenty-four hours after endrin administration, the microviscosity of mitochondrial and microsomal membranes increased by 29% and 51%, respectively, indicating decreases in the fluidity of both membranes and alterations in membrane structure. DNA single strand breaks are an index of oxidative stress and cellular damage. Twenty-four hours post-treatment, a 3.5-fold increase in hepatic DNA single strand breaks was observed relative to the control group (Bagchi and Hassoun et al., 1993). These results indicate that macrophage, mitochondria and microsomes produce ROS following endrin administration to rats, and these ROS may contribute to the toxic manifestations of endrin.

Nitric oxide (NO) (see chapter 1; actual form is ˙NO) has been shown to be an important cellular transmitter in biological systems (Lambert et al., 1991). Kupffer cells, bone marrow and wound macrophages have the capacity to release nitric oxide as nitrite. The synthesis of NO by macrophages produces both cytostatic and cytotoxic effects, and the formation of NO is an indication of macrophage activation and the potential production of an oxidative stress. The effect of the oral administration of endrin to rats on the production of NO by peritoneal macrophages has been investigated (Akubue and Stohs, 1992). Nitric oxide formation was measured as nitrite using Greiss reagent. Dose- and time-dependent increases in endrin-induced NO formation by peritoneal macrophage were observed. At 4.5 mg endrin/kg, the NO secretion by macrophage increased by approximately 300%. Ellagic acid, which has been shown to be a potent anitoxidant, inhibited the elevation of NO production induced by endrin.

Bagchi et al. (1992) have also examined the dose- and time-dependent effects of endrin on hepatic lipid peroxidation, membrane microviscosity and DNA damage in order to further assess the possible role of oxidative stress in the toxicity of this polyhalogenated cyclic hydrocarbon. In these studies, rats were treated with 0, 3.0, 4.5, or 6.0 mg endrin/kg as a single oral dose in corn oil, and the animals were killed, 0, 12, 24, 48 or 72 hours post-treatment. Maximum increases in the three parameters which were measured occurred 24 hours after endrin administration at all three doses, with dose-dependent increases occurring in hepatic mitochondrial and microsomal lipid peroxidation and microviscosity, as well as nuclear DNA single strand breaks. Although the incidence of DNA damage decreased with time after 24 hours, the enhanced lipid peroxidation and microviscosity of microsomal and mitochondrial membranes remained relatively constant. The data suggest that endrin-induced hepatic lipid peroxidation may be responsible for the increased membrane microviscosity which ultimately gives rise to membrane damage. The mechanism by which endrin induces DNA damage is not clear. DNA damage may occur through the induction of ROS, lipid peroxidation, lipid free radicals, and/or reactive endrin metabolites (Bagchi et al., 1992).

The in vitro abilities of rat peritoneal macrophages as well as hepatic mitochondria and microsomes to produce ROS in response to endrin has been investigated (Bagchi and Stohs, 1993). When hepatic mitochondria and microsomes were incubated with endrin for 15 min at a concentration of 100 ng/ml, increases in cytochrome c reduction of 6.5- and 8.6-fold, respectively, were observed. The in vitro incubation of microsomes with endrin also resulted in a 2.1-fold increase in the apparent microviscosity, indicating a significant decrease in membrane fluidity. The results clearly indicate that endrin induces the in vitro production of ROS, and decreases membrane fluidity which may reflect a decrease in membrane function.

The effects of orally administering 1.5, 3.0, 4.5 and 6.0 mg endrin/kg on the urinary excretion of the lipid metabolites MDA, FA, ACT and ACON have been examined (Bagchi et al., 1992a). Urine samples were collected over dry ice for up to 72 hours post-treatment, and the four urinary metabolites were determined by HPLC. Maximum increases in the excretion of the four lipid metabolites occurred at approximately 24 hours post-treatment at all doses with no significant increases in excretion occurring thereafter. The maximum increases in excretion of MDA, FA, ACT and ACON were approximately 160%, 93%, 121%, and 127%, respectively, relative to control values. The simultaneous determination of these four lipid metabolites may be a useful biomarker for assessing exposure to xenobiotics which induce an oxidative stress and enhanced lipid peroxidation.

Bagchi et al. (1992b) have used endrin as a prototypical polyhalogenated cyclic hydrocarbon, and have examined its effects on calcium distribution in hepatic mitochondria, microsomes and nuclei of rats as a function of dose and time. The administration of endrin as a single oral dose increased hepatic mitochondrial, microsomal and nuclear calcium content in a dose- and time-dependent fashion. Seventy-two hours after the administration of 6 mg endrin/kg, the calcium content of nuclei, mitochondria and microsomes increased by 20, 33, and 48%, respectively, as compared to control values. Administration of 100 mg vitamin E succinate/kg for 3 days prior to the administration of endrin resulted in a reversal of the endrin-induced increases in calcium content of mitochondria and microsomes. Membrane perturbation by a wide range of chemical toxicants may result in increased intracellular levels of calcium which activates hydrolytic enzymes resulting in enhanced tissue damage and destruction.

Various studies have demonstrated that the most characteristic pathological lesions produced by TCDD (4.2) and isosteric halogenated aromatic hydrocarbons are mediated by an intracellular receptor protein called the Ah or TCDD receptor (Cook et al., 1987; Safe, 1988). Numerous structurally dissimilar, halogenated cyclic hydrocarbons in addition to TCDD and its bioisosteres induce oxidative stress, and produce a broad spectrum of similar toxic effects including hepatotoxicity, nephrotoxicity, neurotoxicity, lipid mobilization and altered lipid metabolism, DNA damage, and tissue wasting (Matsumura, 1983; Murphy, 1986). In order to determine whether the mechanism of endrin and related cyclodiene insecticides are mediated by the Ah receptor locus, dose-dependent studies on the effect of endrin on hepatic lipid peroxidation, DNA damage and production of NO by peritoneal exudate cells were investigated in two strains of mice, C57BL/6J and DBA/2, which vary at the Ah receptor locus. C57BL/6J mice are TCDD-responsive while DBA/2 mice are TCDD-insensitive. The results indicated that the responsiveness of peritoneal macrophages with respect to both DNA damage and NO production was more dose-dependent in

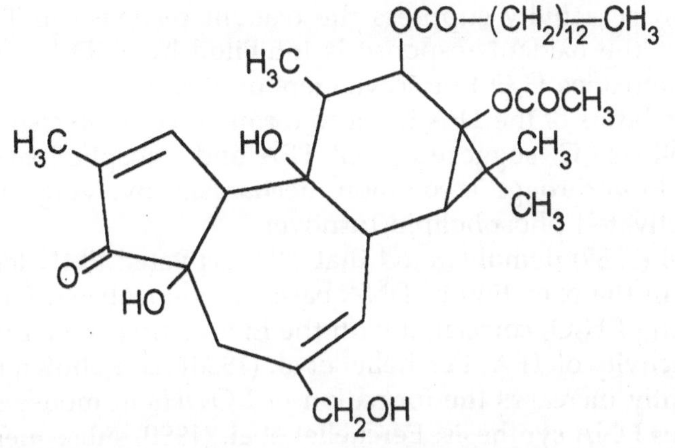

Fig. 4.6   12-O-Tetradecanoylphorbol-13-acetate (TPA)

C57BL/6J mice as compared to DBA/2 mice, while similar results were observed with respect to endrin-induced lipid peroxidation of hepatic mitochondria and microsomes in the two strains of mice (Stohs et al., unpublished). Therefore, endrin is much less reliant on a mechanism involving the Ah receptor system as compared to TCDD and its bioisosteres. Thus, although similar toxicologic effects are produced by structurally dissimilar xenobiotics involving the production of an oxidative stress, a single mechanism may not be involved.

4.5   PHORBOL ESTERS

Extensive evidence has accumulated showing that active oxygen species participate in at least one stage of tumor promotion. Studies have shown that tumor promoters can induce various cell types to undergo an oxidative burst resulting in formation of ROS. The most common group of compounds which have been studied for their tumor promoting activity are the phorbol esters, with 12-O-tetradecanoylphorbol-13-acetate (TPA) (4.6) being the most widely used. Other structurally related phorbol esters have also been studied (Gaudry et al., 1990).

Fischer et al. (1988) have shown that mouse epidermal cells exposed to TPA in vitro results in the production of ROS as detected by chemiluminescence. Agents that inhibit lipoxygenase but not cyclooxygenase activity suppress the oxidant response to TPA. Furthermore, the oxidant response is inhibited by SOD but not CAT (which detoxifies H_2O_2) or scavengers of ˙OH radical or singlet oxygen. Inhibitors of the TPA-induced formation of ROS also inhibited phospholipase C, suggesting that TPA and phospholipase C may produce ROS through a common mechanism involving protein kinase *c* activated phospholipid turnover.

Frenkel (1989) demonstrated that TPA activated PMN leukocytes resulted in the oxidation of DNA bases in co-incubated HeLa cells. Formation of H_2O_2 correlated with the in vivo first-stage tumor promoting activity of TPA. Perchellet et al. (1988) have shown that TPA significantly increases the formation of ROOHs in mouse skin and modulates DNA synthesis. Perchellet et al. (1990) subsequently concluded that the effects of TPA on DNA synthesis were linked to ROOH production, and the effects could be mimicked by the calcium ionophore A23187. Birnboim (1982) extensively studied the factors which affected DNA strand breakage in human leukocytes which had been exposed to TPA. Azide and cyanide greatly increased the level of damage while sulfhydryl compounds (GSH, cysteine and cysteamine) and ascorbate markedly decreased the level of damage. Hydroxyl radical scavengers including dimethylsulfoxide (DMSO) and glycerol also decreased the level of damage by inhibiting the respiratory burst. These results are consistent with a mechanism involving O_2^- and H_2O_2 in the production of TPA-initiated DNA damage.

Robertson et al. (1990) have examined the ability of murine epidermal cells to produce intracellular H_2O_2 in response to TPA. The results were analyzed by flow cytometry and the measurement of 2',7'-dichlorofluorescein (DCFH) oxidation. TPA resulted in increases in DCFH oxidation that were between 2- and 10-fold higher than control levels. The ability of CAT to suppress DCFH oxidation to control levels suggested that intracellular H_2O_2 was responsible for the enhanced rate of DCFH oxidation.

The enzyme involved in the respiratory burst associated with macrophages and other phagocytes is NADPH oxidase. This enzyme is dormant in resting cells but is converted to a catalytically active species when the cells are exposed to any of a large number of stimuli including TPA. The protein kinase *c* system is a signal transduction mechanism involved in the elaboration of ROS by macrophages, and the activation of protein kinase *c* by phorbol esters such as TPA is well documented (Blumberg, 1988). In macrophages, phorbol esters activate the NADPH oxidase system, and have been

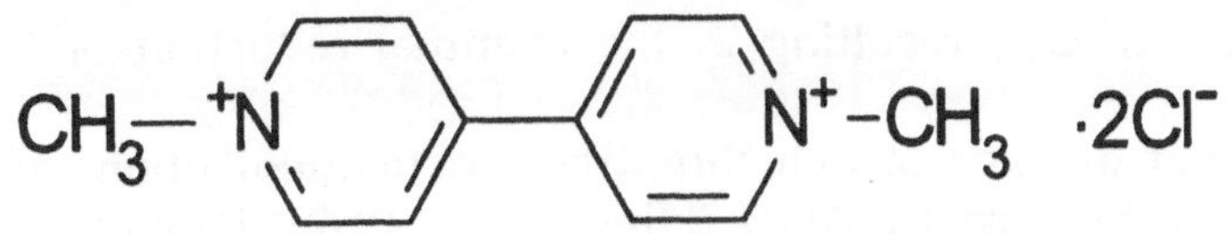

$$CH_3-{}^+N \diagdown \diagup N^+-CH_3 \cdot 2Cl^-$$

Fig. 4.7 1,1'-Dimethyl-4,4'-bipyridinium dichloride (Paraquat)

shown to similarly activate the oxidase in vitro in a reconstituted system which required protein kinase c (Blumberg, 1988). Thus, both the tumor promoting activity as well as the general toxicity of phorbol esters may be due to the production of ROS.

4.6 PARAQUAT AND DIQUAT

Paraquat (1,1'-dimethyl-4,4'-bipyridinium dichloride) (4.7) is a pneumotoxic herbicide which has been used for over thirty years. The herbicide diquat (1,1'-ethylene-2,2'-bipyridinium dichloride) is a structurally related bipyridyl herbicide which primarily produces hepatotoxicity (Smith, 1987a,b; Smith et al., 1985). Gage (1968) first reported that under anaerobic conditions paraquat could be reduced by a NADPH-dependent microsomal system to form its reduced radical. The ability to redox cycle was also demonstrated for diquat. Early studies on the ability of paraquat to produce lipid peroxidation were conducted by Bus et al. (1976). The toxicity of both paraquat and diquat is believed to involve a one-electron reduction to a cation free radical which rapidly reacts with molecular oxygen, regenerating the parent bipyridyl and producing O_2^- radical (Kappus and Sies, 1981). Redox cycling is thought to initiate a chain of events which ultimately lead to diquat (Wolfgang et al., 1991a,b) and paraquat (Smith, 1987a,b) induced toxicity, although the precise mechanisms of toxicity remain controversial.

Smith (1987) has summarized the known mechanism of toxicity of paraquat in the lung. In general, paraquat produces pulmonary edema which progresses to interstitial fibrosis, resulting in anoxia. Paraquat is accumulated in the lung by an energy-dependent, diamine transport process located in the alveolar epithelial cells and the Clara cells. The free radical form of paraquat which is formed by an NADPH-dependent one-electron reduction rapidly reacts with molecular oxygen to reform the cation and produce O_2^-. In addition, oxidation of GSH and other sulfhydryls with the formation of mixed disulfides occurs, resulting in the eventual reduction of NADPH levels.

Paraquat damage of mitochondria is dependent upon molecular oxygen and external NADH (Hirai et al., 1992). Paraquat-induced cell death may result from NADPH and/or sulfhydryl depletion which renders the tissue more susceptible to free radical attack with the resulting peroxidation of vital cellular constituents. Paraquat-induced lung injury is prevented by the free radical spin-trapping agent N-*tert*-butyl-α-phenylnitrone (PBN), and is reduced by CAT, mannitol, ethanol and vitamin E (Sata et al., 1992), strongly suggesting that the toxicity of paraquat involves ROS.

Studies with precision-cut rat liver slices have shown that diquat-induced lipid peroxidation, glutathione depletion and toxicity are effectively reduced by two experimental antioxidants, a 21-aminosteroid (U74,006F) and a troloxamine (U78,517G), as well as the antioxidant N,N'-diphenyl-*p*-phenylenediamine (DPPD) (Wolfgang et al., 1991a). In rat liver microsomes, diquat has been found to be the most potent of the bipyridyls in generating active oxygen species (Sandy et al., 1986, 1987). Smith (1987a,b) has suggested that lipid peroxidation plays a prominant role in diquat-generated, reactive oxygen-mediated hepatic injury.

The effect of an oral dose of 75 mg paraquat/kg to rats has been examined on the urinary excretion of the lipid metabolites MDA, FA, ACT and ACON over 48 hours post-treatment (Bagchi et al., 1993). Time-dependent increases in the urinary excretion of the four metabolites were observed after paraquat administration. Over the 48 hours of the study, paraquat induced urinary excretion of MDA, FA, ACT and ACON increased by approximately 218%, 155%, 331%, and 995%, respectively, relative to control values. Thus, paraquat produces markedly enhanced lipid peroxidation resulting in the excretion of these four lipid metabolites.

Peter et al. (1992) have examined the role of lipid peroxidation and DNA damage in paraquat toxicity using Ehrlich ascites tumor cells. Their results indicate that lipid peroxidation is unlikely as the primary cause for paraquat-induced cell killing, while a good correlation was obtained between cell killing and DNA damage after paraquat treatment. Earlier studies by Combs and Peterson (1983) showed that selenium deficiency in chicks protected against paraquat lethality, while vitamin E supplemented diets had no effect. Selenium deficiency resulted in low levels of GPOX activity. These investigators concluded that lipid peroxidation was not directly involved in paraquat-induced lethality, while depletion of necessary

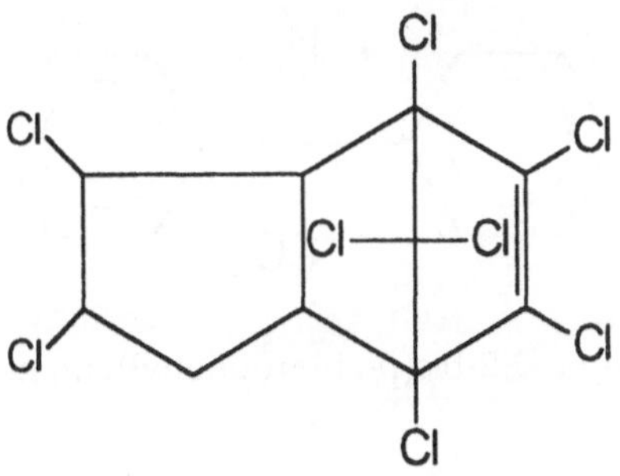

Fig. 4.8   Chlordane

cellular reducing equivalents appeared to be more important in the toxicity of this herbicide. Kitazawa et al. (1991) have concluded that XO plays an important role in paraquat toxicity, although actual tissue damage can not be explained only by the xanthine oxidase activity.

In summary, redox cycling appears to be the important initial step in the toxicity of paraquat and diquat, with the formation of O_2^- radical. Further studies will be required to clearly elucidate the subsequent steps and reaction mechanisms involved in tissue pathogenesis. The tissue specific differences in the toxicities of paraquat and diquat may be related to the toxicokinetics of the two xenobiotics, including differences in rates of metabolism, distribution, metabolic activation, redox cycling, and excretion.

Chlordane (*4.8*) is an organochlorine insecticide which is widely used in the treatment of termites and as a wood preservative. Chlordane produces a variety of toxicologic effects associated with the gastric, hepatic and nervous systems. The effects of chlordane and related compounds including heptachlor and heptachlor epoxide on stimulation of responses of guinea pig PMN leukocytes have been examined (Suzaki et al., 1988). The treatment of PMN with these xenobiotics stimulated O_2^- radical generation, altered membrane potential, and increased intracellular calcium concentrations. The results suggested that the activation of phospholipase C might participate in the generation of O_2^- by these substances. No lag phase was observed in the release of membrane-bound calcium by these substances. A causal relationship between stimulation of O_2^- generation by these xenobiotics and their toxicity was suggested.

Hassoun et al. (1993) have compared the effects of lindane, DDT (*4.9*), chlordane and endrin in rats on the ability to induce hepatic

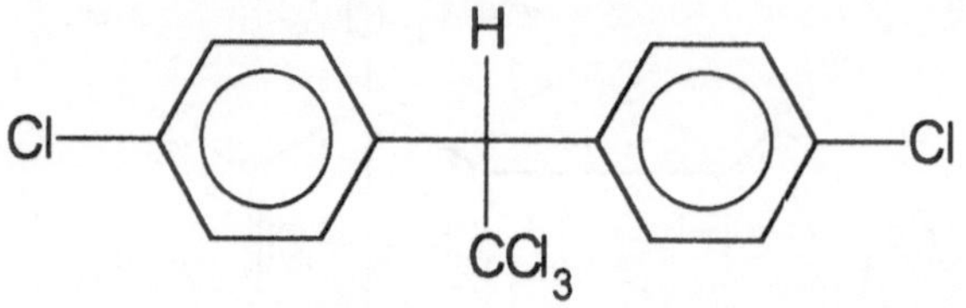

Fig. 4.9   1,1,1-Trichloro-2,2-bis(p-chlorophenyl)ethane (DDT)

lipid peroxidation and DNA single strand breaks. All four xeno-biotics resulted in significant increases in hepatic mitochondrial and microsomal lipid peroxidation and DNA damage. Earliest (6 hr) increases in both lipid peroxidation and DNA damage were observed. Maximum increases in DNA single strand breaks of 2.5- and 2.8-fold were observed 12 hours after chlordane and DDT administration, respectively, while a 4.4-fold increase was observed 24 hours after endrin administration. The results suggest that the four structurally dissimilar polyhalogenated hydrocarbons produce oxidative tissue damage which may contribute to the toxic manifestations of these xenobiotics. The differences in time-course of effects may be due to the different toxicokinetic properties of the xenobiotics.

Alachlor   [2-chloro-N-(2,6-diethylphenyl)-N-(methoxymethyl)-acetamide)] (*4.10*) has been widely used as a herbicide for the control of weeds in corn and soy beans. Although alachlor is cytotoxic to a wide variety of tissues, and is extensively metabolized, little information is available regarding the mechanism of toxicity of this xenobiotic. The effect of administering 800 mg alachlor/kg orally to rats on the urinary excretion of the lipid metabolites MDA, FA, ACT and ACON has been examined (Akubue and Stohs, 1993). These

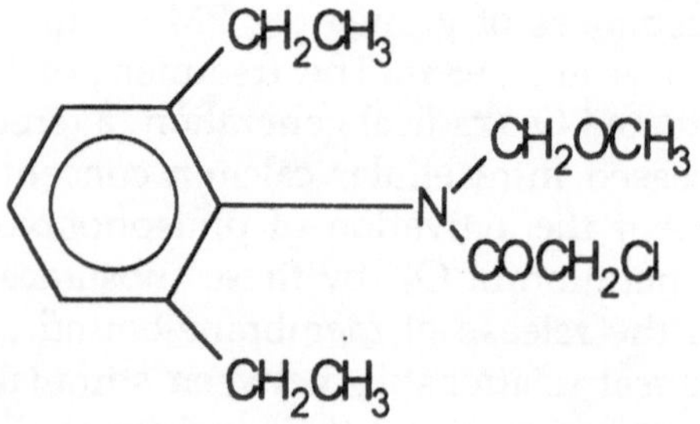

Fig.   4.10   2-Chloro-N-(2,6-diethylphenyl)-N-(methoxymethyl)-acetamide (Alachlor)

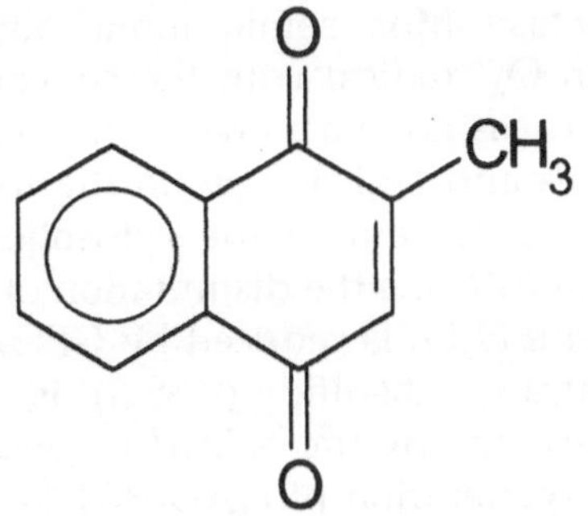

Fig. 4.11   2-Methyl-1,4-naphthaquinone (Menadione)

metabolites were determined simultaneously by HPLC. Urine samples were collected for up to 48 hours post-treatment. Alachlor administration resulted in maximum urinary excretions for MDA, FA, ACT and ACON of approximately 2.0-, 15.5-, 3.3-, and 4.5-fold, respectively. The large increase in FA excretion may be due to markedly enhanced lipid metabolism and/or the formation of formaldehyde as a metabolite of alachlor (Brown et al., 1988). The results clearly indicate that alachlor produces an oxidative stress resulting in the excretion of lipid metabolites.

4.7   QUINONES

Quinone derivatives are believed to produce their cytotoxicity via the induction of an oxidative stress through the formation of ROS. Menadione (2-methyl-1,4-naphthoquinone) (*4.11*) has been the most extensively studied model compound (Min et al., 1992; Comporti, 1989; Bellomo and Orrenius, 1985; Thor et al., 1982). Menadione can undergo both a one- and a two-electron reduction. The two-electron reduction of quinone results in the formation of hydroquinone, and the reaction is catalyzed by the cytosolic enzyme DT-diaphorase (Thor et al., 1985). This reaction is believed to represent a protective mechanism against the conversion of the quinone to the semiquinone (see chapter 7).

The one-electron reduction of menadione to the semiquinone-free radical is catalyzed by several flavoenzymes including microsomal NADPH-cytochrome P450 reductase and mitochondrial NADH-ubiquinone oxidoreductase. Most semiquinones rapidly reduce molecular oxygen to form O_2^- radical with the concomitant regeneration of the quinone, a process known as redox cycling. This process leads to an oxidative stress and toxicity due to the formation of O_2^- and other ROS. SODs are present in the cytosolic and mitochondrial compartments which catalyze the dismutation of O_2^- to O_2 and H_2O_2 (Comporti, 1989). The H_2O_2 is reduced by GPOX, and the oxidized form of GSH, glutathione disulfide (GSSG), is formed which is reduced back to glutathione by the NAD(P)H-dependent glutathione reductase (GR). If regeneration of NAD(P)H becomes rate limiting, the accumulation of ROS results in oxidative tissue damage as lipid peroxidation and DNA single strand breaks.

According to Orrenius and associates (McConkey et al., 1989; DiMonte et al. 1986; Bellomo and Orrenius, 1985), severe cellular damage is associated with depletion of normal protein thiols. The resulting damage to cellular membranes results in an increase in intracellular calcium homeostasis which may lead to cell death. Results obtained by Bellomo et al. (1990), using cultured mammalian cells, have suggested that cytoskeletal structures including actin microfilaments, microtubules, and intermediate-size filaments are targets of menadione-induced oxidative stress. The metabolism of menadione induces microtubule depolymerization and inhibits GTP-induced microtubule assembly from soluble cytosolic components.

The role and importance of oxygen in menadione induced cytotoxicity was demonstrated by Badr et al. (1989) using perfused rat livers. Hepatotoxicity occurred predominantly in the energy-rich zones of liver lobules. In isolated hepatocytes, the mechanism of menadione-mediated toxicity appeared to be concentration-dependent. At low concentrations of menadione, oxidative stress and arylation seemed to be critical mechanisms of toxicity, while at high concentrations of menadione altered membrane fluidity occurred (Shertzer et al., 1992). Dithiolthreitol protected hepatocytes from menadione toxicity at low concentrations, but not at high concentrations. Taken together, the above results indicate that oxidative stress plays an important role in the toxicity of menadione with a variety of macromolecules serving as targets for oxidative damage, although at higher concentrations a mechanical alteration of membrane fluidity, and presumably, ion transport may also contribute to the overall toxic manifestations.

Adriamycin (doxorubicin) (4.12) is a quinone containing anthracycline antibiotic with antineoplastic activity against a wide variety of tumors. Unfortunately, its clinical use has been complicated by

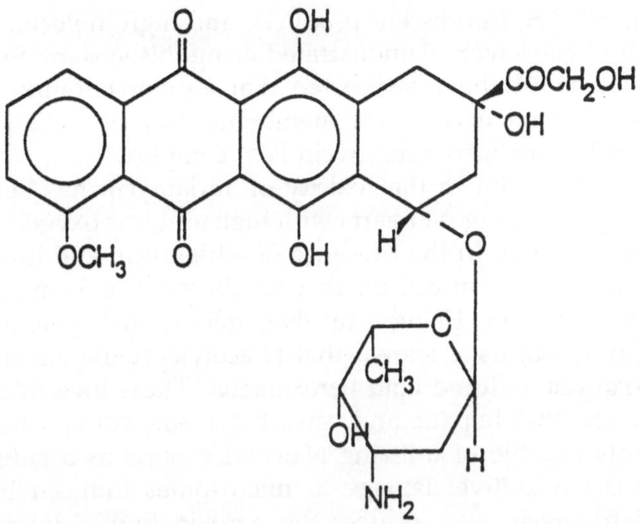

Fig. 4.12 Adriamycin (Doxorubicin)

a dose-limiting, cumulative cardiomyopathy (Jackson et al., 1984). Various studies have provided evidence for the involvement of ROS in the cardiotoxicity of adriamycin (Jackson et al., 1984; Doroshow, 1983). Although adriamycin has been shown to stimulate the production of O_2^-, H_2O_2 and OH radical (Doroshow, 1983), other studies have suggested that H_2O_2 production may not be involved in the oxidative stress caused by the anthracyclines at the cardiac level (D'Alessandro et al., 1988). Thus, although quinones as adriamycin undergo redox cycling with the production of O_2^- and other ROS, the precise mechanisms involved in the production of cardiotoxicity appear to involve multiple effects rather than a single reaction or locus which is responsible for the toxicity.

Recent investigations by Pritsos et al. (1992) have shown that ebselen (PZ-51; 2-phenyl-1,2-benzoisoselenazol-3-[2H-one]) can protect Balb/c mice against adriamycin-induced lipid peroxidation in heart and liver tissue, and adriamycin-induced toxicity in general. Ebselen is a selenoorganic compound with a thiol-dependent, peroxidase-like activity. The results support the contention that adriamycin induces cardiotoxicity via an oxidative mechanism of action. The ability of adriamycin to induce lipid peroxidation has been demonstrated using a rat liver microsomal suspension containing an NADPH-generating system incubated with adriamycin (Fukuda et al., 1992). TBARS, fluorescent products, and high molecular weight protein aggregates were demonstrated using this system. Studies by Miura et al. (1991) have suggested that adriamycin-induced lipid peroxidation of rat erythrocyte membrane may be initiated by an adriamycin-Fe^{3+}-oxygen-adriamycin-Fe^{2+} complex.

The role of oxygen in the toxicity of adriamycin has been demonstrated by perfusing rat hearts with high and low oxygen tensions (Ganey et al., 1991). In the presence of adriamycin and low oxygen tensions, a decrease in cell death was observed as compared to a high oxygen tension. Using a rat liver microsomal system, Powell and McCay (1988) have shown that N-acetylcysteine effectively inhibits adriamycin-induced lipid peroxidation. These investigators also demonstrated that hepatic and cardiac cytosols contain heat-labile components capable of utilizing N-acetylcysteine as a substrate to suppress the oxidative damage to microsomes induced by adriamycin. Adriamycin also induces the peroxidation of lipids in mitochondrial as well as microsomal membranes (Griffin-Green et al., 1988). The process was shown to be enzyme mediated and required iron.

The chronic administration of adriamycin to rats results in persistent biochemical changes in major organs including kidney, heart and liver as well as blood which are consistent with oxidative stress (Thayer, 1988). Different types of lipid peroxidation products are found in different tissues as compared to serum, and may reflect differences in antioxidant and repair mechanisms in these tissues. Adriamycin also enhances the effects of the phorbol ester 12-O-tetradecanoylphorbol-13-acetate (TPA) (4.6) on mouse epidermal GPOX activity, ornithine decarboxylase induction and skin tumor promotion (Perchellet et al., 1985). The enhanced tumor-promoting ability of TPA by adriamycin may be the result of an increased oxidative challenge that exceeds the antioxidant protective mechanisms of the cells.

The calcium channel blocker verapamil has been shown to substantially enhance adriamycin levels in some drug-resistant tumor cells. Furthermore, verapamil effectively inhibits adriamycin-mediated lipid peroxidation in mice, but decreases the survival rate and produces a higher initial peak concentration of adriamycin in the heart (Sridhar et al., 1992). The authors concluded that although verapamil inhibits adriamycin-mediated lipid peroxidation, the cardiac lipid peroxidation is not the major limiting mechanism underlying adriamycin-induced toxicity. The investigators did not exam-

Fig. 4.13   1,4-Benzoquinone

ine changes in membrane fluidity, altered calcium homeostasis or other indicators of oxidative stress.

In recent years, a number of other quinones have been studied for their abilities to redox cycle, produce $O_2^{\cdot-}$ and other ROS, and exhibit a variety of toxicities commensurate with this mechanism. The molecule 1,4-benzoquinone (4.13) is thought to be the ultimate carcinogen of the gasoline additive and industrial solvent, benzene, which is a leukemogen and myelotoxin (Lauriault et al., 1990). Benzene is metabolized to phenol and subsequently to hydroquinone by the liver. The hydroquinone may be transported to the bone marrow where it undergoes oxidative activation to 1,4-benzoquinone by myeloperoxidase and/or prostaglandin synthase. However, benzoquinone is a poor redox cycler, and Rossi et al. (1986) have shown that benzoquinone-induced toxicity in isolated hepatocytes is associated with alkylation of cellular macromolecules and not oxidative stress. The thiol drug diethyldithiocarbamate (an inhibitor of CuZnSOD) protects isolated hepatocytes from benzoquinone-induced alkylation cytotoxicity by changing the initiating cytotoxic mechanism from alkylation to oxidative stress which was less toxic (Lauriault et al., 1990).

Ludewig et al. (1989), using V79 cells, have shown that the genotoxicity of 1,4-benzoquinone (4.13), and 1,4-naphthoquinone, as demonstrated by micronuclei formation and glutathione depletion, are not causally linked. These investigators also concluded that 1,4-benzoquinone induces gene mutations by a mechanism different from oxidative stress and glutathione depletion, and glutathione does not fully protect these cells against the genotoxicity of quinones.

2,3-Dichloro-1,4-naphthoquinone (dichlone, CNQ) (4.14) is a seed fungicide and foliage protectant which has been applied to agricul-

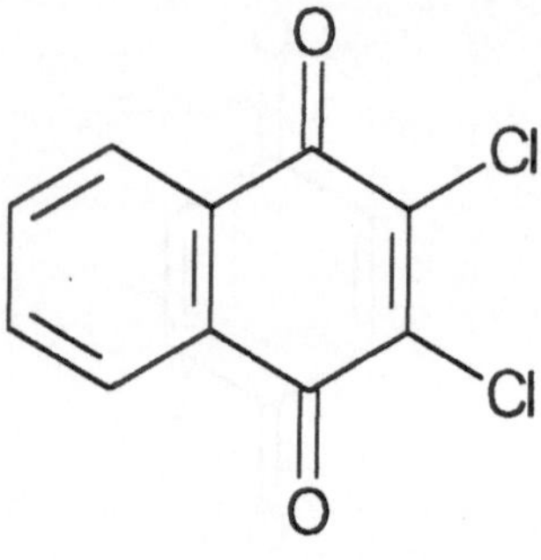

Fig. 4.14   2,3-Dichloro-1,4-naphthoquinone (CNQ)

tural products. Pritsos et al. (1982) have shown that CNQ inhibits beef heart mitochondrial respiration, uncouples oxidative phosphorylation and mediates the generation of O_2^- radical and H_2O_2. In vitro studies demonstrated that the addition of CNQ to isolated mitochondria supplemented with GSSG resulted in a respiratory burst with the production of O_2^- and H_2O_2 , and a decrease in the level of measurable disulfide (Pritsos and Pardini, 1983). CNQ also depleted GSH in rat liver mitochondria, and this depletion could be inhibited by the addition of vitamin E (Pritsos and Pardini, 1984).

Elliot and Pardini (1988) examined the acute and chronic ingestion of CNQ on DT-diaphorase activity and microsomal cytochrome P-450 levels. These investigators demonstrated significant increases in cytochrome P-450 content and the induction of both mitochondrial and cytosolic DT-diaphorase activities. The authors concluded that CNQ exposure leads to increases in hepatic cytochrome P-450 activities which, in turn, result in enhanced formation of a semiquinone intermediate leading to oxidative stress. The above results suggest that the toxicity of CNQ may be associated with the production of an oxidative stress.

2(3)-*tert*-Butyl-4-hydroxyanisole (BHA) is a synthetic antioxidant which is used extensively as a food preservative. Its actions are dependent upon its abilities to inhibit lipid peroxidation. However, the BHA metabolite *tert*-butyl-4-hydroquinone (TBHQ) (4.15) has been shown to stimulate the formation of O_2^- in rat liver microsomes (Kahl et al., 1989). No oxygen activating properties were found for BHA itself. TBHQ was shown to autooxidize to *tert*-butylquinone, and these investigators concluded that *tert*-butylquinone undergoes redox cycling leading to an oxidative burst in the presence of enzymes

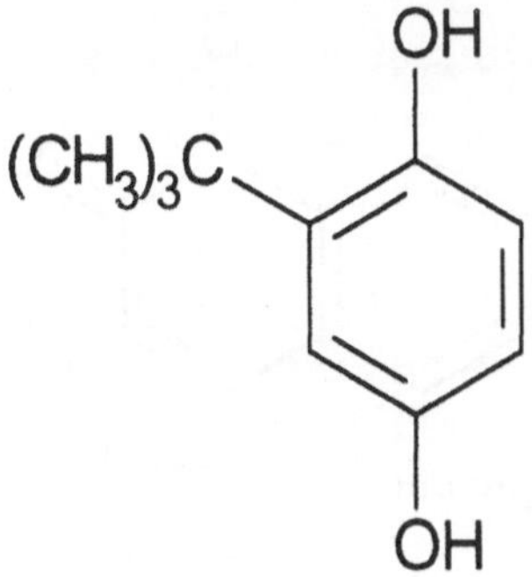

Fig. 4.15   2(3)-tert-Butyl-4-hydroquinone (TBHQ)

capable of one-electron reduction of the quinone with the reformation of the BHA metabolite TBHQ. These studies were conducted using rat liver microsomes and a rat forestomach preparation.

Solveig Walles (1992) has examined the abilities of hydroquinone, catechol, duroquinone and resorcinol to induce DNA single strand breaks in isolated hepatocytes. The DNA damage was assessed using alkaline elution. The results indicated that duroquinone and resorcinol induce DNA damage by oxidative stress presumably involving redox cycling and the production of O_2^-, while hydroquinone and catechol interact with DNA via arylation. When HL-60 cells were treated with the *o*-phenylphenol metabolites *o*-phenylhydroquinone and *o*-phenylbenzoquinone, DNA adducts were formed (Horvath et al., 1992). The authors concluded that peroxidative activation of *o*-phenylphenol may play a role in the carcinogenicity of this compound.

Diaziquinone (2,5-diaziridinyl-3,6-bis[carboethoxyamino]-1,4-benzoquinone) (AZQ) is a quinone containing antineoplastic agent which has been studied for its ability to undergo both one- and two-electron reductions (Fisher and Gutierrez, 1991). Both one- and two-electron enzymatic reductions of AZQ were demonstrated to give rise to the formation of ROS and DNA strand breaks. Autooxidation of the metabolites AZQ semiquinone and hydroquinone in the presence of molecular oxygen is believed to be responsible for these processes.

The ability of a bovine lens protein extract to facilitate the redox cycling of a number of quinones was investigated by Kleber et al. (1991). Cataract induction by naphthalene and its metabolite 1,2-naphthoquinone (*4.16*) are believed to be due to redox cycling. Us-

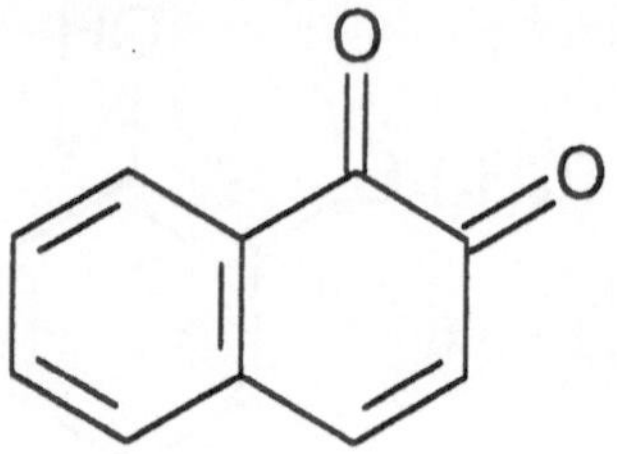

Fig. 4.16   1,2-Naphthoquinone

ing this model system, oxygen consumption could be detected in the presence of pyroloquinoline quinone and juglone (a 1,4-naph-thoquinone), but not in the presence of 1,2-naphthoquinone, men-adione (4.11) or paraquat (4.7). van Ommen et al. (1988) have shown that the covalent binding of tetrachloro-1,4-hydroquinone is depen-dent on molecular oxygen for binding to protein in the metabolic form of tetrachloro-1,4-benzoquinone. The binding could be pre-vented by adding the antioxidants GSH and ascorbic acid. Inhibition of cytochrome P-450 by metyrapone, thus blocking the P-450-me-diated formation of reactive oxygen species, also significantly de-creased binding. Furthermore, the addition of SOD also effectively prevented the covalent binding of the quinone to protein.

Mitozantrone is a novel quinone antineoplastic drug. The toxicity of this quinone has been compared with that of menadione in hu-man Hep G2 hepatoma cells (Duthie and Grant, 1989). Mitozantrone did not cause glutathione depletion, while menadione depleted GSH. The toxicity of mitozantrone appeared to involve activation to an epoxide intermediate rather than redox cycling.

2,4,5-Trichlorophenol (TCP) is an industrial chemical which is used as an intermediate in the manufacture of the herbicide 2,4,5-trich-lorophenoxyacetic acid (2,4,5-T) and its esters. TCP is a major con-taminant of these products. Recent studies have shown that TCP is extensively metabolized, and two of the major metabolic products are the semiquinones 2,5-dichlorohydroquinone (DCH) (4.17) and 3,4,6-trichlorocatechol (TCC) (Juhl et al., 1991). Both DCH and TCC autooxidize to their semiquinone radicals, with the resulting pro-duction of ROS. When DCH and TCC are incubated with DNA, strand breakage occurs which can be differentially inhibited by the scavengers SOD and CAT. The results suggest that redox cycling with the formation of ROS is involved in DNA damage.

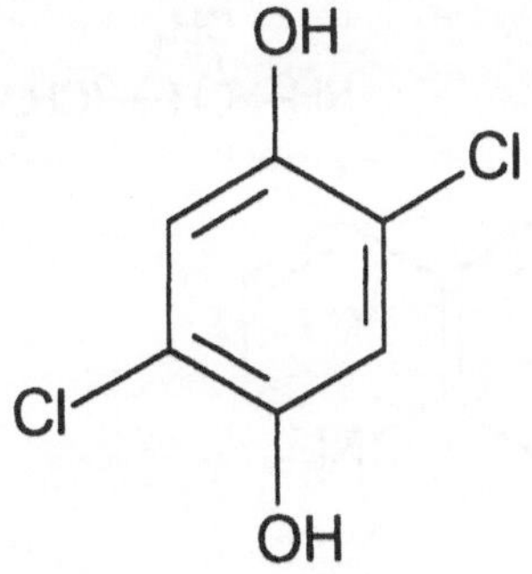

Fig. 4.17 2,5-Dichlorohydroquinone (DHC)

4.8 QUINOLONES

In recent years various fluorinated piperazinyl-substituted quinoline derivatives (quinolones) have been developed and are being used therapeutically as broad spectrum antibacterial agents. Examples include nalidixic acid, ofloxacin, ciprofloxacin (*4.18*), enoxacin and lomafloxacin. Phototoxicity is a common side effect associated with the use of these agents. Ear swelling in mice has been used as a common indicator of phototoxicity induced by oral administration of quinolones followed by ultraviolet-A (UVA) irradiation (Wagai and Tawara, 1991a,b, 1992; Wagai et al., 1989, 1990). The phototoxicity can be inhibited by the use of CAT, SOD, dimethylthiourea, β-carotene and allopurinol, indicating that reactive oxygen species are involved. The authors have concluded that O_2^- and other oxygen

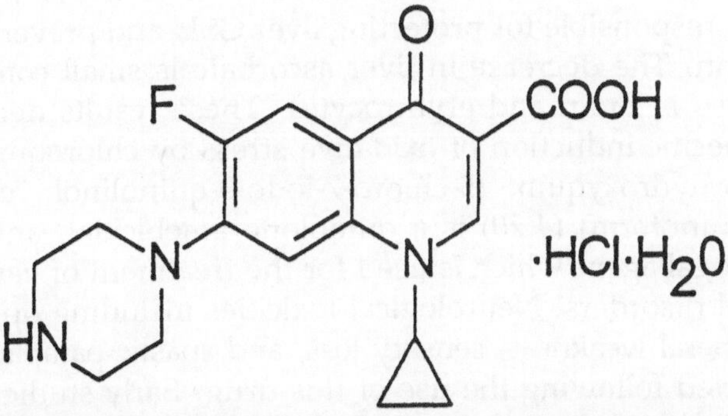

Fig. 4.18 1-Cyclopropyl-6-fluoro-1,4-dihydro-4-oxo-7-(1-piperazinyl)-3-quinolinecarboxylic acid hydrochloride monohydrate (Ciprofloxacin)

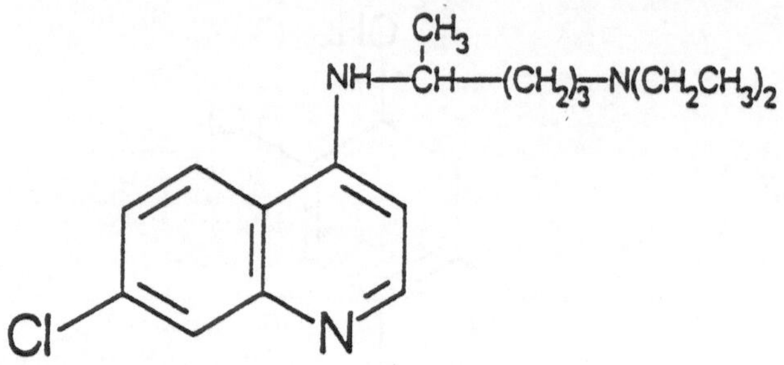

Fig. 4.19   Chloroquine

metabolites generated by the xanthine oxidase pathway are involved in the toxicity. Bailly et al. (1990) have studied the effects of quinolone antibacterial agents on production of tumor necrosis factor α-TNF in human monocytes. Various quinolones were found to alter TNF production in a dose-dependent manner, possibly through the induction of an accumulation of intracellular cAMP.

Chloroquine (*4.19*) belongs to a group of 4-aminoquinolone-containing drugs which are effective against malaria and other parasitic infections. Related quinolone drugs include amodiaquine, primaquine and quinacrine. Chloroquine produces an oxidative stress in brain and erythrocytes of rats as evidenced by decreases in GSH and ascorbate content and an increase in lipid peroxidation (Abdel-Gayoum et al., 1992). A decrease in ascorbate is also observed in the liver. However, GSH content remains unchanged in the liver, and lipid peroxidation does not occur. In the liver, a significant increase in the activity of glucose-6-phosphate dehydrogenase activity occurs which results in the increased availability of NADPH which presumably is responsible for protecting liver GSH and preventing lipid peroxidation. The decrease in liver ascorbate is small compared to the decrease in brain and erythrocytes. These results demonstrate a tissue specific induction of oxidative stress by chloroquine.

Iodochlorhydroxyquin (5-chloro-7-iodo-8-quinolinol; clioquinol; vioform; chinoform) (*4.20*) is a quinolone amebicidal, antibacterial and antifungal agent which is used for the treatment of various dermatological disorders. Neurological toxicities including optical atrophy, peripheral weakness, sensory loss, and spastic paraparesis have been reported following the use of this drug. Early studies by Yagi et al. (1985) demonstrated that an iodochlorhydroxyquin-ferric chelate incubated with cultured neural retinal cells from chick embryos resulted in significant time-dependent increases in lipid peroxide

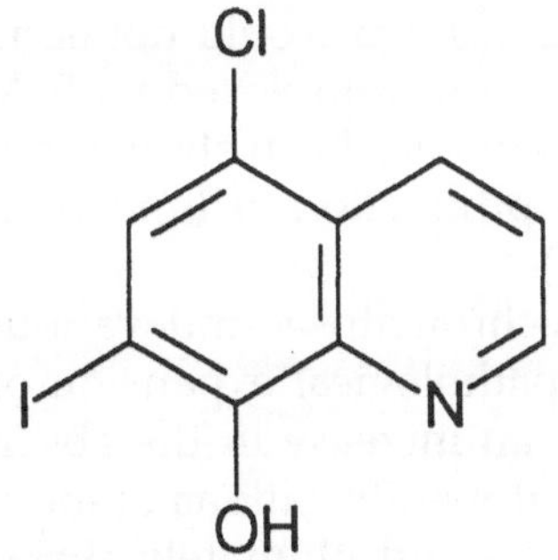

Fig. 4.20   5-Chloro-7-iodo-8-quinolinol (Iodochlorohydroxyquin)

levels in the cells. The data indicated that the iodochlorhydroxyquin served as a carrier for the entry of iron into the cells, and the incorporated iron induced lipid peroxidation, resulting in neural cell degeneration. Yagi et al. (1990) have also shown that an iron chelate of iodochlorhydroxyquin significantly increases the lipid peroxidation associated with rat liver microsomes. The iodochlorhydroxyquin-ferric chelate more effectively enhanced microsomal lipid peroxidation than the use of ADP-ferric chelate. The investigators have speculated that lipid peroxidation induced by iodochlorhydroxyquin may be responsible for the neurotoxicity of this xenobiotic.

4.9   TRANSITION METALS AND CATION COMPLEXES

A growing body of evidence has demonstrated that transition metals act as catalysts in the oxidative deterioration of biological macromolecules and the toxicities of many xenobiotics. The two most commonly studied transition metals are the cations iron and copper. A variety of studies have demonstrated the ability of iron chelates or complexes to catalyze the formation of ROS and stimulate lipid peroxidation.

Aust (1989) has reviewed the relationship between metal ions, oxygen radicals and tissue damage. The role of iron in the initiation of lipid peroxidation has also been reviewed (Minotti and Aust, 1987). These investigators have presented evidence that lipid peroxidation requires both Fe^{3+} and Fe^{2+}, probably as a dioxygen-iron complex. Iron is capable of catalyzing redox reactions between oxygen and biological macromolecules that would not occur if catalytically active iron were not present. Iron complexed with ADP, histidine, EDTA, citrate and other chelators has been shown to facilitate the formation of ROS and enhance production of lipid peroxidation (Aust, 1989; Ryan and Aust, 1992).

In hereditary hemochromatosis and various forms of secondary hemochromatosis, a pathological expansion of body iron stores occurs primarily due to an increase in the absorption of dietary iron. The major pathological manifestations associated with chronic iron overload include fibrosis and ultimately cirrhosis. The mechanisms associated with liver injury in chronic iron overload are believed to include increased lysosomal membrane fragility mediated by iron-induced lipid peroxidation, and peroxidative damage of organelles as microsomes and mitochondria (Bacon and Britton, 1989).

Upon ingestion of iron, iron is either oxidized and stored in the iron storage protein ferritin or associates with the iron transport protein transferrin in the blood stream. In order for iron to facilitate the formation of ROS via the Fenton reaction, the iron must be in a free or catalytically active form. A variety of xenobiotics have been shown to facilitate the release of iron from ferritin, including, paraquat, diquat, nitrofurantoin, adriamycin, daunomycin and diaziquone (Ryan and Aust, 1992). Thus, a variety of xenobiotics may enhance the formation of ROS not only by undergoing redox cycling, but may also facilitate the release of free iron which catalyzes the formation of ROS.

The role of iron in the initiation of lipid peroxidation is well known (Minotti and Aust, 1987; Alleman et al., 1985). Evidence indicates that chelated iron acts as a catalyst for the Fenton reaction, facilitating the conversion of O_2^- and H_2O_2 to $\cdot OH$ radical, a species frequently proposed to initiate lipid peroxidation (Imlay et al., 1988; Halliwell and Gutteridge, 1986). Studies using iron nitrilotriacetate have shown that this complex exhibits a prooxidant activity, resulting in enhanced lipid peroxidation of hepatocyte cultures with the leakage of lactate dehydrogenase and transaminase enzymes, as well as a substantial increase in Trypan blue staining which indicates membrane damage (Morel et al., 1990; Carini et al., 1992). However, in these studies, the protein and non-protein thiol content of the cells was not affected, although a dramatic increase in the conjugated diene content of mitochondria of iron nitrilotriacetate treated hepatocytes was observed (Carini et al., 1992). Lipid peroxidation was shown to occur in both inner membranes and plasma membrane. Free radical scavengers including SOD, vitamin

E and mannitol reduced lipid peroxidation intracellularly in response to iron nitrilotriacetate, while CAT and thiourea seemed to protect plasma membranes as evidenced by a decrease in enzyme leakage (Morel et al., 1990).

Studies on the interaction of paraquat with microsomes and ferric complexes have demonstrated an increase in oxygen radical generation (Puntarulo and Cederbaum, 1989). Ferric ion complexed with citrate, ATP, EDTA, and diethylenetriamine penta-acetic acid (DETAPAC) increased the catalytic effectiveness of paraquat in promoting microsomal generation of oxygen radicals. Thus, the interaction of iron complexes with paraquat may contribute to the oxidative stress and toxicity produced by paraquat in biological systems. Redox cycling of adriamycin has been shown to result in a parallel reductive release of membrane-bound nonheme iron (Minotti, 1990). Lipid peroxidation was shown to occur in the presence of low levels of adriamycin which favored only partial Fe^{2+} autooxidation and a Fe^{2+}/Fe^{3+} ratio of approximately 1:1, providing further evidence that both forms of iron are required for the initiation of lipid peroxidation.

Iron has also been shown to play a role in the toxicity of TCDD. Microsomal lipid peroxidation induced by TCDD requires iron (Al-Bayati and Stohs, 1987). TCDD markedly alters the distribution of iron as well as copper, zinc and magnesium (Wahba et al., 1988). Endrin also alters iron distribution in hepatic mitochondria and microsomes of rats in a manner similar to the results observed for TCDD. Wahba et al. (1990) have shown that TCDD administration to rats results in an increase in the amount of free or catalytically active iron associated with microsomes. Endrin may produce a similar effect with respect to the amount of catalytically active or available iron. The ability of iron to act as a synergist for hepatocellular carcinoma induced by PCBs in Ah-responsive mice has been demonstrated by Smith et al. (1990). The precise role of iron in potentiating carcinomas in mice is not clear. However, since iron facilitates the formation of ROS, and reactive oxygen species are believed to play a central role in tumor formation (Cerutti, 1989), the iron may act to facilitate carcinogenesis in this manner.

The iron chelator DFX has been used to assess the role of iron in the toxicity of various xenobiotics. Wahba et al. (1990) have shown that DFX can modulate the toxicity of TCDD. DFX blocks the depletion of hepatic GSH and the formation of MDA by tert-butyl-hydroperoxide (t-BOOH) in isolated perfused rat livers (Younes and Strubelt, 1990). Furthermore, DFX inhibits t-BOOH-induced GSH depletion, lipid peroxidation and hepatotoxicity in rats. These results emphasize the importance of iron in t-BOOH-induced lipid peroxidation and hepatotoxicity, as well as the important role of iron in propagation of lipid peroxidation reactions (Younes and Wess, 1990). Due to the short half-life of DFX, it can not readily be used to prevent oxidative stress associated with chemical-induced toxicities. However, the use of hydroxyethyl starch conjugated DFX has been shown to modulate the neurologic injury which occurs during brain ischemia and reperfusion in rats (Rosenthal et al., 1992). This complex may also be useful in modulating the neurotoxicity and hepatotoxicity associated with a wide range of toxicants.

The induction of DNA single strand breaks in macrophage by Fe^{3+} under cool-white fluorescent light in the presence and absence of low molecular weight chelators was investigated by Chao and Aust (1993). The results indicate that the photochemical reduction of Fe^{3+} to Fe^{2+} by chelators including citrate, nitrilotriacetate or EDTA results in the formation of DNA single strand breaks without the addition of H_2O_2. The results further indicate that Fe^{2+} and, OH radical or similarly reactive species, are involved in the induction of the DNA damage.

The redox properties of iron and copper complexes of adriamycin, bleomycin and thiosemicarbazones have been investigated by electron spin resonance spectroscopy (ESR) (Antholine et al., 1985). A common property of these metal complexes is their ready ability to be reduced by thiol compounds and oxidized by iron or reduced species of iron to produce radicals. Copper is a common cofactor for many enzymes including oxidases and oxygenases (Gutteridge, 1984). Similar to iron, copper acts as a catalyst in the formation of ROS and catalyzes peroxidation of membrane lipids (Chan et al., 1982). Studies have shown that when cupric acetate is added with adriamycin in the Ames *salmonella* mutagenicity test, the presence of copper results in more than a 700% increase in the mutagenicity of adriamycin. The results support the contention that drug-metal ion-DNA associations might contribute to genotoxicity (Yourtee et al., 1992).

The role of copper in the oxidation of hydroquinone to benzoquinone (*4.10*) as well as the cytotoxicity of hydroquinone have been examined (Li and Trush, 1993a). Copper was shown to significantly accelerate the oxidation of hydroquinone to benzoquinone in a concentration-dependent manner. Furthermore, copper added to primary bone marrow stromal cell cultures significantly enhanced hydroquinone-induced cytotoxicity. Antioxidants including GSH and dithiothreitol but not CAT completely prevented the enhanced cytotoxicity of hydroquinone by copper. Further studies have demonstrated that copper markedly enhances the formation of DNA strand breaks in the presence of hydroquinone (Li and Trush, 1993b). The presence of singlet oxygen ($^1O_2$) scavengers but not ˙OH radical scavengers provides partial protection, indicating that $^1O_2$ rather than ˙OH radical may play a role in the induction of DNA strand breaks. Other metal ions including Fe^{3+}, Mn^{2+}, Cd^{2+} and Zn^{2+} did not significantly enhance oxidation of hydroquinone or the induction of DNA strand breaks (Li and Trush, 1993b). Thus, copper may be an important factor in the generation of reactive oxygen species, the cytotoxicity, and the formation of DNA damage in target cells by hydroquinone.

Cadmium (Cd) is an abundant, non-essential element which is generating concern due to its accumulation in the environment as a result of industrial practices. Soluble Cd salts accumulate and result in toxicity to liver, kidneys, brain, lungs, heart, testes and the central nervous system. The mechanisms responsible for the toxicity of Cd are not well understood. Cadmium does not appear to generate free radicals (Ochi et al., 1987), but does elevate lipid peroxidation in tissues soon after exposure (Muller, 1986). Studies by Fariss (1991) have shown that free radical scavengers and antioxidants are useful in protecting against cadmium toxicity.

Manca et al. (1991) examined the susceptibility of liver, kidneys, brains, lungs, heart and testes of rats given intraperitoneal doses of cadmium chloride. The animals received from 25-1250 μg Cd/kg as $CdCl_2$, and the animals were sacrificed 24 hours post-treatment. Greatest increases in lipid peroxidation were demonstrated in lungs and brain as well as liver, based on the formation of TBARS. The authors concluded that lipid peroxidation is an early and sensitive consequence of cadmium exposure.

Treatment of rats with a single carcinogenic dose of $CdCl_2$ (30 mμmol/kg) has been shown to cause severe hemorrhagic damage in testis within 12 hours after administration of the metal. Furthermore, lipid peroxidation levels, iron content, and cellular production of H_2O_2 were markedly elevated in testicular Leydig cells, the target population for Cd carcinogenesis (Koizumi and Li, 1992). In addition, GPOX activity increased while GR and CAT activities decreased. The authors concluded that oxygen species such as H_2O_2 may play an important role in the initiation of carcinogenesis within the target cell population.

The influence of ascorbic acid on lipid peroxidation in guinea pigs treated with Cd (1 mg Cd/animal/day in drinking water) has been examined by Hudecova and Ginter (1992). Cadmium administration in conjunction with a low intake of ascorbic acid (2 mg/animal/day) increased lipid peroxidation (lipid peroxides) in kidney, liver and serum. A high intake of ascorbic acid (100 mg/animal/day) decreased formation of MDA in these same tissues. Thus, ascorbic acid levels appear to play a role in the ability of cadmium to induce lipid peroxidation in the guinea pig.

Hussain et al. (1987) examined the in vitro and in vivo effects of Cd on SOD and lipid peroxidation in liver and kidney of rats. Cadmium acetate administered in vivo inhibited the activity of SOD and increased lipid peroxidation in liver and kidneys. Addition of Cd in vitro also inhibited SOD in both tissues. Furthermore, lipid peroxidation was markedly increased after addition of Cd to fresh homogenates of both tissues. Thus, the results demonstrate both in vitro and in vivo the ability of cadmium to induce lipid peroxidation.

By manipulating the Cd concentrations in the media of V79 Chinese hamster fibroblasts in culture, some of the cells developed cadmium resistance and cross-resistance to oxidative stress (Chubatsu et al., 1992). Most of the cross-resistance to oxidative stress in Cd-challenged cells can be accounted for by a parallel increase in GSH content. Metallothionein content did not seem to exert a major effect against oxidative stress in Cd challenged cells. Exposure of HeLa and HL60 cells to $CdCl_2$ or sodium arsenite leads to a marked increase in the synthesis of a stress protein (Taketani et al., 1989). Further studies have demonstrated that this protein in these human cells is heme oxygenase, suggesting that an oxidative stress is involved in the toxicity of Cd.

Chromium (Cr) is widely known to cause allergic dermatitis as well as toxic and carcinogenic effects in humans and animals. The role of physiological antioxidants in Cr (VI)-induced cellular injury has been reviewed by Sugiyama (1992). This author has reviewed recent in vitro and in vivo effects of oxygen scavengers, GSH, vitamin B2, vitamin E, and vitamin C on chromate-induced injuries including DNA damage, lipid peroxidation, enzyme inhibition, cytotoxicity and mutagenesis. Vitamin E was shown to dramatically decrease chromate-induced cytotoxicity, lipid peroxidation and DNA damage, while vitamin B2 did not exhibit these effects (Sugiyama, 1991). Chromium occurs in the workplace primarily in the valence forms Cr(VI) and Cr(III). The chromate ion $[CrO_4]^{-2}$, the dominant form of Cr(VI) in neutral aqueous solutions, can readily cross cellular membranes via non-specific anion carriers (Danielsson et al., 1982).

While it has been postulated that Cr(V) is the ultimate carcinogenic form of Cr compounds, Kawanishi et al. (1986) have demonstrated that it is not Cr(V) itself that is carcinogenic, but rather oxygen free radicals such as O_2^- anion, $^1O_2$ and OH radicals. These investigators examined the mechanism of DNA cleavage induced by Cr(VI) in the presence of H_2O_2. Reactive oxygen species were produced by the decomposition of Cr(V) $(O_2)_4^{-3}$ ion, resulting in DNA damage. The generation of OH radical was detected by ESR.

Shi and Dalal (1989) have also used ESR to demonstrate the formation of long-lived Cr(V) intermediates in the reduction of Cr(VI) by GR in the presence of NADPH, and the generation of OH radical. Hydrogen peroxide suppresses Cr(V) and enhances the formation of OH radicals through a Cr(V) catalyzed Fenton-like reaction. Subsequent investigations with SOD showed no significant participation of O_2^- in the generation of OH radicals (Shi and Dalal, 1990). Their results indicated that the Cr(V) complexes produced in the reduction of Cr(VI) by cellular reductants react with H_2O_2 to generate OH radicals, which may be the initiators of the primary events in Cr(VI) cytotoxicity. Related studies by Jones et al. (1991) have provided evidence which suggests that OH radicals are generated from a Cr(V) intermediate which is responsible for causing DNA strand breaks.

Until recently, Cr(III) was thought to be relatively non-toxic. However, Ozawa and Hanaki (1990) demonstrated that Cr(III) can be reduced to Cr(II) by the biological reductants L-cysteine and NADH, and in turn, the newly formed Cr(II) reacts with H_2O_2 to produce OH radical which can be detected by ESR and HPLC. The resulting OH radical is presumably responsible for tissue damaging effects.

Sugden et al. (1992) have employed the *Salmonella* reversion assay to identify mutagenic Cr(III) complexes. Relaxation of supercoiled DNA was used to show in vitro interactions with plasmid DNA. These investigations demonstrated that mutagenic Cr(III) complexes display characteristics of reversibility and positive shifts of the Cr(III)/Cr(II) redox couple consistent with the ability of these Cr(III) complexes to serve as cyclical electron donors in a Fenton-like reaction. These same mutagenic Cr(III) complexes relaxed supercoiled DNA, presumably by the induction of single-strand breaks. The results suggest that the mechanism involved in the potentiation of mutagenesis by Cr complexes involves an oxygen radical as an active intermediate. Furthermore, Cr(III) may be one of the most biologically active oxidation states of Cr.

In summary, the literature indicates that both Cr(VI) and Cr(III) are biologically active oxidation states of Cr. Furthermore, both oxidation states of Cr are involved in redox cycling with the production of ROS. Further studies are required to elucidate the mechanisms involved in the regulation of tissue damaging effects by Cr complexes.

The toxicity of mercury (Hg) and its ability to react with and deplete free sulfhydryl groups is well known. The decrease in free sulfhydryl groups may lead to the formation of an oxidative stress, resulting in tissue damaging effects. The subcutaneous administration of mercuric chloride ($HgCl_2$) to rats results in nephrotoxic acute renal failure (Gstraunthaler et al., 1983). Mercury causes a depletion of GSH in the renal tubules, and also a reduction in the activities of SOD, CAT, and GPOX, enzymes responsible for the protection of cells against the peroxidative action of O_2^- and ROOHs. Thus, nephrotoxicity may be due to Hg-induced alterations in membrane integrity via the formation of ROS and the perturbation of antioxidant defense mechanisms.

Lipid peroxidation occurs in rat kidney as early as 12 hours after Hg administration to rats (Fukino et al., 1984). A decrease in vitamin C and vitamin E contents in the kidney are also observed 12 hours post-mercury administration. When rats are pretreated with zinc, a decrease in kidney lipid peroxidation occurs with a concomitant increase in the GSH content. When rats are pretreated with the thiol antidotes 2,3-dimercapto-1-propanesulfonic acid (DMPS) and D-penicillamine (PA) prior to the administration of mercury, protection against Hg-induced lipid peroxidation is afforded by both thiols in the liver, while in the kidneys only PA exhibited a protective effect (Benov et al., 1990). In in vitro experiments, these same investigators demonstrated that both antidotes can act as oxygen radical scavengers and inhibitors of lipid peroxidation. However, PA was significantly more effective. Thus, the antioxidant properties of these chelating agents may be beneficial in metal intoxications.

Exposure of catfish to $HgCl_2$ (0.20 mg/L) for up to 30 days resulted in significant increases in lipid peroxidation in liver, brain and muscle (Bano and Hasan, 1989). When rat kidney mitochondria are incubated with mercuric ion in vitro, approximately a 4-fold increase in H_2O_2 formation occurs at the ubiquinone-cytochrome *b* (antimycin A inhibited) region, and a 2-fold increase occurs in the

NADH dehydrogenase (rotenone inhibited) region (Lund et al., 1991). Concomitantly, a 3.5-fold increase in iron-dependent lipid peroxidation occurs at the NADPH dehydrogenase region with a small increase at the ubiquinone-cytochrome *b* region. In addition, mitochondrial GR concentrations decrease as a function of both Hg concentration and incubation time. Thus, at low concentrations (12-30 nmol/mg protein), Hg depletes mitochondrial GSH and enhances H_2O_2 formation under conditions of impaired respiratory chain electron transport. The increased H_2O_2 may lead to oxidative tissue damage, including lipid peroxidation, resulting in mercury-induced nephrotoxicity.

More recent studies have shown that exposure of mice to Hg, Cr or silver results in enhanced production of MDA in liver and kidneys (Rungby and Ernst, 1992). MDA levels remain elevated for approximately 60 minutes and return to normal in kidneys, whereas MDA levels in liver are elevated for over 48 hours after exposure. Furthermore, synergistic lipid peroxidation and toxicity was observed between inorganic Hg and carbon tetrachloride, while pre-exposure to Cr, organic mercury or silver did not enhance carbon tetrachloride toxicity as measured by organ content of MDA.

DNA damage has also been demonstrated in response to the in vivo administration of mercuric acetate (Williams et al., 1987). These investigators demonstrated that in cultured human KB cells, the inhibition of dUTPase and DNA polymerase alpha activities and the activation of uracil-DNA glycosylase activity correlated with the induction of DNA single strand breaks and the decrease in cell viability.

LeBel et al. (1992) have examined the hypothesis that methyl mercury may exert its neurotoxicity by way of iron-mediated oxidative damage. The formation of ROS in rat brain was estimated based on the rate of oxidation of fluorescent probes. A significant increase in the rate of formation of ROS was observed 7 days after a single interperitoneal injection of methyl mercury (5 mg/kg i.p.), while pretreatment with the potent iron chelator DFX (500 mg/kg) completely prevented this effect. These findings provide evidence that iron-catalyzed oxygen radical production plays a role in methyl mercury neurotoxicity, and provides support for the use of the iron chelator DFX in xenobiotic-induced oxidative damage. When cerebellar granule cells from rats are incubated with methyl mercury, time- and concentration-dependent cell killing occurs (Sarafian and Verity, 1991). Significant protection from methyl mercury-induced cell death was observed by the addition of the chelators EGTA and DFX, or KCN. The results indicate that oxidative processes contribute to the cytotoxicity of methyl mercury in isolated cerebellar granular neurons.

Suda et al. (1991) have examined the degradation of methyl mercury and ethyl mercury into inorganic mercury by oxygen free radical producing systems in vitro. Copper ascorbate, xanthine oxidase-hypoxanthine, and iron-EDTA plus H_2O_2 were used to produce $\cdot OH$ radical. Both methyl mercury and ethyl mercury were readily degraded by these three systems. The degradation appeared to be unrelated to either O_2^- production or H_2O_2 production, indicating that $\cdot OH$ radical might be the oxygen free radical primarily responsible for degradation of the two forms of organic mercury.

The administration of nickel (Ni) to rats results in enhanced lipid peroxidation, decreased GPOX activity, and increased tissue iron levels (Athar et al., 1987). The authors proposed that the carcinogenicity of Ni compounds may be related to enhanced production of ROS. Furthermore, the Ni-induced accumulation of iron may be directly responsible for the formation of ROS and the subsequent enhancement of lipid peroxidation. Based on the fact that inorganic nickel chloride ($NiCl_2$) forms DNA adducts, induces hepatic DNA strand breaks, produces chromosome aberrations and induces lipid peroxidation, Stinson et al. (1992) have examined the relationship between $NiCl_2$-induced lipid peroxidation and DNA strand breaks in rat liver. At a dose of 0.52 nmol/kg, $NiCl_2$ subcutaneously induced DNA strand breakage at 4 hours and lipid peroxidation at 12 hours post-treatment in rat liver. DFX (1 g/kg, i.p., 15 minutes prior to $NiCl_2$ injection) completely inhibited DNA strand breakage, but had no effect on lipid peroxidation. The authors concluded that lipid peroxidation is not causally related to genetic damage. Nickel chloride-induced DNA strand breakage may be caused by the induction of the Fenton reaction, generating $\cdot OH$ radicals. However, the rapid excretion of DFX may prevent it from inhibiting the lipid peroxidation which was observed 12 hours after treatment with Ni.

Shi et al. (1992) have shown that the incubation of nickel with cumene hydroperoxide or t-BOOH in the presence of GSH, carnosine, homocarnosine and anserine resulted in the formation of alkyl, alkoxyl, and peroxyl radicals. Glutathione, carnosine, homocarnosine and anserine are normally considered to be cellular antioxidants. These studies suggest that instead of protecting against oxidative damage, these oligopeptides may facilitate Ni-mediated free radical production, and may thus participate in both the carcinogenicity and toxicity of Ni.

Younes and Strubelt (1991) have shown a strong correlation between vanadate-induced hepatotoxicity and the induction of lipid peroxidation. Both processes are inhibited in parallel by antioxidants, suggesting a role for lipid peroxidation in vanadate-induced hepatotoxicity. Previous studies had shown that vanadate induces lipid peroxidation in isolated hepatocytes (Stacey and Klaassen, 1981), and depletes GSH (Stacey and Kappus, 1982). Furthermore, vanadate stimulates the exhalation of ethane in rats (Harvey and Klaassen, 1983), providing evidence that vanadate induces an oxidative stress and lipid peroxidation.

Various investigators have proposed that a biochemical function of zinc (Zn) is the maintenance of membrane structure and function (Bettger and O'Dell, 1981). Dietary Zn deficiency has been shown to increase the susceptibility of rat hepatic microsomes to lipid peroxidation both in vivo (Sullivan et al., 1980) and in vitro (Sullivan et al., 1980; Burke and Fenton, 1985). The ability of a dietary Zn deficiency to stimulate the production of carbon-centered free radicals in lung microsomes was reported by Bray et al. (1986). Hammermueller et al. (1987) have demonstrated that a Zn deficiency causes leakage of H_2O_2 from a NADPH-dependent cytochrome P450 enzyme system. Xu and Bray (1992) have examined the effects of increased microsomal oxygen radicals on the function and stability of cytochrome P450 in dietary Zn deficient rats. Based on their results, they concluded that severe dietary Zn deficiency in rats causes a functional and structural impairment of liver microsomal cytochrome P450 function via a free radical mediated mechanism. In summary, although most cations facilitate the formation of an oxidative stress, Zn appears to act as a membrane stabilizer and prevents the formation of ROS through a mechanism which may involve protection of sulhydryl groups against oxidation, and/or the inhibition of the production of ROS by transition metals (Bray and Bettger, 1990). However, other critical antioxidant functions for Zn may still be found.

4.10 MISCELLANEOUS INDUCERS OF OXIDATIVE STRESS

A wide range of structurally dissimilar xenobiotics has been shown to induce an oxidative stress. Studies involving a number of these compounds are summarized. Nilutamide (4.21) is a nitrofluorinated dimethylimidazolidinedione antiandrogen drug which acts as a competitive inhibitor of the androgen receptor, and has been used

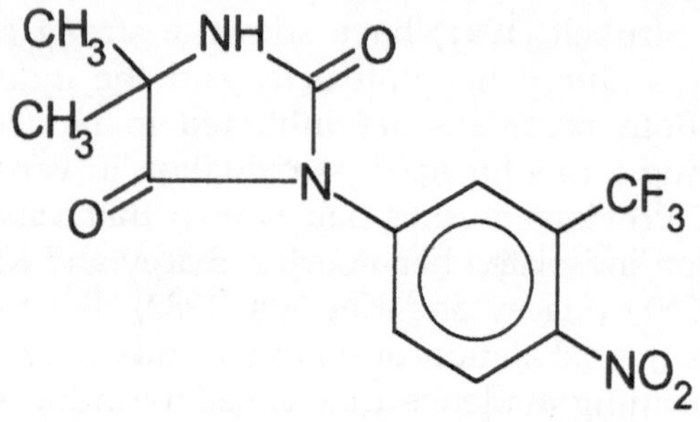

Fig. 4.21   5,5-Dimethyl-3-[4-nitro-3-(trifluoromethyl)phenyl]-2,4-imidazoli-dinedione (Nilutamide)

in the treatment of metastatic prostatic carcinoma. Berson et al. (1991) have shown that nilutamide undergoes redox cycling in aerobic rat liver microsomes, being reduced to a nitro anion-free radical which reacts with oxygen to regenerate the parent drug and form O_2^{-} which dismutates to H_2O_2. Using an isolated hepatocyte system, Fau et al. (1992) have shown that the toxicity of nilutamide is associated with depletion of GSH, increased levels of GSSG, increased calcium-dependent phosphorylase a activity, and oxidation and accumulation of cytoskeleton-associated proteins and formation of blebs. Toxicity was prevented by GSH precursors, thiol reductants and vitamin E, indicating that the mechanism of toxicity was consistent with an oxidative stress.

Warren and Reed (1991) have demonstrated that the treatment of rats with 40 mg/kg methyl ethyl ketone peroxide (2-butanone peroxide) (4.22) rapidly induced lipid peroxidation, and from one to four hours post-treatment depressed plasma content of vitamin E and the liver contant of GSH while increasing serum levels of alanine and aspartate aminotransferases. Adam et al. (1990) have assessed the role of redox cycling in the toxicity of nitrofurantoin [N-(5-nitro-2-furfurylidene)-1-aminohydantoin] (4.23). Nitrofurantoin is an antibiotic used in the treatment of urinary tract infections, and the lung appears to be a primary target of toxicity of this compound. The results of this study indicated that nitrofurantoin was a poor redox cycler, similar to paraquat, but an order of magnitude less potent than either diquat or menadione. A rat lung tissue slice system was used for these studies. Although nitrofurantoin is a poor redox cycler, the results do not eliminate or preclude the possibility

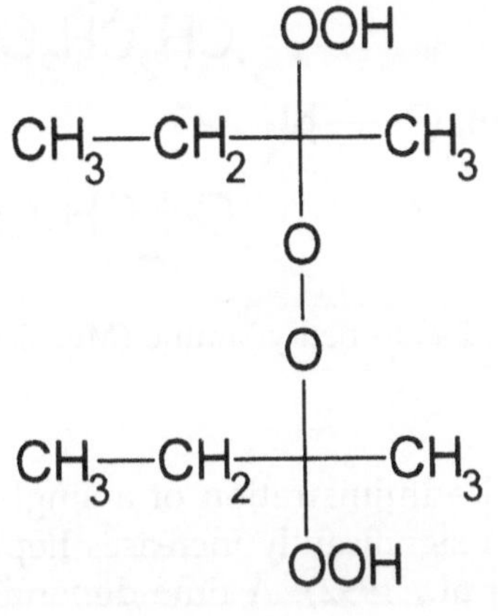

Fig. 4.22 Methyl ethyl ketone peroxide (2-Butanone peroxide)

that redox cycling and oxidative stress are involved in the toxicity of nitrofurantoin.

The cytotoxicity of alkylating antineoplastic drugs as well as the acute toxicity of neoplastic alkylating agents may be associated with the depletion of GSH and the induction of lipid peroxidation. The alkylating antineoplastic drugs mechloethamine (4.24), chlorambucil, cyclophosphamide, carmustine (BCNU) and lomustine (CCMU) rapidly deplete glutathione from isolated rat hepatocytes (Khan et al., 1992). These investigators have demonstrated that lipid peroxidation occurs following GSH depletion and prior to the onset of observable cytotoxicity. Furthermore, cytotoxicity can be delayed by the antioxidants BHA, vitamin E, the iron chelator DFX and the radical trapping agent 4-hydroxy-2,2,6,6-tetramethylpiperidine-N-oxyl (TEMPO).

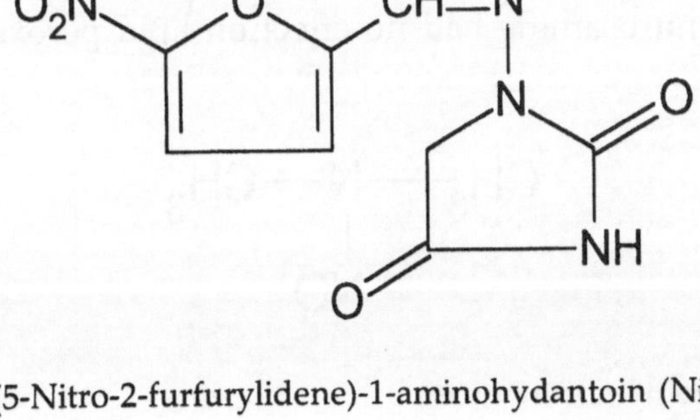

Fig. 4.23 N-(5-Nitro-2-furfurylidene)-1-aminohydantoin (Nitrofurantoin)

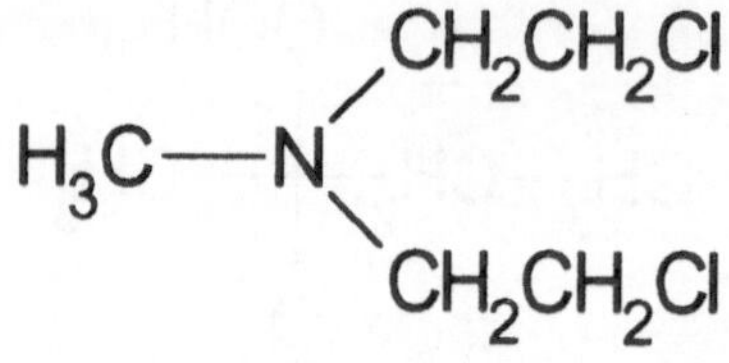

Fig. 4.24   N-Methylbis(2-chloroethyl)amine (Mechlorethamine)

The intraperitoneal administration of a single dose of cyclophos-phamide (200 mg/kg) significantly increases hepatic microsomal lipid peroxidation (Lear et al., 1992). A time-dependent increase in lipid peroxidation was observed. The administration of adriamycin (10 mg/kg) resulted in a 175% increase in rat hepatic microsomal lipid peroxidation seven days post-treatment (Lear et al., 1992). The sub-cutaneous injection of BCNU to rats produced dose- and time-de-pendent effects with respect to the inhibition of GR activity in lung and liver, and also induced hepatic lipid peroxidation (Nakagawa, 1987). The extent of lipid peroxidation in the lung as determined by MDA content was not affected by BCNU throughout three days of the study.

Nitrosamines are chemical carcinogens. The tumor promoting ca-pability of these agents may be linked to their ability to induce ox-idative stress. Dimethylnitrosamine (4.25) administered intraperi-toneally to Wistar rats at doses of 13.3 and 133 μmol/kg produced time-dependent increases in ethane exhalation, and time-dependent increases in hepatic lipid peroxidation as demonstrated by increases in conjugated dienes, fluorescent products, increased TBARS, and increased chemiluminescence (Ahotupa et al., 1987). Diethylnitro-samine also increased ethane exhalation and hepatic lipid peroxi-dation, whereas methylbenzylnitrosamine had no effect. Thus, he-patocarcinogenic nitrosamines induced lipid peroxidation while a non-carcinogenic nitrosamine had no effect on lipid peroxidation.

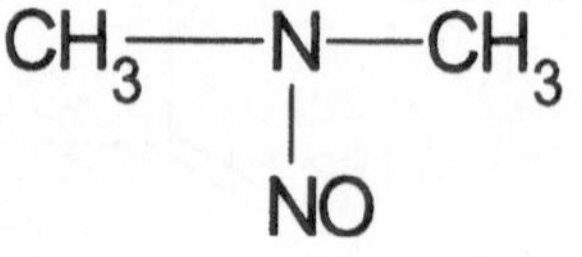

Fig. 4.25   Dimethylnitrosamine

Butyl-2-chloroethyl sulfide (butyl mustard, BCS) is a vesicant which produces local tissue injury apparently by alkylation and cross-linking of purine nucleotides and proteins. One and 24 hours after injection subcutaneously in mice, BCS produced significant decreases in kidney GSH levels (Omaye et al., 1991). Kidney GPOX activity was markedly increased while NADPH-isocitrate dehydrogenase activity decreased. The investigators concluded that these results were consistent with changes associated with oxidative stress or detoxification mechanisms for BCS.

ST789 is a novel immunomodulator characterized by an amino acidic group (L-arginine) joined to the N9 position of the hypoxanthine ring. This drug has been reported to protect immunosuppressed mice against tumor growth and microbial infections. Foresta et al. (1992) has shown that the in vitro addition of this compound to splenocytes from immunosuppressed mice markedly enhances phagocytic activity. Furthermore, this compound activates peritoneal exudate cells (primarily macrophages) as evidenced by the enhanced release of nitric oxide, and significantly increases the production of interleukin 1 and TNF. Superoxide anion and other ROS are implicated in the responses to this potent phagocyte activator.

4.11 SUMMARY AND CONCLUSIONS

The preceding discussion provides insight into the broad range of structurally dissimilar xenobiotics which are capable of inducing an oxidative stress in biological systems. The precise role of reactive oxygen species (ROS) in the toxicity of many xenobiotics is unclear, although it is apparent that ROS contribute to the overall toxic manifestations. With some xenobiotics, the production of an oxidative stress is an early event which leads to the compromise of antioxidant defense mechanisms, resulting in a cascading response resulting in cell damage and death. With respect to other xenobiotics, the production of ROS may be a late occurring event in the biological processes associated with toxicity, with the initial toxic insult being unrelated to the production of ROS. However, in such situations, the ultimate formation of ROS still may ultimately contribute to the final demise and disintegration of the cell.

Increasing evidence indicates that multiple mechanisms are involved in the production of ROS. Furthermore, a single xenobiotic may initiate formation of ROS by more than a single mechanism, involving more than one cell type or organelle. Phagocytic cells, and membranous fractions as mitochondria, microsomes and peroxisomes constitute major sources of production of ROS which include H_2O_2, O_2^-, ˙OH radical and $^1O_2$.. Most membranous fractions or cells may possess the potential for generating ROS. In addition, enzyme systems such as xanthine oxidase and myeloperoxidase, which are not associated with mitochondria or microsomes, are also active sources of ROS.

The involvement of ROS and/or other free radicals may be common to most if not all toxic mechanisms of most xenobiotics, although differences will exist in whether the formation of ROS is an early, intermediate or late event in the sequence of events leading to irreversible cell damage and death. Much information is still required before a clear understanding of the roles of ROS in the toxicity of xenobiotics, in general, is satisfactorily understood.

Pharmacokinetics (toxicokinetics) may explain many of the inconsistencies and contradictions which appear to exist with respect to the roles of ROS, lipid peroxidation, and antioxidants on the toxicities of various xenobiotics. The tissue distribution of xenobiotics, compartmentation within cells, microenvironments, localization of enzyme systems, and the cellular distribution of antioxidant defense mechanisms all contribute to our inability to establish generalized mechanisms for all xenobiotics. These considerations may explain why a xenobiotic is toxic in one tissue, as for example the lung, and not highly toxic in another tissue, as for example the liver. Toxicokinetic properties may also explain differences in toxicities of structurally related compounds as well as xenobiotics, which possess structurally similar moieties but apparently exhibit somewhat different mechanisms of toxicity. For example, xenobiotics which undergo redox cycling do not all seem to possess a single, well delineated mechanism of toxicity.

Information is rapidly accumulating regarding the roles of divalent cations as iron, copper and calcium in the cascade of events responsible for toxicity. The roles of iron and copper are becoming well recognized for their abilities to play a crucial role in the formation of ROS, while the role of calcium in the activation of various hydrolytic enzymes may be a critical but late occurring event in irreversible cell damage and death.

Evidence suggests that many lipid soluble xenobiotics can interact directly and in an unmetabolized state with plasma membranes to sufficiently alter membrane integrity, and thus enhance membrane permeability and alterations in calcium homeostasis. Furthermore, some xenobiotics, for example polyhalogenated cyclic hydrocarbons, can interact directly with various membranous fractions including mitochondria and microsomes resulting in the production of ROS. However, it is not clear whether the production of ROS occurs as a result of the physical presence of the xenobiotic in an unmetabolized form or a metabolic activation is required.

Our knowledge of the ability of a wide range of structurally dissimilar xenobiotics to induce oxidative stress has increased tremendously in recent years. However, numerous questions remain unanswered regarding ROS, their transport, tissue damaging effects, and relative roles in the mechanistic sequences associated with toxicity and carcinogenicity. The advent of new technologies will greatly aid in addressing these questions.

References

Abdel-Gayoum, A.A., Haider, S.S., Tash, F.M. and Ghawarsha, K. (1992) Effect of chloroquine on some carbohydrate metabolic pathways: Contents of GHS, ascorbate and lipid peroxidation in the rat. *Pharmacol. Toxicol.* **71**, 161–164.

Adam, A., Smith, L.L. and Cohen, G.M. (1990) An assessment of the role of redox cycling in mediating the toxicity of paraquat and nitrofurantoin. *Environ. Hlth. Persp.* **85**, 113–117.

Ahotupa, M., Bussacchini-Griot, V., Bereziat, Camus, A.M. and Bortsch, H. (1987) Rapid oxidative stress induced by N-nitrosamines. *Biochem. Biophys. Res. Commun.* **146**, 1047–1054.

Akubue, P.I. and Stohs, S.J. (1992) Endrin-induced production of nitric oxide by rat peritoneal macrophages. *Toxicol. Lett.* **62**, 311–316.

Akubue, P.I. and Stohs, S.J. (1993) The effect of alachlor on the urinary excretion of malondialdehyde, formaldehyde, acetaldehyde and acetone in serum of rats. *Bull. Environ. Contam. Toxicol.* **50**, 565–571.

Al-Bayati, Z.A.F. and Stohs, S.J. (1987) The role of iron in 2,3,7,8-tetrachlorodibenzo-p-dioxin (TCDD)-induced lipid peroxidation in rat liver microsomes. *Toxicol. Lett.* **38**, 115–121.

Alleman, M.A., Koster, J.F., Wilson, J.H.P., Edixhoven-Bosdijk, A., Slee, R.G., Kroos, M.J. and von Eijk, H.G. (1985) The involvement of iron and lipid peroxidation in the pathogenesis of HCB-induced porphyria. *Biochem. Pharmacol.* **34**, 161–166.

Alsharif, N.Z., Hassoun, E., Bagchi, M., Lawson, T. and Stohs, S.J. (1994) The effects of anti-TNF-alpha antibody and dexamethasone on TCDD-induced oxidative stress. *Pharmacology* **48**, 127–136.

Alsharif, N.Z., Lawson, T. and Stohs, S.J. (1990) TCDD-induced production of superoxide anion and DNA single strand breaks in peritoneal macrophage of rats. *The Toxicologist* **10**, 276.

Antholine, W.E., Kalyanaraman, B. and Petering, D.H. (1985) ESR of copper and iron complexes with antitumor and cytotoxic properties. *Environ. Hlth. Persp.* **64**, 19–35.

Athar, M., Hasan, S.K. and Srivastava, R.C. (1987) Evidence for the involvement of hydroxyl radicals in nickel mediated enhancement of lipid peroxidation: Implications for nickel carcinogenesis. *Biochem. Biophys. Res. Commun.* **147**, 1276–1281.

Aust, S.D. (1989) Metal ions, oxygen radicals and tissue damage. *Bibl. Nutr. Diet.* **43**, 266–277.

Bacon, B.R. and Britton, R.S. (1989) Hepatic injury in chronic iron overload. Role of lipid peroxidation. *Chem.-Biol. Interact.* **70**, 183–226.

Badr, M.Z., Ganey, P.E., Yoshihara, H., Kauffman, F.C. and Thurman, R.G. (1989) Hepatotoxicity of menadione predominates in oxygen-rich zones of the liver lobule. *J. Pharmacol. Exp. Therap.* **248**, 1317–1322.

Bagchi, D., Bagchi, M., Hassoun, E.A., Moser, J. and Stohs, S.J. (1992) Effects of carbon tetrachloride, menadione and paraquat on the urinary excretion of malondialdehyde, formaldehyde, acetaldehyde and acetone in rats. *J. Biochem. Toxicol.* **8**, 101–106.

Bagchi, D., Bagchi, M., Hassoun, E. and Stohs, S.J. (1992a) Endrin-induced urinary excretion of formaldehyde, acetaldehyde, malondialdehyde and acetone in rats. *Toxicology* **75**, 81–89.

Bagchi, D., Bagchi, M., Hassoun, E. and Stohs, S.J. (1992b) Effect of endrin on the hepatic distribution of iron and calcium in female Sprague-Dawley rats. *J. Biochem. Toxicol.* **7**, 37–42.

Bagchi, D., Dickson, P.H. and Stohs, S.J. (1992) The identification and quantitation of malondialdehyde, formaldehyde, acetaldehyde and acetone in serum of rats. *Toxicology Methods* **2**, 270–279.

Bagchi, M., Hassoun, E.A., Bagchi, D. and Stohs, S.J. (1993) Endrin-induced increases in hepatic lipid peroxidation, membrane microviscosity, and DNA damage in rats. *Arch. Environ. Contam. Toxicol.* **23**, 1–5.

Bagchi, M., Hassoun, E.A., Bagchi, D. and Stohs, S.J. (1993) Production of reactive oxygen species by peritoneal macrophages, and hepatic mitochondria and microsomes from endrin-treated rats. *Free Rad. Biol. Med.* **14**, 149–155.

Bagchi, M. and Stohs, S.J. (1993) *In vitro* induction of reactive oxygen species by 2,3,7,8-tetrachlorodibenzo-*p*-dioxin, endrin, and lindane in rat peritoneal macrophages, and hepatic mitochondria and microsomes. *Free Rad. Biol. Med.* **14**, 11–18.

Bailly, S., Fay, M., Roche, Y. and Gougerot-Pocidali, M.A. (1990) Effects of quinolones on tumor necrosis factor production by human monocytes. *Int. J. Immunopharmacol.* **12**, 31–36.

Bano, Y. and Hasan, M. (1989) Mercury induced time-dependent alterations in lipid profiles and lipid peroxidation in different body organs of cat-fish Heteropneustes fossilis. *J. Environ. Sci. Health B.* **24**, 145–166.

Barros, S.B.M., Simizu, K. and Junqueira, V.B.C. (1991) Liver lipid peroxidation-related parameters after short-term administration of hexachlorocyclohexane isomers to rats. *Toxicol. Lett.* **56**, 137–144.

Barros, S.B.M., Videla, L.A., Simizu, K., Halsema, L.V. and Junqueira, V.B.C. (1988) Lindane-induced oxidative stress. II. Time course of changes in hepatic glutathione status. *Xenobiotica* **18**, 1305–1310.

Bellomo, G., Mirabelli, F. and Vairetti, M., Monte, D., Richemi, P, (1990) Cytoskeleton as a target in menadione-induced oxidative stress in cultured mammalian cells. I. Biochemical and immunocytochemical features. *J. Cell. Physiol.* **143**, 118–128.

Bellomo, G. and Orrenius, S. (1985) Altered thiol and calcium homeostasis in oxidative hepato cellular injury. *Hepatology* **5**, 876–882.

Benov, L.C., Benchev, I.C. and Monovich, O.H. (1990) Thiol antidotes effect on lipid peroxidation in mercury-poisoned rats. *Chem.-Biol. Interact.* **76**, 321–332.

Berson, A., Wolf, C., Berger, V., Fau, D., Chachaty, C., Fromenty, B. and Pessayre, D. (1991) Generation of free radicals during the reductive metabolism of the nitroaromatic compound, nilutamide. *J. Pharmacol. Exp. Therap.* **257**, 714–719.

Bettger, W.J. and O'Dell, B.L. (1981) A critical physiological role of zinc in the structure and function of biomembranes. *Life Sci.* **28**, 1425–1438.

Birnboim, H.C. (1982) Factors which affect DNA strand breakage in human leukocytes exposed to a tumor promoter, phorbol myristate acetate. *Can. J. Physiol. Pharmacol.* **60**, 1359–1366.

Blumberg, P.M. (1988) Protein kinase C as the receptor for the phorbol ester tumor promoters: Sixth Rhoads Memorial Award Lecture. *Cancer Res.* **48**, 1–8.

Bombick, D.W. and Matsumura, F. (1987) 2,3,7,8-Tetrachlorodibenzo-p-dioxin causes elevation of the levels of the protein tyrosine kinase pp60^{c-src}. *J. Biochem. Toxicol.* **2**, 141–154.

Bray, T.M. and Bettger, W.J. (1990) The physiological role of zinc as an antioxidant. *Free Rad. Biol. Med.* **8**, 281–291.

Bray, T.M., Kubow, S. and Bettger, W.J. (1986) Effect of dietary zinc on endogenous free radical production in rat lung microsomes. *J. Nutr.* **116**, 1054–1060.

Brown, M.A., Kimmel, E.C. and Casida, J.E. (1988) DNA adduct formation by alachlor metabolites. *Life Sci.* **43**, 2087–2094.

Burke, J.P. and Fenton, M.R. (1985) Effect of a zinc-deficient diet on lipid peroxidation in liver and tumor subcellular membranes. *Proc. Soc. Exp. Biol. Med.* **179**, 187–191.

Bus, J.S., Cagen, M., Olgaard, M. and Gibson, J.E. (1976) A mechanism of paraquat toxicity in men and mice. *Toxicol. Appl. Pharmacol.* **35**, 501–513.

Carini, R., Parola, M., Dianzani, M.U. and Albano, E. (1992) Mitochondrial damage and its role in causing hepatocyte injury during stimulation of lipid peroxidation by iron nitriloacetate. *Arch. Biochem. Biophys.* **297**, 110–118.

Cerutti, P.A. (1989) Response modification in carcinogenesis. *Environ. Hlth. Persp.* **81**, 39–43.

Chan, P.C., Peller, O.G. and Kesner, L. (1982) Copper(II)-catalyzed lipid peroxidation in liposomes and erythrocyte membrane. *Lipids* **17**, 331–337.

Chao, C.C. and Aust, A.E. (1993) Photochemical reduction of ferric iron by chelators results in DNA strand breaks. *Arch. Biochem. Biophys.* **300**, 544–550.

Chubatsu, L.S., Gennari, M. and Meneghini, R. (1992) Glutathione is the antioxidant responsible for the resistance to oxidative stress in V79 Chinese hamster fibroblasts rendered resistant to cadmium. *Chem.-Biol. Interact.* **82**, 99–110.

Clark, G.C., Taylor, M.J., Tritscher, A.M. and Lucier, G.W. (1991) Tumor necrosis factor involvement in 2,3,7,8-tetrachlorodibenzo-p-dioxin-mediated endotoxin hypersensitivity in C57BL/6J mice congenic at the Ah locus. *Toxicol. Appl. Pharmacol.* 111, 422–431.

Combs, G.F., Jr. and Peterson, F.J. (1983) Protection against acute paraquat toxicity by dietary selenium in the chick. *J. Nutr.* 113, 538–545.

Comporti, M. (1989) Three models of free radical-induced cell injury. *Chem.-Biol. Interact.* 72, 1–56.

Cook, J.C., Gaido, K.W. and Greenlee, W.F. (1987) Ah receptor: Relevance of mechanistic studies to human risk assessment. *Environ. Hlth. Persp.* 76, 71–77.

Cook, J.C. and Greenlee, W.F. (1989) Characterization of a specific binding protein for 2,3,7,8-tetrachlorodibenzo-p-dioxin in human thymic epithelial cells. *Mol. Pharmacol.* 35, 713–719.

Cox, J.R. and Volp, R.F. (1993) Liver microsomal lipid peroxidation induced by chloropropanes. *The Toxicologist* 13, 335.

Cuthill, S., Wilhelmsson, A., Mason, G.G.F., Gillner, M., Poellinger, L. and Gustafson, J.A. (1988) The dioxin receptor: A comparison with the glucocorticoid receptor. *J. Steroid Biochem.* 30, 277–280.

D'Alessandro, N., Rausa, L. and Crescimanno, M. (1988) *In vivo* effects of doxorubicin and isoproterenol on reduced glutathione and hydrogen peroxide production in mouse heart. *Res. Commun. Chem. Pathol. Pharmacol.* 62, 19–30.

Danielsson, B.R.G., Hassoun, E. and Dencker, L. (1982) Embryo toxicity of chromium: Distribution in pregnant mice and effects on embryonic cells *in vitro*. *Arch. Toxicol.* 51, 233–245.

Danni, O., Chiarpotto, E. Aragno, M., Biasi, F., Comoglio, A., Belliardo, F., Dianzani, M.U. and Poli, G. (1991) Lipid peroxidation and irreversible cell damage: Synergism between carbon tetrachloride and 1,2-dibromoethane in isolated rat hepatocytes. *Toxicol. Appl. Pharmacol.* 110, 216–222.

Dargel, R. (1992) Lipid peroxidation—a common pathogenetic mechanism? *Exp. Toxic. Pathol.* 44, 169–181.

DiMonte, D., Sandy, M.S., Ekstrom, G. and Smith, M.T. (1986) Comparative studies on the mechanisms of paraquat and 1-methyl-4-phenylpyridine (MPP$^+$) cytotoxicity. *Biochem. Biophys. Res. Commun.* 137, 303–309.

Doroshow, J.H. (1983) Effect of anthracycline antibiotics on oxygen radical formation in rat heart. *Cancer Res.* 43, 4543–4551.

Duthie, S.J. and Grant, M.H. (1989) The toxicity of menadione and mitozantrone in human liver-derived HEP G2 hepatoma cells. *Biochem. Pharmacol.* 38, 1247–1255.

Elliott, A.J. and Pardini, R.S. (1988) Modulation of hepatic cytochrome P-450 and DT-diaphorase by oral and sub-cutaneous administration of the pro-oxidant fungicide dichlone (2,3-dichloro-1,4-naphthoquinone). *Bull. Environ. Contam. Toxicol.* 41, 164–171.

El Sisi, A.E.D., Earnest, D.L. and Sipes, I.G. (1993a) Vitamin A potentiation of carbon tetrachloride hepatotoxicity: Enhanced lipid peroxidation without enhanced biotransformation. *Toxicol. Appl. Pharmacol.* 119, 289–294.

El Sisi, A.E.D., Earnest, D.L. and Sipes, I.G. (1993b) Vitamin A potentiation of carbon tetrachloride hepatotoxicity: Role of liver macrophages and active oxygen species. *Toxicol. Appl. Pharmacol.* **119**, 295–301.

El Sisi, A.E.D., Hall, P., Sim, W.L.W., Earnest, D.L. and Sipes, I.G. (1993) Characterization of vitamin A potentiation of carbon tetrachloride-induced liver injury. *Toxicol. Appl. Pharmacol.* **119**, 280–288.

Emerit, I. and Chance, B. (eds.) (1992) *Free Radicals and Aging.* Birkhauser Verlag, Basel.

Fariss, M.W. (1991) Cadmium toxicity: Unique cytoprotective properties of alpha tocopheryl succinate in hepatocytes. *Toxicology* **69**, 63–77.

Fau, D., Berson, A., Eugene, D., Fromenty, B., Fisch, C. and Pessayre, D. (1992) Mechanism for the hepatotoxicity of the antiandrogen, nilutamide. Evidence suggesting that redox cycling of this nitroaromatic drug leads to oxidative stress in isolated hepatocytes. *J. Pharmacol. Exp. Therap.* **263**, 69–77.

Fischer, S.M., Cameron, G.S., Baldwin, J.K., Jashaway, D.W. and Patrick, K.E. (1988) Reactive oxygen in the tumor promotion stage of skin carcinogenesis. *Lipids* **23**, 592–597.

Fisher, G.R. and Gutierrez, P.L. (1991) Free radical formation and DNA strand breakage during metabolism of diaziquone by NAD(P)H quinone-acceptor oxidoreductase (DT-diaphorase) and NADPH cytochrome c reductase. *Free Rad. Biol. Med.* **11**, 597–607.

Foresta, P., Ruggiero, V., Albertoni, C., Leoni, B. and Arrigoni-Martelli B. (1992) *In vitro* activation of murine peritoneal exudate cells (PEC) and peritoneal macrophages by ST 789. *Int. J. Immunopharmacol.* **14**(6), 1061–1068.

Fraga, C.G., Leibovitz, B.E. and Tappel, A.L. (1987) Halogenated compounds as inducers of lipid peroxidation in tissue slices. *Free Rad. Biol. Med.* **3**, 119–123.

Frenkel, K. (1989) Oxidation of DNA bases by tumor promoter-activated processes. *Environ. Hlth. Persp.* **81**, 45–54.

Frenkel, K. (1992) Carcinogen-mediated oxidant formation and oxidative DNA damage. *Pharmacol. Therap.* **53**, 127–166.

Fukino, H., Hirai, M., Hsueh, Y.M. and Yamane, Y. (1984) Effect of zinc pretreatment on mercuric chloride-induced lipid peroxidation in the rat kidney. *Toxicol. Appl. Pharmacol.* **73**(3), 395–401.

Fukuda, F., Kitada, M., Horie, T. and Awazu, S. (1992) Evaluation of adriamycin-induced lipid peroxidation. *Biochem. Pharmacol.* **44**, 755–760.

Gage, J.C. (1968) The action of paraquat and diquat on the respiration of liver cell fractions. *Biochem. J.* **109**, 757–761.

Ganey, P.E., Carter, L.S., Mueller, R.A. and Thurman, R.G. (1991) Doxorubicin toxicity in perfused rat heart. Decreased cell death at low oxygen tension. *Circ. Res.* **68**, 1610–1613.

Gaudry, M., Combadiere, C., Marquetty, C. and Hakim, J. (1990) A comparison of the priming effect of phorbol myristate acetate and phorbol dibutyrate in fMet-Leu-Phe-induced oxidative burst in human neutrophils. *Immunopharmacol.* **20**, 45–56.

Goel, M.R., Shara, M. and Stohs, S.J. (1988) Induction of lipid peroxidation by hexachlorocyclohexane, dieldrin, TCDD, carbon tetrachloride and hexachlorobenzene. *Bull. Environ. Contam. Toxicol.* **40**, 255–262.

Gribble, G.W. (1992) Naturally occurring organohalogen compounds—a survey. *J. Nat. Prod.* **55**(10), 1353–1395.

Griffin-Green, E.A., Zaleska, M.M. and Erecinska, M. (1988) Adriamycin-induced lipid peroxidation in mitochondria and microsomes. *Biochem. Pharmacol.* **37**, 3071–3077.

Gstraunthaler, G. Pfaller, W. and Kotanko, P. (1983) Glutathione depletion and *in vitro* lipid peroxidation in mercury or maleate induced acute renal failure. *Biochem. Pharmacol.* **32**(19), 2969–2972.

Gutteridge, J.M.C. (1984) Tissue damage by oxy-radicals: The possible involvement of iron and copper complexes. *Med. Biol.* **62**, 101–104.

Halliwell, B. and Gutteridge, J.M.C. (1986) Iron and free radical reactions: Two aspects of antioxidant protection. *Trends Biochem. Sci.* **11**, 372–375.

Hammermueller, J.D., Bray, T.M. and Bettger, W.J. (1987) Effect of zinc and copper deficiency on microsomal NADPH-deficient active oxygen generation in rat lung and liver. *J. Nutr.* **116**, 894–901.

Harvey, M.J. and Klaassen, C.D. (1983) Interaction of metals and carbon tetrachloride on lipid peroxidation and hepatotoxicity. *Toxicol. Appl. Pharmacol.* **71**, 316–321.

Hassan, M.Q., Numan, I.T., Al-Nasiri, N. and Stohs, S.J. (1991) Endrin-induced histopathological changes and lipid peroxidation in livers and kidneys of rats, mice, guinea pigs and hamsters. *Toxicol. Pathol.* **19**, 108–114.

Hassoun, E., Bagchi, M., Bagchi, D. and Stohs, S.J. (1993) Comparative studies on lipid peroxidation and DNA single strand breaks induced by lindane, DDT, chlordane and endrin in rats. *Comp. Biochem. Physiol.* **104C**, 427–431.

Hirai, K.-I., Ikeda, K. and Wang, G.-Y. (1992) Paraquat damage of rat liver mitochondria by superoxide production depends on extramitochondrial NADH. *Toxicology*, **72**, 1–16.

Horvath, E., Levay, G., Pongracz, K. and Bodell, W.J. (1992) Peroxidative activation of O-phenylhydroquinone leads to the formation of DNA adducts in HL-60 cells. *Carcinogenesis* **13**, 1937–1939.

Hudecova, A. and Ginter, E. (1992) The influence of ascorbic acid on lipid peroxidation in guinea pigs intoxicated with cadmium. *Fd. Chem. Toxic.* **30**, 1011–1013.

Hussain, T., Shukla, G.S. and Chandra, S.F. (1987) Effects of cadmium on superoxide dismutase and lipid peroxidation in liver and kidney of growing rats: *In vivo* and *in vitro* studies. *Pharmacol. Toxicol.* **60**, 355–358.

Imlay, J.A., Chin, S.M. and Linn, S. (1988) Toxic DNA damage by hydrogen peroxide through the Fenton reaction *in vivo* and *in vitro*. *Science* **240**, 640–642.

Jackson, J.A., Reeves, J.P., Muntz, K.H., Kruk, D., Prough, R.A., Willerson, J.T. and Buja, L.M. (1984) Evaluation of free radical effects and catecholamine alterations in adriamycin cardiotoxicity. *Am. J. Pathol.* **117**, 140–153.

Jones, P., Kortenkamp, A., O'Brien, P., Wang, G. and Yang, G. (1991) Evidence for the generation of hydroxyl radicals from a chromium(V) intermediate isolated from the reaction of chromate with glutathione. *Arch. Biochem. Biophys.* **286**, 652–655.

Juhl, U., Blum, J.K., Butte, W. and Witte, I. (1991) The induction of DNA strand breaks and formation of semiquinone radicals by metabolites of 2,4,5-trichlorophenol. *Free Rad. Res. Commun.* **11**(6), 295–305.

Junqueira, V.B.C., Simizu, K., Pimentel, R., Azzalis, L.A., Barros, S.B.M., Koch, O. and Videla, L.A. (1991) Effect of phenobarbital and 3-methylcholanthrene on the early oxidative stress component induced by lindane in rat liver. *Xenobiotica* **21**(8), 1053–1063.

Junqueira, V.B.C., Simizu, K., Van Halsema, L.V., Koch, O.R., Barros, S.B.M. and Videla, L.A. (1988) Lindane-induced oxidative stress. I. Time course of changes in hepatic microsomal parameters, antioxidant enzymes, lipid peroxidative indices and morphological characteristics. *Xenobiotica* **18**(11), 1297–1304.

Kahl, R., Weinke, S. and Kappus, H. (1989) Production of reactive oxygen species due to metabolic activation of butylated hydroxyanisole. *Toxicology*, **59**, 179–194.

Kappus, H. and Sies, H. (1981) Toxic drug effects associated with oxygen metabolism: Redox cycling and lipid peroxidation. *Experientia* **37**, 1233–1241.

Kawanishi, S., Inoue, S. and Sano, S. (1986) Mechanism of DNA cleavage induced by sodium chromate (VI) in the presence of hydrogen peroxide. *J. Biol. Chem.* **261**, 5952–5958.

Khan, S., Ramwani, J.J. and O'Brien, P.J. (1992) Hepatocyte toxicity of mechlorethamine and other alkylating anticancer drugs. Role of lipid peroxidation. *Biochem. Pharmacol.* **43**, 1963–1967.

Kitazawa, Y., Matsubara, M., Takeyama, N. and Tanaka, T. (1991) The role of xanthine oxidase in paraquat intoxication. *Arch. Biochem. Biophys.* **288**(1), 220–224.

Kleber, E., Kroner, R. and Elstner, E.F. (1991) Cataract induction by 1,2,-naphthoquinone. I. Studies on the redox properties of bovine lens proteins. *Z. Naturforsch.* **46c**, 280–284.

Kociba, R.J. and Schwetz, B.A. (1982) Toxicology of 2,3,7,8-tetrachlorodibenzo-p-dioxin (TCDD). *Drug Metab. Rev.* **13**, 387–406.

Koizumi, T. and Li, Z.G. (1992) Role of oxidative stress in single-dose, cadmium-induced testicular cancer. *J. Toxicol. Environ. Hlth.* **37**(1), 25–36.

Lambert, L.E., Whitten, J.P., Baron, B.M., Cheng, H.C., Doherty, M.S. and McDonald, I.A. (1991) Nitric oxide synthesis in the CNS, endothelium and macrophages differs in its sensitivity to inhibition by arginine analogues. *Life Sci.* **48**, 69–75.

Lauriault, V.V., McGirr, L.G., Wong, W.W.C. and O'Brien, P.J. (1990) Modulation of benzoquinone-induced cytotoxicity by diethyldithiocarbamate in isolated hepatocytes. *Arch. Biochem. Biophys.* **282**, 26–33.

Lear, L., Nation, R.L. and Stupans, I. (1992) Effects of cyclophosphamide and adriamycin on rat hepatic microsomal glucuronidation and lipid peroxidation. *Biochem. Pharmacol.* **44**, 747–753.

LeBel, C.P., Ali, S.F. and Bondy, S.C. (1992) Deferoxamine inhibits methyl mercury-induced increases in reactive oxygen species formation in rat brain. *Toxicol. Appl. Pharmacol.* **112**(1), 161–165.

Li, Y. and Trush, M.A. (1993a) Oxidation of hydroquinone by copper: Chemical mechanism and biological effects. *Arch. Biochem. Biophys.* **300**, 346–355.

Li, Y. and Trush, M.A. (1993b) Reactive oxygen-mediated DNA damage initiated by copper-dependent hydroquinone (HQ) oxidation. *The Toxicologist*, **13**, 338.

Ludewig, G., Dogra, S. and Glatt, H. (1989) Genotoxicity of 1,4-benzoquinone and 1,4-naphthoquinone in relation to effects on glutathione and NAD(P)H levels in V79 cells. *Environ. Hlth. Persp.*, **82**, 223–228.

Lund, B.O., Miller, D.M. and Woods, J.S. (1991) Mercury-induced H_2O_2 production and lipid peroxidation *in vitro* in rat kidney mitochondria. *Biochem. Pharmacol.* **42**(S), S181–S187.

Manca, D., Ricard, A.C., Trottier, B. and Chevalier, G. (1991) Studies on lipid peroxidation in rat tissues following administration of low and moderate doses of cadmium chloride. *Toxicology* **67**(3), 303–323.

Masuda, Y. and Nakamura, Y. (1990) Effects of oxygen deficiency and calcium omission on carbon tetrachloride hepatotoxicity in isolated perfused livers from phenobarbital-pretreated rats. *Biochem. Pharmacol.* **40**(8), 1865–1876.

Matsumura, F. (1983) Biochemical aspects of action mechanisms of 2,3,7,8-tetra-chlorodibenzo-p-dioxin (TCDD) and related chemicals in animals. *Pharmacol. Therap.* **19**, 195–209.

McConkey, D.J., Harzell, P., Duddy, S.K., Hahansson, H. and Orrenius, S. (1989) 2,3,7,8-Tetrachlorodibenzo-p-dioxin kills immature thymocytes by Ca^{2+}-mediated endonuclease activation. *Science* **242**, 256–259.

Min, K.-S., Terano, Y., Onosaka, S. and Tanaka, K. (1992) Induction of metallo-thionein synthesis by menadione or carbon tetrachloride is independent of free radical production. *Toxicol. Appl. Pharmacol.* **113**, 74–79.

Minotti, G. (1990) NADPH- and adriamycin-dependent microsomal release of iron and lipid peroxidation. *Arch. Biochem. Biophys.* **277**, 268–276.

Minotti, G. and Aust, S.D. (1987) The role of iron in the initiation of lipid peroxidation. *Chem. Phys. Lipids* **44**, 191–208.

Miura, T., Muraoka, S. and Ogiso, T. (1991) Lipid peroxidation of rat erythrocyte membrane induced by adriamycin-Fe^{3+}. *Pharmacol. Toxicol.* **69**, 296–300.

Morehouse, L.A. and Aust, S.D. (1988) Reconstituted microsomal lipid peroxidation: ADP-Fe^{+3}-dependent peroxidation of phospholipid vesicles containing NADPH-cytochrome P-450 reductase and cytochrome P-450. *Free Rad. Biol. Med.* **4**, 269–277.

Morel, I., Lescoat, G., Cillard, J., Pasdeloup, N., Brissot, P. and Cillard, P. (1990) Kinetic evaluation of free malondialdehyde and enzyme leakage as indices of iron damage in rat hepatocyte cultures. *Biochem. Pharmacol.* **39**(11), 1647–1655.

Muller, L. (1986) Consequences of cadmium toxicity in rat hepatocytes: Mitochondrial dysfunction and lipid peroxidation. *Toxicology*, **40**, 285–292.

Murphy, S.D. (1986) Toxic effects of pesticides. In *Toxicology: The Basic Science of Poisons*, (M.O. Amdur, J. Doull, and C.D. Klaassen, eds.), McMillan Publishing Company, New York, pp. 519–581.

Nakagawa, Y. (1987) Effects of 1,3-Bis(2-chloroethyl)-1-nitrosourea (BCNU) on the levels of glutathione and lipid peroxidation and the activity of glutathione reductase in liver and lung. *Toxicol. Lett.* **35**, 269–275.

Nebert, D.W., Petersen, D.D. and Fornace, A.J., Jr. (1990) Cellular responses to oxidative stress: The [Ah] gene battery as a paradigm. *Environ. Health Perspect.* **88**, 13–25.

Numan, I.T., Hassan, M.O. and Stohs, S.J. (1990a) Protective effects of antioxidants against endrin-induced lipid peroxidation, glutathione depletion, and lethality in rats. *Arch. Environ. Contam. Toxicol.* **19**, 302–306.

Numan, I.T., Hassan, M.Q. and Stohs, S.J. (1990b) Endrin-induced depletion of glutathione and inhibition of glutathione peroxidase activity in rats. *Gen. Pharmacol.* **21**, 625–628.

Ochi, T., Takahashi, K. and Ohsawa, M. (1987) Indirect evidence for the induction of a prooxidant state by cadmium chloride in cultured mammalian cells and a possible mechanism for the induction. *Mutat. Res.* **180**, 257–266.

Omaye, S.T., Elsayed, N.M., Klain, G.J. and Korte, D.W. (1991) Metabolic changes in the mouse kidney after subcutaneous injection of butyl 2-chloroethyl sulfide. *J. Toxicol. Environ. Hlth.* **33**, 19–27.

Orrenius, S., McCabe, M.J. and Nicotera, P. (1992) Ca^{2+}-dependent mechanisms of cytotoxicity and programmed cell death. *Toxicol. Lett.* **64-65**, 357–364.

Ozawa, T. and Hanaki, A. (1990) Spin-trapping studies on the reactions of Cr(III) with hydrogen peroxide in the presence of biological reductants: Is Cr(III) non-toxic? *Biochem. Int.* **22**, 343–352.

Paraidathathu, T., De Groot, H. and Kehrer, J.P. (1992) Production of reactive oxygen by mitochondria from normoxic and hypoxic rat heart tissue. *Free Rad. Biol. Med.* **13**, 289–297.

Parola, M., Leonarduzzi, G., Biasi, F., Albano, E., Biocca, M.E., Poli, G. and Dianzani, M.U. (1992) Vitamin E dietary supplementation protects against carbon tetrachloride-induced chronic liver damage and cirrhosis. *Hepatology* **16**, 1014–1021.

Perchellet, E.M., Abney, N.L. and Perchellet, J.-P. (1988) Stimulation of hydroperoxide generation in mouse skins treated with tumor-promoting or carcinogenic agents *in vivo* and *in vitro*. *Cancer Lett.* **42**, 169–177.

Perchellet, E.M., Jones, D. and Perchellet, J.-P. (1990) Ability of the Ca^{2+} ionophores A23187 and ionomycin to mimic some of the effects of the tumor promoter 12-O-tetradecanoylphorbol-13-acetate on hydroperoxide production, ornithine decarboxylase activity, and DNA synthesis in mouse epidermis *in vivo*. *Cancer Res.* **50**, 5806–5812.

Perchellet, J.P., Kishore, G.S. and Perchellet, E.M. (1985) Enhancement by adriamycin of the effects of 12-O-tetradecanoylphorbol-13-acetate on mouse epidermal glutathione peroxidase activity, ornithine decarboxylase induction and skin tumor promotion. *Cancer Lett.* **29**, 127–137.

Peter, B., Wartena, M., Kampinga, H.H. and Konigs, A.W.T. (1992) Role of lipid peroxidation and DNA damage in paraquat toxicity and the interaction of paraquat with ionizing radiation. *Biochem. Pharmacol.* **43**, 705–715.

Poland, A. and Knutson, J.C. (1982) 2,3,7,8-Tetrachlorodibenzo-p-dioxin and related halogenated aromatic hydrocarbons: Examination of the mechanism of toxicity. *Ann. Rev. Pharmacol. Toxicol.* **22**, 517–554.

Powell, S.R. and McCay, P.B. (1988) Inhibition of doxurubicin-initiated membrane damage by N-acetylcysteine: Possible mediation by a thiol-dependent, cytosolic inhibitor of lipid peroxidation. *Toxicol. Appl. Pharmacol.* **96**, 175–184.

Pritsos, C.A., Jensen, D.E., Pisani, D. and Pardini, R.S. (1982) Involvement of superoxide in the interaction of 2,3-dichloro-1,4-naphthoquinone with mitochondrial membranes. *Arch. Biochem. Biophys.* **217**, 98–109.

Pritsos, C.A. and Pardini, R.S. (1983) A role for glutathione disulfide as a scavenger of oxygen radicals produced by 2,3-dichloro-1,4-naphthoquinone. *Res. Commun. Chem. Pathol. Pharmacol.* **42**, 271–280.

Pritsos, C.A. and Pardini, R.S. (1984) A redox cycling mechanism of action for 2,3-dichloro-1,4-naphthoquinone with mitochondrial membranes and the role of sulfhydryl groups. *Biochem. Pharmacol.* **33**, 3771–3777.

Pritsos, C.A., Sokoloff, M. and Gustafson, D.L. (1992) PZ-51 (Ebselen) *in vivo* protection against adriamycin-induced mouse cardiac and hepatic lipid peroxidation and toxicity. *Biochem. Pharmacol.* **44**, 839–841.

Puntarulo, S. and Cederbaum, A.I. (1989) Interactions between paraquat and ferric complexes in the microsomal generation of oxygen radicals. *Biochem. Pharmacol.* **38**, 2911–2918.

Recknagel, R.O., Glende, Jr., E.A. and Dolak, J.A. (1989) Mechanisms of carbon tetrachloride toxicity. *Pharmac. Therap.* **43**, 139–154.

Robertson, F.M., Beavis, A.J., Oberyszyn, T.M., O'Connel, S.M., Dokidos, A., Laskin, D.L., Laskin, J.D. and Ruiners , J.J. (1990) Production of hydrogen peroxide by murine epidermal keratinocytes following treatment with the tumor promoter 12-*O*-tetradecanoylphorbol-13-acetate. *Cancer Res.* **50**, 6062–6067.

Rosenthal, R.E., Chanderbhan, R., Marshall, G. and Fiskum, G. (1992) Prevention of post-ischemic brain lipid conjugated diene production and neurological injury by hydroxyethyl starch-conjugated deferoxamine. *Free Rad. Biol. Med.* **12**, 29–33.

Rossi, L., Moore, G.A., Orrenius, S. and O'Brien, P.J. (1986) Quinone toxicity in hepatocytes without oxidative stress. *Arch. Biochem. Biophys.* **251**, 25–35.

Rungby, J. and Ernst, E. (1992) Experimentally induced lipid peroxidation after exposure to chromium, mercury or silver: Interactions with carbon tetrachloride. *Pharmacol. Toxicol.* **70**, 205–207.

Ryan, T.P. and Aust, S.D. (1992) The role of iron in oxygen-mediated toxicities. *Crit. Rev. Toxicol.* **22**, 119–141.

Safe, S.H. (1988) The aryl hydrocarbon (Ah) receptor. *ISI Atlas Sci. Pharmacol.*, 78–83.

Sandy, M.S., Moldeus, P., Ross, D. and Smith, M.T. (1986) Role of redox cycling and lipid peroxidation in bipyridal herbicide cytotoxicity. Studies with a compromised isolated hepatocyte model system. *Biochem. Pharmacol.* **35**, 3095–3101.

Sandy, M.S., Moldeus, P., Ross, D. and Smith, M.T. (1987) Cytotoxicity of the redox cycling compound diquat in isolated hepatocytes: Involvement of hydrogen peroxide and transition metals. *Arch. Biochem. Biophys.* **259**, 29–37.

Sarafian, T. and Verity, M.A. (1991) Oxidative mechanisms underlying methyl mercury neurotoxicity. *Int. J. Dev. Neurosci.* **9**, 147–153.

Sata, T., Kubota, E., Misra, H.P., Mojarad, M., Pakbaz, H. and Said, S.J. (1992) Paraquat-induced lung injury: Prevention by N-*tert*-butyl α-phenylnitrone, a free radical spin-trapping agent. *Am. J. Physiol.* **262**, L147–L152.

Shara, M.A., Dickson, P.H., Bagchi, D. and Stohs, S.J. (1992) Excretion of formaldehyde, malondialdehyde, acetaldehyde and acetone in the urine of rats in response to 2,3,7,8-tetrachlorodibenzo-*p*-dioxin, paraquat, endrin and carbon tetrachloride. *J. Chromatogr.* **576**, 221–233.

Shertzer, H.G., Lastbom, L., Sainsbury, M. and Muldeus, P. (1992) Menadione-mediated membrane fluidity alterations and oxidative damage in rat hepatocytes. *Biochem. Pharmacol.* **43**, 2135–2141.

Shi, X. and Dalal, N.S. (1989) Chromium (V) and hydroxyl radical formation during the glutathione reductase-catalyzed reduction of chromium (VI). *Biochem. Biophys. Res. Commun.* **163**, 627–634.

Shi, X. and Dalal, N.S. (1990) On the hydroxyl radical formation in the reaction between hydrogen peroxide and biologically generated chromium(V) species. *Arch. Biochem. Biophys.* **277**, 342–350.

Shi, X., Dalal, N.S. and Kasprzak, K.S. (1992) Generation of free radicals from lipid hydroperoxides by NI^{2+} in the presence of oligopeptides. *Arch. Biochem. Biophys.* **299**, 154–162.

Smith, A.G., Francis, J.E. and Carthew, P. (1990) Iron as a synergist for hepatocellular carcinoma induced by polychlorinated biphenyls in Ah-responsive C57BL/10ScSn mice. *Carcinogenesis* **11**, 437–444.

Smith, C.V. (1987a) Effect of BCNU pretreatment on diquat-induced oxidant stress and hepatotoxicity. *Biochem. Biophys. Res. Commun.* **144**, 415–421.

Smith, C.V. (1987b) Evidence for participation of lipid peroxidation and iron in diquat-induced hepatic necrosis *in vivo*. *Mol. Pharmacol.* **32**, 417–422.

Smith, C.V., Hughes, H., Lauterburg, B.H. and Mitchell, J.R. (1985) Oxidant stress and hepatic necrosis in rats treated with diquat. *J. Pharmacol. Exper. Therap.* **235**, 172–177.

Smith, L.L. (1987) Mechanism of paraquat toxicity in lung and its relevance to treatment. *Human Toxicol.* **6**, 31–36.

Solveig Walles, S.A. (1992) Mechanisms of DNA damage induced in rat hepatocytes by quinones. *Cancer Lett.* **63**, 47–52.

Sridhar, R., Dwivedi, C., Anderson, J., Baker, P.B., Sharma, H.M., Desai, P. and Engineer, F.N. (1992) Effects of verapamil on the acute toxicity of doxorubicin *in vivo*. *J. Natl. Cancer Inst.* **84**, 1653–1660.

Stacey, N.H. and Kappus, H. (1982) Comparison of methods of assessment of metal-induced lipid peroxidation in isolated rat hepatocytes. *J. Toxicol. Environ. Hlth.* **9**, 277–281.

Stacey, N.H. and Klaassen, C.D. (1981) Inhibition of lipid peroxidation without prevention of cellular injury in isolated rat hepatocytes. *Toxicol. Appl. Pharmacol.* **58**, 8–15.

Stinson, T.J., Jaw, S., Jeffery, E.H. and Plewa, M.J. (1992) The relationship between nickel chloride-induced peroxidation and DNA strand breakage in rat liver. *Toxicol. Appl. Pharmacol.* **117**, 98–103.

Stohs, S.J. (1990) Oxidative stress induced by 2,3,7,8-tetrachlorodibenzo-*p*-dioxin (TCDD). *Free Rad. Biol. Med.* **9**, 79–90.

Stohs, S.J., Shara, M.A., Alsharif, N.Z., Wahba, Z.Z. and Al-Bayati, Z.Z. (1990) 2,3,7,8,-Tetrachlorodibenzo-*p*-dioxin (TCDD)-induced oxidative stress in female rats. *Toxicol. Appl. Pharmacol.* **106**, 126–135.

Suda, I., Totoki, S. and Takahashi, H. (1991) Degradation of methyl and ethyl mercury into inorganic mercury by oxygen free radical-producing systems: Involvement of hydroxyl radical. *Arch. Toxicol.* **65**, 129–134.

Sugden, K.D., Geer, R.D. and Rogers, S.J. (1992) Oxygen radical-mediated DNA damage by redox-active Cr(III) complexes. *Biochemistry* **31**, 11626–11631.

Sugiyama, M. (1991) Effects of vitamins on chromium(VI)-induced damage. *Environ. Hlth. Persp.* **92**, 63–70.

Sugiyama, M. (1992) Role of physiological antioxidants in chromium(VI)-induced cellular injury. *Free Rad. Biol. Med.* **12**, 397–407.

Sullivan, J.F., Jetton, M.M., Hahn, H.K.J. and Birch, R.E. (1980) Enhanced lipid peroxidation in liver microsomes of zinc-deficient rats. *Am. J. Clin. Nutr.* **33**, 51–56.

Suzaki, E., Inoue, B., Okimasu, E., Ogata, M. and Utsumi, K. (1988) Stimulative effect of chlordane on the various functions of the guinea pig leukocytes. *Toxicol. Appl. Pharmacol.* **93**, 137–145.

Taketani, S., Kohno, H., Yoshinaga, T. and Tokunaga, R. (1989) The human 32-dKa stress protein induced by exposure to arsenite and cadmium ions is heme oxygenase. *FEBS Lett.* **245**, 173–176.

Taylor, M.J., Lucier, G.W., Mahler, J.F., Luster, M.I. Mahler, J.F., Thompson, M. and Clark, G.C. (1992) Inhibition of acute TCDD toxicity by treatment with anti-tumor necrosis factor antibody or dexamethasone. *Toxicol. Appl. Pharmacol.* **117**, 126–132.

Thayer, W.S. (1988) Evaluation of tissue indicators of oxidative stress in rats treated chronically with adriamycin. *Biochem. Pharmacol.* **37**, 2189–2194.

Thor, H., Hartzell, P., Svensson, S.-A. Orrenius, S., Mirabelli, F., Marinoni, V. and Bellomo, G. (1985) On the role of thiol groups in the inhibition of liver microsomal Ca^{2+} sequestration by toxic agents. *Biochem. Pharmacol.* **34**, 3717–3723.

Thor, H., Smith, M.T., Hartzell, P., Bellomo, G., Jewell, S.A. and Orrenius, S. (1982) The metabolism of menadione (2-methyl-1,4-naphthoquinone) by isolated hepatocytes. *J. Biol. Chem.* **257**, 12419–12425.

van Kuijk, F.J.G.M., Sevanian, A., Handelman, G.J. and Dratz, E.A. (1987) A new role for phospholipase A_2: Protection of membranes from lipid peroxidation. *Trends Biochem. Sci.* **12**, 31–34.

van Ommen, B., Voncken, J.W., Muller, F. and van Bladeren, P.J. (1988) The oxidation of tetrachloro-1,4-hydroquinone by microsomes and purified cytochrome P-450b. Implications for covalent binding to protein and involvement of reactive oxygen species. *Chem.-Biol. Interact.* **65**, 247–259.

Videla, L.A., Barros, S.B.M. and Junqueira, V.B.C. (1990) Lindane-induced liver oxidative stress. *Free Rad. Biol. Med.* **9**, 169–179.

Videla, L.A., Simizu, K., Barros, S.B.M. and Junqueira, V.B.C. (1989) Lindane-induced liver oxidative stress: Respiratory alterations and the effect of desferrioxamine in the isolated perfused rat liver. *Cell Biochem. Funct.* **7**, 179–183.

Videla, L.A., Simizu, K., Barros, S.B.M. and Junqueira, V.B.C. (1991) Mechanisms of lindane-induced hepatotoxicity: Alterations of respiratory activity and sinusoidal glutathione efflux in the isolated perfused rat liver. *Xenobiotica* **21**, 1023–1032.

Wagai, N. and Tawara, K. (1991a) Important role of oxygen metabolites in quinolone antibacterial agent-induced cutaneous phototoxicity in mice. *Arch. Toxicol.* **65**, 495–499.

Wagai, N. and Tawara, K. (1991b) Quinolone antibacterial-agent-induced cutaneous phototoxicity: Ear swelling reactions in Balb/c mice. *Toxicol. Lett.* **58**, 215–223.

Wagai, N. and Tawara, K. (1992) Possible direct role of reactive oxygens in the cause of cutaneous phototoxicity induced by five quinolones in mice. *Arch. Toxicol.* **66**, 392–397.

Wagai, N., Yamaguchi, F., Sekiguchi, M. and Tawara, K. (1990) Phototoxic potential of quinolone antibacterial agents in Balb/c mice. *Toxicol. Lett.* **54**, 299–308.

Wagai, N., Yamaguchi, F., Tawara, K. and Onodera, T. (1989) Studies on experimental conditions for detecting phototoxic potentials of drugs in Balb/c mice. *J. Toxicol. Sci.* **14**, 197–204.

Wahba, Z.Z., Al-Bayati, Z.A.F. and Stohs, S.J. (1988) Effect of 2,3,7,8-tetrachlorodibenzo-p-dioxin on the hepatic distribution of iron, copper, zinc, and magnesium in rats. *J. Biochem. Toxicol.* **3**, 121–129.

Wahba, Z.Z., Murray, W.J. and Stohs, S.J. (1990) Altered hepatic iron distribution and release in rats after exposure to 2,3,7,8-tetrachlorodibenzo-p-dioxin (TCDD). *Bull. Environ. Contam. Toxicol.* **45**, 436–445.

Warren, D.L. and Reed, D.J. (1991) Modification of hepatic vitamin E stores *in vivo*. *Arch. Biochem. Biophys.* **285**, 45–52.

Whitlock, J.P. (1989) The control of cytochrome P-450 gene expression by dioxin. *Trends Pharmacol. Sci.* **10**, 285–288.

Wilhelmsson, A., Wikstrom, A.C. and Poellinger, L. (1986) Polyanionic-binding properties of the receptor for 2,3,7,8-tetrachlorodibenzo-p-dioxin. A comparison with the glucocorticoid receptor. *J. Biol. Chem.* **261**, 13456–13463.

Williams, M.V., Winters, T. and Waddell, K.S. (1987) *In vivo* effects of mercury (II) on deoxyuridine triphosphate nucleotidohydrolase, DNA polymerase (alpha, beta), and uracil-DNA glycosylase activities in cultured human cells: Relationship to DNA damage, DNA repair, and cytotoxicity. *Mol. Pharmacol.* **31**, 200–207.

Wolfgang, G.H.I., Jolly, R.A., Donarski, W.J. and Petry, T.W. (1991a) Inhibition of diquat-induced lipid peroxidation and toxicity in precision-cut rat liver slices by novel antioxidants. *Toxicol. Appl. Pharmacol.* **108**, 321–329.

Wolfgang, G.H.I., Jolly, R.A. and Petry, T.W. (1991b) Diquat-induced oxidative damage in hepatic microsomes: Effects of antioxidants. *Free Rad. Biol. Med.* **10**, 403–411.

Xu, Z. and Bray, T.M. (1992) Effects of increased microsomal oxygen radicals on the function and stability of cytochrome P450 in dietary zinc deficient rats. *J. Nutr. Biochem.* **3**, 326–332.

Yagi, K., Ishida, N. and Ohtsuka, K. (1990) Induction by chinoform-ferric chelate of lipid peroxidation in rat liver microsomes. *J. Clin. Biochem. Nutr.* **9**, 11–15.

Yagi, K., Ohtsuka, K. and Ohishi, N. (1985) Lipid peroxidation caused by chinoform-ferric chelate in cultured neural retinal cells. *Experientia* **41**, 1561–1563.

Younes, M. and Strubelt, O. (1990) The role of iron and glutathione in t-butyl hydroperoxide-induced damage towards isolated perfused rat livers. *J. Appl. Toxicol.* **10**, 319–324.

Younes, M. and Strubelt, O. (1991) Vanadate-induced toxicity towards isolated perfused rat livers: The role of lipid peroxidation. *Toxicology* **66**, 63–74.

Younes, M. and Wess, A. (1990) The role of iron in t-butyl hydroperoxide-induced lipid peroxidation and hepatotoxicity in rats. *J. Appl. Toxicol.* **10**, 313–317.

Yourtee, D.M., Elkins, L.L., Nalvarte, E.L. and Smith, R.E. (1992) Amplification of doxorubicin mutagenicity by cupric ion. *Toxicol. Appl. Pharmacol.* **116**, 57–65.

CHAPTER 5

Metabolic Detoxification of Plant Prooxidants

May R. Berenbaum

5.1 INTRODUCTION

Whereas oxidants are chemicals capable of causing direct oxidative damage to living systems, prooxidants are compounds that must first undergo activation to cause oxidative damage (Ahmad, 1992). While oxidants tend to be small inorganic molecules, such as nitrogen oxides or ozone, a broad range of naturally occurring organic compounds have prooxidant properties. Prooxidant compounds are produced in great abundance by plants; because they upon activation initiate chain reactions that magnify their biological effects relative to their initial concentrations, they represent energy-efficient investments in defense against environmental stresses (Ahmad, 1992). Among those organisms susceptible to the toxic effects of plant prooxidants are insect and mammalian herbivores as well as microbial phytopathogens and plant competitors (Downum, 1992).

Two major forms of prooxidant activation are known (Ahmad, 1992). Photosensitization is a mechanism by which prooxidants can absorb photons of light energy and form an unstable (short-lived) singlet excited state. This excited state molecule can rearrange to form a longer lived triplet state, which can then proceed to interact with ground state oxygen to form singlet oxygen ($^1O_2$) by energy transfer, or superoxide anion O_2^- (and eventually hydroxyl radical; $\cdot$OH) by charge or electron transfer (see chapter 1). Photoionization, the mechanism by which electron transfer to oxygen occurs, results in a free radical or radical ion as the other product; these species too may react with oxygen to produce a peroxyl radical, another potentially reactive and damaging agent for biomolecules. Toxic oxygen molecules can disrupt living systems by reacting with proteins, nucleic acids, lipids, and other biomolecules critical to cell function. The excited state sensitizer can also interact, via charge or electron transfer, not with oxygen but with these biomolecules directly and thus cause damage to living systems without the participation of molecular oxygen. A second form of activation, redox cycling, occurs in phenolic compounds, particular dihydroxy derivatives. The dihydroxy compound can be converted by one-electron reduction into a quinone or semiquinone radical. A one-electron addition regenerates the dihydroxy compound, and further interactions between the semiquinone radical and ground state oxygen produce the quinone as well as O_2^- radical (Appel, 1993). Phenolic compounds can undergo oxidation enzymatically as well as nonenzymatically. Phenolic-oxidizing enzymes include tyrosinases, catechol oxidases, laccases, and peroxidases; among the small molecules capable of activating phenolics are metal cations, organic peroxides, tocopherols, and photosensitizers (Appel, 1993).

Irrespective of the manner in which they are generated, toxic oxyradicals can interfere with a broad range of biological processes. Virtually all organisms that consume plants possess defense mechanisms against plant prooxidants. Resistance to plant prooxidants in plant consumers takes many forms. In the absence of light or oxygen, the toxicity of photosensitizers or redox-active compounds may be greatly reduced; accordingly, many organisms consume and process plant tissues containing these compounds in the absence of light or oxygen (Berenbaum, 1987). Aerobic organisms in general possess a battery of antioxidant compounds and enzymes that target the toxic oxyradicals produced by activated prooxidants or photosensitizers (Appel, 1993). Small antioxidant molecules include, among others, carotenoid pigments, tocopherols, ascorbic acid, amines, furans, and uric acid, all of which are widely distributed constituents of consumer organism (see chapter 6). Antioxidant enzymes include superoxide dismutase, catalase, glutathione peroxidase, glutathione transferase, and glutathione reductase (see chapter 7). With the exception of glutathione peroxidase, which appears to be absent in insects (Ahmad, 1992), these enzymes are widespread among organisms. Finally, many organisms have developed catabolic pathways for specifically altering the structure of prooxidants so as to render them incapable of undergoing activation. In contrast with antioxidants, which appear to be highly conserved among species, detoxificative enzyme systems are extremely diverse and idiosyncratically distributed among species.

Metabolic detoxification is generally a two-phase process. Phase I transformations involve structural alteration of the xenobiotic, usually to a more hydrophilic, less toxic form; Phase II transformations are synthetic reactions generally involving conjugation to a more polar carrier to facilitate export from the body (Berenbaum, 1991a). Prooxidants are structurally and biosynthetically an extremely varied group of compounds; thus, no single system suffices to process all prooxidants and the relative importance of phase I and phase II transformations varies according to the structural type and the nature of activation. Moreover, for some forms of prooxidants, virtually nothing is known of metabolic pathways for detoxification; such transformations may be secondary to detoxification systems geared at disabling toxic oxygen species generated by activated prooxidants. Thus, in this review, representative prooxidants will be examined individually according to their biosynthetic classification and mode of activation.

5.2 DETOXIFICATION OF PHOTOSENSITIZERS

5.2.1 *Furanocoumarins*

Metabolic detoxification of prooxidant allelochemicals is perhaps most thoroughly characterized for the prooxidant phototoxic furanocoumarins. Furanocoumarins, benz-2-pyrone compounds with a furan ring attached at the 6,7 (linear) or 7,8 (angular) positions, are known from a handful of higher plant families but are produced in great quantity and diversity principally in the Apiaceae and Rutaceae (Murray et al., 1982). Over 200 individual furanocoumarins have been isolated and characterized from natural sources. In the presence of ultraviolet light, these compounds form excited states that can react directly with pyrimidine bases in DNA to form covalent cyclobutane cycloadducts as well as crosslink. Such cycloadducts can also form with unsaturated lipids (Specht et al., 1988). In addition, furanocoumarins are capable of interacting with oxygen to produce $^1O_2$ and to a lesser extent O_2^- and OH radicals (Joshi and Pathak, 1983; Berenbaum, 1991b; Larson and Marley, 1994). Damage to protein as well as DNA cleavage may be mediated by a free-radical mechanism.

The ability of an organism to metabolize furanocoumarins is closely associated with its ecological and evolutionary association with furanocoumarins; this pattern holds true across a broad range of taxa. Plant-derived strains of the fungal species *Gibberella pulicaris* can metabolize as many as 13 different furanocoumarins (Desjardins et al., 1989). Furanocoumarin tolerance is highly correlated with rate of metabolic detoxification among strains of this fungus, although virulence of these strains in furanocoumarin-containing hostplants does not appear to be. Whereas 88% of plant-derived strains examined were capable of metabolizing furanocoumarins, only 5% (2/38) of soil-derived strains demonstrated any metabolic activity against these plant secondary metabolites. O-demethylation and other oxidative transformations do not appear to be involved in xanthotoxin metabolism in this fungus (Desjardins et al., 1989). Rather, the principal metabolite, 5-(2-carboxyethyl)-6-hydroxy-7-methylbenzofuran, results from a reductive transformation (Spencer et al., 1990).

Furanocoumarins are metabolized oxidatively in a broad range of organisms via cytochrome P450 monooxygenases, membrane-bound enzymes catalyzing a variety of oxidative reactions that convert lipophilic substrates into more hydrophilic metabolites (Gonzalez and Nebert, 1990). Cytochrome P450-mediated metabolism is the principal mechanism for the disposition of furanocoumarins in herbivorous insects. The black swallowtail, *Papilio polyxenes* (Lepidoptera: Papilionidae) is an oligophage that feeds almost exclusively on furanocoumarin-containing plants (Berenbaum, 1981). Metabolism of furanocoumarins is localized in the midgut, is NADPH dependent and is inhibited by piperonyl butoxide (Ivie et al., 1983); these attributes are characteristic of P450 involvement. Two principal metabolites are excreted by black swallowtails in frass (Fig. 5.1); the structure of these carboxylic acid metabolites is consistent with the P450-catalyzed epoxidation of the furan ring, followed by ring opening. O-demethylation appears to be a minor pathway for detoxifying 5 or 8-substituted furanocoumarins (Ivie et al., 1983; Bull et al., 1986; Ivie, 1987). The location of the furan ring influences the suitability of furanocoumarins as P450 substrates. Ivie and Bull et al. (1986) confirmed that angelicin, an angular furanocoumarin, is also metabolized by P450s, albeit only 1/3 as rapidly as are linear furanocoumarins. In *P. polyxenes*, metabolism of xanthotoxin (a linear furanocoumarin) is induced up to eightfold by prior exposure to xanthotoxin but is not induced by standard synthetic inducers or even by related coumarin derivatives (such as coumarin itself or the angular furanocoumarin angelicin) (Berenbaum et al., 1990). The fact that a fourfold increase in xanthotoxin metabolic activity results in no significant change in total P450 content in response to dietary xanthotoxin is consistent with induction of only a small subset of P450s (Cohen et al., 1992).

Two full-length cDNAs encoding P450s induced by xanthotoxin and responsible for xanthotoxin metabolism in *P. polyxenes* have been cloned and sequenced (Cohen et al., 1992). They share 32% identity with previously characterized *CYP6A1*, a phenobarbital inducible cDNA isolated from insecticide resistant house flies (Feyereisen et al., 1989) and thus have been classified as members of the same P450 family, although in a separate subfamily, *CYP6B*. The two cDNAs appear to be allelic variants, sharing 98% amino acid identity; when expressed in a baculovirus-infected recombinant insect cell system, these isozymes were capable of metabolizing substantial amounts of linear, but not angular, furanocoumarins (Ma et al., 1994). These findings suggest that there are other, as yet unidentified, P450s principally responsible for angular furanocoumarin metabolism.

Xanthotoxin metabolism has been investigated in other lepidopteran species as well (Table 1). Generally, the ability to metabolize xanthotoxin is proportional to the probability of dietary exposure to this furanocoumarin: (1) whereas generalists, which rarely encounter these compounds, can metabolize xanthotoxin at rates of approximately 0.1 nmol/min/mg protein or less; (2) oligophages such as the black swallowtail and giant swallowtail *P. cresphontes* can metabolize xanthotoxin at rates of 1 to 8 nmol/min/mg protein; and (3) ecological monophages such as *Depressaria pastinacella*, the parsnip webworm, which feeds exclusively throughout much of its range on the furanocoumarin-rich flowers and fruits of wild parsnip *Pastinaca sativa*, can metabolize xanthotoxin at rates in excess of 30 nmol/min/mg protein (Berenbaum et al., 1990). Thus, detoxicative abilities appear to evolve in response to selection pressure exerted by hostplant allelochemistry.

Patterns of furanocoumarin biosynthesis may also reflect selection pressures on plants exerted by herbivorous insects. Angular furanocoumarins are far more restricted in distribution than are linear furanocoumarins, appearing principally, in several genera of Fabaceae and Apiaceae (Murray et al., 1982). As well, they tend to be

Table 5.1.
Cytochrome P450-Mediated Xanthotoxin-Metabolic Activity in Lepidoptera (Activity Expressed as nmol/mg protein/min)

Insect species	Constitutive	Induced	Reference
Oligophages on furanocoumarin-containing plants			
Depressaria pastinacella	21.3	31.5	Nitao, 1989
Papilio polyxenes	1.1	8.33	Cohen et al., 1989
Papilio cresphontes	3.4	6.6	Cohen et al., 1992
Polyphages rarely encountering furanocoumarins			
Papilio glaucus	0.5	2.8	Cohen, 1991
Spodoptera frugiperda	0.11	ND	Yu, 1987
Trichoplusia ni	0.15	0.15	Lee and Berendaum, 1989
Oligophages rarely encountering furanocoumarins			
Papilio troilus	ND	ND	Cohen, 1991
Eurytides marcellus	ND	ND	Cohen, 1991
Battus philenor	ND	ND	Cohen, 1991

ND = not detected.

produced in lower concentrations in plants. Due to the positioning of the furan ring at the 7,8 positions, these compounds are less effective at forming DNA crosslinks and thus are generally less phototoxic (Berenbaum, 1991b). Despite their decreased capability for causing photobiological damage, they have been implicated as resistance factors in plants against several lepidopterans oligophagous on furanocoumarin-containing plants, including *Depressaria pastinacella* (Berenbaum et al., 1986) and *Papilio polyxenes* (Berenbaum and Feeny, 1981). This resistance is attributable in part to the fact that angular furanocoumarins inhibit metabolism of the more phototoxic furanocoumarins (Berenbaum and Zangerl, 1993; Ma et al., 1994).

In mammals, furanocoumarins are also metabolized by P450s. In rodents, as many as five pathways are involved; these include O-demethylation (in the case of 5 or 8 substituted furanocoumarins), aryl hydroxylation at the 5 position, pyrone ring hydrolysis, furan ring oxidation, and oxidation of the 5,8 dihydroquinone to the quinone (Figure 5.1) (Ivie, 1987). Products of primary metabolism may subsequently undergo sulfate or glucuronide conjugation prior to export out of the body. Virtually all metabolites are excreted in urine. Furanocoumarin metabolism is phenobarbital-inducible, as would be expected of P450-mediated transformations. Furanocoumarin metabolism in humans resembles that in rodents. Metabolism

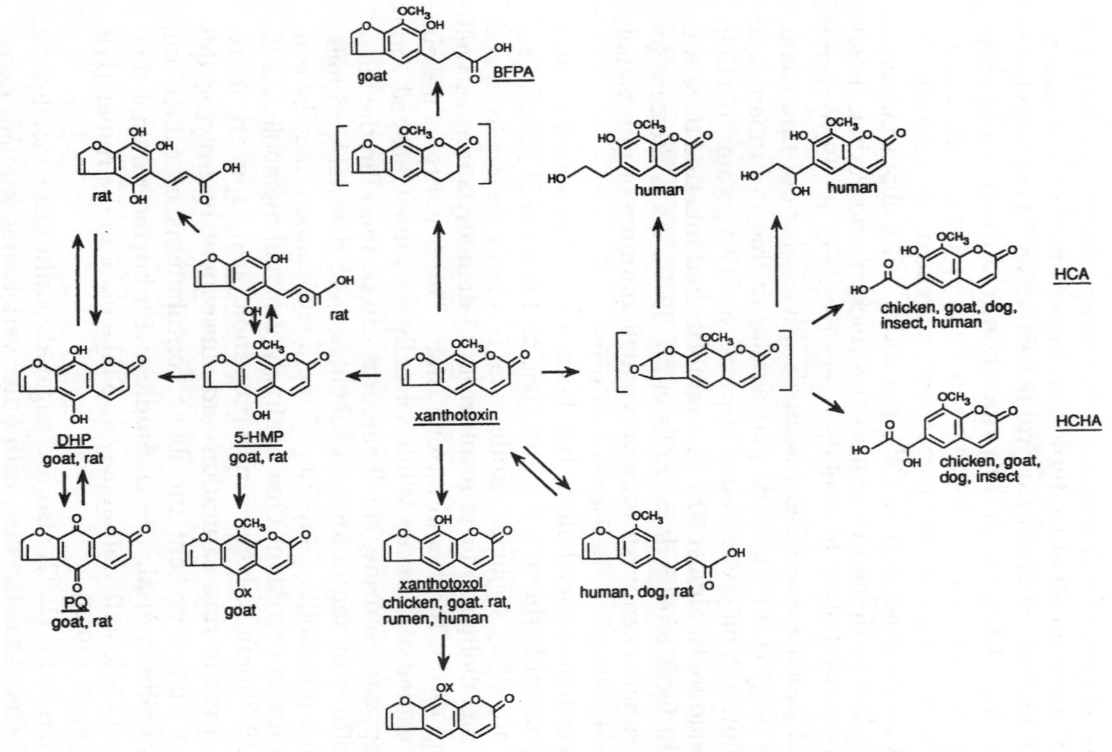

Fig. 5.1 Metabolic detoxification of xanthotoxin (from Ivie, 1987 and Pangilinan, 1992). Brackets indicate proposed intermediates. X = conjugated moieties; HCA = 6-(7-hydroxy-8-methoxycoumaryl)-acetic acid; HCHA = 6-(7-hydroxy-8-methoxycoumaryl)-hydroxyacetic acid; BFPA = 3-[5-(6-hydroxy-7-methoxybenzofuryl)]-propanoic acid; 5-HMP = 5-hydroxy-8-methoxypsoralen; DHP = dihydroxypsoralen; PQ = psoralenquinone

is rapid, with almost 80% of a 40 mg oral dose excreted (primarily in urine) within 24 hours (Schmid et al., 1980). At least five metabolites are produced in human urine after ingestion of xanthotoxin; these result from O-demethylation (8-hydroxypsoralen), opening of the lactone ring (furanocoumaric acid), and epoxidation of the furan ring (4', 5' dihydrodiol of xanthotoxin). Ingestion of xanthotoxin results in an increase in total cytochrome P450 as well as two- to three-fold induction of hepatic aryl hydrocarbon hydroxylase and ethylmorphine N-demethylase (Bickers and Pathak, 1984). Oral administration of the angular furanocoumarin angelicin, in contrast, did not effect an increase in total P450 nor did it bring about induction of hydroxylase or demethylase activities (Bickers and Pathak, 1994). Xanthotoxin metabolism in miniature pigs (Ikeda et al., 1990) is similar to that described in other mammals. Metabolism of furanocoumarins in dogs is for the most part similar to that in humans and rodents although biliary excretion appears to play a greater role than in other mammals; almost 40% of a radiolabelled ingested dose was excreted in feces after 3 days (Kolis et al., 1979). As is the case for other mammals examined, only very small amounts of unmetabolized furanocoumarins are excreted in urine.

Furanocoumarin metabolism has been studied in detail in only one mammalian herbivore, the goat (Ivie and Beier et al., 1986; Pangilinan et al., 1992). In goats, xanthotoxin is rapidly metabolized and excreted; metabolic products result from O-demethylation, as well as opening of the furan and lactone rings. The majority of metabolites are excreted in urine. While virtually no unmetabolized xanthotoxin appears in urine, small amounts can be found in the feces; no xanthotoxin or xanthotoxin metabolites were detectable in milk. At least one metabolite, 5-hydroxy-8-methoxypsoralen, may be conjugated prior to excretion. One relatively unusual metabolite, 3-[5-(6-hydroxyl-7-methoxybenzofuryl)]-propanoic acid, appears to result from pyrone ring saturation and subsequent hydrolysis; this conversion may represent metabolic contributions of rumen microbes (Pangilinan et al., 1992). Studies with bovine ruminal fluid suggest that O-demethylation may also take place in the rumen (Ivie and Beier et al., 1986).

Pangilinan et al. (1992) also investigated xanthotoxin metabolism in laying hens. As is the case with other vertebrates studied, excretion of metabolites is rapid, with 90% of an oral dose eliminated within one day. Of the six metabolites characterized, all had been isolated previously from other vertebrates, including xanthotoxol (the product of O-demethylation), 6-(7-hydroxy-8-methoxycoumary)hydroxyacetic acid, and 6-(7-hydroxy-8-methoxycoumaryl)-acetic acid (products of furan ring epoxidation). A small amount of xanthotoxin is excreted unmetabolized in excreta as well. Unmetabolized xanthotoxin, albeit at concentrations less than 1 ppm, constituted over 85% of labelled material isolated in egg yolks and whites.

5.2.2 Coumarin Metabolism

Coumarins, known from over 70 plant families, are capable of limited photosensitization. UVA exposure leads to free radical formation; coumarin as well as 5,7 dimethoxycoumarin can form cycloadducts with alkenes and pyrimidine bases (Song, 1984; Larson and Marley, 1994). Like furanocoumarins, coumarin metabolism is mediated by cytochrome P450 monoxygenases in a variety of taxa. Coumarin itself is metabolized in humans principally via hydroxylation to 7-hydroxycoumarin, which is excreted in urine. This transformation is catalyzed by CYP2A6. Whereas 7-hydroxycoumarin production is independent of coumarin dose within a range of 5 mg to 50 mg, formation of the metabolite is inhibited in the presence of the linear furanocoumarin, xanthotoxin (Kraul et al., 1992). Excretion rates in the presence of xanthotoxin are unchanged, which suggests that, while phase I transformations are inhibited by xanthotoxin, phase II transformations are unaffected.

Hydroxycoumarins differ in distribution and in biosynthetic origin from coumarin derivatives lacking a 7-oxygen functionality; whereas the former are derived from p-coumaric acid, the latter are derived from cinnamic acid (Murray et al., 1982). They also differ in the way that they are metabolized. In many instances, hydroxycoumarins are excreted unchanged, or as glucuronide or sulfate conjugates. Although metabolism of hydroxycoumarins can also be P450-mediated, different isozymes are involved than are those catalyzing detoxification of coumarin derivatives. Dealkylation is catalyzed by CYP1A2 (Furuya et al., 1991; Mayuzumi et al., 1993). The methylether of umbelliferone, herniarin, is demethylated in rats and rabbits to yield umbelliferone as a metabolite (Scheline, 1978). Hydroxycoumarins can also be hydroxylated, as is the case for herniarin, which is hydroxylated, principally at the 3 position but also at the 6 position, to yield phenylacetic acid derivatives (Fig. 5.2). Umbelliferone (7-hydroxycoumarin) can also undergo ring cleavage in rats, yielding 2,4-dihyroxyphenylpropionic acid, a reaction thought to be mediated by the intestinal microflora (Scheline, 1978). In rats,

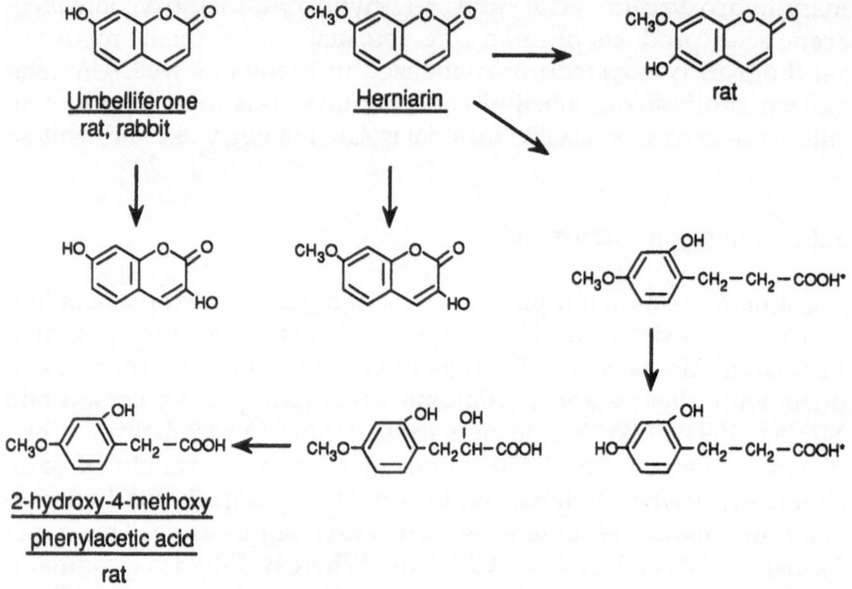

Fig. 5.2 Metabolism of herniarin in rats and rabbits (adapted from Scheline, 1978). Metabolites marked with an asterisk are absent in urine of germ-free rats ingesting herniarin; most likely these compounds are derived from microbial metabolism

scoparone (6,7-dimethoxycoumarin) is selectively O-demethylated to yield scopoletin or isoscopoletin. Cytochrome P450s from several gene families, (including P4501A1, P4501A2, P4502B1, P4502B2, and P4503A2) may play a role in these transformations (Witkamp et al., 1993).

5.2.3 *Thiophenes and Polyacetylenes*

While polyacetylenes in general are widespread among flowering plants, photosensitizing polyacetylenes, including the sulfur-containing thiophenes, are known only from the plant family Asteraceae (Downum, 1992). In contrast with furanocoumarins, the excited states of which can interact directly with biomolecules, polyacetylenes owe their phototoxicity primarily to interaction with oxygen and subsequent production of toxic oxygen molecules, particularly $^1O_2$, which causes damage by oxidizing cell membrane components (Larson and Marley, 1994). A considerable amount of effort has gone into characterizing the metabolism of thiophenes in general in a number of organisms (rat—Murphy et al., 1992; Jacob et al., 1991: bacteria—Alam and Clark, 1991; Abdulrashid and Clark, 1987), due to the fact that they are anthropogenic pollutants, like the polycyclic aromatic hydrocarbons they resemble structurally. However, investigations of the detoxification of phototoxic polyacetylenes have apparently been restricted to insects. Iyengar et al. (1987) compared rates of clearance of ingested α-terthienyl, a phototoxic thiophene, in three species of herbivorous lepidopterans. While *Heliothis virescens* (tobacco budworm) (Noctuidae) and *Ostrinia nubilalis* (European corn borer) (Pyralidae) cleared the ingested chemical rapidly, through excretion in feces, *Manduca sexta*, the tobacco hornworm (Sphingidae), retained sufficient quantities of unmetabolized α-terthienyl in the body to cause toxicity. Subsequent studies (Iyengar et al., 1990) revealed that a substantial amount (as much as 40%) of ingested α-terthienyl is excreted unmetabolized in these three species. Metabolites were produced in midgut microsomes of *H. virescens* and *O. nubilalis*, however, 16 and 30 times faster, respectively, than in midgut microsomes of *M. sexta*.

This pattern is consistent with the feeding habits and host associations of these species. *Ostrinia nubilalis* is a generalist feeder frequently associated with Asteraceae; *H. virescens*, also a generalist, is known to feed on a wide variety of asteraceous hosts. In contrast, *Manduca sexta* is restricted to feeding on species in the family Solanaceae, none of which is known to contain phototoxic polyacetylenes. Inhibition of microsomal formation of metabolites by piperonyl butoxide implicates the cytochrome P450 monooxygenase system in thiophene processing (Iyengar et al., 1990). Although metabolites were not characterized structurally, they were tested for phototoxicity in a yeast bioassay; the absence of phototoxicity of these metabolites suggests that the conjugated double bond system is disrupted during detoxification. Total cytochrome P450 levels do not appear to be inducible by exposure to α-terthienyl; indeed, at high doses, there is a slight but significant decrease in total P450 content of *H. virescens*. That this decrease in total P450 is not accompanied by a decrease in the rate of disappearance of α-terthienyl suggests that only a small subset of P450s is involved in α-terthienyl processing.

Hasspieler et al. (1991) demonstrated that α-terthienyl metabolism is mediated by cytochrome-P450 monooxygenases in the detritivorous larvae of the mosquito *Culex tarsalis*. This conclusion was based on the fact that piperonyl butoxide increased susceptibility to the toxin in vivo and decreased rates of elimination of α-terthienyl. Virtually no α-terthienyl is excreted unmetabolized; export from the body is apparently dependent upon conversion to a more hydrophilic product, which is then conjugated and excreted. That pretreatment of mosquito larvae with the P450 inducer β-naphthoflavone resulted in a reduction in the susceptibility of larvae to α-terthienyl but did not affect rates of α-terthienyl excretion suggests that conjugation of hydrophilic products may be the limiting step in clearing α-terthienyl from the system. β-naphthoflavone is known to induce metabolism of the structurally similar polycyclic aromatic hydrocarbons.

Some insights into metabolism of α-terthienyl and other phototoxic polyacetylenes may be gained by examining pathways leading to detoxification of nonphototoxic thiaarenes with central or peripheral thiophene moieties (Jacob et al., 1991). In rat liver, compounds with a central thiophene ring are converted to sulfoxides and sulfones via S-oxidation; compounds with a peripheral thiophene ring form hydroxylated derivatives such as phenols, diols, thiols, and diol epoxides. Phenobarbital induces both sulfone and sulfoxide formation, consistent with a P450-mediated mechanism; differential rates of product formation suggest, however, that different P450s are involved in sulfoxide and sulfone formation.

5.2.4   Citral Metabolism

Citral, a monoterpene aldehyde, was recently demonstrated to possess phototoxic properties (Asthana et al., 1992). In nature it exists as a mixture of two isomers, neral and geranial. Citral is widely distributed within the family Rutaceae and is a common essential oil component; it has been isolated from the essential oils of species in the eucalyptus (Myrtaceae) and grass (Poaceae) families as well. As is the case for polyacetylenes, citral owes its phototoxicity to its ability to interact with oxygen and produce toxic oxygen species, particularly hydrogen peroxide (H_2O_2) (Asthana et al., 1992). Many alpha, beta unsaturated aldehydes are capable of undergoing free radical reactions and thus citral may act not only as a phototoxin but as a redox-cycling prooxidant as well.

Fungi are capable of metabolizing citral (Moleyar and Narasimham, 1987); in the bacteria *Pseudomonas aeruginosa* and *P. convexa*, citral is converted to geranic acid (Diliberto et al., 1990). In rats, citral is metabolized rapidly and metabolites produced mainly in urine; expired air and feces each account for approximately 7 to 16% of an ingested dose. Biliary excretion increases in importance with repeated exposure. In rats, as in other mammals, the primary metabolites of citral are mono- or dicarboxylic (C1 and C8) acids (Fig. 5.3). Metabolism involves oxidation at the aldehyde, reduction of the 2,3 unsaturated bond, and allylic oxidation at C-8 or C-9. Attachment of an oxygen functionality at C-8 or C-9 involves cytochrome P450, via omega-hydroxylation (Parke and Rahman, 1969), with subsequent oxidation by aldehyde dehydrogenase to yield carboxylic acid moieties at these positions.

Hydrophilic metabolites of citral can be conjugated prior to export; at least one biliary metabolite is excreted as a glucuronide derivative. Despite the reactivity of alpha and beta unsaturated aldephydes toward sulfhydryl groups, there is no evidence for a role of glutathione in citral metabolism or export (e.g., no mercapturic acid metabolites); in general, 1,4 addition reactions do not figure prominently in citral metabolism. Pretreatment with citral results in increases in total cytochrome P450 content, as well as increases in activities of biphenyl 4-hydroxylase, 4-nitrobenzoate reductase, and glucuronyl transferase, in rat liver (Diliberto et al., 1990).

Despite the plausibility of aldehyde dehydrogenase-catalyzed oxidation of the aldehyde functionality, Boyer and Petersen (1991) could find no evidence for this role of aldehyde dehydrogenase (ALDH) in citral metabolism. In fact, citral actually inhibits ALDH-mediated acetaldehyde oxidation. The role of aldehyde dehydrogenase may be restricted to secondarily processing metabolites produced by cytochrome P450 activity. Cytosolic alcohol dehydrogenase, however, does play a central role in citral metabolism, catalyzing the reduction of citral to the corresponding alcohol. This reaction is NADH-dependent. The reduction is biphasic, possibly corresponding to successive reduction of the two isomers of citral, neral and geranial. Boyer and Peterson (1991) raise the possibility that bacterial hydrogenases may contribute to citral metabolism in mammals (oxidizing the monoterpene aldehyde to the acid form).

5.2.5 Quinine and Related Quinoline Alkaloids

The phototoxicity of the alkaloid quinine, known exclusively from plants in the Rubiaceae, was first observed in humans consuming the substance as a therapeutic agent against malaria (Moore, 1980).

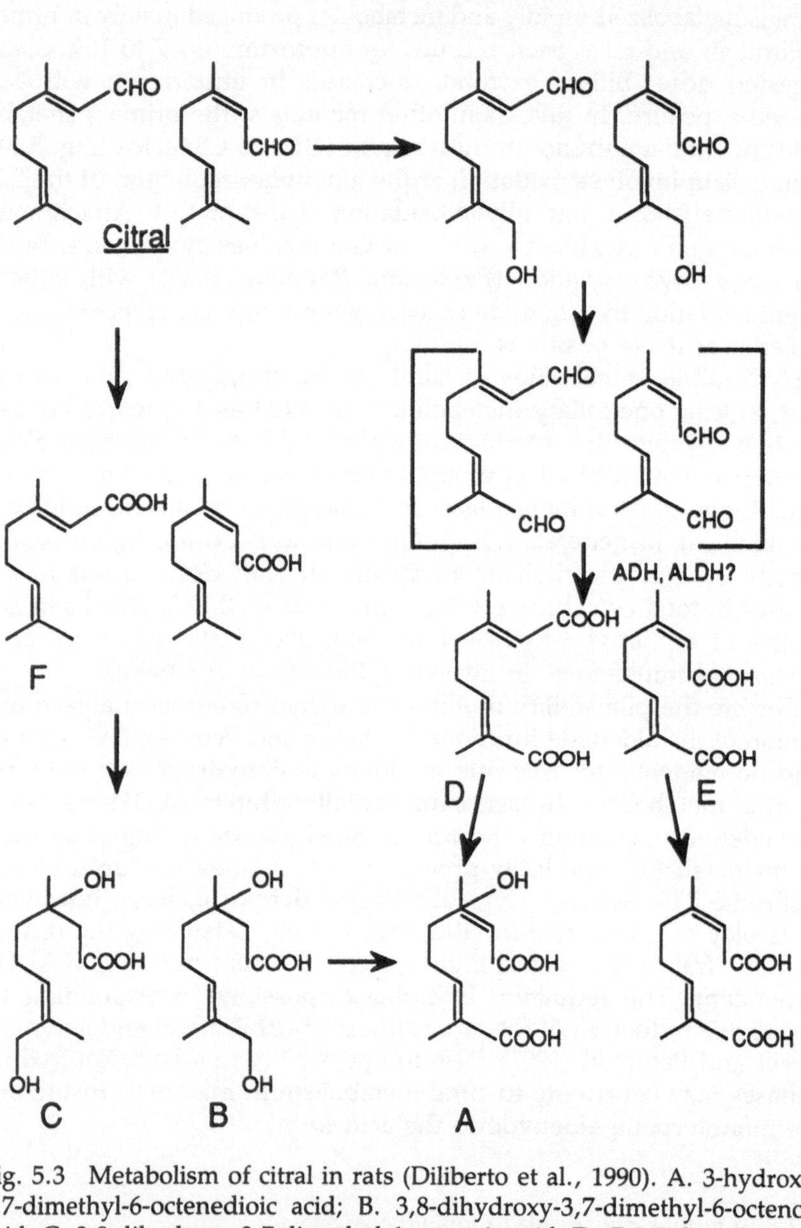

Fig. 5.3 Metabolism of citral in rats (Diliberto et al., 1990). A. 3-hydroxy-3,7-dimethyl-6-octenedioic acid; B. 3,8-dihydroxy-3,7-dimethyl-6-octenoic acid; C. 3,9-dihydroxy-3,7-dimethyl-6-octenoic acid; D. *E* 3,7-dimethyl-2,6-octadienedioic acid; E. *Z* 3,7-dimethyl-2,6-octadienedioic acid; F. *E* and *Z* 3,7-dimethyl-2,6-octadienoic acid

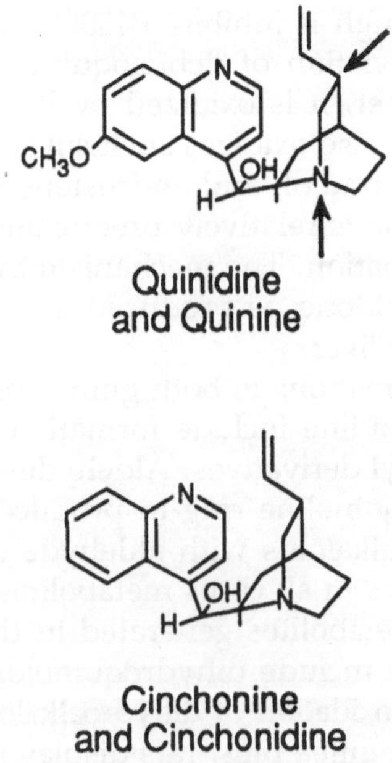

Quinidine
and Quinine

Cinchonine
and Cinchonidine

Fig. 5.4   Structure of cinchona alkaloids

Phototoxicity was noted early as a complication of treatment. In the presence of sunlight, quinine can intercalate DNA and induce mutation; although these reactions may be mediated principally by $^1O_2$, the related quinoline can photoionize and form O_2^- as well (Larson and Marley, 1993). The so-called Cinchona alkaloids consist of the stereoisomers quinine and quinidine, and the stereoisomers cinchonine and cinchonidine, which lack the 6'-methoxy substituent present in quinine and quinidine (Figure 5.4).

Quinine metabolism was first investigated over 50 years ago. The major metabolite in human urine, 2' quininone, was thought to result from the action of aldehyde oxidase (Knox et al., 1987 cited in Beedham et al., 1992). Other quinine metabolites include 3' hydroxyquinine and O-desmethylquinine; minor metabolites are quinine 10,11 epoxide and dihydrodiol. In rabbits, 2' quinolone is a major metabolite. Metabolism of quinidine, the diastereoisomer of quinine, has been studied extensively at least in part due to the fact that it has been used as an antiarrhythmic for decades. It is rapidly metabolized in humans; among the major metabolites are 3-hydroxyquinidine and quinidine N1' oxide, produced by cytochrome

P450 oxidation. Although it inhibits P450DB, the isozyme responsible for the 4-hydroxylation of debrisoquine, it is not a substrate for this isozyme. Rather, it is oxidized by P450NF, nifedipine oxidase (CYP3A4), which also oxidizes such substrates as aldrin, benzphetamine, cortisol, estradiol, and androstenedione (Guengerich et al., 1987). This isozyme is relatively unique among P450s in that it catalyzes N-oxide formation. The mechanism by which another major metabolite, 2' quinidone, is produced is thought to involve aldehyde oxidase in the liver.

Microsomal transformations in both guinea pig and rabbit liver of both quinine and quinidine include formation of 3-hydroxy derivatives and O-desmethyl derivatives. Aldehyde oxidase catalyzes oxidation at C-2 of the quinoline ring to produce a cyclic lactam. Incubation of cinchona alkaloids with aldehyde oxidase from guinea pig and rabbit produces in all cases metabolites with a 2' quinoline ring system. Minor metabolites generated in the presence of aldehyde oxidase probably include dihydroquinolones. These products are likely the result of oxidation of dihydroalkaloids. In contrast with humans, rabbits, and guinea pigs, rats display low levels of hepatic aldehyde oxidase activity; 2' quinoline is not detectable and only small quantities of 2' quininone is produced after quinine ingestion (Beedham et al., 1992).

5.2.6 β-Carboline Alkaloids

β-carboline alkaloids are tricyclic compounds that are widely distributed throughout the plant kingdom, although they are most diverse and abundant in the Rutaceae and Simaroubaceae (Downum, 1992). Their phototoxicity is associated with their ability to bind to DNA; while a free radical mechanism has been proposed, the harmala alkaloid harmane and its derivatives can produce toxic oxygen species, including $^1O_2$ and O_2^- (Chae and Ham, 1986; Larson et al., 1988). Phototoxicity to bacteria and insects, however, does not correspond to production of either of these oxygen species rather, lipophilicity is a better predictor of phototoxicity (Larson et al., 1988).

Metabolic detoxification of phototoxic β-carboline alkaloids has not yet been investigated in insects. In contrast, while the phototoxicity of harmala alkaloids to mammals has not been extensively studied, the metabolic disposition of these compounds in mammals has been elucidated (Figure 5.5). Studies of harmala alkaloid metabolism are summarized by Scheline (1978). In contrast with other phototoxic

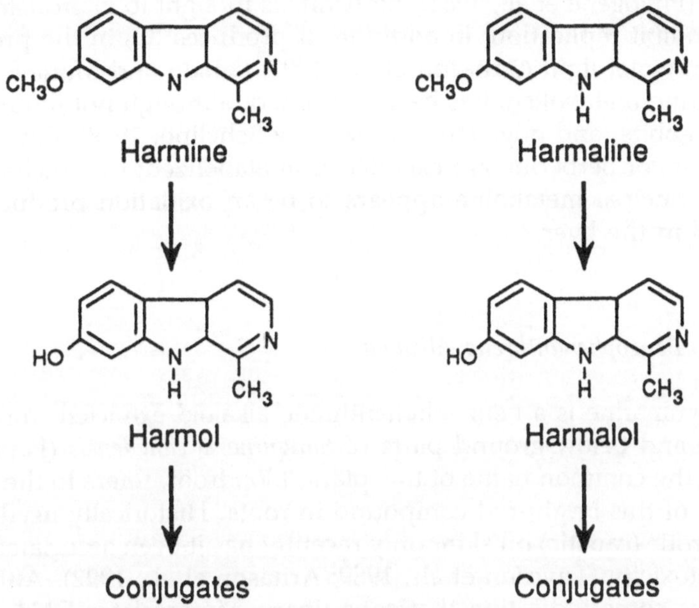

Fig. 5.5 Mammalian metabolism of harmala alkaloids (from Scheline, 1978)

alkaloids, biliary excretion rather than urine, is a major pathway in harmine metabolism; rats excrete almost 3/4 of an ingested dose in bile, and the remaining amount in urine. Very little unmetabolized harmine can be found in either bile or urine. In at least six mammalian species, harmine is demethylated by liver microsomes, to yield harmol, which is subsequently conjugated, primarily with sulfate. In humans, the major excretory product is also harmol, although glucuronate conjugation predominates over sulfate conjugation. In addition to undergoing microsomal demethylation, as does harmine, harmaline and harmalol undergo dehydrogenation at the 3,4 position to yield harmol. Harmalol can also be conjugated and excreted directly, principally in the form of the glucuronide in rats.

5.2.7 Isoquinoline Alkaloids

The benzylisoquinoline alkaloids are known from a wide variety of plants, particularly in the woody Ranales. Of the many compounds known, only berberine is reported to have phototoxic properties (Philogene et al., 1984). Berberine is thought to intercalate DNA and inhibit replication; in addition, it produces $^1O_2$ in the presence of ultraviolet light (Arnason et al., 1992; Larson and Marley, 1994). Berberine metabolism has been examined, although not in detail, in rats, rabbits, and dogs (summarized by Scheline, 1978). Only small amounts of berberine are excreted unmetabolized; the uncharacterized principal metabolite appears to be an oxidation product generated in the liver.

5.2.8   Benzophenanthrene Alkaloids

Sanguinarine is a benzophenanthrene alkaloid extracted from both above and below-ground parts of *Sanguinaria canadensis* (Papaveraceae); the common name of this plant, bloodroot, refers to the abundance of this bright-red compound in roots. Historically used as an escharotic (caustic) on skin, only recently has it been associated with phototoxicity (Tuveson et al., 1989; Arnason et al., 1992). Although there is some indication that sanguinarine intercalates DNA (Maiti et al., 1982), other studies show that in aqueous solution sanguinarine produces H_2O_2 via photoionization (Tuveson et al., 1989; Arnason et al., 1992). Efficiency of $^1O_2$ production varies with solvent (Larson and Marley, 1994). Scheline (1978) cites only a single study of sanguinarine metabolism. The major metabolite detected in blood, milk, and urine, is 3,4 benzacridine. The process by which this metabolite is produced was not elucidated.

5.2.9   Hydroxycinnamic Acid Derivatives

Hydroxycinnamic acid derivatives are virtually universal constituents of plants, involved in the biosynthesis of a vast array of secondary metabolites, including lignans, coumarins, phenolics, and alkaloids. They frequently are found as methyl esters or as esters of glucose or acids. Recent studies (Ashwood-Smith et al., 1993; Mohammad et al., 1991) indicate that methyl esters of a number of cinnamic and hydroxycinnamic acid derivatives are phototoxic to DNA repair-deficient mutants of *Escherichia coli*, suggesting intercalation with DNA as a possible mechanism. In vitro studies also demonstrate UV-mediated covalent cycloadduct formation with DNA by p-methoxymethylcinnamic acid (Mohammad et al., 1991). At least

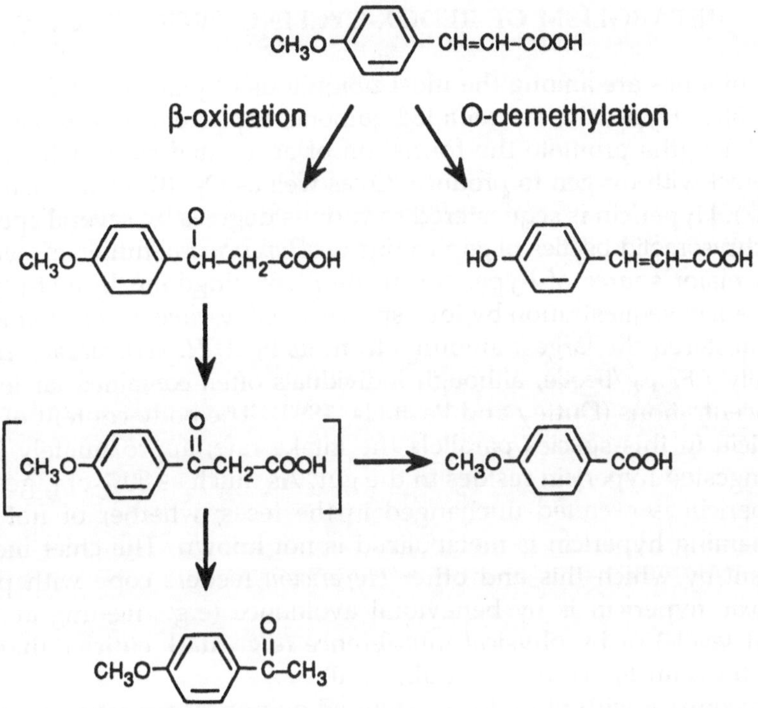

Fig. 5.6   Metabolism of 4-methoxycinnamic acid in rats (from Scheline, 1978)

one compound, 2-ethylhexyl-p-methoxycinnamate, is used medici-
nally, in sunscreen preparations, and is thought to be mutagenic
(Ashwood-Smith et al., 1993).

Extensive studies of the metabolism of cinnamic acid itself have
been conducted (reviewed in Scheline, 1978); over 80 years ago, the
major metabolite excreted by cats and dogs was characterized as
hippuric acid (Figure 5.6). Excretion in the form of metabolites is
extensive in all mammals examined. Conjugation of this material
with glucuronic acid has been documented. Other metabolites that
have been found in urine of rats following ingestion of cinnamic
acid to rats include 4-hydroxy and 3,4-dihydroxy- derivatives, as well
as the monomethyl ethers of the 3,4 dihydroxyderivatives (in Scheline, 1978). Many of these compounds are produced by plants (including p-coumaric, ferulic, and caffeic acids) and may be excreted
unmetabolized. No information was found on the processing of
methylated cinnamic acid derivatives.

5.3   METABOLISM OF REDOX-CYCLING PROOXIDANTS

Quinones are among the most ubiquitous of plant secondary metabolites. Hypericin, an extended quinone, is phototoxic—visible light
wavelengths promote the formation of an excited state, which can
interact with oxygen to produce $^1O_2$ as well as O_2^- (Knox and Dodge,
1987). Hypericin is sequestered to various degrees by several species
of chrysomelid beetles oligophagous on *Hypericum* (Guttiferae), which
is a major source of hypericin in the plant kingdom. In a study of
hypericin sequestration by four species of *Chrysolina*, *C. brunsvicensis*
sequestered the largest amounts from its host, *H. perforatum*—maximally 1.87 µg/beetle, although individuals often contained far lower
concentrations (Duffey and Pasteels, 1993). The body content of hypericin in this species parallels the intake rate; approximately 80%
of ingested hypericin resides in the gut. As much as 80% of ingested
hypericin is excreted unchanged in the feces; whether or not the
remaining hypericin is metabolized is not known. The chief mechanism by which this and other *Hypericum* feeders cope with phototoxic hypericin is by behavioral avoidance (e.g., feeding at low
light levels) or by physical interference (e.g., dark cuticles that do
not transmit light) (Fields et al., 1990).

In contrast with phototoxic extended quinones, activation of most
other quinones and related phenolics into oxidants is independent
of light energy. Phenolics are essentially universal constituents of
flowering plants. They possess a wide range of biological properties;
generation of toxic oxyradicals is but one of several mechanisms underlying those properties (Appel, 1993). Activation via one-electron
transfer leads to the formation of reactive semiquinones, which can
then pass on an electron to oxygen to form O_2^- and subsequently
other oxygen radicals. Metabolism of these compounds differs dramatically from metabolism of phototoxins, which are generally oxidized by cytochrome P450s or related enzymes. The principal mode
of detoxifying oxidized phenolics is via reduction reactions, catalyzed by DT-diaphorase (quinone reductase or NAD(P)H oxidoreductase [quinone acceptor]); these enzymes facilitate the transfer of
two electrons to convert quinones into more stable hydroquinones
(Atallah et al., 1986). The importance of DT-diaphorase in detoxification of quinones was recognized over 25 years ago (Ernstner,
1986). DT-diaphorase, an electron-transferring flavoprotein, has since
been characterized from a broad range of animals; activity levels vary
not only between species but also between individuals within the
same species (Ernstner, 1986). In mammals, highest levels are gen-

erally found in liver tissues. Basically, the enzyme effects a two-electron transfer from nicotinamide nucleotides (either NADH or NADPH) to a quinone acceptor. The resulting hydroquinone, in addition to being inherently less reactive (i.e., less likely to participate in free radical chain reactions), is also more likely to undergo conjugation reactions and subsequent excretion. Among quinones, unsubstituted benzoquinones and naphthoquinones are particularly reactive. Whereas a one-electron reduction (as catalyzed by NADPH-cytochrome P450 reductase, NADH-cytochrome b_5 reductase, mitochondrial NADH dehydrogenase, or other flavoprotein oxidoreductases) produces a semiquinone free radical, which can generate O_2^- radical when oxidized back to the starting quinone by molecular oxygen, a two-electron reduction produces a stable hydroquinone, effectively terminating redox cycling. Thus, DT-diaphorase competes with other NAD(P)H-oxidizing flavoproteins for substrates. DT-diaphorase activity is inducible by many plant constituents, including quercetin and coumarin, as well as by various synthetic materials, including polycyclic hydrocarbons, β-naphthoflavone, BHA and other antioxidants, and, to some extent, phenobarbital (Ernstner, 1986). In some cases, induction may be linked to cytochrome-P450 mediated activation of the inducing agent, as evidenced by involvement of the *Ah*, arylhydrocarbon hydroxylase, gene (Ernstner, 1986).

Quinone reductase activity (same as DT-diaphorase) has been documented in mammals, plants, bacteria, and insects (Yu, 1987). In herbivorous lepidopteran larvae, juglone reductase activity is concentrated in midgut microsomes and is inducible by flavone, indole-3-carbinol, and several other inducers of P450 activity; in contrast with P450s, however, juglone reductase activity is not inhibited by piperonyl butoxide. The level of reductase activity is not related to host breadth; the highly polyphagous *Heliothis* (=*Helicoverpa*)*zea*, with over 100 recorded hostplants, has reductase activities only 2/3 that of *Anticarsia gemmatalis*, the velvetbean caterpillar, with only a dozen recorded hosts.

In addition to reductive reactions, phenolics can be detoxified via hydrolases (although hydrolysis can, as in the case of quercetin glycosides, lead to activation—Lindroth, 1991) or esterases. Lindroth et al. (1989a,b,c) demonstrated that dietary phenolic glycosides from salicaceous hosts of the tiger swallowtail *Papilio glaucus* cause significant induction of soluble esterase activity; inhibition of these esterases leads to enhanced toxicity. In contrast, elevated levels of ingested phenolic glycosides did not affect cytochrome P450 activity in this insect.

Detoxification of oxidized phenolics can also involve conjugation reactions. In vertebrates, glucuronic acid conjugates of phenols predominate, while in invertebrates glutathione or glucose conjugates are reported (Appel, 1993). Lindroth et al. (1989c) reported that supplementing hostplant foliage of *Papilio glaucus* with hostplant-derived phenolic glycosides increased glutathione-S-transferase activity by approximately 50%. Interestingly, Lee (1991) demonstrated that several plant-derived phenolics actually inhibit glutathione-S-transferase activity in the cabbage looper, *Trichoplusia ni*.

Flavonols, which consist of two aromatic rings linked by a 3-carbon unit that is hydoxylated at the 3 position, are ubiquitous constituents of angiosperm foliage and flowers (Goodwin and Mercer, 1974). Like other phenolics, they are activated by redox cycling, particularly those possessing a catechol moiety (ortho-dihydroxy subtituents on the B ring). The disposition of the flavonol quercetin and several of its glycosides has been examined in a variety of organisms. In insects, these water-soluble compounds are often excreted in unaltered form. Approximately 90% of an ingested dose of rutin (quercitin-3-rutinoside), for example, is excreted unchanged in frass of *Heliothis* (=*Helicoverpa*)*zea* (Lepidoptera: Noctuidae) (Isman and Duffey, 1983). Scheline (1978) summarized studies of quercetin metabolism in mammals. In rats, rabbits, guinea pigs, and humans, ingestion of quercetin results in production of homoprotocatechuic acid (3,4-dihydroxyphenylacetic acid), homovanillic (4-hydroxy-3-methoxyphenylacetic) acid, and m-hydroxyphenylacetic acid in urine. These metabolites are thought to result from cleavage of the B ring at the 1,2 and 3,4 bonds. The appearance of a variety of other acids in urine, as well as appearance of radiolabel in expired carbon dioxide following ingestion of a radiolabelled dose, indicates that quercetin metabolism is undoubtedly complex. Indications are that many of these transformations are mediated by intestinal microbes; substantial amounts of radiolabel are recovered in feces and anaerobic incubations of rutin with rat caecal or fecal microorganisms produced the characteristic C6-C3 urinary metabolites. Bovine rumen microbes also contribute to quercetin metabolism; end products of detoxification, however, include phloroglucinol and 3,4 dihydroxybenzaldehyde in addition to 3,4-dihydroxyphenylacetic acid (homoprotocatechuic acid), suggesting that a different detoxification pathway may be involved.

5.4 SUMMARY AND CONCLUSIONS

Identifying unifying themes in the disposition of prooxidants is a not an easy task. Few general patterns are readily discernible. One pattern that does emerge is that oxidative mechanisms predominate in the disposition of prooxidant photosensitizers. In a sense, this pattern is not surprising in that oxidation reactions are the most widespread and diverse of phase I metabolic transformations (Scheline, 1978). However, oxygen may be even more likely to play a role than is usually the case in xenobiotic detoxification because prooxidants are by their very nature highly reactive toward oxygen. If a distinction is to be made, it appears that metabolism of prooxidants that are activated by redox recycling, rather than by photosensitization, tends to be reductive rather than oxidative. In some cases, these reductive reactions are attributable to actions of microflora associated with the digestive tract.

A second pattern that is suggested by an examination of metabolic pathways for the disposition of prooxidants is that efficacy of detoxification may be a function of frequency of exposure. Species that as specialists consume relatively large quantities of particular prooxidant compounds tend to have either specialized detoxification reactions or, at the very least, general detoxification reactions that are more rapid or efficient than those found in nonspecialists. This appears to be the case for insect metabolism of furanocoumarins and thiophenes. Such has also been demonstrated for antioxidant enzyme activity in specialist and nonspecialist insects (Ahmad, 1992). This pattern may also apply to degradation of plant prooxidants by mammalian herbivores in comparison with mammalian omnivores or carnivores, although no specific comparisons have been made in any single study and comparing results in different studies is difficult.

One additional observation is that detoxification enzymes involved in structurally altering phototoxins appear to target those portions of the molecule critical for photoactivation. Cycloadduct formation by linear furanocoumarins, for example, involves the 4',5' double bond of the furan ring; in humans, insects, goats, and dogs, the principal metabolites of xanthotoxin are thought to result from epoxidation of that furan ring double bond (Ivie, 1987). Similarly, metabolites of alpha-terthienyl lack phototoxicity (Iyengar et al., 1990). It is evidently a high priority for organisms ingesting light-activated compounds to render those materials incapable of absorbing light energy as quickly as possible. In general, export of phototoxins from the body is quite rapid and complete; in most instances, organisms ranging from insects to mammals excrete over 60% of ingested doses within a day. Because of the tremendous potential of these toxins to participate in regenerating free radical chain reactions, rapid export would appear to be of even greater importance than it would be for other types of toxins the effects of which tend to be more localized and restricted. It is probably not a coincidence that even those species with highly efficient detoxification systems for phototoxins have comparatively high constitutive levels of antioxidant activities, to protect against oxidative stress resulting from phototoxin molecules that escape detoxification (Lee and Berenbaum, 1989, 1993).

The one commonality that characterizes studies to date of metabolic detoxification of prooxidants is that not much is known about the process. With few exceptions, studies of detoxification have arisen out of an interest in human drug disposition rather than ecological or evolutionary associations between plants and their consumers. Even those studies with an ecological focus tend to be very limited; furanocoumarin metabolism, for example, is well characterized in lepidopterans yet insects in this order make up only a portion of the herbivorous insect community associated with furanocoumarin-containing plants. Until comprehensive studies are undertaken, designed explicitly to take into account ecological levels of exposure to plant prooxidants and evolutionary associations between plants and their consumers, it will be, to say the least, difficult to reconstruct the evolutionary process by which metabolic resistance to prooxidants has evolved.

ACKNOWLEDGMENTS

I thank Drs. Arthur Zangerl, Richard Larson and Sami Ahmad for helpful comments on the manuscript. This manuscript was prepared with support from NSF DEB 91-19612.

References

Abdulrashid, N. and Clark, D.P. (1987) Isolation and genetic analysis of mutations allowing the degradation of furans and thiophenes by *Escherichia coli*. *J. Bacteriol.* **169**, 1267–1271.

Ahmad, S. (1992) Biochemical defence of pro-oxidant plant allelochemicals by herbivorous insects. *Biochem. Syst. Ecol.* **20**, 269–296.

Alam, K.Y. and Clark, D.P. (1991) Molecular cloning and sequence of the *thdF* gene, which is involved in thiophene and furan oxidation by *Escherichia coli*. *J. Bacteriol.* **173**, 6018–6024.

Appel, H.M. (1993) Phenolics in ecological interactions: the importance of oxidation. *J. Chem. Ecol.* **19**, 1521–1552.

Arnason, J.T., Guerin, B., Kraml, M.M., Mehta, B., Redmond, R.W. and Scaiano, J. (1992) Phototoxic and photochemical properties of sanguinarine. *Photochem. Photobiol.* **55**, 35–38.

Ashwood-Smith, M., Stanley, C., Towers, G.H.N. and Warrington, P.J. (1993) UV-A-mediated activity of *p*-methoxymethylcinnamate. *Photochem. Photobiol.* **57**, 814–818.

Asthana, A., Larson, R.R.A., Marley, K.A. and Tuveson, R.W. (1992) Mechanisms of citral phototoxicity. *Photochem. Photobiol.* **56**, 211–222.

Atallah, A., Landolph, J.R. and Hochstein, P. (1986) DT-diaphorase and quinone detoxification. *Chemica Scripta* **27A**, 141–144.

Beedham, C., Al-Tayib, Y. and Smith, J.A. (1992) Role of guinea pig and rabbit hepatic aldehyde oxidase in oxidative in vitro metabolism of cinchona antimalarials. *Drug Metab. Disp.* **20**, 889–895.

Berenbaum, M. (1981) Effects of linear furanocoumarins on an adapted specialist insect *(Papilio polyxenes)*. *Ecol. Entomol.* **6**, 345–351.

Berenbaum, M.R. (1987) Charge of the light brigade: insect adaptations to phototoxins. In *Light-Activated Pesticides* (J.R. Heitz and K.R. Downum, eds.), Washington, D.C.: *ACS Symp. Ser.* **339**, 206–216.

Berenbaum, M.R. (1991a) Comparative processing of allelochemicals in Papilionidae. *Arch. Insect Biochem. Physiol.* **17**, 213–222.

Berenbaum, M.R. (1991b) Coumarins. In *Herbivores. Their Interactions with Secondary Plant Metabolites*, (G.A. Rosenthal and M.R. Berenbaum, eds.), Academic Press, New York, vol. 1: 221–249.

Berenbaum, M.R., Cohen, M.B. and Schuler, M.A. (1990) Cytochrome P450 in plant-insect interactions: inductions and deductions. In *Molecular Insect Science*, (H. Hagedorn, J. Hildebrand, M.G. Kidwell, and J. Law, eds.), Plenum Pub. Co., New York, pp. 257–262.

Berenbaum, M.R. and Feeny, P.P. (1981) Toxicity of angular furanocoumarins to swallowtails: escalation in the coevolutionary arms race. *Science* **212**, 927–929.

Berenbaum, M.R., Nitao, J.K. and Zangerl, A.R. (1986) Constraints on chemical coevolution: wild parsnips and the parsnip webworm. *Evolution* **40**, 1215–1228.

Berenbaum, M.R. and Zangler, A.R. (1993) Fumarocoumarin metabolism in *Papilio polyxenes*: genetic variability, biochemistry, and ecological significance. *Oecologia* **95**, 370–375.

Bickers, D.R. and Pathak, M.A. (1984) Psoralen pharmacology: studies on metabolism and enzyme induction. In *Photobiologic, Toxicologic, and Pharmacologic Aspects of Psoralens* (M.A. Pathak, ed.), Bethesda: National Cancer Institute Monograph NIH Publication No. 84-2692, pp. 77–83.

Boyer, C.S. and Petersen, D.R. (1991) The metabolism of 3,7-dimethyl-2,6-octadienal (citral) in rat hepatic mitochondrial and cytosolic fractions. Interactions with aldehyde and alcohol dehydrogenases. *Drug Met. and Disp.* **19**, 81–86.

Bull, D.L., Ivie, G.W., Beier, R.C. and Pryor, N.W. (1986) In vitro metabolism of a linear furanocoumarin (8-methoxypsoralen, xanthotoxin) by mixed-function oxidases of larvae of black swallowtail butterfly and fall armyworm. *J. Chem. Ecol.* **12**, 885–892.

Chae, K.H. and Ham, H.S. (1986) Production of singlet oxygen and superoxide anion radicals by b-carbolines. *Bull. Korean Chem. Soc.* **7**, 478–479.

Cohen, M.B., Berenbaum, M.R. and Schuler, M.A. (1989) Induction of cytochrome P450-mediated detoxification in the black swallowtail. *J. Chem. Ecol.* **15**, 2347–2355.

Cohen, M.B., Schuler, M.A. and Berenbaum, M.R. (1992) A host-inducible cytochrome P450 from a host-specific caterpillar: molecular cloning and evolution. *Proc. Natl. Acad. Sci. USA* **89**, 10920–10924.

Desjardins, A.E., Spencer, G.F. and Plattner, R.D. (1989) Tolerance and metabolism of furanocoumarins by the phytopathogenic fungus *Gibberella pulicaris (Fusarium sambucinum)*. *Phytochemistry* **28**, 2963–2969.

Desjardins, A.E., Spencer, G.R., Plattner, R.D. and Beremand, M.N. (1989) Furanocoumarin phytoalexins, trichothecene toxins, and infection of *Pastinaca sativa* by *Fusarium sporotrichoides*. *Phytopathology* **79**, 170–175.

Diliberto, J.J., Srinivas, P., Overstreet, D., Usha, G., Burka, L.T. and Birnbaum, L.S. (1990) Metabolism of citral, an α,β-unsaturated aldehyde in male F344 rates. *Drug Metab. Disp.* **18**, 866–875.

Downum, K.R. (1992) Tansley Review No. 43. Light-activated plant defence. *New Phytol.* **122**, 401–420.

Duffey, S.S. and Pasteels, J.M. (1993) Transient uptake of hypericin by chrysomelids is regulated by feeding behaviour. *Physiol. Entomol.* **18**, 119–129.

Ernster, L. (1986) DT-Diaphorase: a historical review. *Chemica Scripta* **27A**, 1–14.

Feyereisen, R., Koerner, J.F., Farnsworth, D.E. and Nebert, D.W. (1989) Isolation and sequence of a cDNA encoding a cytochrome P450 from an insecticide-resistant strain of the house fly, *Musca domestica*. *Proc. Natl. Acad. Sci. USA* **88**, 4558–4562.

Fields, P.G., Arnason, J.T. and Philogene, B.J.R. (1990) Behavioural and physical adaptations of three insects that feed on the phototoxic plant *Hypericum perforatum*. *Can. J. Zool.* **68**, 339–346.

Furuya, H., Shimizu, T., Hirano, K., Hatano, M., Fujii-Kuriyama, Y., Raag, R. and Poulos, T.L. (1989) Site-directed mutagenesis of rat liver cytochrome P-450d: catalytic activities toward benzphetamine and 7-ethoxycoumarin. *Biochemistry* **28**, 6848–6857.

Gonzalez, F.J. and Nebert, D.W. (1990) Evolution of the P450 gene superfamily: animal-plant "warfare", molecular drive, and human genetic differences in drug oxidation. *Trends in Genetics* **6**, 182–186.

Goodwin, T.W. and Mercer, E.I. (1972) *Introduction to Plant Biochemistry*. Pergamon Press, New York.

Guengerich, F.P., Muller-Enoch, D. and Blair, I.A. (1987) Oxidation of quinidine by human liver cytochrome P450. *Mol. Pharmacol.* **30**, 287–295.

Hasspieler, B.M., Arnason, J.T. and Downe, A.E.R. (1991) Metabolism of the phototoxin α-terthienyl in *Culex tarsalis* larvae: involvement of the polysubstrate monooxygenase system. *Pestic. Biochem. Physiol.* **40**, 191–197.

Ikeda, G.J., Sapienza, P.P., Sager, A.O. and Kornhauser, A. (1990) Exretion and tissue distribution of 14C-labelled 8-methoxypsoralen in beagle dogs and miniature pigs. *Food Chem. Toxicol.* **28**, 333–338.

Isman, M.B. and Duffey, S.S. (1983) Pharmacokinetics of chlorogenic acid and rutin in larvae of *Heliothis zea*. *J. Insect Physiol.* **29**, 295–300.

Ivie, G.W. (1987) Biological actions and metabolic transformations of furanocoumarins. In *Light-Activated Pesticides* (J.R. Heitz and K.R. Downum, eds.), Washington, D.C., *ACS Symp. Ser.* **339**, 217–230.

Ivie, G.W., Beier, R.C., Bull, D.L. and Oertli, E.H. (1986) Fate of [^{14}C]xanthotoxin (8-methoxypsoralen) in a goat and in bovine ruminal fluid. *Am. J. Vet. Res.* **47**, 799–803.

Ivie, G.W., Bull, D.L., Beier, R.C., Pryor, N.W. and Oertli, E.H. (1983) Metabolic detoxification: mechanism of insect resistance to psoralens. *Science* **221**, 374–376.

Ivie, G.W., Bull, D.L., Beier, R.C. and Pryor N.W. (1986) Comparative metabolism of [^{3}H] psoralen and [^{3}H]-isopsoralen by black swallowtail (*Papilio polyxenes* Fabr.) caterpillars. *J. Chem. Ecol.* **12**, 871–884.

Iyengar, S., Arnason, J.T., Philogene, B.J.R., Morand, P., Werstiuk, N.H. and Timmins, G. (1987) Toxicokinetics of the phototoxic allelochemical α-terthienyl in three herbivorous Lepidoptera. *Pestic. Biochem. Physiol.* **29**, 1–9.

Iyengar, S., Arnason, J.T., Philogene, B.J.R., Werstiuk, N.H. and Morand, P. (1990) Comparative metabolism of the phototoxic allelochemical α-terthienyl in three species of lepidopterans. *Pestic. Biochem. Physiol.* **37**, 154–164.

Jacob, J., Schmoldt, A., Augustin, C., Raab, G. and Grimmer, G. (1991) Rat liver microsomal ring- and S-oxidation of thiaarenes with central or peripheral rings. *Toxicology* **67**, 181–194.

Joshi, P.C. and Pathak, M.A. (1983) Production of singlet oxygen and superoxide radicals by psoralens and their biological significance. *Biochem. Biophys. Res. Commun.* **112**, 638–646.

Knox, J.P., Samuels, R.I. and Dodge, A.D. (1987) Photodynamic action of hypericin. In *Light-Activated Pesticides* (J.R. Heitz and K.R. Downum, eds.), Washington, D.C., *ACS Symp. Ser.* **339**, 265–270.

Kolis, S.J., Williams, T.H., Postma, E.J., Sasso, G.J., Confalone, P.N. and Schwartz, M.A. (1979) The metabolism of ^{14}C-methoxsalen by the dog. *Drug Metab. Disp.* **7**, 220–225.

Kraul, H., Brix, R., Otto, A. and Hoffman, A. (1992) Dose-dependence of coumarin elimination and inhibitory effect of 8-methoxypsoralen. *Pharmazie* **47**, 389–390.

Larson, R.A. and Marley, K.A. (1994) Oxidative mechanisms of phototoxicity. In *Oxidants in the Environment. Advances in Environmental Science and Technology* (J.O. Nriagu and M.S. Simmons, eds.), New York, J. Wiley and Sons, 269–317.

Larson, R.A., Marley, K.A., Tuveson, R.W. and Berenbaum, M.R. (1988) β-carboline alkaloids: mechanisms of phototoxicity to bacteria and insects. *Photochem. Photobiol.* **48**, 665–674.

Lee, K. (1991) Glutathione-S-transferase activities in phytophagous insects: induction and inhibition by plant phototoxins and phenols. *Insect Biochem.* **21**, 353–361.

Lee, K. and Berenbaum, M.R. (1989) Action of antioxidant enzymes and cytochrome P450 monooxygenases in the cabbage looper in response to plant phototoxins. *Arch. Insect Biochem. Physiol.* **10**, 151–162.

Lee, K. and Berenbaum, M.R. (1993) Food utilization and antioxidant enzyme activities of black swallowtail in response to plant phototoxins. *Arch. Insect Biochem. Physiol.* **23**, 79–89.

Lindroth, R. (1989a) Host plant alteration of detoxication activity in *Papilio glaucus glaucus*. *Entomol. Exp. Appl.* **50**, 29–35.

Lindroth, R. (1989b) Biochemical detoxication: mechanism of differential tiger swallowtail tolerance to phenolic glycosides. *Oecologia* **81**, 219–224.

Lindroth, R. (1989c) Differential esterase activity in *Papilio glaucus* subspecies: absence of cross-resistance between allelochemicals and insecticides. *Pestic. Biochem. Physiol.* **35**, 185–191.

Lindroth, R.L. (1991) Differential toxicity of plant allelochemicals to insects: role of enzymatic detoxification systems. In *Insect-Plant Interactions* (E.A. Bernays, ed.), CRC Press, Boca Raton, FL, III, 2–33.

Ma, R., Cohen, M.B., Berenbaum, M.R. and Schuler, M.A. (1994) Black swallowtail (*Papillo polyxenes*) encode cytochrome P450s that selectively metabolize linear furanocoumarins. *Arch. Biochem. Biophys.* **310**, 332–340.

Maiti, M., Nandi, R. and Chaudhuri, K. (1982) Sanguinarine: a monofunctional intercalating alkaloid. *FEBS Letts.* **142**, 280–284.

Mayuzumi, H., Sambongi, C., Kiroya, H., Shimizu, T., Tateishi, T. and Hatano, M. (1993) Effect of mutation of ionic amino acids of cytochrome P450 1A2 on catalytic activities toward 7-ethoxycoumarin and methanol. *Biochemistry* **32**, 5622–5628.

Mohammad, T., Baird, W.M. and Morrison, H. (1991) Photochemical covalent binding of p-methoxycinnamic acid to calf thymus DNA. *Bioorg. Chem.* **19**, 88–100.

Moleyar, V. and Narasimham, P. (1987) Detoxification of essential oil component (citral and menthol) by *Aspergillus niger* and *Rhizopus stolonifer*. *J. Sci. Food Agric.* **39**, 240–246.

Moore, D.E. (1980) Photosensitization by drugs: quinine as a photosensitizer. *J. Pharm. Pharmacol.* **32**, 216–218.

Murphy, S.E., Amin, S., Coletta, K. and Hoffman, D. 1992. Rat liver metabolism of benzo[b]naphtho[2,1-d]thiophene. *Chem. Res. Toxicol.* **5**, 491–495.

Murray, R.D.H., Mendez, J. and Brown, S.A. (1982) *The Natural Coumarins: Occurrence, Chemistry and Biochemistry*. J. Wiley and Sons, Chichester.

Nitao, J.K. (1989) Enzymatic adaptation in a specialist herbivore for feeding on furanocoumarin-containing plants. *Ecology* **70**, 629–635.

Pangilinan, N.C., Ivie, G.W., Clement, B.A., Beier, R.C. and Uwayjan, M. (1992) Fate of [^{14}C]xanthotoxin (8-methoxypsoralen) in laying hens and a lactating goat. *J. Chem. Ecol.* **18**, 253–270.

Parke, D. and Rahman, M. 1969. The effects of some terpenoids and other dietary nutrients on hepatic drug metabolizing enzymes. *Biochem. J.* **113**, 12.

Philogene, B.J.R., Arnason, J.T., Towers, G.H.N., Abramowski, Z., Campos, F., Champagne, D. and McLachlan, D. (1984) Berberine: a naturally occurring phototoxic alkaloid. *J. Chem. Ecol.* **10**, 115–123.

Scheline, R.R. (1978) *Mammalian Metabolism of Plant Xenobiotics*. Academic Press, London.

Schmid, J., Prox, A., Reuter, A., Zipp, H. and Koss, F.W. (1980) The metabolism of 8-methoxypsoralen in man. *Eur. J. Drug Metab. Pharmacokinetics* 5, 81–92.

Song, P.S. (1984) Photoreactive states of furanocoumarins. *Natl. Cancer Inst. Monogr.* 66, 15–19.

Specht, K.G., Kittler, L. and Midden, W.R. (1988) A new biological target of furanocoumarins: photochemical formation of covalent adducts with unsaturated fatty acids. *Photochem. Photobiol.* 47, 537–541.

Spencer, G.F., Desjardins, A.E. and Plattner, R.F. (1990) 5-(2-carboxyethyl)-6-hydroxy-7-methoxybenzofuran, a fungal metabolite of xanthotoxin. *Phytochemistry* 29, 2495–2497.

Tuveson, R.W., Larson, R.A., Marley, K.A., Wong, G.-R. and Berenbaum, M.R. (1989) Sanguinarine, a phototoxic H_2O_2-producing alkaloid. *Photochem. Photobiol.* 50, 733–738.

Waterman, P.G. and Grundon, M.F. (1975) Chemistry and chemical taxonomy of the Rutales. *Ann. Proc. Phytochem. Soc. Europe* 22. Academic Press, New York.

Witkamp, R.F., Nijmeijer, S.M., Mennes, W.C., Rozema, A.Z., Noordhoek, J. and Van Miert, A.S.J.P.A.M. (1993) Regioselective O-demethylation of scoparone (6,7-dimethoxycoumarin) to assess cytochrome P450 activities *in vitro* in rat. Effects of gonadal steroids and the involvement of constitutive P450 enzymes. *Xenobiotica* 23, 401–410.

Yu, S.J. (1987) Quinone reductase of phytophagous insects and its induction by allelochemicals. *Comp. Biochem. Physiol.* 87B, 621–624.

CHAPTER 6

Antioxidant Mechanisms of Secondary Natural Products

Richard A. Larson

6.1 INTRODUCTION

Atmospheric oxygen is unusual among stable molecules in that its ground state has two unpaired electrons. This permits it to enter into energetically favorable reactions (eq. 6.1) with many organic free radicals.

$$RH \xrightarrow{\text{init.}} R^{\cdot} + {}^{\cdot}O{-}O^{\cdot} \xrightarrow{O_2} ROO^{\cdot} \xrightarrow{R'H} ROOH + R'' \quad (6.1)$$

These more or less spontaneous reactions, which take part in the spoilage of foodstuffs and the degradation of many other materials, are called autooxidations. Autooxidative reactions, especially when they occur by chain mechanisms (see section 6.2), permit the rapid conversion of a chemically reduced material into oxygenated forms. These mechanisms are involved in the progressive decomposition, or weathering, of many synthetic and natural substances, ranging from petroleum and rubber to biologically important membrane lipids.

Nonchain, or stoichiometric, mechanisms of autooxidation are also possible in materials or tissues where significant quantities of reduced substances are not present; these autooxidative reactions, however, may still produce sufficient oxidation products to accumulate to damaging levels. Biologically relevant examples include nucleic acids and proteins.

One potentially important nonchain oxidative process involves reactions with singlet oxygen, $^1O_2$. This compound is produced by energy transfer from photochemically excited donor molecules, called sensitizers, to ground-state (triplet) molecular oxygen. The product of the photoreaction, $^1O_2$, is more reactive than ground-state O_2 toward many biologically important compounds, including several protein amino acids (cysteine, methionine, tryptophan, and histidine) which react with it at quite rapid rates. Polyunsaturated fatty acids (PUFAs) also react at much slower rates; these peroxides are

$$-CH=CH-CH_2-CH=CH- \rightarrow -CH-CH=CH-CH=CH- \atop \underset{OOH}{|} \qquad (6.2)$$

likely contributors to damage and dysfunction in cell and organelle membranes.

Synthetic organic chemists have created many effective inhibitors of oxidative damage for rubber, hydrocarbon fuels, plastics, foodstuffs, and many other materials. In principle, free radical chain reactions within a material can be inhibited either by adding chemicals that would retard the formation of free radicals (preventive antioxidants), or by introducing substances that would compete for the existing radicals and remove them from the reaction (chain-breaking antioxidants). The first mechanism is exemplified by the addition of carbon black to rubber to prevent the penetration of light into the product. Most research in the field of antioxidants, however, is concerned with the second mechanism; designing chemicals which, when added in small quantities to a material, react rapidly with the free radical intermediates of an autooxidation chain and stop it from progressing.

Many naturally occurring substances, including those found in animals and plants, also have antioxidant activity. Because of increasing interest in oxygen-containing free radicals in biological systems (and their implied roles as causative agents in the etiology of a variety of chronic disorders), the protective biochemical functions of such naturally occurring antioxidants have become subjects of great interest. A better understanding of the mechanisms of biological autooxidation reactions and their inhibition should help us comprehend and treat not only human health disorders, but also detrimental effects on crop plants and other organisms.

6.2   KINETICS OF ANTIOXIDATION

Many studies of purported antioxidant compounds are reported in qualitative terms; varying amounts of the test substance are added to an oxidizable mixture, often a very complex material such as linseed oil, and the extent of oxidation is measured after a fixed period of time. Such data, while undoubtedly useful in a practical sense, tell little about the kinetic or mechanistic aspects of the reactions under study. For this information, it is necessary to work under more controlled conditions, and particularly to follow the time course of the autooxidation reaction.

Measurements of autooxidation in solution often show characteristic sigmoidal plots, in which a slow initial rate of oxygen uptake is followed by a rapid uptake phase that eventually slows or stops. In the presence of heterogeneous systems incorporating surfaces, cells, or micelles, more complex kinetics are often observed. Rates of diffusion of antioxidants into heterogeneous media, for example, may take on overwhelming kinetic importance. These complications are probably responsible for the lack of correlation sometimes encountered between the effectiveness of an antioxidant in solution to that of the same antioxidant in a food product, living organism, etc. (Doba et al., 1985, Pryor et al., 1988).

6.2.1   *Chain Reactions: Initiation*

Autooxidation in solution is almost always a series of free-radical reactions, begun by the reaction of some free-radical initiator with the substrate being oxidized. The formation of organic (usually carbon-centered) free radicals from non-radical precursors is called the *initiation* phase of an autooxidative chain reaction. This process, which is often quite slow, results in the characteristic "lag period" of a radical chain reaction. The initiating free radical (or first productive radical in the chain), R·, can be formed from a precursor such as RH by a variety of mechanisms such as gamma radiation, UV illumination, sonication, reaction with transition metal ions or pre-existing free radicals, mechanical disintegration, or heat (eq. 6.3).

$$RH \xrightarrow{\;k_i\;} R^{\cdot} \tag{6.3}$$

For many years, studies of autooxidation and antioxidation kinetics were hindered by the difficulty of initiating the radical chain reaction at a reproducible rate. However, most investigators now use the thermal decomposition of a lipophilic or hydrophilic azo compound in the presence of excess oxygen to perform kinetic studies (Boozer et al., 1955; also see Barclay et al., 1984 and references therein). These compounds decompose by emitting nitrogen to form alkyl radicals, which then react at a nearly diffusion-controlled rate with oxygen to give peroxyl radicals as products (eq. 6.4).

$$R_3C-N{=}N-CR_3 \rightarrow N_2 + 2\,R_3C^{\cdot} \xrightarrow{\;O_2\;} R_3COO^{\cdot} \tag{6.4}$$

6.2.2 Reactivities of Free Radicals: Propagation

In the *propagation* phase of a free-radical chain reaction, there is a buildup of peroxyl radicals, ROO$^{\cdot}$, and the subsequent reaction of these radicals with other compounds (R'H) having extractable hydrogen atoms. The new radicals, R'', are then available for further reaction with molecular oxygen.

Initiators R$^{\cdot}$, when carbon-centered, react with molecular oxygen at rates k_{ox} that are virtually diffusion-controlled (eq. 6.5).

$$R^{\cdot} + {}^{\cdot}O{-}O^{\cdot} \xrightarrow{\;k_{ox}\;} ROO^{\cdot} \tag{6.5}$$

The resulting peroxyl radicals are then able to continue the chain process by reacting with another reactive hydrogen atom donor, R'H, with a rate k_p (eq. 6.6).

$$ROO^{\cdot} + R'H \xrightarrow{\;k_p\;} ROOH + R'' \tag{6.6}$$

6.2.3 Termination Processes: External Inhibitors

In the final stages of a chain reaction, when all the oxygen or active hydrogen species are used up, the *termination* phase begins. In this phase, the radicals recombine with each other to produce inactive, nonradical products.

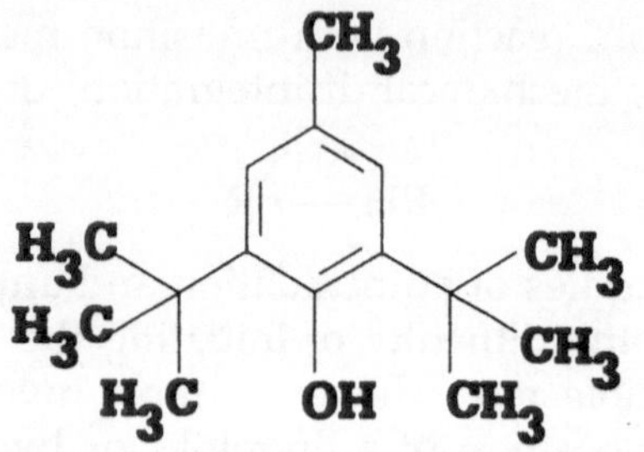

Fig. 6.1

Without an antioxidant present, a free-radical chain reaction ends when peroxyl radicals recombine with each other to produce non-radical oxidation products as shown in eq. 6.7.

$$ROO^. + ROO^. \xrightarrow{k_t} \text{nonradical products} + O_2 \qquad (6.7)$$

Reaction 6.7 is usually much slower than typical radical-radical coupling reactions. Under these conditions, the rate law for oxygen uptake by an autooxidizing system is given by eq. 6.8.

$$-\frac{d[O_2]}{dt} = \frac{[RH] \, (k_i)^{1/2} \, k_p}{(2k_t)^{1/2}} \qquad (6.8)$$

However, in the presence of an external antioxidant AH, one or more peroxyl radicals, rather than reacting with each other, are scavenged by the antioxidant with a rate constant k_{inh}:

$$n \, ROO^. + AH \xrightarrow{k_{inh}} \text{products} \qquad (6.9)$$

For the case in which $k_{inh} \gg k_t$, the rate law becomes:

$$-\frac{d[O_2]}{dt} = \frac{[RH] k_i \, k_p}{n \, k_{inh}[AH]} \qquad (6.10)$$

For many common antioxidants, both natural and synthetic, including phenols such as BHT (Fig. 6.1) and α-tocopherol or vitamin E (Fig. 6.2), the stoichiometry of n has been shown to be equal to 2. This is a consequence of the reaction of $ROO^.$ with the antioxidant to give a free radical $A^.$ capable of further reaction:

$$ROO^. + AH \rightarrow ROOH + A^. \rightarrow; ROO^. + A^. \rightarrow A\!-\!OOR \quad (6.11)$$

The most reliable way of measuring k_{inh} is the so-called "lag period

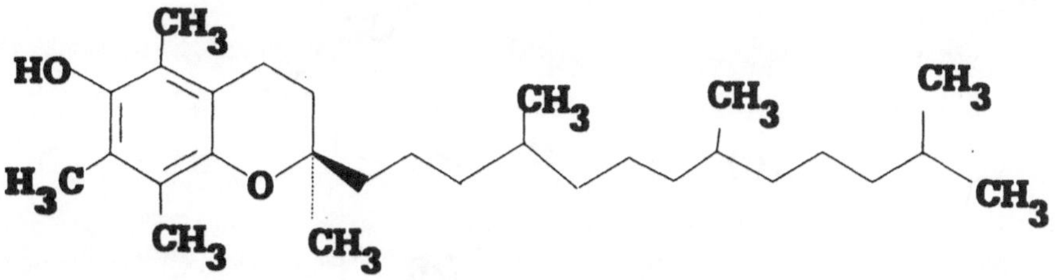

Fig. 6.2

method," in which the characteristic delay in the onset of a standard radical chain reaction is measured in the presence and absence of antioxidant. The lag period is directly proportional to k_{inh} and the concentration of the inhibitor.

6.3 ANTIOXIDANTS AS REDUCING AGENTS

Many compounds with antioxidant activity are, as might be expected, readily oxidizable materials. This property allows them both to intercept primary oxidants such as transition metal ions, molecular oxygen, hydroxyl radical ($\cdot$OH), or hydrogen peroxide (H_2O_2), and also to compete with chain-carrying free radicals to terminate autooxidation processes. Both stoichiometric and catalytic mechanisms are known for this class of antioxidants.

6.3.1 *Redox Potentials*

In principle, it should be possible to correlate the antioxidative effectiveness of a compound toward an oxidizing species such as an electrophilic free radical with its ease of oxidation by an electron-transfer process in vitro. Possible complicating factors might include differences in mechanism between the standard oxidation reaction whose potential is being measured and the reaction being quenched by the antioxidant in the living cell or other biologically related system, or competing intracellular processes that could affect the availability or speciation of the purported antioxidant. For whatever reason, this approach has not often been tried, although data that may be useful are available. In one study of a series of indoles, most of which were synthetic, an excellent correlation was shown between

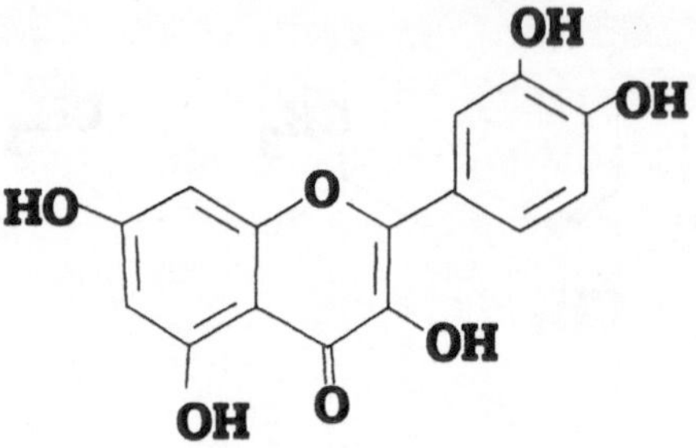

Fig. 6.3

the measured electrochemical redox potential for the compound and its inhibition of iron-ascorbate-initiated lipid peroxidation in liposomes (Brown et al., 1991).

Using a pulse radiolysis technique, the one-electron redox potentials of various phenols and amines of biological interest (including many which have been reported to show antioxidant activity) were determined (Steenken and Neta, 1982). The redox potentials of many flavonoids (polyphenols such as quercetin, Fig. 6.3), and chalcones have been determined by cyclic voltammetry (Hodnick et al., 1988) and discontinuous titration (Loth and Diedrich, 1967). The half-wave potentials of 17 Vitamin E derivatives have also been tabulated (Mukai et al., 1991).

6.3.2   Reactions with Peroxides

In organisms a variety of enzymes act as peroxide-destroying agents. Catalase, for example, decomposes H_2O_2 and peroxidases reduce hydroperoxides (ROOH) (see chapter 7). The mechanism of action of these molecules is beyond the scope of this chapter. However, many small molecules found in cells also have the property of reacting with peroxides or other cellular oxidants and removing them from autooxidative reactions. Sulfur compounds, for example, such as thiols and dihydrolipoic acid (Fig. 6.4), are rather easily oxidized by many electrophiles such as ROOH, H_2O_2, superoxide (O_2^-), $^1O_2$, and ·OH radical (Asada and Kanematsu, 1976; Suzuki et al., 1991). Typical products of these reactions are, in the case of thiols, the corresponding disulfides and sulfonic acids; and in the case of sulfides, the corresponding sulfoxides and sulfones.

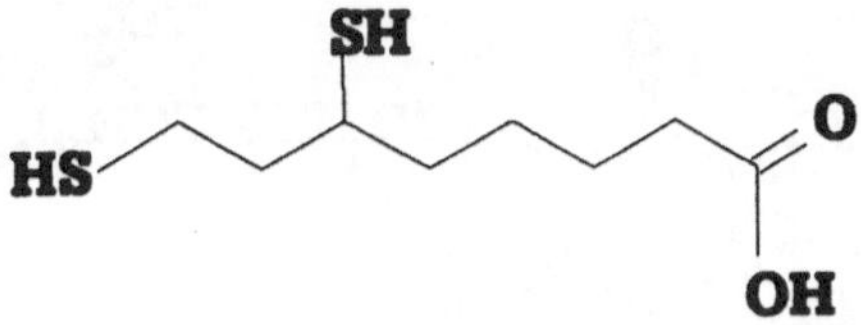

Fig. 6.4

6.4 ANTIOXIDANTS AS RADICAL QUENCHERS

Different free radicals have differing reactivities; for example, ˙OH is so unselective that it normally finds a partner within a very few molecular collisions, but peroxyl radicals, ROO˙ are normally several orders of magnitude less reactive with most compound types, and may "live" long enough to seek out a readily oxidized species such as an antioxidant. A general ability to enter into rapid reactions with free radicals is a great advantage for a potential antioxidant compound. Many synthetic antioxidants are specifically designed to react with oxygen radicals and to form sterically hindered or otherwise inactive radical products that effectively terminate radical chains. Natural antioxidants such as vitamin E have similar mechanisms of action.

Polyenes such as β-carotene (Fig. 6.5) react rapidly with many free radicals under certain conditions. The derived radicals are rendered less energetic (more stable) by extensive resonance contributions from the long conjugated system of double bonds, making it less likely for the radical to take part in chain processes. β-Carotene reacts rapidly with the triplet states of certain photosensitizing agents, especially ketonic sensitizers whose excited states are diradical-like. This process reduces the rate of chain initiation and is therefore an example of preventive antioxidation (see section 6.1).

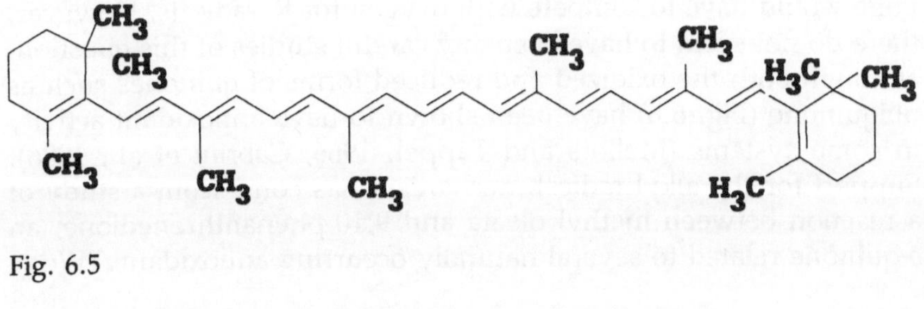

Fig. 6.5

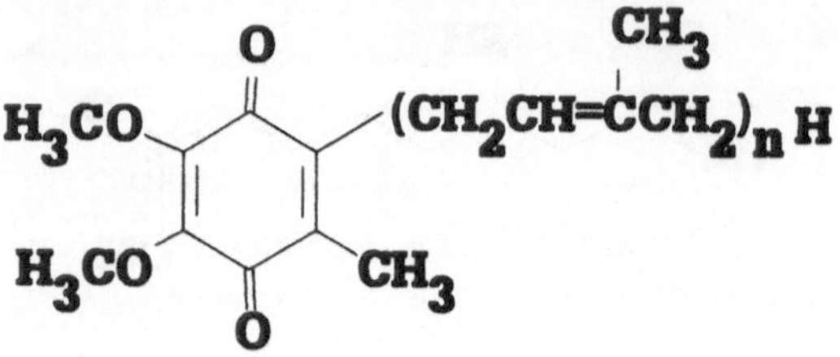

Fig. 6.6

Many naturally occurring compounds could have mechanisms of action that are similar to those that have been established for synthetic materials. For example, sterically hindered secondary amines have been synthesized as antioxidants for polyolefins. Their effectiveness appears to depend on their capacity to form stable nitroxyl radicals, which then react further with polymer free radicals to stop the chain reaction (Allen, 1980):

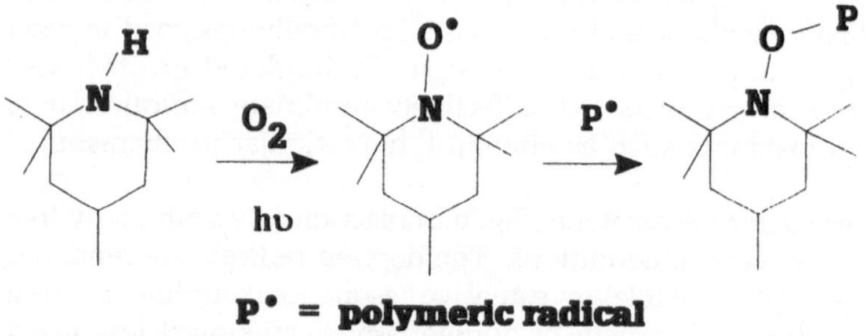

(6.12)

P• = polymeric radical

It is not known whether naturally occurring nitrogenous compounds could have similar mechanisms of action.

In another example, quinones are known to act as vinyl polymerization inhibitors in synthetic polymers, due to their interactions with carbon-centered free radicals. It has been proposed that biological quinones could act in the same way in the presence of low oxygen concentrations (Katbab et al., 1985). In other words, the quinone would have to compete with oxygen for R• radicals. However, there do not seem to have been any careful studies of this question, although both the oxidized and reduced forms of quinones such as ubiquinone (Fig. 6.6) have been shown to have antioxidant activity in some systems (Mellors and Tappel, 1966; Cabrini et al., 1986). Support for the alkyl radical mechanism has come from a study of a reaction between methyl oleate and 9,10-phenanthrenedione, an o-quinone related to several naturally occurring antioxidants (Weng

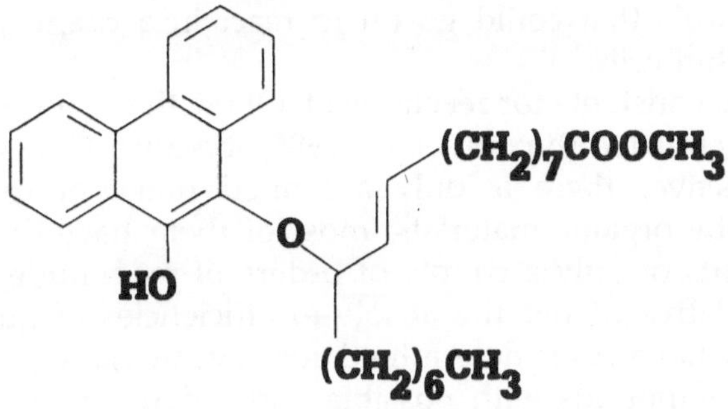

Fig. 6.7

and Gordon, 1992). When the reactants were heated together at 50°C for a prolonged period, an addition product (Fig. 6.7) was isolated in unspecified yield.

Uric acid (Fig. 6.8) also appears to have high activity toward a variety of types of free radicals including ROO˙ and ˙OH radicals (Ames et al., 1981; Cohen et al., 1984; Niki et al., 1986). Purines with similar structural characteristics were virtually inactive radical radical scavengers. Uric acid exists at rather high ($>10^{-4}$ M) concentrations in human plasma (Ames et al., 1981). Its activity appears to be restricted to regions of the cell that are high in water.

6.4.1 Hydroxyl Radicals

Hydroxyl radicals, if generated in the immediate vicinity of an important biomolecule, could produce direct damage. Another mechanism of their activity could be to initiate the formation of sec-

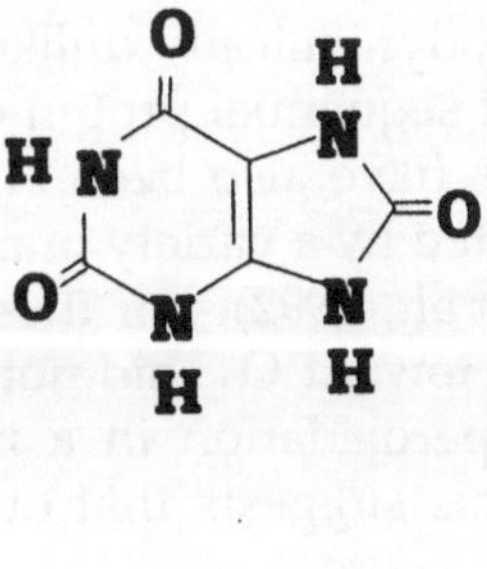

Fig. 6.8

ondary radicals that could go on to react in a chain or nonchain fashion within cells.

Many rate constants for reactions of ˙OH with organic compounds have been tabulated (Buxton et al., 1988). Because ˙OH is so reactive and unselective, there is only a limited range of reactivity for quenching by organic materials; most of them have differences of rate constants of only a couple of orders of magnitude. Nevertheless, the relative (if not the absolute) efficiencies of quenching of ˙OH radical have been determined for several classes of naturally occurring compounds with possible antioxidant activity, including polyphenols (Husain et al., 1987; Chimi et al., 1991; Ricardo da Silva et al., 1991) and phenylpropanoids (Taira et al., 1992).

6.4.2   Superoxide and Hydroperoxyl Radicals

Superoxide is the one-electron reduction product of molecular oxygen. It is protonated (at pHs below 6 or so) to the hydroperoxyl radical, ˙OOH. Although not a good oxidizing agent on its own, superoxide is rapidly converted to powerful oxidants by a self-redox reaction (eq. 6.13), that has been given the special name of *disproportionation*. The uncatalyzed disproportionation reaction is quite fast

$$2 \, O_2^- + 2H^+ \rightarrow H_2O_2 + O_2 \tag{6.13}$$

under normal physiological and environmental pHs, but in aerobic organisms is further catalyzed by the enzyme superoxide dismutase. (see chapter 7). The product, H_2O_2, is a moderately active oxidizing agent, but is further converted to even more potent oxidants such as the ˙OH radical by metal ion-catalyzed processes, such as the Fenton reaction (see chapter 1, eq. 1.16).

Phenolic antioxidants such as Vitamin E interact with O_2^- to produce an intermediate shown by ESR studies to be the phenoxyl radical (Tajima et al., 1983). Rather than a direct oxidation of the phenol by O_2^-, which is thermodynamically unlikely, the observation may represent an example of sequential proton-electron transfer (Sawyer et al., 1985). Flavonoids have also been shown repeatedly to react rapidly with O_2^- generated by a variety of methods (Robak and Gryglewski, 1988; Cotelle et al., 1992). For these compounds, however, their scavenging activity toward O_2^- did not correlate at all with their ability to inhibit lipid peroxidation in a mouse liver homogenate (Yuting et al., 1990). This suggests that other mechanisms of inhi-

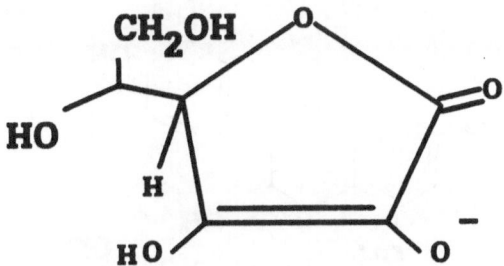

Fig. 6.9

bition could be more important for polyphenols, at least in this system.

Vitamin C (ascorbic acid, AH: Fig. 6.9) is found in quite high (millimolar and up) concentrations in many animal and plant tissues and reacts quite rapidly with O_2^- and $\cdot OOH$ radicals, as shown by pulse radiolysis and flash photolysis studies (Cabelli and Bielski, 1983). The product, ascorbate radical ($A\cdot$), can terminate the chain reaction by disproportionating into the nonradical products, ascorbate and dehydroascorbate. However, the reaction efficiency is partly ameliorated by the ability of ascorbate to produce O_2^- upon its own oxidation:

$$AH + O_2^- \rightarrow A^- + O_2 + H^+ \tag{6.14}$$

$$AH + O_2 \rightarrow A\cdot + O_2^- + H^+ \tag{6.15}$$

The presence of trace amounts of transition metal ions as catalysts for these reactions appears to govern the relative extents of the two processes (Bendich et al., 1986; Buettner, 1988).

The interactions of β-carotene with O_2^- are complicated, and of uncertain importance for determining its antioxidant capacity. It has been shown that the radical anion of β-carotene is capable of transferring an electron to oxygen to form O_2^-, but the reverse reaction does not occur. However, spectroscopic evidence suggests the formation of an unstable β-carotene-superoxide radical addition complex (Conn et al., 1992).

6.4.3 Peroxyl Radicals

Vitamin E has been shown to react very rapidly with peroxyl radicals; one measurement indicates a second-order rate constant of 2.4

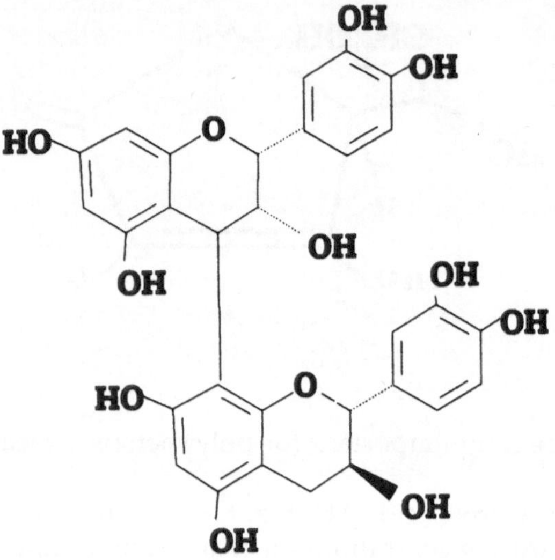

Fig. 6.10

$\times$ 10^6 L/mol sec, making it the most reactive chain-breaking anti-oxidant known (Burton and Ingold, 1981). In other systems, widely varying values for this rate constant have been determined, but all are in agreement that the compound and related tocopherol deriv-atives have very high affinities for reaction with peroxyl and other free radicals.

Naturally occurring polyphenols have repeatedly been shown to scavenge peroxyl radicals (Torel et al., 1986), although few rate con-stants for inhibition have been measured. Using the induction-pe-riod method, Ariga and Hamano (1990) measured k_{inh} in aqueous solution of 6×10^4, 5.9×10^4, and 2.7×10^4 L/mol sec for pro-cyanidin B-1 (Fig. 6.10), procyanidin B-3 (Fig. 6.11), and (+)-cate-chin (Fig. 6.12), respectively. For comparison, k_{inh} for vitamin E was 1.1×10^5 in their system. Since polyphenols may have numerous potential sites for reaction with peroxyl radicals, they may have high stoichiometries, far above the typical 2.0 radicals per antioxidant molecule for a synthetic antioxidant such as BHT; values of 8 or more were measured for the two procyanidins depicted in Fig. 6.10 and Fig. 6.11 (Ariga and Hamano, 1990).

Carotenoids, particularly β-carotene, scavenge free radicals under some conditions. Peroxyl radicals, in particular, have been shown to add rapidly to the long chain of conjugated double bonds present

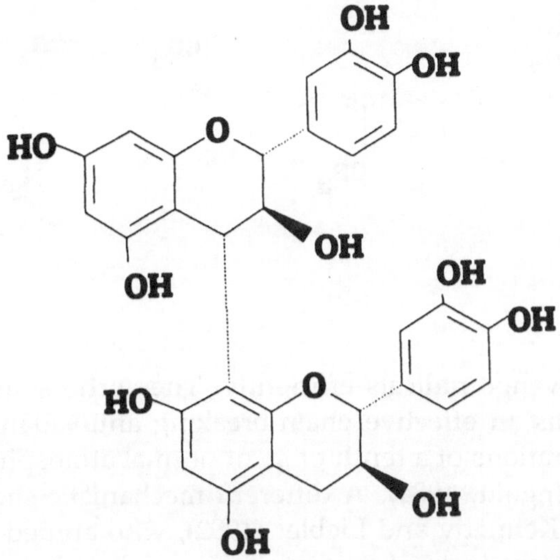

Fig. 6.11

in β-carotene and other carotenoids. For example, Packer et al., (1981) showed that the second-order rate constant for the reaction of β-carotene with trichloromethyl peroxyl radical ($Cl_3COO^{\cdot}$) was 1.5×10^9/M sec, about a tenth of the diffusion-controlled rate. Others, however, have criticized this value as unrealistic for natural systems, due to the heightened reactivity of the $Cl_3COO^{\cdot}$ radical (Burton and Ingold, 1984). In the presence of high oxygen tensions, however, it has been suggested that molecular oxygen, too, adds in a reversible manner to the carotenoid chain, producing peroxyl radicals with chain-carrying ability and limiting the ability of the com-

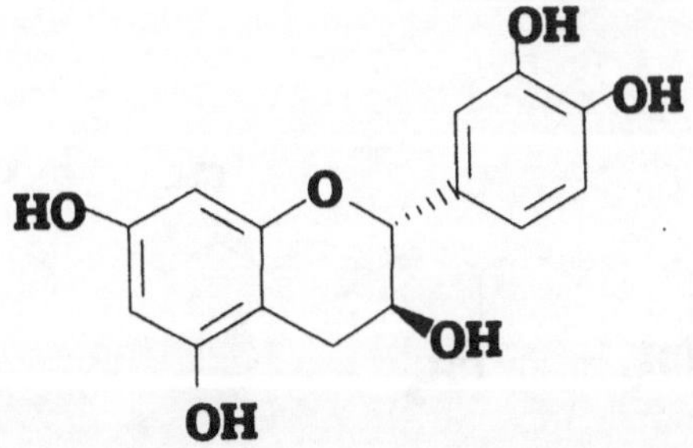

Fig. 6.12

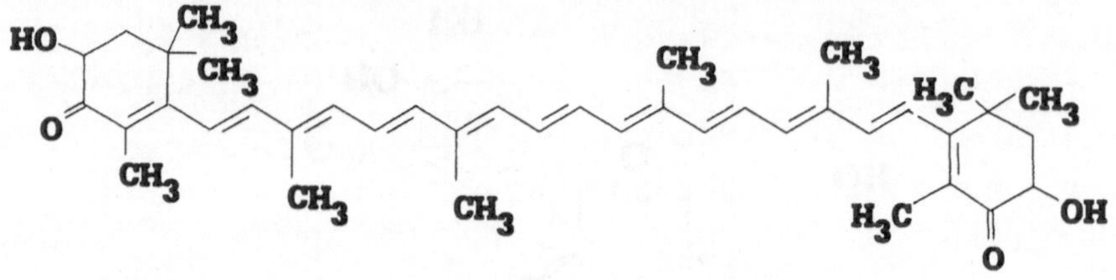

Fig. 6.13

pound to scavenge radicals efficiently. The authors suggested that β-carotene was an effective chain-breaking antioxidant only at oxygen concentrations of a tenth or so of normal atmospheric pressure (Burton and Ingold, 1984). A different mechanistic suggestion was advanced by Kennedy and Liebler (1992), who argued that inactive autooxidation products of β-carotene accumulated in the medium, and were formed at faster rates at higher oxygen pressures.

In some studies, oxygenated carotenoids such as astaxanthin (Fig.6.13) and canthaxanthin (Fig. 6.14) appear to be more effective antioxidants toward radical-initiated peroxidation reactions than β-carotene (Palozza and Krinsky, 1992). The greater efficiency of the keto carotenoids in lipid peroxidation has been attributed to a reduced activity of the stabilized polyene peroxyl radical toward propagation of the chain reaction (Terao, 1989).

Ascorbic acid reacts with peroxy radicals in water with a rate constant (k_{inh}) of 5×10^4 L/mol sec (Ariga and Hamano, 1990). The number of peroxyl radicals trapped per mole of ascorbate ranged from near 2 at low ascorbate concentrations to almost 0 at high concentrations (Bendich et al., 1986), apparently because of self-termination reactions that compete with radical trapping at higher ascorbate levels.

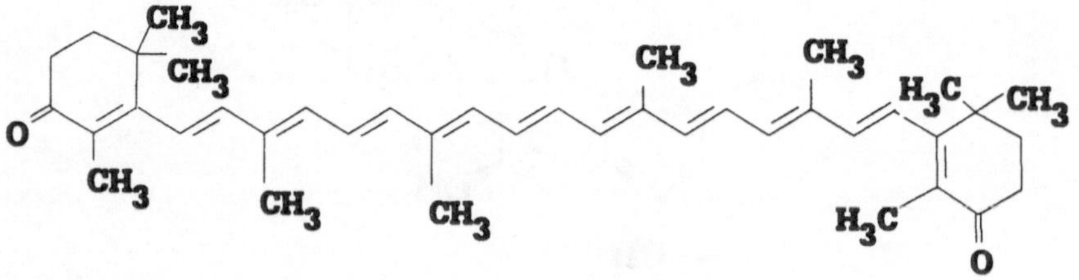

Fig. 6.14

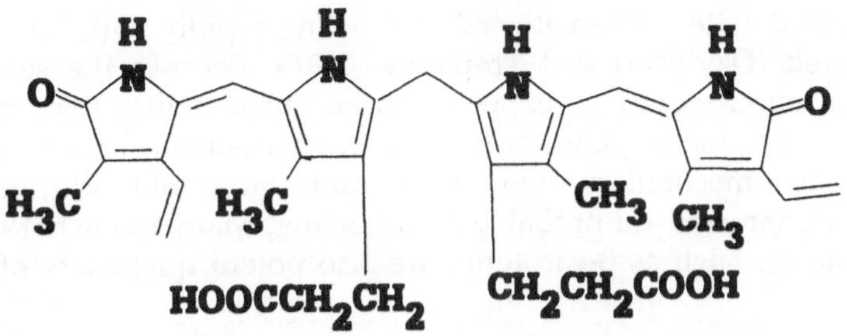

Fig. 6.15

Tertiary amines such as trimethylamine have been shown to be very effective quenchers of peroxyl radicals. The radicals derived from the amines have very large termination rate constants, contributing to the efficient cessation of radical chain reactions (Howard and Yamada, 1981). Very few studies of this type with naturally occurring nitrogenous bases such as alkaloids have been reported to date (Speisky et al., 1991); however, it has been pointed out that these compounds, being protonated at physiological pHs, should be strongly attracted to intracellular anionic sites such as DNA or membranes (Terce et al., 1982). Such a coulombic interaction might permit such a compound to function more productively due to its increased effective concentration in the vicinity of a potential target molecule. Of course, it is also beyond doubt that the protonation state of a nitrogenous compound may also strongly affect its ability to function as an antioxidant.

The bile pigment bilirubin (Fig. 6.15) has been shown to be an efficient quencher of peroxyl radicals both in homogeneous solution and in liposomes (Stocker et al., 1987). The antioxidant activity of bilirubin increases as the oxygen concentration of the medium is decreased, in analogy with β-carotene. The precise mechanism of radical quenching by bilirubin is not understood fully, but it has been suggested that a hydrogen atom could be donated to ROO˙ from one of the compound's carbon atoms, resulting in a resonance-stabilized product radical which might be able to scavenge other radicals to form nonradical products.

6.5 ANTIOXIDANTS AS SINGLET OXYGEN QUENCHERS

Singlet oxygen-induced damage in tissues could theoretically be minimized either by preventing the $^1O_2$ from forming (intercepting the excited states of sensitizers) or reacting rapidly with $^1O_2$ once it is formed (Davidson and Trethewy, 1976). Because the sensitizer concentration would be expected to be much higher than the $^1O_2$ steady-state concentration (at least in an environment high in water), the former mechanism ought to be quite important. Within cells, both mechanisms are probably co-occurring. Many quenchers of excited states, such as β-carotene, are also potent quenchers of $^1O_2$.

6.5.1 Physical Quenching

Collisional quenching of an energetically excited molecule such as $^1O_2$ occurs with transfer of the excess energy to the quencher. The quencher may then dissipate its newly acquired energy by a variety of mechanisms, most usually by emitting it to the medium as heat, without change in structure. This property may permit the quencher to deactivate many molecules of $^1O_2$. Quenching by solvents usually follows this type of mechanism. Water, for example, rapidly deactivates $^1O_2$, which has a lifetime of only a few microseconds in aqueous solution; in some organic solvents, however, the lifetime may be several orders of magnitude greater (Wilkinson and Brummer, 1981).

β-Carotene is the most reactive naturally occurring singlet oxygen quencher known (Foote et al., 1970), with a rate constant near that of diffusion control, exceeding 10^{10}/M sec. This rate constant exceeds that for the reaction of $^1O_2$ with most biologically important PUFAs by 4–5 orders of magnitude, thus apparently allowing a relatively low concentration of β-carotene to effectively protect membrane lipids from reactions of $^1O_2$ leading to peroxidation. The quenching process appears to be mostly physical, with excess energy being dissipated either thermally to the medium or by reversible isomerization of one of the double bonds of the compound.

Amines have long been known to be physical quenchers of $^1O_2$. Among naturally occurring compounds, several alkaloids of various structural types have been found to be potent inhibitors of $^1O_2$. Particularly effective are indole alkaloids such as strychnine (Fig. 6.16) and brucine (Fig. 6.17) that incorporate a basic nitrogen atom in a rigid, cage-like structure (Larson and Marley, 1984). Such alkaloids appear to be strictly physical quenchers, and are not destroyed chemically by the process of quenching (Gorman et al., 1984). Therefore, in principle, they could inactivate many molecules of $^1O_2$ per molecule of alkaloid. Similarly, ergothioneine (Fig. 6.18), a his-

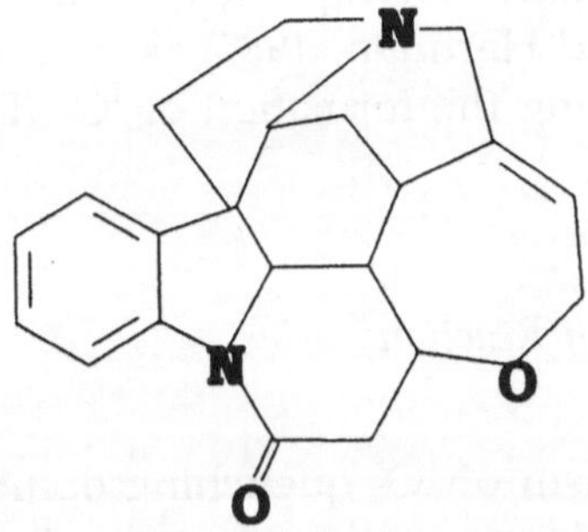

Fig. 6.16

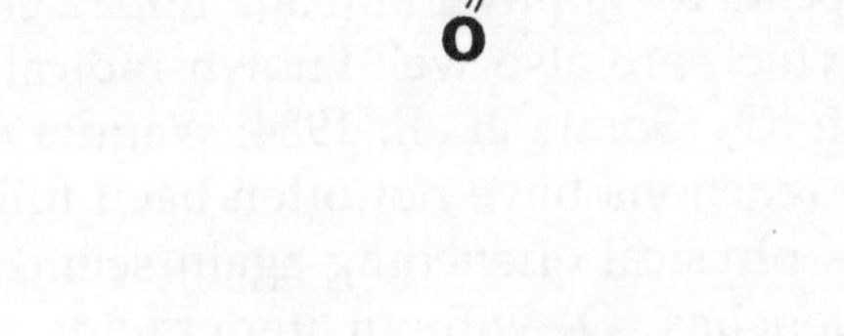

Fig. 6.17

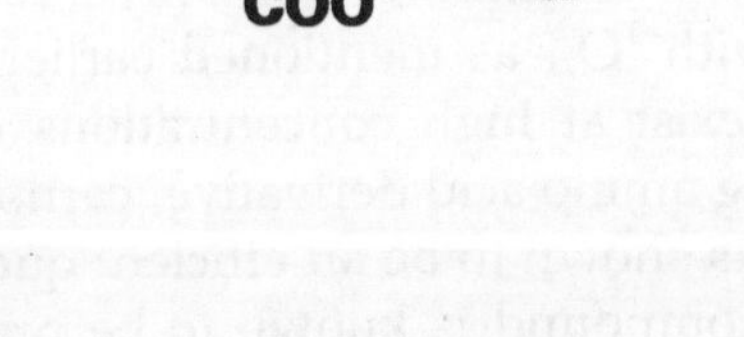

Fig. 6.18

tidine derivative incorporating a thiol function, is found in large amounts in many fungi, and is preserved in red cells and the liver, as well as in other body locations, after ingestion by mammals. It reacts readily with many electrophilic oxidants, including H_2O_2 and ROOH (Hartman and Hartman, 1987) as well as with sensitizer excited states, preventing the formation of $^1O_2$ (Dahl et al., 1988).

6.5.2   Quenching with Reaction

A second mechanism of $^1O_2$ quenching occurs by the reaction of an acceptor with the incorporation of oxygen into the quencher molecule; in some cases, the initially formed oxidation product decomposes spontaneously into new, more stable products.

Many phenolic compounds, in addition to being potent quenchers of free radical reactions, also react quite rapidly with $^1O_2$. Vitamin E, for example, has been determined by several authors (Wilkinson and Brummer, 1981) to have a rate constant (in alcoholic solvents) of roughly 5×10^8 L/mol sec (about 5% of that of β-carotene) with $^1O_2$. Interestingly, there is a high degree of correlation between the $^1O_2$ quenching activity and radical-quenching activities of 17 Vitamin-E related compounds (Mukai et al., 1991). Although reaction products of Vitamin E with $^1O_2$ have been determined, the bulk of the quenching appears to be physical (Fahrenholtz et al., 1974). Some flavonoids, too, which are also well-known radical scavengers, are quite reactive with $^1O_2$ (Sorata et al., 1984; Wagner et al., 1988). The products of these reactions have not often been fully characterized, but in some cases physical quenching again seems to predominate (the flavonoid quenches $^1O_2$ without undergoing much change in concentration) whereas in others the flavonoid is removed rapidly from solution.

Ascorbic acid quenches singlet oxygen with a rate constant of about 10^7 L/mol sec (Chou and Kahn, 1983). Only a few free amino acids are very reactive with $^1O_2$, as mentioned earlier, and not many of these compounds exist at high concentrations in most tissues. A commonly occurring amino acid derivative, carnosine (the dipeptide alanylhistidine), was shown to be an efficient quencher of $^1O_2$ (Dahl et al., 1988). This compound is known to be present in quite high concentrations (up to 40 mM) in muscle.

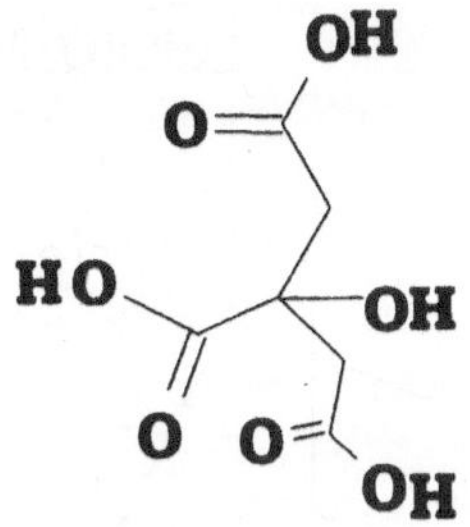

Fig. 6.19

6.6 ANTIOXIDANTS AS METAL ION COMPLEXING AGENTS

Several oxidized transition metal ions such as iron(III) and copper(II) have readily accessible reduced states and, furthermore, are present in high enough abundances in many tissues to make them plausible reactants for one-electron oxidations or reductions that could generate reactive free radicals. For example, a well-known route to ·OH and other radicals is the Fenton reaction, shown in simplified form in eq. 6.16.

$$Fe(II) + H_2O_2 \rightarrow Fe(III) + \cdot OH + HO^- \qquad (6.16)$$

Agents that could bind reactive transition metal cations by complexation could decrease their biological effects dramatically. Commercial synthetic antioxidants such as erythorbic acid and EDTA appear to function largely by this mechanism. In nature, there are many agents which probably could chelate metal ions strongly enough to make them effective antioxidants. Citric acid (Fig. 6.19), one such compound, is in fact used as an additive for foods. Flavonoids, although they may have other antioxidant mechanisms, have also been demonstrated to form stable complexes with copper(II) cations (Thompson et al., 1976, Hudson and Lewis, 1983a). A second effect of complexed metal ions could be to change the redox properties of their ligands, and therefore either promote or inhibit their antioxidant capabilities.

Phytic acid (inositol hexaphosphate; Fig. 6.20) is an abundant constituent of many plants, making up 1–5% of the dry weight of a variety of seeds and pollens. Iron and phytic acid form stable complexed phytates which are ineffective catalysts for the Fenton reaction and are potent inhibitors of lipid peroxidation (Graf et al., 1987).

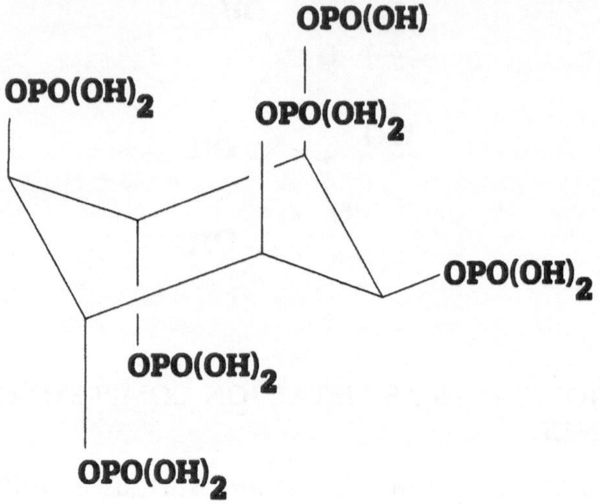

Fig. 6.20

The metal complexing ability of β-diketones is well known in the organometallic literature. The number of examples of these products known from natural sources has been increasing. Curcumin (Fig. 6.21), a well-known pigment isolated from the common spice, turmeric, has been shown to have antioxidant activity (Toda et al., 1985), although its mechanism of action is not known with certainty. Furthermore, many leaf lipids and plant waxes contain these diketones, having long-chain alkyl substituents, which have also been demonstrated to be effective antioxidants (Osawa et al., 1992) in model food-type systems. Although it is possible that the enol form of these molecules could be contributing to their activity, by analogy with phenols, acetylacetone (Fig. 6.22) has virtually no protective effect

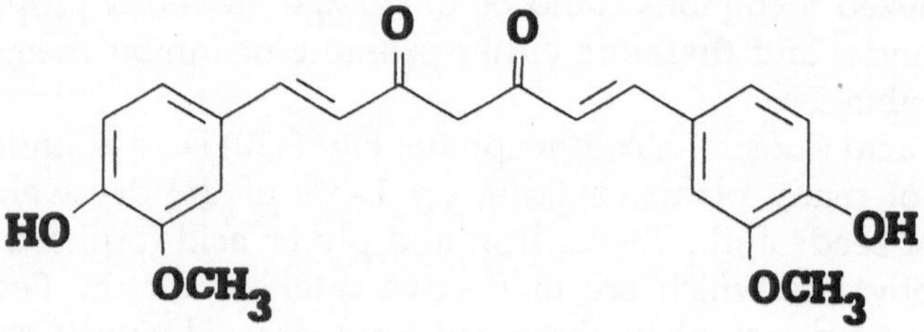

Fig. 6.21

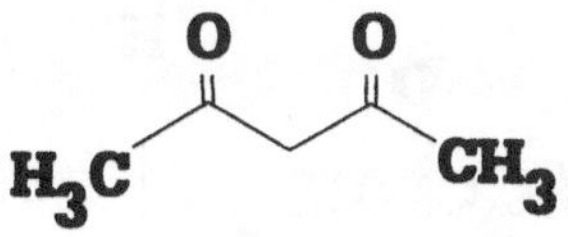

Fig. 6.22

for PUFA oxidation in vitro, suggesting that the long alkyl groups are also essential contributing factors. Human and mammalian tissues and urine also contain significant quantities of β-diketones, derived from fatty acids. The complexing ability of these compounds for calcium ion has led to suggestions that they have roles in intracellular ionic interactions (Douglas, 1991). It is also quite possible that these compounds could be effective sequestrants for transition metal ions with redox activity.

6.7 SYNERGISTIC EFFECTS

It has often been noted that combinations of antioxidants are more effective than would be expected if they were acting independently of one another. For example, vitamins C and E have long been known to be highly effective in combination, although vitamin C is significantly less effective when used alone. The mechanism for this particular interaction has been established, although many others remain to be elucidated. Vitamin E radicals react with vitamin C, abstracting a hydrogen atom to give the nonradical form of vitamin E and the radical of vitamin C (Packer et al., 1979; Niki et al., 1982, 1983; Barclay et al., 1985). This reaction has been demonstrated to have a comparable rate to that of the reaction of vitamin E radicals with peroxyl radicals (Burton and Ingold, 1981), and because vitamin C concentrations are normally manyfold higher than those of ROO˙, regeneration of vitamin E can proceed effectively. Several other naturally occurring compounds, including some amino acid and indole derivatives, may react with vitamin E in a similar electron-transfer mechanism (Cadenas et al., 1989).

A complex lipid, phosphatidylethanolamine (Fig. 6.23), was shown to be a very effective synergist of polyphenol antioxidation in a test in which the induction period for lard autooxidation was measured (Hudson and Lewis, 1983b). It was similarly shown that phosphatidylcholine (Fig. 6.24) as well as phosphatidylethanolamine syn-

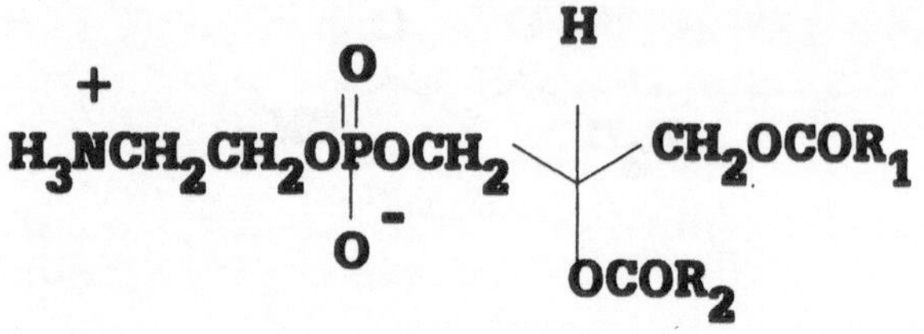

$$\underset{H_3NCH_2CH_2OPOCH_2}{\overset{+}{}} \quad \underset{O^-}{\overset{O}{\underset{\|}{P}}} \quad \underset{OCOR_2}{\overset{H}{\underset{|}{C}}} CH_2OCOR_1$$

Fig. 6.23

ergized the antioxidant effect of α-tocopherol (Hudson and Mah-goub, 1981). The presence of nitrogen, together with phosphorus, in these complex lipids presumably increases their metal-complexing ability and thus their antioxidant activity, but no studies have addressed this problem.

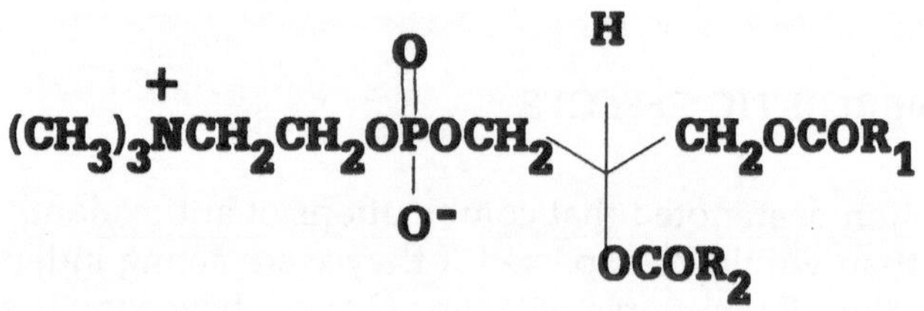

$$\underset{(CH_3)_3NCH_2CH_2OPOCH_2}{\overset{+}{}} \quad \underset{O^-}{\overset{O}{\underset{\|}{P}}} \quad \underset{OCOR_2}{\overset{H}{\underset{|}{C}}} CH_2OCOR_1$$

Fig. 6.24

6.8 SUMMARY

Many naturally occurring compounds have demonstrable antioxidant activity in vitro, and plausible mechanisms for their action have been demonstrated in many cases. These mechanisms include chemical reduction of peroxides or free radicals, deactivation of free radicals by covalent addition reactions, complexation of redox-active metal ions, and singlet oxygen quenching or product formation. However, the mechanisms of action of these and other compounds in vivo and their importance for potentially damaging intracellular processes remain to be fully elucidated.

ACKNOWLEDGMENTS

I thank the U.S. Department of Agriculture for financial assistance (Grant # AF89-372804-897), and May Berenbaum, Karen Marley,

Penney Stackhouse Miller, and Mike Walker for helpful comments on the manuscript.

References

Allen, N.S. (1980) Interaction of phenolic antioxidants with hindered piperidine compounds: a spectrophotometric study. *Makromol. Chem. Rapid Commun.* **1**, 235–24.

Ames, B. N., Cathcart, R., Schwiers, E. and Hochstein, P. (1981) Uric acid provides an antioxidant defense in humans against oxidant- and radical-caused aging and cancer: a hypothesis. *Proc. Natl. Acad. Sci. USA* **78**, 6858–6862.

Ariga, T. and Hamano, M. (1990) Radical scavenging action and its mode in procyanidins B-1 and B-3 from azuki beans to peroxyl radicals. *Agric. Biol. Chem.* **54**, 2499–2504.

Asada, K. and Kanematsu, S. (1976) Reactivity of thiols with superoxide radicals. *Agric. Biol. Chem.* **40**, 1891–1892.

Barclay, L.R.C., Locke, S.J. and MacNeil, J.M. (1985) Autoxidation in micelles. Synergism of vitamin C with lipid-soluble vitamin E and water-soluble Trolox. *Can. J. Chem.* **63**, 366–374.

Barclay, L.R.C., Locke, S.J., MacNeil, J.M., Van Kessel, J., Burton, G.W. and Ingold, K.U. (1984) Autoxidation of micelles and model membranes. Quantitative kinetic measurements can be made by using either water-soluble or lipid-soluble initiators with water-soluble or lipid-soluble chain-breaking antioxidants. *J. Am. Chem. Soc.* **106**, 2479–2481.

Bendich, A., Machlin, L.J., Scandurra, O., Burton, G.W. and Ingold, K.U. (1986) The antioxidant role of vitamin C. *Adv. Free Rad. Biol. Med.* **2**, 419–444.

Boozer, C.E., Hammond, G.S., Hamilton, C.E. and Sen, J.N. (1955) Air oxidation of hydrocarbons. II. The stoichiometry and fate of inhibitors in benzene and chlorobenzene. *J. Am. Chem. Soc.* **77**, 3233–3237.

Brown, D.W., Graupner, P.R., Sainsbury, M. and Shertzer, H.G. (1991) New antioxidants incorporating indole and indoline chromophores. *Tetrahedron* **47**, 4383–4408.

Buettner, G.R. (1988) In the absence of catalytic metals ascorbate does not autoxidize at pH 7. Ascorbate as a test for catalytic metals, *J. Biochem. Biophys. Meth.* **16**, 27–40.

Burton, G.W. and Ingold, K.U. (1981) Autoxidation of biological molecules. I. The antioxidant activity of vitamin E and related chain-breaking phenolic antioxidants *in vitro*. *J. Am. Chem. Soc.* **103**, 6472–6477.

Burton, G.W. and Ingold, K.U. (1981) Beta-carotene: an unusual type of lipid antioxidant. *Science* **224**, 569–573.

Buxton, G.V., Greenstock, C.L., Helman, W.P. and Ross, A.B. (1988) Critical review of rate constants for reactions of hydrated electrons, hydrogen atoms, and hydroxyl radicals in aqueous solution. *J. Phys. Chem. Ref. Data* **17**, 514–886.

Cabelli, D.E. and Bielski, B.H.J. (1983) Kinetics and mechanism for the oxidation of ascorbic acid/ascorbate by HO_2/O_2^- radicals. A pulse radiolysis and stopped-flow photolysis study. *J. Phys. Chem.* **87**, 1809–1817.

Cabrini, L., Pasquali, P., Tadolini, B., Sechi, A.M. and Landi, L. (1986) Antioxidant behavior of ubiquinone and β-carotene incorporated in model membranes. *Free Rad. Res. Commun.* **2**, 85–92.

Cadenas, E., Simic, M.G. and Sies, H. (1989) Antioxidant activity of 5-hydroxy-tryptophan, 5-hydroxyindole, and DOPA against microsomal lipid peroxidation and its dependence on vitamin E. *Free Rad. Res. Commun.* **6**, 11–17.

Chimi, H., Cillard, J., Cillard, P. and Rahmani, M. (1991) Peroxyl and hydroxyl radical scavenging activity of some natural phenolic antioxidants. *J. Am. Oil Chem. Soc.* **68**, 307–312.

Chou, P.-T. and Khan, A.U. (1983) L-ascorbate quenching of singlet delta molecular oxygen in aqueous media: a generalized antioxidant property of vitamin C. *Biochem. Biophys. Res. Commun.* **115**, 932–937.

Cohen, A.M., Aberdroth, R.E. and Hochstein, P. (1984) Inhibition of free radical-induced DNA damage by uric acid. *FEBS Lett.* **174**, 147.

Conn, P.F., Lambert, C., Land, E.J., Schalch, W. and Truscott, T.G. (1992) Carotene-oxygen radical interaction. *Free Rad. Res. Commun.* **16**, 401–408.

Cotelle, N., Bernier, J.L., Henichart, J.P., Catteau, J.P., Guydou, E. and Wallet, J.C. (1992) Scavenger and antioxidant properties of ten synthetic flavones. *Free Rad. Biol. Med.* **13**, 211–219.

Dahl, T.A., Midden, W.R. and Hartman, P.E. (1988) Some prevalent biomolecules as defenses against singlet oxygen damage. *Photochem. Photobiol.* **47**, 357–362.

Davidson, R.S. and Trethewey, K.R. (1976) The role of the excited singlet state of dyes in dye-sensitized photooxidation reactions. *J. Am. Chem. Soc.* **98**, 4008–4009.

Doba, T., Burton, G.W. and Ingold, K.U. (1985) Antioxidant and co-oxidant activity of vitamin C. The effect of vitamin C, either alone or in the presence of vitamin E or a water-soluble vitamin E analogue, upon the peroxidation of aqueous multilamellar phospholipid liposomes. *Biochim. Biophys. Acta* **835**, 298–303.

Douglas, D.E. (1991) Higher aliphatic 2,4-diketones: a ubiquitous lipid class with chelating properties, in search of a physiological function. *J. Lipid Res.* **32**, 553–558.

Fahrenholtz, S.R., Doleiden, F.H., Trozzolo, A.M. and Lamola, A. (1974) On the quenching of singlet oxygen by α-tocopherol. *Photochem. Photobiol.* **20**, 505–509.

Foote, C.S., Denny, R.W., Weaver, L., Chong, Y. and Peters, J. (1970) Chemistry of singlet oxygen. X. Carotenoid quenching parallels biological protection. *J. Am. Chem. Soc.* **92**, 5216–5219.

Gorman, A.A., Hamblett, I., Smith, K. and Standen, M.C. (1984) Strychnine: a fast physical quencher of singlet oxygen. *Tetrahedron Lett.* **25**, 581–584.

Graf, E., Empson, K.L. and Eaton, J.W. (1987) Phytic acid, a natural antioxidant. *J. Biol. Chem.* **262**, 11647–11650.

Hartman, Z. and Hartman, P.E. (1987) Interception of some direct-acting mutagens by ergothioneine. *Environ. Molec. Mutagen.* **10**, 3–15.

Hodnick, W.F., Milosavljevic, E.B., Nelson, J.H. and Pardini, R.S. (1988) Electrochemistry of flavonoids. Relationships between redox potentials, inhibition of mitochondrial respiration, and production of oxygen radicals by flavonoids. *Biochem. Pharmacol.*, **37**, 2607–2611.

Howard, J.A. and Yamada, T. (1981) Absolute rate constants for hydrocarbon autoxidation. 31. Autoxidation of cumene in the presence of tertiary amines. *J. Am. Chem. Soc.* **103**, 7102–7106.

Hudson, B.J.F. and Lewis, J.I. (1983a) Polyhydroxy flavonoid antioxidants for edible oils. Structural criteria for activity. *Food Chem.* **10**, 47–55.

Hudson, B.J.F. and Lewis, J.I. (1983b) Polyhydroxy flavonoid antioxidants for edible oils. Phospholipids as synergists. *Food Chem.* **10**, 111–120.

Hudson, B.J.F. and Mahgoub, S.E.O. (1981) Synergism between phospholipids and naturally-occurring antioxidants in leaf lipids. *J. Sci. Food Agric.* **32**, 208–210.

Husain, S.R., Cillard, J. and Cillard, P. (1987) Hydroxyl radical scavenging activity of flavonoids. *Phytochemistry* **26**, 2489–2491.

Katbab, A.A., Ogunbanjo, A. and Scott, G. (1985) Mechanisms of antioxidant action: antidegradant activities of phenols and quinones derived from phenolic sulfides in a peroxide vulcanizate. *Polym. Degr. Stabil.* **12**, 333–347.

Kennedy, T.A. and Liebler, D.C. (1992) Peroxyl radical scavenging by β-carotene in lipid bilayers. *J. Biol. Chem.* **267**, 4658–4663.

Larson, R.A. and Marley, K.A. (1984) Quenching of singlet oxygen by alkaloids and related nitrogen heterocycles. *Phytochemistry* **23**, 2351–2354.

Loth, H. and Diedrich, H. (1967) Redoxpotentiale einiger 3,4-dihydroxylierter Flavon- und Flavonolderivate. *Arch. Pharm.* **301**, 103–110.

Mellors, A. and Tappel, A.L. (1966) The inhibition of mitochondrial peroxidation by ubiquinone and ubiquinol. *J. Biol. Chem.* **241**, 4353–4356.

Mukai, K., Daifuku, K., Ozabe, K., Tanigaki, T. and Inouye, K. (1991) Structure-activity relationships in the quenching reaction of singlet oxygen by tocopherol (vitamin E) derivatives and related phenols. Finding of linear correlation between the rates of quenching os singlet oxygen and scavenging of peroxyl and phenoxyl radicals in solution. *J. Org. Chem.* **56**, 4189–4192.

Niki, E., Saito, T. and Kamiya, Y. (1983) The role of vitamin C as an antioxidant. *Chem. Lett.*, 631–632.

Niki, E., Saito, M., Yoshikawa, Y., Yamamoto, Y. and Kamiya, Y. (1986) Oxidation of lipids. XII. Inhibition of oxidation of soybean phosphatidylcholine and methyl linoleate in aqueous dispersions by uric acid. *Bull. Chem. Soc. Japan* **59**, 471–477.

Niki, E., Tsuchiya, J., Tanamura, R. and Kamiya, Y. (1982) Oxidation of lipids. II. Rate of inhibition of oxidation by α-tocopherol and hindered phenols measured by chemiluminescence. *Bull. Chem. Soc. Japan* **55**, 1551–1555.

Osawa, T., Ramanatham, M., Kawakishi, S. and Namiki, M. (1992) Antioxidant defense systems generated by phenolic plant constituents. *Am. Chem. Soc. Symp.* **507**, 122–134.

Packer, J.E., Mahood, J.S., Mora-Arellano, V.O., Slater, T.F., Willson, R.L. and Wolfenden, B.S. (1981) Free radicals and singlet oxygen scavengers: reaction of a peroxy radical with beta-carotene, diphenylfuran, and 1,4-diazabicyclo [2.2.2]octane. *Biochem. Biophys. Res. Commun.* **98**, 901–906.

Packer, J.E., Slater, T.F. and Willson, R.L. (1979) Direct observation of a free radical intermediate between vitamin E and vitamin C. *Nature* **278**, 737–738.

Palozza, P. and Krinsky, N.I. (1992) Astaxanthin and canthaxanthin are potent antioxidants in a membrane model. *Arch. Biochem. Biophys.* **297**, 291–295.

Pryor, W.A., Strickland, T. and Church, D.F. (1988) Comparison of the efficiencies of several natural and synthetic antioxidants in aqueous sodium dodecyl sulfate micelle solutions. *J. Am. Chem. Soc.* **110**, 2224–2229.

Ricardo da Silva, J.M., Darmon, N., Fernandez, Y. and Mitjavila, S. (1991) Oxygen free radical scavenger capacity in aqueous models of different procyanidins from grape seeds. *J. Agric. Food Chem.* **39**, 1549–1552.

Robak, J. and Gryglewski, R.J. (1988) Flavonoids are scavengers of superoxide anions. *Biochem. Pharmacol.* **37**, 837–841.

Sawyer, D.T., Calderwood, J.S., Johlman, C.L. and Wilkins, C.L. (1985) Oxidation by superoxide ion of catechols, ascorbic acid, dihydrophenazine, and reduced flavins to their respective anion radicals. A common mechanism via a sequential proton-hydrogen atom transfer. *J. Org. Chem.* **50**, 1409–1412.

Sorata, Y., Takahama, U. and Kimura, M. (1984) Cooperation of quercetin with ascorbate in the protection of photosensitized lysis of human erythrocytes in the presence of hematoporphyrin. *Biochim. Biophys. Acta* **799**, 313–317.

Speisky, H., Cassels, B.K., Lissi, E. and Videla, L.A. (1991) Antioxidant properties of the alkaloid boldine in systems undergoing lipid peroxidation and enzyme inactivation. *Biochem. Pharmacol.* **41**, 1575–1581.

Steenken, S. and Neta, P. (1982) One-electron redox potentials of phenols. Hydroxy- and aminophenols and related compounds of biological interest. *J. Phys. Chem.* **86**, 3661–3667.

Stocker, R., Yamamoto, Y., McDonagh, A.F., Glazer, A.N. and Ames, B.N. (1987) Bilirubin is an antioxidant of possible physiological importance. *Science* **235**, 1043–1046.

Suzuki, Y.J., Tsuchiya, M. and Packer, L. (1991) Thioctic acid and dihydrolipoic acid are novel antioxidants which interact with reactive oxygen species. *Free Rad. Res. Commun.* **15**, 255–263.

Taira, J., Ikemoto, T., Yoneya, T., Hagi, A., Murakami, A. and Makinko, K. (1992) Essential oil phenyl propanoids: useful as ˙OH scavengers? *Free Rad. Res. Commun.* **16**, 197–204.

Tajima, K., Sakamoto, M., Okada, K., Mukai, K., Ishizu, K., Sakurai, H. and Mori, H. (1983) Reaction of biological phenolic antioxidants with superoxide generated by cytochrome P-450 model system. *Biochem. Biophys. Res. Commun.* **115**, 1002–1008.

Terao, J. (1989) Antioxidant activity of β-carotene-related carotenoids in solution. *Lipids* **24**, 659–661.

Terce, F., Tocanne, J.-F. and Laneelle, G. (1982) Interaction of ellipticine with model or natural membranes. A spectrophotometric study. *Eur. J. Biochem.* **125**, 203–207.

Thompson, M., Williams, C.R. and Elliott, G.E.P. (1976) Stability of flavonoid complexes of copper(II) and flavonoid antioxidant activity. *Anal. Chim. Acta* **85**, 375–381.

Toda, S., Miyase, T., Arichi, H., Tanizawa, H. and Takino, Y. (1985) Natural antioxidants. III. Antioxidative components isolated from rhizome of *Curcuma longa* L. *Chem. Pharm. Bull.* **33**, 1725–1728.

Torel, J., Cillard, L. and Cillard, P. (1986) Antioxidant activity of flavonoids and reactivity with peroxy radical. *Phytochemistry* **25**, 383–385.

Wagner, G.R., Youngman, R.J. and Elstner, E.F. (1988) Inhibition of chloroplast photo-oxidation by flavonoids and mechanisms of the antioxidative action. *J. Photochem. Photobiol. B* **1**, 451–460.

Weng, X.C. and Gordon, M.H. (1992) Antioxidant activity of quinones extracted from tanshen (*Salvia miltiorrhiza* Bunge). *J. Agric. Food Chem.* **40**, 1331–1336.

Wilkinson, F. and Brummer, J.G. (1981) Rate constants for the decay and reactions of the lowest electronically excited singlet state of molecular oxygen in solution. *J. Phys. Chem. Ref. Data* **10**, 809–999.

Yuting, C., Rongliang, Z., Zhongjian, J. and Yong, J. (1990) Flavonoids as superoxide scavengers and antioxidants. *Free Rad. Biol. Med.* **9**, 19–21.

CHAPTER 7

Antioxidant Mechanisms of Enzymes and Proteins

Sami Ahmad

7.1 INTRODUCTION

While the aerobic life-style offers great advantages, it is subject to oxygen toxicity due to endogenous activation of ground-state molecular oxygen to the superoxide anion radical ($O_2^{\cdot-}$), hydrogen peroxide (H_2O_2) and hydroxyl radical ($^{\cdot}OH$). Thus, "any organism that avails itself of the benefits of oxygen does so at the cost of maintaining an elaborate system of defenses against these intermediates," (Fridovich, 1983).

7.1.1 The Oxygen Radical Cascade

The oxygen activation follows a series of electron transfer reactions (see chapter 1), and the pertinent ones are as follows.

$$O_2 + e^- \longrightarrow O_2^{\cdot-} \tag{7.1}$$

$$O_2^{\cdot-} + H^+ \longrightarrow HO_2^{\cdot} \tag{7.2}$$

$$HO_2^{\cdot} + H^+ + e^- \longrightarrow H_2O_2 \tag{7.3}$$

$$H_2O_2 + e^- \longrightarrow {}^{\cdot}OH + HO^- \tag{7.4}$$

The endogenous sources for production of $O_2^{\cdot-}$ range from small to large autoxidizable molecules such as catecholamines, ubihydroquinone, and oxidoreductases such as hemoproteins and flavin enzymes. Thus, O_2^- is generated in virtually all subcellular compartments including cytosol, mitochondria, endoplasmic reticulum, nuclei, and chloroplasts in plants. Superoxide radical produced is in equilibrium with its protonated form, the hydroperoxyl radical $HO_2^{\cdot}$ (eq. 7.2). Both forms dismutate to H_2O_2 which in the presence of metals such as iron decomposes to $^{\cdot}OH$ radical and hydroxyl anion. Ground-state O_2 is also activated in photosensitization and other reactions to singlet O_2 ($^1O_2$) (Singh, 1989; see chapter 1).

7.1.2   The Lipid Peroxidation Chain Reaction

While O_2^- and H_2O_2 are toxic to cells, the extremely high reactivity of $^{\cdot}OH$ and $^1O_2$ renders these activated forms most cytotoxic due to deleterious peroxidation reactions with lipids, proteins and DNA. Briefly, this occurs as exemplified for lipid peroxidation (eqs. 7.5–7.7; see also chapter 1 for more details). The $^{\cdot}OH$ radical initiates a lipid peroxidation chain reaction with a variety of lipids such as a

$$LH + {^{\cdot}OH} \longrightarrow L^{\cdot} + H_2O \qquad (7.5)$$

$$L^{\cdot} + O_2 \longrightarrow LO_2^{\cdot} \qquad (7.6)$$

$$LO_2^{\cdot} + LH \longrightarrow L^{\cdot} + LOOH \qquad (7.7)$$

polyunsaturated fatty acid (PUFA; shown above as LH) to form a lipid radical ($L^{\cdot}$) and a lipid hydroperoxide (LOOH). Singlet O_2 attack results in a direct insertion of O_2 in a LH molecule with the formation of LOOH or a cyclic endoperoxide. The $L^{\cdot}$ radical can self-react to form LL (a chain termination process), but unless LOOH's are removed $L^{\cdot}$ will continue the lipid peroxidation chain reaction by reacting with O_2 and regenerating the peroxidizing lipid peroxyl radical, $LO_2^{\cdot}$ (eq. 7.6).

Oxidative stress is exerted by all peroxides which directly damage cells and tissues, or from their more reactive breakdown products such as malondialdehyde and hydroxynonenals (Mannervik, 1985). Moreover, LOOH's are catalytically decomposed by metals such as iron or copper to free radicals such as $LO_2^{\cdot}$ which will contribute to the propagation of the lipid peroxidation chain reaction (Borg and Schaich, 1988). It is evident then that "oxidative stress is a chain-event, and a single initiating event caused by a prooxidant may cascade into a widespread chain reaction that produces many deleterious products in concentrations many magnitudes greater than the initiator" (Ahmad, 1992). This is exemplified by the fact that thousands of PUFA molecules may be destroyed by a lipid peroxidation chain reaction initiated by a single free radical (McCord, 1985). It is imperative that in order to prevent this vicious chain reaction, the O_2 radical cascade to O_2^- and H_2O_2 must be attenuated, and the peroxides converted to innocuous metabolites. Consequently, all aerobic organisms possess elaborate defense mechanisms to prevent the formation of toxic forms of oxygen and to remove peroxides formed.

As a first line of defense against oxygen toxicity, many endogenous non-protein small antioxidant molecules are important in quenching free radicals and reacting with peroxides (see chapters 1 and 6 for details). However, the quenching mechanisms *per se* are often insufficient because the reactions involved may decrease the tissue levels of such key antioxidants as α-tocopherol and ascorbic acid. Furthermore, under conditions of oxidative stress arising from exposure to a prooxidant, the supply of antioxidant molecules may be more drastically curtailed. Therefore, along with antioxidant molecules, antioxidant enzymes are considered essential for alleviating oxidative stress by acting as chain-breakers of the O_2 radical cascade and lipid peroxidation chain reaction.

This chapter serves as an introduction and overview of the antioxidant enzymatic defenses and some proteins with antioxidant properties.

7.2 PRIMARY ANTIOXIDANT ENZYMES

The primary antioxidant enzymes that minimize the O_2 radical cascade and remove cytotoxic peroxides are listed (Table 7.1).

7.2.1 *Superoxide Dismutase*

At pH 7.8, the spontaneous dismutation of either two O_2^- radicals or two HO_2 radicals is slow, while that between O_2^- and HO_2 radicals is a rapid process (eq. 7.8).

$$O_2^- + O_2^- + 2H^+ \longrightarrow H_2O_2 + O_2 \quad k_2 = 8 \times 10^4 \, M^{-1} \, sec^{-1} \quad (7.8)$$

This fast spontaneous dismutation raises the paradox as to why an

Table 7.1
Primary Antioxidant Enzymes

Enzymes	Systematic Name	EC Number
Superoxide dismutase	Superoxide: superoxide dismutase	1.15.1.1
	Hydroperoxidases	
Cytochrome-*c* peroxidase	Ferrocytochrome-*c*: hydrogen-peroxide oxidoreductase	
Catalase	Hydrogen-proxide: hydrogen-peroxide oxidoreductase	1.11.1.6
Alkyl hydroperoxidase	NE*	NE*
Peroxidase	Donor: hydrogen-peroxide oxidoreductase	1.11.1.7
Ascorbate peroxidase	L-Ascorbate: hydrogen-peroxide oxidoreductase	1.11.1.11
	Glutathione peroxidases	
Glutathione peroxidase (Selenium dependent)	Glutathione: hydrogen-peroxide oxidoreductase	1.11.1.9
Glutathione transferase	RX: glutathione s-transferase	2.5.1.18
	Glutathione reductase	
Glutathione reductase	NAD(P)H: oxidized-glutathione oxidoreductase	1.6.4.2.

* NE; not established.

enzyme such as superoxide dismutase (SOD) should be at all required to dismutate O_2^-. Fridovich (1983) provided a logical solution to this paradox. First, the rate constant for the spontaneous dismutation is second order (eq. 7.8), because the first half-life of O_2^- is an inverse function of its initial concentration. The first half-life of O_2^- at 1×10^{-4} M is about 0.05 seconds, while at 1×10^{-10} M the first half-life is as long as 14 hours. The tissue concentration of SOD averages 1×10^{-5} M, whereas that of O_2^- is 1×10^{-11} M. The possibility of collision of a O_2^- with SOD is 10^6-fold greater than the possibility of collision with another O_2^-; the enzymatic dismutation therefore is 10^6-fold faster than the spontaneous dismutation. On the other hand, the reaction between SOD and O_2^- is first order with a rate constant of 2×10^9 M^{-1} sec^{-1}, and this rate is 10^4 times faster than the spontaneous rate (eq. 7.8). From these calculations, Fridovich (1983) concluded that SOD eliminates O_2^- 10^{10}-fold faster than spontaneous dismutation. This generalization, however, is very much a function of pH 7.8 and the steady-state concentration of O_2^-; the spontaneous dismutation is maximal about pH 5.0 and is very slow above pH 9.0. This cautionary statement is perhaps of least importance in relation to dismutation rates near physiological pH and O_2^- concentration. Nonetheless, the need for an enzyme to rapidly remove O_2^- from tissues is abundantly clear.

There are three types of SOD's and each type appears different, although based on amino acid sequence homology they represent two evolutionary groups (Stallings et al., 1984). The manganese containing enzyme (MnSOD) and iron containing enzyme (FeSOD) are somewhat similar and of ancient origin having evolved in procaryotes, while the conspicously different copper/zinc containing enzyme (CuZnSOD) of eucaryotes represents a more recent evolutionary origin. Procaryotes possess both MnSOD and FeSOD, and some species such as *Escherichia coli* contain a heterodimer composed of both MnSOD and FeSOD (see chapter 8). MnSOD has been well characterized from the fungus, yeast, and plants possess all three types of SOD's. In vertebrates CuZnSOD is the predominant form present in cell cytosol (about 80% of total SOD), while the remainder is MnSOD present in the mitochondrial matrix (Fridovich, 1983). A similar pattern of SOD occurs in invertebrates. For example, in insects the CuZnSOD in cytosol is in the 50–70% range, but at least in one species, *Trichoplusia ni*, the CuZnSOD is only 25% of the total SOD (Ahmad et al., 1988a,b 1990). Thus higher amounts of MnSOD exist in some animals but the reason for this anomaly is at present obscure. Another feature of eucaryotic SOD distribution is a complete absence of MnSOD from erythrocytes and leukocytes, and as recently discovered insect blood cells, hemocytes, are also devoid of MnSOD (Ahmad et al., 1991). Similarly in plants, the CuZnSOD is a cytosolic enzyme and MnSOD is a mitochondrial enzyme with the exception that chloroplasts contain FeSOD, CuZnSOD, or both forms, and in some species MnSOD may also be present (see chapter 9).

CuZnSOD is a dimeric protein whose molecular weight averages 32 kDa (Paoletti and Mocali, 1990). Many isoenzymes have been reported. MnSOD is a tetrameric protein with an average molecular weight of 86 kDa. FeSOD is a dimer with a molecular weight of 41 kDa (Salin 1987). Regardless of the metal co-factor, all forms of SOD's share the same mechanism and speed in dismutating O_2^-. The catalytic mechanism of CuZnSOD is well characterized (Tainer et al., 1983; Getzoff et al., 1983; Klapper et al., 1986; Sharp et al., 1987) (see chapter 8 for details).

The SOD-O_2^- interaction is promoted by electrostatic charges of SOD for substrate guidance and charge compatibility. The channel leading to the active site has an extensive electrostatic field for rapid passage of the negatively charged O_2^- without charge hinderance until it reaches the highly positive binding site. The sequence conserved amino acids are responsible for directing the long-range approach of O_2^-, and there is a striking positive region that extends from the long deep channel to the enzyme surface, which no doubt serves as the recognition/attraction site for O_2^-. These features of enzyme topography, and electrostatic field are ideal for rapid interception and fast dismutation of O_2^-, by SOD. The catalytic mechanisms of MnSOD and FeSOD have not been studied in as much depth as CuZnSOD, but apparently the mechanisms are the same including the active site configuration and conserved amino acids, with alternate cycles of metal reduction and reoxidation to complete the reaction cycle (see chapter 8). Thus, a generalized scheme of the catalytic cycle of SOD's emerges as follows. In this scheme, $M(n+1)$ represents oxidized states of the metals, i.e., Cu(II), Mn(III) and

$$E + M(n+1) + O_2^- + H^+ \longrightarrow E-M(n) + O_2 \qquad (7.9)$$

$$E-M(n) + O_2^- + H^+ \longrightarrow E - M(n-1) + H_2O_2 \qquad (7.10)$$

$$\text{Overall: } 2O_2^- + 2H^+ \longrightarrow O_2 + H_2O_2 \qquad (7.11)$$

Fe(III), and the corresponding reduced forms, i.e., Cu(I), Mn(II) and Fe(II).

The SOD types can be distinguished by their differential response to H_2O_2 and KCN (Bowler et al., 1989). MnSOD is not affected by either H_2O_2 or KCN, FeSOD is sensitive to H_2O_2 but not KCN, while CuZnSOD is inhibited by both H_2O_2 and KCN. In addition, diethyldithiocarbamate is an excellent Cu chelator and can be used both in vivo and in vitro to inhibit CuZnSOD, and the portion not inhibited in case of animal preparations by substraction yields a fairly good estimate of the amount of MnSOD (Misra, 1979). Another feature of these enzymes is their inducibility. Under excessive oxidative stress such as by prooxidants, the eucaryotic CuZnSOD is usually rapidly induced (Bowler et al., 1989; Ahmad and Pardini, 1990a). Prooxidants usually do not induce the MnSOD or FeSOD; however under intrinsic conditions which may lead to increased mitochondrial or chloroplast metabolism (Bowler et al., 1989) or from stimulus of interleukin 1, MnSOD is induced (Masuda et al., 1988). Thus, endogenous metabolic cues rather than exogenous factors are more important in inducing the activity of MnSOD, and both factors may induce FeSOD. However, this cautious proposal requires clarity. Thus SOD appears to be the most important antioxidant enzyme since its point of attack is O_2^- which is the initiator of the oxygen radical cascade that feeds into the lipid peroxidation chain reaction. Thus, it is not surprising that almost all aerobic organisms possess this enzyme. In this regard a few anomalies should be pointed out. A CuZnSOD was isolated from a bacterium, *Photobactor leiognathi* (Puget and Michelson, 1974). As it lives as a symbiont in the light organ of the pony fish, Martin and Fridovich (1981) suggested that the bacterium obtained this enzyme by gene transfer from the host fish. The presence of CuZnSOD in a free-living bacterium *Caulobacter crecentus* is hard to explain (Steinman, 1982). If this occurred by gene transfer then it must have been long time ago; the donor which is unknown may have transferred the gene to some symbiotic/parasitic ancestor of *C. crecentus*. Since FeSOD is also a procaryotic enzyme and its occurrence in plants is not ubiquitous, gene transfer again may be responsible. Lastly, a number of microbials have been found to lack this enzyme, and at least in one case of *Lactobacillus planatarum*, an explanation of this anamoly exists. This bacterium lives in Mn-rich media and accumulates up to 25 mM Mn. Although less efficient than MnSOD, Mn can catalyse the dismutation of O_2^-. If the bacteria are deprived of Mn by growing them in Mn-free media, then they become sensitive to O_2-dependent toxicity of quinones (Archibald and Fridovich, 1981, 1982).

7.2.2 Hydroperoxidases

The dismutation of O_2^- radicals results in the production of H_2O_2 which represents a higher state of oxidation than O_2^-. Hydroperoxidases require for catalytic activity one molecule of either H_2O_2 or an LOOH (or any organic hydroperoxide; ROOH) as an oxidant and another substrate (DH_2) as a hydrogen (H) donor; there are many types of a DH_2 substrate including H_2O_2 and ROOH's.

i. Catalase

Catalase (CAT) acts sequentially to SOD to dismutase two identical substrate molecules to lower the oxidation state (eq. 7.11) (Chance et al., 1979). This overall reaction proceeds in two steps.

$$H_2O_2 + H_2O_2 \rightarrow 2H_2O + O_2 \tag{7.11}$$

$$CAT\text{-}Fe(III) + H_2O_2 \rightarrow Complex\ I \tag{7.12}$$

$$Complex\ I + H_2O_2 \rightarrow CAT\text{-}Fe(III) + 2H_2O + O_2 \tag{7.13}$$

$$Complex\ I + DH_2 \rightarrow CAT\text{-}Fe(III) + 2H_2O + D \tag{7.14}$$

As characteristic of hydroperoxidases, CAT exhibits dual activities, i.e., "catalatic" in which two molecules of H_2O_2 are dismutated, one molecule acting as an oxidant while the other as a reductant (H donor) in two consecutive steps (eqs. 7.12 and 7.13). The other reaction is called "peroxidative" in which the reductant (DH_2) is the H donor; examples of DH_2 are methanol, ethanol, formic acid, phenols and amines (eq. 7.14) (Chance et al., 1979; Aebi, 1984). In both reactions, an active enzyme-H_2O_2 intermediate called complex I is formed first. In the typical catalatic activity of CAT, the reaction between complex I and another molecule of H_2O_2 proceeds extremely rapidly, and the rate constant k for this reaction (eq. 7.13) equals 10^7 liters $mol^{-1}\ sec^{-1}$, while the peroxidative reaction (eq. 7.13) proceeds slowly at a rate constant $k = 10^2 - 10^3$ liters $mol^{-1}\ sec^{-1}$ (Aebi, 1984).

CAT is an unusual enzyme in that it does not obey the classical Michelis-Menton kinetics. The enzyme exhibits non-saturation kinetics, and enzyme activity increases linearly with the available H_2O_2 over a wide range of its concentration. Thus, K_m can not be determined, and even at saturation concentration the enzyme decomposes H_2O_2 in a first order reaction. Therefore, to avoid a rapid decrease in the initial rate of reaction, Aebi's (1984) recommendation is followed to conduct the CAT assay at a relatively low substrate concentration, i.e., 10 mM H_2O_2. There is only slight dependence on either temperature or pH for this fast-acting enzyme, and most assays are conducted at pH 7.0 and 20–25 °C. Also, only that portion of activity is computed that represents a steep slope (0–30 sec) of a first order reaction. Only at H_2O_2 concentration exceeding 0.1 M, CAT is inactivated (called autoinactivation) due to the conversion of complex I to inactive complexes II and III (Aebi, 1984).

The primary role of CAT is to decompose H_2O_2 although some activity occurs with methyl- and ethyl hydroperoxides. The mammalian CAT has no activity toward *tert*-butylhydroperoxide or other ROOH's. CAT is present in most eucaryotic cells and the enzyme is primarily localized in the peroxisomes, although its biogenesis is from extraperoxisomally synthesized precursors. Peroxisomes or the microbodies of Rouiller originate from the golgi-endoplasmic reticulum-lysosome (GERL) complex (Ross and Reith, 1985). The peroxisomes are generally continuous with the smooth endoplasmic reticulum (Weiss, 1983). The peroxisomal location of CAT is considered strategic in that H_2O_2 is produced by many peroxisomal enzymes by a direct two-electron reduction of O_2, and without the intermediacy of O_2^- radical. Peroxisomal production of H_2O_2 is from flavin enzymes such as urate oxidase, glucose oxidase and D-amino acid oxidases.

Peroxisomal turnover is rapid and its measured half-life is approximately 1.5–2 days. During the fiftees and sixtees conflicting reports had existed for the occurrence of intact extraperoxisomal CAT because of the fragility of peroxisomes (Chance et al., 1979). More recently, Nohl and Jordan (1980) have confirmed that in the rat heart mitochondria nearly 85% of H_2O_2 generated is destroyed by a CAT located in the mitochondrial matrix, while only 15% of the H_2O_2 was decomposed by the Se-dependent glutathione peroxidase. Further studies by Jones et al. (1981) have demonstrated that CAT in the endoplasmic reticulum becomes more important than a Se-dependent glutathione peroxidase when H_2O_2 concentration exceeds nM concentrations.

In our studies of insects using a very gentle subcellular fraction protocol, CAT's of three insect species were found widely distributed in mitochondria, microsomes and cytosol, with negligible activity in nuclei (Ahmad et al., 1988a, 1988b, 1990). In these studies, peroxisomes were associated with the endoplasmic reticulum and all vesicles were found intact upon electron microscopy. Moreover, in all three insect species examined, the mitochondrial CAT level was higher than that of the microsome-peroxisome complex and cytosol; the latter two organelles possessed nearly equal amounts of CAT. Another striking feature of insect CAT is its high basal activity in cell-free homogenates, about 200–300+ units (Ahmad and Pardini, 1990a; Ahmad, 1992), and this feature has been confirmed for additional insect species (Del Vicchio, 1988; Aucoin et al., 1991).

The role of CAT in plants in contrast to other peroxidases is relatively minor (see chapter 9). In a study of cytosolic H_2O_2 arising from mitochondrial O_2^- dismutation in soybean embryonic axes, H_2O_2 elimination was found in part due to CAT along with other hydroperoxidases (Puntarulo et al., 1991). In procaryotes, however, CAT is an important enzyme, and in *E. coli* there are two forms called HPI and HPII (see chapter 8).

CAT's are homotetrameric hemoproteins (in some exceptional cases homodimeric and homohexameric forms have been reported) with one protoheme IX group per monomer. The average molecular weight of the native protein in mammalian species is 250–270 kDa. At least in two insect species *Drosophila melanogaster* (Nahmias and Bewley, 1984) and *T. ni* (Mitchell et al., 1991), the purified enzyme resembles the bovine CAT. Recently purified CAT of the insect *Spodoptera eridania* is also a tetramer, and the presence of four protoheme IX groups, one per monomer, has been confirmed (Weinhold, L.C., Ahmad, S. and Pardini, R.S., 1993, unpublished). Unlike any other CAT, however, this enzyme is a unique polymer of four heterogenous subunits (42–65 kDa) and the native enzymes molecular weight is about 270 kD. The catalatic activity of this enzyme is strongly inhibited by KCN, and the peroxidative activity is maximally inhibited by NaN_3 in slightly acidic pH, which are inhibitions characteristic of all CATs. Whereas, CAT's are also inhibited by 3-amino-1,2,4-triazole (AT), the *S. eridania* enzyme is minimally affected by AT. Explanation for this anomaly will be sought, especially with respect to the heteropolymeric nature of the protein which may be causing a steric hinderance for AT interaction.

The paradoxically high activity and wide subcellular distribution of CAT in insects deserves further comment. At least three explanations can be made at this time: (1) the absence of a Se-dependent glutathione peroxidase; (2) higher rate of steady state accumulation of H_2O_2 than in other organisms; and (3) the demand for energy for these short-lived animals during their more rapid embryogenesis, tissue differentiation and metamorphosis. In this regard, CAT involvement in gluconeogenesis has been suggested (Giorgi and Deri, 1976; Best-Belpomme and Ropp, 1982). This may involve rapid deamination of D-amino acids to α-keto acids, and formation of glucose from lipids and other noncarbohydrate precursors (Weiss, 1983). High demand for ATP would generate large amounts of H_2O_2 in mitochondria and cytosol. Thus, H_2O_2 concentration may far exceed the normal nM range for efficient diffusion into peroxisomes for destruction.

The two CATs of *E. coli*, HPI and HPII, catalyze the same basic reaction but their gene regulation and compartmentation are different; HPI appears confined to the periplasmic and cell membrane, whereas HPII is confined to the cytoplasm (Heimberger and Eisenstark 1988). In particular, HPII is produced in response to conditions of stationary growth and starvation. Another difference between the two HPs is that HPI exhibits a broad peroxidase activity, while HPII is essentially devoid of any peroxidative activity (Claiborne et al., 1979; Heimberger and Eisenstark, 1988). CAT of the insect *S. eridania* is induced four-fold under conditions of starvation, but not from prooxidants (Pritsos et al., 1990). An explanation for this behavior of HPII of *E. coli* and *S. eridania* CAT, which are functionally similar to other CAT's including HPI, remains obscured.

Although CAT decomposes H_2O_2 very efficiently, the enzyme is turned on at rather high H_2O_2 concentrations (Halliwell, 1982). Thus, CAT has a relatively minor role in catabolism of H_2O_2 at low rates of H_2O_2 generation (Jones et al., 1981), but the CAT's role increases and becomes indispensable when the H_2O_2 production is enhanced from oxidative stress.

ii.   Alkyl hydroperoxidase

Alkyl hydroperoxidase (AHP), also known as AHP reductase (see chapter 8) is a relatively new discovery from procaryotes *E. coli* and *Salmonella typhyimurium* (Farr and Kogoma, 1991). This is a flavo-protein enzyme and depends on NADH or NADPH for catalyzing the reduction of alkyl hydroperoxides to their corresponding alcohols. In *S. typhimurium*, AHP upon isolation yields a larger 57 kDa subunit and a smaller 22 kDa; the exact stoichiometry of the enzyme has not yet been confirmed since the larger subunit has a tendency to dimerize in solutions, and the smaller subunit forms polymeric aggregates ranging from 65 to 300 kDa. This may be an artifact and the enzyme may consist only of two subunits. This aspect requires resolution. The larger component behaves both as a NAD(P)H dependent oxidase and a diaphorase. The enzyme is capable of reducing a variety of electron acceptors such as O_2 and cytochrome *c*. When both subunits are reconstituted, AHP acts as a hydroperoxidase reducing aryl-, thymine-, and linoleic acid hyroperoxides to their corresponding innocous alcohols (Jacobson et al., 1989). AHP is not a selenoprotein, yet there is a remarkable similarity in its catalytic activity toward ROOHs with the Se-dependent glutathione peroxidase found in vertebrates (Harris, 1992).

A unique feature of this enzyme is that based on sequence analysis, the larger subunit resembles thioredoxin reductase (EC 1.6.4.5), an enzyme similar to glutathione reductase which reduces disulfides to monosulfides. In fact, the *oxyR* (H_2O_2 inducible) regulon which is responsible for the induction of many proteins, controls the expression of CAT, AHP and glutathione reductase. Thus, AHP appears to be an important hydroperoxidase, its thioredoxin reductase activity is well suited as a quencher of α-tocopheroxyl radical, whereas glutathione reductase has no such activity (Packer, 1987, Tartaglia et al., 1990). Although the endogenous substrates of AHP are unknown, based on its reactions with the ROOHs named above, it may participate in the prevention of lipid peroxidation chain reaction, and may also protect DNA from peroxidative damage (Jacobson et al., 1989).

iii. Peroxidase

Peroxidase (HRP) is an enzyme of ancient evolutionary origin, and its acronym HRP represents the initial name given to this enzyme as horseradish peroxidase. It is also abbreviated as POD (see chapter by G.W. Felton), and recently it has also been called guaiacol peroxidase (see chapter 9), since guaiacol is an excellent substrate (Hochman and Shemesh, 1987). Peroxidases, including HRP, occur ubiquitously in aerobic organisms and are hemoproteins (protoheme IX), but unlike CAT these enzymes are not polymeric (Hochman and Shemesh, 1987). HRP catalyzes the basic peroxidative reaction depicted in eq. 7.12 and 7.14. It differs from the peroxidative activity of CAT in two respects. First, the range of DH_2 substrates is wider and includes mono- and dihydroxy alcohols and phenols, dihydroquinones, amines and many other H donors. Despite its conspicuous presence in plants, its physiological role is still obscure.

The enzyme may act as a scavenger of H_2O_2 in plants and animals, where the reaction involved, depending on the DH_2 substrate, does not generate a free radical as depicted in eq. 7.14. This antioxidant function has been demonstrated for *D. melanogaster*, where the role of HRP was reported to surpass that of CAT (Armstrong et al., 1978; Nickla et al., 1983). In insects, HRP uses H_2O_2 and a number of phenolic substrates, *o*-catechols in particular, to form semiquinone radicals and quinones. These products are used in the cuticular protein cross-linking process (Hasson and Sugumaran, 1987). Recently, the HRP-catalyzed reaction between H_2O_2 and arbutin (PhOH; 4-hydroxyphenyl-β-D-glucopyranoside) was shown to generate an arbutin free radical, which reacted with α-tocopherol (α-T-

OH) to regenerate PhOH and produce α-tocopheroxyl radical (α-T-O˙) (Mehlhorn et al., 1990).

$$H_2O_2 + 2PhOH \longrightarrow 2H_2O_2 + PhO˙ \tag{7.15}$$

$$PhO˙ + α\text{-}T\text{-}OH \longrightarrow PhOH + α\text{-}T\text{-}O˙ \tag{7.16}$$

On balance then, HRP may have a limited role as an antioxidant enzyme but may play a significant role in plants, fungi and procaryotes which synthesize toxic prooxidant aromatic metabolites. In animals such as insects, phenolic substances acquired from the diet, may be used for cuticular protein tanning with the use of HRP, and there is evidence that the necessary co-substrate H_2O_2 is generated *in situ* by peroxisomal enzymes such as glucose oxidase (Hasson and Sugumaran, 1987).

iv. Ascorbate peroxidase

Discovered in 1979 from spinach chloroplasts (Grodon and Beck, 1979), the occurrence of this specific peroxidase (AP) has since been firmly established in almost all plant cells (see chapter 9 for details). AP removes H_2O_2 at the expense of ascorbate (AH_2) which is reduced to dehydroascorbate (DHA).

$$H_2O_2 + AH_2 \longrightarrow 2H_2O + DHA \tag{7.17}$$

This enzyme works in a sequential team which regenerates AH_2 from DHA as discussed in section III of this chapter (Dalton et al., 1986). AP is a typical hemoprotein peroxidase with a molecular weight in the 28–34 kDa range, and represents a very early evolutionary divergence from HRP (Dalton, 1990; Asada, 1992). In addition to being present in the chloroplast, an isozyme is also present in the cell cytosol. There are some differences among the two forms of AP. The H donor for both forms is AH_2, but the specificity is higher for the chloroplast than the cytosolic enzyme. Upon isolation, the cytosolic AP is more stable than the chloroplastic AP and this is related to AH_2 depletion in the isolation medium (Dalton et al., 1987; Chen and Asada, 1989). There are differences also in the relative sensitivity to inhibition, e.g. thiol inhibitors. The importance of such differences in vivo is not clear, and perhaps both forms are equally important in scavenging H_2O_2 from chloroplasts as well as the cytosol.

There is no evidence of the occurrence of AP in animals or fungi. There are scant reports of existence of AP-like enzymes in bovine eye tissue, the parasite *Trypnosomo cruzi* and the insect *Heliothis*

(=Helicoverpa) zea, but the enzymes *per se* have yet to be character-ized (see chapter 10).

v. Cytochrome-c peroxidase

In fungi, cytochrome-c peroxidase (CCP) is the main H_2O_2 and other peroxide scavenging peroxidase. The reaction catalyzed is as follows.

$$H_2O_2 + 2\text{ Ferrocytochrome } c = 2\text{ }H_2O + 2\text{ Ferricytochrome } c \quad (7.18)$$

As any other peroxidase, CCP is a hemoprotein and although it pos-sesses some homology with the plant AP, neither AP nor HRP are known to function as peroxide scavengers in fungal species such as yeast (see chapter 9). Mitochondrial location of this enzyme is stra-tegic in that it is the main source of O_2^- and H_2O_2 in fungi (Boveris, 1978) and the co-substrate, cytochrome *c*.

7.2.3 Glutathione Peroxidases

A cytosolic Se-dependent glutathione peroxidase (GPOX) present in the cell cytosol and mitochondrial matrix has long been regarded as a crucial enzyme for the removal of cytotoxic hydroperoxides (Chance et al., 1979). Present in vertebrate species, this enzyme has an absolute requirement for the hydrogen donor GSH for catalytic activity. In the reaction catalyzed, the enzyme reduces H_2O_2 to two molecules of H_2O, and membrane peroxides (ROOHs/LOOHs) to one molecule of H_2O and one molecule of the corresponding in-nocuous alcohol (ROH/LOH; eqs. 7.19 and 7.20).

$$H_2O_2 + 2GSH \longrightarrow 2H_2O + GSSG \quad (7.19)$$

$$LOOH + 2GSH \longrightarrow H_2O + LOH + GSSG \quad (7.20)$$

Under identical incubation conditions, GPOX catalyzes the reduc-tion of H_2O_2, ethyl hydroperoxide, *tert*-butylhydroperoxide, cumene hydroperoxide, linoleic acid hydroperoxide (Flohé and Günzler, 1973), and nucleotide or steroid derived hydroperoxides (Chance et al., 1979) at nearly comparable rates.

A critical requirement for the cytosolic GPOX activity for action on membrane hydroperoxides is the enzyme phospholipase A_2 which liberates ROOHs/LOOHs such as PUFA into the cytosol (van Kuijk et al., 1987). GPOX is a tetrameric protein with the molecular weight ranging from 76–105 kDa depending upon the isolation sources of vertebrate tissues, with four gram atoms of Se per molecule. Selenium is present at the catalytic center as selenocysteine selenolate in the resting (or reduced) form of the enzyme. Catalysis of hydroperoxides occurs in several steps. First, the enzyme's catalytic center reduces the hydroperoxide and, in turn, the enzyme is oxidized. Second, the oxidized enzyme forms a complex with one molecule of GSH. Third, the complex uses another molecule of GSH resulting in the oxidation of two molecules of GSH to GSSG with the liberation of the enzyme once again in its reduced state. The dependence of the enzyme for reducing equivalents from GSH and not other thiols is crucial. In fact, GPOX is inhibited by mercaptocarboxylic acids and other mercaptans. For example, mercaptosuccinate is one of the three most potent inhibitory mercaptans reported by Chaudiere et al. (1984).

It should be evident that for the maintenance of optimal GPOX in vivo activity, regeneration of GSH by reduction of GSSG is required. An NAD(P)H-dependent glutathione reductase activity has a similar subcellular distribution to that of GPOX. As discussed later GPOX and the reductase act sequentially, and the oxidation of NAD(P)H establishes the operation of the two enzymes with the NAD(P)H linked substrates. Lastly, there exists a fairly good correlation for the GPOX activity/Se level ratio among monkeys, sheep and rats (1.2–1.6), although for humans the ratio approximates 0.2 and higher amounts of Se are toxic (Ahmad et al., 1989). The point is that dietary Se level does affect GPOX activity of vertebrates, although the levels required vary among mammalian and other vertebrate species. There is no evidence for the occurrence of this enzyme in either procaryotes, plants or invertebrates, and some trivial activity detected in some species remains uncharacterized and is likely due to the peroxidative activity of glutathione-S-transferase (Ahmad et al. 1989, Ahmad, 1992). For more information on comparative aspects of GPOX activity, especially on the fish and fowl enzyme, the reader is referred to chapter 10.

In recent years, at least three other Se-dependent GPOXs have been discovered. The second enzyme is the extracellular plasma GPOX (PL-GPOX). According to Takahashi et al. (1987), in humans the PL-GPOX is a monomeric selenoglycoprotein of about 22.5 kDa. An identical protein has been found in the kidney suggesting that this organ is both the site for synthesis and distribution of this enzyme (Yoshimura et al., 1991). PL-GPOX like activity has been recently observed in the human milk (Avissar et al., 1991). Unlike the cytosolic GPOX, however, the PL-GPOX has high K_m for GSH approximating the mM range, whereas the plasma concentration of GPOX is in the μM range (Halliwell and Gutteridge, 1990a). Because of this anomaly, at the present time the antioxidant role of this enzyme remains confounding.

The third novel intracellular GPOX was discovered from pig liver, and was found to preferentially reduce esterified phospholipid hydroperoxides (Ursini et al., 1982). The selenoprotein nature of this phospholipid hydroperoxide GPOX (PH-GPOX) was established in an elegant study by Weitzel et al. (1990) in several mouse organs during both dietary Se deficiency and supplementation. Although PH-GPOX content seldom exceeds 20% of the overall intracellular total GPOX activity, it is considered a physiologically important enzyme for the following reasons: (1) the enzyme is of broad tissue distribution; (2) it directly attacks membrane phospholipid hydroperoxides without the necessity of a phospholipase to liberate the hydroperoxide from the membrane (a requisite for the classical cytosolic GPOX); (3) the lack of a requirement for activating/mobilizing the phospholipase (i.e., A_2) reduces the risk of liberating excessive amounts of substrates of phospholipase A_2 for prostanoid synthesis (Krinsky, 1992); and (4) resistance to the loss of activity over as long a period of 250 days of Se deficiency, which in contrast is a period nearly twofold for complete depletion of the main cytosolic GPOX. Nonetheless, the enzyme does require rather high levels of Se for maintenance of full activity. The resistance of PH-GPOX to loss of activity from Se deficiency has been interpreted as either a tighter retainment of Se than by GPOX, or that the residual Se is better competed for by the PH-GPOX than GPOX. More work is needed to further characterize this most interesting type of selenoprotein enzyme.

A fourth form of Se-GPOX most recently reported is a tetrameric cytosclic enzyme which is distinct from other GPOXs because it does not cross-react with their antisera (Chu et al., 1993). This enzyme is active against H_2O_2 and most of the LOOHs, but does not accept phosphatidylcholine hydroperoxide as a substrate. This enzyme is currently abbreviated as GPOX-GI (see chapter 10), where GI stands for gastrointestinal. Although found in other organs of humans, this enzyme is primarily located in the gastrointestinal tract of rodents. As suggested by Felton in chapter 10, GPOX-GI may have a prominent role in the removal of dietary hydroperoxides. I suspect however, that additionally or alternatively, it may be a strategic enzyme for protecting the GI tract's epithelial/mucosal cells from lipid-per-oxidizing dietary prooxidants.

7.2.4   Glutathione Transferase

Glutathione transferase (GST; glutathione-S-transferase) is of ubiquitous occurrence, and is comprised of a superfamily of iso-zymes of multiple functionality. The majority of the isoenzymes are involved in detoxification of toxic xenobiotics via conjugation with GSH. The prototype conjugative reaction is depicted with a model substrate, 1-chloro-2,4-dinitrobenzene (CDNB) as follows (Habig et al., 1974).

$$CDNB + GSH \longrightarrow 1\text{-GS-DNB} + H^+ + Cl^- \qquad (7.21)$$

The relevance of GST in the mitigation of oxidative stress is its per-oxidase activity (GST_{px}). Based on mammalian work, Prohaska (1980) demonstrated that GST_{px} reduces LOOH or ROOH in two steps.

$$LOOH + GSH \longrightarrow LOH + [GSOH] \qquad (7.22)$$
$$[GSOH] + GSH \longrightarrow H_2O + GSSG \qquad (7.23)$$

Overall reaction: $LOOH + 2GSH \longrightarrow LOH + H_2O + GSSG$. In the first step, the prototype LOOH, cumene hydroperoxide (cumOOH) is reduced to the innocuous alcohol cumenol, and GSH to a highly reactive intermediate, the sulfenic acid GSOH. In the second step, GSOH instantly reacts with another molecule of GSH to form GSSG and H_2O. As with the Se-GOPXs the GSSG formed during the second step (eq. 7.23) is reduced back to GSH by glu-tathione reductase as discussed later. Prohaska (1980) used cysteam-ine and cyanide to distinguish the similar overall reaction of GST_{px} from that of Se-dependent GPOX. While cysteamine reacts with GSOH to form a mixed disulfide, KCN rapidly reacts to form GSCN. The mixed disulfide is not readily reduced to GSH by the enzyme glutathione reductase, and GSCN is not a substrate for this reduc-tase.

Extensive work on GSTs has been focused in the past on the sol-uble isoenzymes which fall into four families, α, μ, π (Mannervik et al., 1985) and θ (Meyer et al. 1991; Pemble and Taylor, 1992). At least 13 subunits have been characterized—more awaiting charac-terization—which form homodimers as well as heterodimers com-prised of subunits of the four families (Ketterer and Coles, 1991). GSTs perform several types of catalysis with the participation of GSH;

while some forms carry out most reactions, albeit at different rates, other isozymes depict a marked preference for specific substrates (Jakoby, 1985). For example, these enzymes are involved in numerous GSH-transferal reactions including some isomerization reactions and disulfide interchange. GSH conjugation is a very important reaction for the deactivation of toxic lipophilic xenobiotics, and their facile excretion as polar GS-conjugates (Ahmad et al., 1986). However, it is also apparent that this basic reaction of GSTs is important in the detoxification of endogenous substances. For example, 4-hydroxyalkenals which are highly reactive and deleterious breakdown products of LOOHs, are conjugated with GSH by GST, and rendered innocuous and excretal (Danielson et al., 1987). Specific isoenzymes also dehydrochlorinate the insecticide DDT to DDE, detoxify cytotoxic α, β-unsaturated ketones, and epoxides by forming GSH conjugates while opening the epoxide ring. The GST_{px} activity of GSTs is highest for θ isoenzyme, it is moderate for α, low for μ, and least for the π isoenzymes.

Regardless of the wide array of reactions catalyzed, the enzymatic mechanism involves GST-mediated attack on compounds that are sufficiently lipophilic and possess strong electrophilic centers of either carbon, oxygen or nitrogen (Ketterer and Coles, 1991). Thus a wide range of LOOHs are good substrates for GST_{px} activity. H_2O_2 is a known inducer of GST but the reaction with this peroxide is negligible (Mannervik, 1985) presumably due to the high polarity of H_2O_2.

Mammalian species have been known to possess microsomal and mitochondrial enzymes (Ahmad et al., 1986). Microsomes contain strongly adsorbed soluble GSTs, and another isoenzyme which shows GST_{px} activity. The latter isoenzyme is 17.2 kDa and is not homologous with the known amino acid sequences of soluble GSTs (Morgenstern and DePierre, 1988). In insects, soluble GSTs have been extensively studied for the deactivation of insecticides via GSH conjugation (Ahmad et al., 1986). Our laboratory recently demonstrated that in the insect *T. ni*, GST activity is also present in the microsomes and nuclei (Ahmad and Pardini, 1988; 1989). Further our studies along the protocols established by Prohaska (1980) demonstrated a marked association of this GST activity with that of GST_{px}. A surprising feature was that the GST_{px} activity was highest in nuclei, moderate in microsomes, and although low in cytosol, the activity was substantial compared to the mammalian levels. Subsequently, a GST_{px}-like GST activity was reported from other eucaryotic nuclei, with activity towards lipid hydroperoxides (Ketterer and Meyer, 1989). In another insect species, the black swallowtail butterfly, *Papilio polyxenes*, the highest GST_{px} activity was not only found once again in the nucleus, activity was also detected in the mitochondria which in level was second to the nuclei, followed by levels in microsomes and the cytosol (Ahmad et al., 1990). Earlier, evidence had been provided that thymine hydroperoxide was a substrate for rat GST_{px} (Tan et al., 1986). DNA is well-known for its susceptibility to peroxidation and for containing hydroperoxides such as thymine hydroperoxide and 5-hydroperoxymethyluracil residues. The high nuclear GST_{px} activity may have a pivotal role in the reduction of these hydroperoxides to respective glycols and 5-hydroxymethyl residues. These derivatives may then be released by a DNA glycosylase. "The resulting apyrimidinic DNA should then undergo repair," Ketterer and Coles (1991).

GST activity is also crucial in the biogenesis and regulation of eicosanoid cell mediators, involving both GSH-conjugative and GST_{px} activity, since many intermediates possess peroxyl moieties (Ketterer and Coles, 1991). In the leukotriene pathway originating from arachidonic acid, leukotriene A_4 (LTA_4) is converted to LTC_4. Moreover, GST_{px} attack on the precursor of LTA_4 HPETE ([S]-5-hydroperoxy-6-*trans*-8,11,14-*cis*-eicosatetraenoic acid) causes its reduction to HETE. In the biogenesis of prostaglandins, arachidonic acid is converted by cyclooxygenase to the hydroperoxyendoperoxide prostaglandin PGG_2, and then reduces it to hydroxyendoperoxide PGH_2. GST_{px} apparently converts PGH_2 to a non-endoperoxide PGF_{2a}, and GST isomerase activity converts PGH_2 to PGE_2 and PGD_2.

Generally the mammalian GSTs relative to Se-GPOXs exhibit higher K_ms, and this seems also to be the case with insect GSTs. Based on crude enzyme extracts of several insect species, the V_{max}s for CDNB conjugation and cumOOH reduction were found to be unusually very high, indicating the enzyme possesses high catalytic power K_{cat} (Weinhold et al., 1990). This aspect requires clarification with the use of purified isoenzymes. However, lower affinity but higher catalytic power of GSTs is apparent at this time. These features have resemblance to the catalytic properties of CAT, and perhaps both enzymes are crucial only when severe oxidative stress leads to high steady-state concentrations of H_2O_2 and LOOHs, respectively. Lastly, GST along with its various non-conjugative activities (including GST_{px} activity) is inhibited by a number of phenolics such as the flavonoid, quercetin. Quercetin also inhibits a wide range of enzymes including glutathione reductase; therefore, phenolics are not specific inhibitors of GSTs. The inhibition by phenolics presumably occurs via hydrogen bonding of catechols or dihydroquinones, or covalent binding with quinones or semiquinone radicals (Ahmad and Pardini, 1990b). Other useful and more specific inhibitors are hematin and bromosulfophthalein (BSP) (Singh et al., 1987). However, in GST_{px} assay that is coupled to that of glutathione reductase involving the oxidation of NADPH to NAD^+P, hematin can not be used since it directly and very rapidly draws electrons from NADPH. On the other hand, BSP is an excellent inhibitor of GST for both in vitro and in vivo studies and useful in characterizing GSTs' peroxidative activity from that of other GSH-dependent peroxidases (Meyers et al., 1992).

7.2.5 Glutathione Reductase

In all aerobic organisms the redox status of cells heavily relies on the maintenance of glutathione in its reduced state, GSH, and prevention of its oxidation to the oxidized state, GSSG. GSH is the most abundant intracellular thiol (5–10 mM range) and is crucial in many biological processes (Meister and Anderson, 1993). For example, it serves as a co-enzyme for many enzymes and forms conjugates with endo- and exobiotics/xenobiotics. It is also involved in the transport of amino acid and in the synthesis of DNA precursors (Anderson, 1987). Thus, it is not surprising that much of this thiol is present in tissues as GSH (Sies and Akerboom, 1984). However, under oxidative stress, the normal ratio of GSH: GSSG which is 99+: <1% shifts, causing a decrease in GSH level and a concomitant rise in the GSSG level. As reviewed in Ahmad et al. (1988a), the oxidation of free thiols such as GSH is very deleterious to cells; for example, if 12% of the mitochondrial thiols are oxidized mitochondrial swelling occurs, and if the level of oxidized thiol reaches 15% then lipid peroxidation is initiated. The activities of H_2O_2 and LOOHs/ROOHs detoxifying enzymes, e.g., Se-GPOX and GST_{px} are in part responsible for the formation of GSSG above the physiologically acceptable levels. The cytosolic enzyme glutathione (GSSG) reductase (GR) is the crucial enzyme for the regeneration of GSH from GSSG (eq. 7.24).

$$GSSG + NAD(P)H + H^+ \longrightarrow 2GSH + NAD^+(P) \qquad (7.24)$$

GR is a FAD-containing flavoprotein and is distributed in the cytosol, mitochondrial matrix, and some activity is also associated with other membrane-bound organelles such as the endoplasmic reticulum. It has been purified from several mammalian species including humans, a few invertebrates (i.e., the mollusks, sea urchin and mussels), plants and fungi (see chapters 9 and 10). Thus, a great deal is known about the nature of this enzyme and its kinetics. In mammalian species, the molecular weight of the enzyme is 100 kDa, and the enzyme is a dimer consisting of two equal subunits of 50 kDa each with one FAD molecule per subunit (Carlberg and Mannervik, 1985). The enzyme seems specific for the reduction of GSSG because mixed-disulfides (GSSR) are extremely poor substrates. High affinity for the substrates is exhibited by k_ms in the μM range; for GSSG these values are 101 and 26, and for NADPH 21 and 8 μM for calf and rat liver GR, respectively (Carlberg and Mannervik, 1985).

The plant GRs are of higher molecular weights and tetrameric forms have been reported. The K_m value for GSSG is about 52 μM, and for hGSSG (derived from homoglutathione, hGSH) the K_m is higher (see chapter 9 for more details). Fungi contain high amounts of both GSH and the enzyme GR. Fungal GRs have been thoroughly characterized (Schirmer et al., 1989; Williams, 1992). The best commercial source of GR is from the Baker's yeast, and is routinely used in assays of GPOX and GST_{px} activities coupled to that of exogenously added GR and its required co-factor NADPH. Fungal GRs are dimeric proteins and resemble closely with mammalian GRs. The subunit is about 55 kDa containing one molecule of FAD.

The enzymes from all sources are remarkably stable when refrigerated (and not frozen) at a concentration of about 2 mg protein/mL. Furthermore, from the literature cited NADPH rather than NADH seems a better co-substrate for most GRs. lastly, for characterization studies the best inhibitor of GR is bischloronitrosourea (BCNU), which is useful for both in vitro and in vivo studies (Michiels and Ramacle, 1988).

The GR activity is at the expense of NAD(P)H (see eq. 7.24), however, the cytosolic glucose-6-phosphate dehydrogenase reduces back $NAD^+(P)$ to NAD(P)H returning the cell to its fully reduced state. Locally, i.e., in mitochondria and microsomes a number of NAD(P)H trans/de/hydrogenases may also participate in the regeneration of NAD(P)H from its oxidized form. This premise is supported by the fact that during induced oxidative stress all hydrogenases mentioned herein are induced (see chapter 8).

The function of the primary antioxidant enzymes is to work sequentially to prevent O_2 radical cascade and to attenuate the lipid peroxidation chain reaction. This is achieved by successive attacks, first by SOD to destroy O_2^- radicals, second by hydroperoxidases to decompose H_2O_2, third by peroxidases to decompose LOOHs/ROOHs, and fourth by a specific reductase, GR, to regenerate GSH

Table 7.2
The Ancillary Antioxidant Enzymes

Enzyme	Systematic Name	EC Number
NAD(P)H dehydrogenase (quinone); DT-diaphorase	NAD(P)H: (quinone acceptor) oxido-reductase	1.6.99.2
Ascorbate recycling enzymes		
Monodehydroascorbate reductase (NADH); AFR reductase	NADH: monodehydro-ascorbate oxido-reductase	1.6.5.4
Glutathione dehydrogenase (ascorbate); DHA reduc-tase	Glutathione: dehydroascorbate oxidoreductase	1.8.5.1
Antioxidant proteases		
Intracellular neutral proteases	NE*; multicatalytic against oxidized proteins	
Macroxyproteinase	NE*; active against oxidized proteins of red blood cells	

* NE = not established.

from GSSG which is an essential but undesirable by-product of the peroxidase reactions.

7.3 ANCILLARY ANTIOXIDANT ENZYMES

The ancillary antioxidant enzymes are summarized in Table 7.2.

7.3.1 *NAD(P)H Dehydrogenase (Quinone); or DT-diaphorase*

Some quinones undergo non-enzymatic redox-cycling to semi-quinone radical, while others are reduced to semiquinones by the microsomal reductase of the cytochrome P-450, the one-electron transfer flavoprotein called NADPH-ferrihemoprotein reductase (Ahmad, 1992). Semiquinones can readily transfer an electron to O_2 forming O_2^- radical which can then initiate oxygen radical cascade. The dicumarol-sensitive cytosolic flavoprotein enzyme DT-diapho-

rase can directly reduce quinones (Q) to more stable dihydroquinones (QH$_2$) by a direct two-electron reduction (eq. 7.25).

$$Q + NAD(P)H + H^+ \longrightarrow QH_2 + NAD^+(P) \qquad (7.25)$$

Apparently, the enzyme (quinone reductase) is comprised of two
flavoproteins, each of which accepts one e$^-$ at a time from NAD(P)H
and, in turn, is partially reduced. This facilitates the conversion of
Q to a semiquinone first, and then a rapid reduction to QH$_2$, while
the flavoproteins return to their respective oxidized states (Cadenas
et al., 1992). The exact stoichiometry of this reaction is still obscure,
and possibly the source of another proton as shown in eq. 7.25 is
a flavoprotein. DT-diaphorase is distinguished from other NAD(P)H
diaphorases based on a direct interaction of dicumarol with NAD(P)H,
which prevents the flow of e$^-$ for enzymatic activity.

A similar enzyme has been reported to be active towards the di-
pyridinium compound, paraquat. Our recent work has provided
preliminary evidence that paraquat may indeed be stabilized by this
enzyme (Pardini and Ahmad, 1991). More work is required to clarify
the various competing systems in mitigating oxidative stress from
quinones and, in particular, from DT-diaphorase. Until then as stated
by Brunmark et al. (1987) DT-diaphorase appears to be an important
"cellular control device against semiquinone and O$_2^-$ generation."

7.3.2 Ascorbate Recycling Enzymes

Ascorbic acid (AH$_2$) is an outstanding soluble antioxidant. The
ascorbate anion (AH$^-$) readily transfers an e$^-$ to reactive O$_2$ species
which results in the production of monohydroascorbate, or ascorbyl
free radical (A$^{\cdot-}$). This free radical disproportionates to AH$_2$ and dehydroascorbate (DHA = A). DHA can also form by a 1−e$^-$ reduction of the A$^{\cdot-}$ radical, or as shown below in eq. 7.26 by a direct
2−e$^-$ reduction of AH$^-$.

$$H_2O_2 + AH^- + H^+ \longrightarrow 2H_2O + A \qquad (7.26)$$

A rapid depletion of AH$_2$ can occur under conditions of severe
oxidative stress. This is a cataclysmic event requiring aerobic organisms to obtain it from a dietary source, or possess the ability to regenerate from the oxidized forms. As briefly discussed below two
separate enzymes have crucial roles in the recovery of AH$_2$.

i. Monohydroascorbate reductase (NADH), or ascorbate free radical reductase

The ascorbate free radical reductase (AFR) converts $A^{\cdot-}$ to AH_2 as shown below (eq. 7.27), and it is a well-characterized enzyme from

$$A^{\cdot-} + NADH \longrightarrow AH^- + NAD^+ \qquad (7.27)$$

plants (Borracino et al., 1989). A functionally similar enzyme has been reported from mammalian tissues by Rose and Bode (1992). However, at this juncture it is not clear whether this enzyme is same as in plants; the NADH dependence has not been established and the mammalian enzyme may depend on other sources of reducing equivalents. The occurrence of an AFR-like enzyme was recently claimed for an insect species *Helicoverpa zea* (Felton and Duffy, 1992). Clearly, more work is needed to characterize the animal vs. plant enzyme.

ii. Glutathione dehydrogenase (ascorbate), or dehydroascorbate reductase

Dehydroascorbate reductase (DHA reductase) is a GSH dependent enzyme (Dipierro and Borracino, 1991) and converts A (=DHA) to AH_2 as follows.

$$A + 2GSH \longrightarrow AH_2 + GSSG \qquad (7.28)$$

The necessity of this enzyme in the presence of AFR or AFR-like enzymes for the regeneration of AH_2 is not quite clear. Nonetheless, the two enzymes may be complementing each other in recovering ascorbate from its two oxidized states (for more details see chapters 9 and 10). The occurrence of DHA reductase in animal species has been speculated, and the first reliable claim for the presence of this enzyme in the insect species, *H. zea*, was made very recently (Felton and Duffy, 1992). The enzyme has since been isolated and its characteristics are that of the plant enzyme (Summers and Felton, 1993).

These enzymes are not only important in regulating the steady-state concentration of AH_2, they also participate in scavenging H_2O_2. H_2O_2 is reduced to H_2O by ascorbate with the formation of dehydroascorbate (eq. 7.26), which is then reduced to AH^- by DHA reductase (eq. 7.28); the GSSG formed from the latter reaction is then reduced back to GSH by the enzyme GR (eq. 7.24). This phenomenon has been demonstrated for both plants (Dalton et al., 1986) and phytophagous insects (Felton and Duffy, 1992).

7.3.3   Antioxidant Proteases

Under oxidative stress a wide variety of proteins are induced including heat shock proteins (see chapter 8). Among the proteins induced only a few have been characterized, and these include antioxidant proteases.

i.   Intracellular neutral proteases

All amino acid residues of a protein are attacked by ˙OH; the oxidized proteins demonstrate toxic/pathologic consequences (Davies, 1986). In animal species, intracellular proteolytic systems (Davies, 1986) recognize oxidized proteins from unoxidized ones, and degrade only the former (Stadtman, 1992).

ii.   Macroxyproteinase

Macroxyproteinase (M.O.P.) is comprised of a heavy molecular weight proteinase complex with a molecular weight of 670 kDa (Pacifici et al., 1989). These proteins selectively degrade oxidatively denatured proteins.

Together, these proteases and hitherto uncharacterized battery of proteins that are induced under oxidative stress, may constitute an important ancillary antioxidant enzymatic defenses.

7.4   ANTIOXIDANT PROTEINS

In addition to the antioxidant enzymes, it is becoming increasingly clear that several constitutive/proteins possess antioxidant properties (Table 7.3). Only a brief discussion is provided here (see chapter 10 for an in-depth discussion).

The metal-catalyzed Haber-Weiss or Fenton reaction is responsible for generating lipid-peroxidizing ˙OH radical (eq. 7.4), whereas the ˙OH radical causes lipid peroxidation of unsaturated organic molecules i.e., PUFA (eqs. 7.5–7.7). The implicated transitional metals catalyzing these reactions are copper (Cu [I] $\leftrightarrow$ Cu [II]) and iron (Fe [II] $\leftrightarrow$ Fe [III]).

7.4.1   Albumin

Albumin has the ability to rapidly bind with Cu. Thus, it acts as a good scavenger of Cu ions and, in turn, it could minimize the

Table 7.3.
Antioxidant Proteins

Protein	Functionality
Albumin	Cu scavenging
Ceruloplasmin	Ferro-O_2-oxidoreductase (EC 1.16.3.1)
Heat shock proteins	Stress repair proteins
Metallothioneins	Metal binding proteins
Transferrins	Fe transport proteins
Ferritins (phytoferritins)	Fe storage proteins

generation of the potent ˙OH radical and reduce the threat for lipid peroxidation (Halliwell and Gutteridge, 1990a,b).

7.4.2 Transferrins

Transferrins are iron transport proteins found in plasma. These glycoproteins are monomeric with a molecular weight of 80 kDa. In mammals transferrin carries two Fe molecules per protein molecule, but in insects despite considerable homology, it carries only one Fe molecule (Law et al., 1992). Their antioxidant role is considered together with ferritins.

7.4.3 Ferritins

Ferritins (phytoferritins in plants; see chapter 9) are Fe storage proteins. In mammals ferritin is primarily stored in the liver cytosol, and is also found in µM range in the serum. The serum protein is glycosylated. Typically, the molecular weight of ferritins is around 500 kDa. In the insect the tobacco hornworm, *Maduca sexta*, a 490 kDa mammalian-like ferritin exists as polymer of two subunits of 24 and 26 kDa (Law et al., 1992). In the butterfly *Calpodes ethlius*, the protein is larger, about 660 kDa, and comprised of larger subunits, 24 and 31 kDa (Law et al., 1992). The 24 kDa subunit upon glycosylation yields the 31 kDa subunit. The ferritin of *C. ethlius* is abundant in the hemolymph and in midgut tissues; in the latter tissue it is not cytosolic and is stored in special vacoules, but is abundant in the hemolymph (Law et al., 1992).

Both transferrins and ferritins are critical proteins to minimize the occurrence of free Fe in plasma/serum as well as intracellulary. Iron is therefore subject to controlled transport and release, and this is crucial in mitigating oxidative stress (Gutteridge and Quinlan, 1993).

7.4.4   Metallothioneins

These proteins are notable for scavenging a large variety of metals, including metalloids. As such they serve antioxidant function (Halliwell and Gutteridge, 1990a), but there is increasing evidence that these proteins can also act as direct scavengers of free radicals (Sato and Bremer, 1993).

7.4.5   Ceruloplasmin

Ceruloplasmin (CP) is a multifunctional Cu-protein (Samokyszyn et al., 1989). The ferrioxidase activity of CP which converts Fe[II] to Fe [III] that promotes incorporation of ferric into the ferritin, is the antioxidant property of this protein. As the protein is of multiple functionality (action as a protein rather than enzyme) the readers of this chapter are referred to the more detailed account of CP in chapter 10.

7.4.6   Heat Shock Proteins

As briefly mentioned in chapter 8, during oxidative stress as many as 40 proteins are induced. Among them are antioxidant enzymes, DNA repair enzymes, and other metabolic enzymes such as glucose-6-phosphate dehydrogenase. Others include inducible intracellular proteases and metallothioneins. Still a large number of proteins or enzymes remain uncharacterized. However, there is considerable interest now with the discovery that ROS induce the "stress," or heat shock proteins (HSP). The synthesis of HSPs in response to stress and heat shock, is a universal phenomenon ranging from bacteria to humans (Schlesinger, 1990). Oxidant-induced injury such as during ischemia and reperfusion induces a 72kDa HSP (Donnelly et al., 1992). HSP72 is in fact a marker HSP since it is often induced in most organisms, although the number of proteins with lower and some higher molecular weights is substantial.

As reviewed by Schlesinger (1990), HSPs possess protective or salvaging role, e.g., HSP70 proteins are necessary for the import of folded polypeptides (by unfolding them) into eukaryotic cell organelles, and in the dissociation of protein aggregates. For example, under the oxidant-induced stress, nuclear proteins and immunoglobins may become aggregated and insoluble. ROS-mediated induction of HSPs is now well established from the well-known prooxidant Hg^{2+}, which generates H_2O_2 and other ROS (Lee and Kim, 1992); the protein induced is the HSP72. Future work will clarify the role of HSPs in alleviating oxidative stress.

7.5 SUMMARY

All aerobic organisms are subject to cytotoxicity from reactive oxygen species (ROS). Although ROS are endogenously generated as part of routine physiological processes, their production is exacerbated by exogenous prooxidants. Therefore, all aerobic organisms have evolved a battery of enzymatic and non-enzymatic protein defenses to prevent the free radical cascade of O_2, and deleterious lipid, protein, and DNA oxidations. The role of primary antioxidant enzymes, superoxide dismutase, hydroperoxidases such as catalase, is to destroy O_2^- and H_2O_2. Many glutathione (GSH) dependent and independent peroxidases destroy organic peroxides. The oxidized glutathione (GSSG) resulting from catalytic activities of the GSH-dependent proxidases is reduced back by glutathione reductase to GSH. Ancillary antioxidant systems such as DT-diaphorase prevent the formation of O_2^- from quinoid compounds, and other enzymes regenerate a most crucial antioxidant, ascorbic acid, from its oxidized forms. Proteases have been identified as important ancillary enzymes in degrading oxidized proteins. Resultant amino acids and peptides are re-used in the synthesis of normal proteins. Lastly, a number of proteins have been found to possess antioxidant properties.

References

Aebi, H. (1984) Catalase *in vitro*. *Methods Enzymol.* **105**, 121–126.

Ahmad, S. (1992) Biochemical defence of pro-oxidant allelochemicals by herbivorous insects. *Biochem. Syst. Ecol.* **20**, 269–296.

Ahmad, S., Beilstein, M.A. and Pardini, R.S. (1989) Glutathione peroxidase activity in insects: a reassessment. *Arch. Insect Biochem. Physiol.* **12**, 31–49.

Ahmad, S., Brattsten, L.B., Mullin, C.A. and Yu, S.J. (1986) Enzymes involved in the metabolism of plant allelochemicals. In *Molecular Aspects of Insect-plant Associations* (L.B. Brattsten and S. Ahmad, eds.), Plenum Press, New York, pp. 73–153.

Ahmad, S., Duval, D.L., Weinhold, L.C. and Pardini, R.S. (1991) Cabbage looper antioxidant enzymes: tissue specificity. *Insect Biochem.* **21**, 563–572.

Ahmad, S. and Pardini, R.S. (1988) Evidence for the presence of glutathione peroxidase activity towards an organic hydroperoxide in larvae of the cabbage looper moth, *Trichoplusia ni. Insect Biochem.* **18**, 861–866.

Ahmad, S. and Pardini, S. (1989) Corrigendum, *Insect Biochem.* **19**, 109.

Ahmad, S. and Pardini, R.S. (1990a) Mechanisms for regulating oxygen toxicity in phytophagous insects. *Free Rad. Biol. Med.* **8**, 401–413.

Ahmad, S. and Pardini, R.S. (1990b) Antioxidant defense of the cabbage looper, *Trichoplusia ni*: enzymatic responses to the superoxide-generating flavonoid, quercetin, and photodynamic furanocoumarin, xanthotoxin. *Photochem. Photobiol.* **51**, 305–311.

Ahmad, S., Pritsos, C.A., Bowen, S.M., Heisler, C.R., Blomquist, G.J. and Pardini, R.S. (1988a) Subcellular distribution and activities of superoxide dismutase, catalase, glutathione peroxidase and glutathione reductase in the southern armyworm, *Spodoptera eridania. Arch. Insect Biochem. Physiol.* **7**, 173–186.

Ahmad, S., Pritsos, C.A., Bowen, S.M., Heisler, C.R., Blomquist, G.J. and Pardini, R.S. (1988b) Antioxidant enzymes of the cabbage looper moth, *Trichoplusia ni*: subcellular distribution and activities of superoxide dismutase, catalase and glutathione reductase. *Free Rad. Res. Commun.* **4**, 403–408.

Ahmad, S., Pritsos, C.A. and Pardini, R.S. (1990) Antioxidant enzyme activities in subcellular fractions of the black swallowtail butterfly, *Papilio polyxenes. Arch. Insect Biochem. Physiol.* **15**, 101–109.

Anderson, M.E. (1987) Tissue glutathione. In *CRC Handbook of Methods for Oxygen Radical Research* R.A. Greenwald, (ed.), CRC Press, Boca Raton, FL, pp. 317–323.

Archibald, F.S. and Fridovich, I. (1981) Defense against oxygen toxicity in *Lactobacillus planatarum. J. Bacteriol.* **145**, 442–451.

Armstrong, D., Rinehart, R., Dixon, L. and Reigh, D. (1978) Changes of peroxidase with age in *Drosophila. Age* **1**, 8–12.

Asada, K. (1992) Ascorbate peroxidase—a hydrogen peroxide-scavenging enzyme in plants. *Physiol. Plant* **85**, 235–241.

Aucoin, R.R., Philogene, B.J.R. and Arnason, J.T. (1991) Antioxidant enzymes as biochemical defenses against phototoxin-induced oxidative stress in three species of herbivorous Lepidoptera. *Arch. Insect Biochem. Physiol.* **16**, 139–152.

Avissar, N., Slemonnon, J.R., Palmer, I.S. and Cohen, H.J. (1991) Partial sequence of human plasma glutathione peroxidase and immunologic identification of milk glutathione peroxidase as the plasma enzyme. *J. Nutr.* **121**, 1243–1249.

Best-Belpomme, M. and Ropp, M. (1982) Catalase is induced by ecdysterone and ethanol in *Drosophila* cells. *Eur. J. Biochem.* **121**, 349–355.

Borg, D.C. and Schaich, K.M. (1988) Iron and hydroxyl radicals in lipid peroxidation: Fenton reactions in lipid and nucleic acids co-oxidized with lipids. In *Oxy-*

radicals in Molecular Biology and Pathology (P.A. Cerutti, I. Fridovich and J.M. McCord eds.), Allan R. Liss, New York, pp. 427–441.

Borracino, G., Dipierro, S. and Arrigoni, O. (1989) Interaction of ascorbate free radical with sulfhydryl reagents. *Phytochemistry* **28**, 715–717.

Boveris, A. (1978) Production of superoxide anion and hydrogen peroxide in yeast mitochondria. In *Biochemistry and Genetics of Yeasts*, (M. Bacila, B.L. Horecker and A.O.M. Stoppani, eds.), Academic Press, New York, pp. 65–80.

Bowler, C., Alliotte, A., DeLoose, M., Van Montagu, M. and Inze, D. (1989) The induction of manganese superoxide dismutase in response to stress in *Nicotiana plumbaginifolia*. *EMBO J.* **8**, 31–38.

Brunmark, A., Cadenas, E., Lind, C., Segura-Aguilar, J. and Ernster, L. (1987) DT-diaphorase catalyzed two-electron reduction of quinone epoxides. *Free Rad. Biol. Med.* **3**, 181–188.

Cadenas, E., Hochstein, P. and Ernster, L. (1992) Pro- and antioxidant function of quinones and quinine derivatives in mammalian cells. *Adv. Enzymol.* **65**, 97–146.

Carlberg, I. and Mannervik, B. (1985) Glutathione reductase. *Method Enzymol.* **113**, 484–490.

Chance, B., Sies, H. and Boveris, A. (1979) Hydroperoxide metabolism in mammalian organs. *Physiol. Rev.* **59**, 527–605.

Chaudiere, J., Wilhemsen, E.C. and Tappel, A.L. (1984) Mechanism of selenium-glutathione peroxidase and its inhibition by mercaptocarboxylic acids and other mercaptans. *J. Biol. Chem.* **259**, 1043–1050.

Chen, G.-X. and Asada, K. (1989) Ascorbate peroxidase in tea leaves: occurrence of two isozymes and the differences in their enzymatic and molecular properties. *Plant Cell Physiol.* **30**, 987–998.

Chu, F.F., Doroshow, J.H. and Esworthy, R.S. (1993) Expression, characterization, and tissue distribution of a new selenium-dependent glutathione peroxidase, GSHPx-GI. *J. Biol. Chem.* **268**, 2571–2576.

Claiborne, A., Malinowski, D.P. and Fridovich, I. (1979) Purification and characterization of hydroperoxidase II of *Escherichia coli* B. *J. Biol. Chem.* **254**, 11664–11668.

Dalton, D.A. (1990) Ascorbate peroxidase. In *Peroxidases in Chemistry and Biology*, Vol. II, (J. Everse, K. Everse and M.B. Grisham, eds.), CRC Press, Boca Raton, FL, pp. 139–153.

Dalton, D.A., Hanus, F.J., Russell, F.J. and Evans, H.J. (1987) Purification, properties and distribution of ascorbate peroxidase in legume root nodules. *Plant Physiol.* **83**, 789–794.

Dalton, D.A., Russell, S.A., Hanus, F.J., Pascoe, G.A. and Evans, E.H. (1986) Enzymatic reactions of ascorbate and glutathione that prevent peroxide damage in soybean root nodules. *Proc. Natl. Acad. Sci. USA* **83**, 3811–3815.

Danielson, U.H., Esterbauer, H. and Mannervik, B. (1987) Structure-activity relationships of 4-hydroxyalkenals in the conjugation catalyzed by mammalian glutathione transferases. *Biochem. J.* **247**, 707–713.

Davies, K.J.A. (1986) Intracellular proteolytic systems may function as secondary defenses: an hypothesis. *J. Free Rad. Biol. Med.* **2**, 155–173.

Del Vicchio, R.J. (1988). Some physiological effects of gamma radiation on larvae of the navel orangeworm (*Amyelosis transitella*). Ph.D. dissertation, University of California, Davis, CA.

Dipierro, S. and Borraccino, G. (1991) Dehydroascorbate reductase from potato tubers. *Phytochemistry* 30, 427–429.

Donnelly, T.J., Sievers, R.E., Vissern, F.L.J., Welch, W.J. and Wolfe, C.L. (1992) Heat shock protein induced in rat hearts. *Circulation* 85, 769–778.

Farr, S.B. and Kogoma, T. (1991) Oxidative stress responses in *Escherichia coli* and *Salmonella typhimurium*. Sequence and homology to thioredoxin reductase and other flavoprotein disulfide oxidoreductases. *Microbiol. Rev.* 55, 561–585.

Felton, G.W. and Duffy S.S. (1992) Ascorbate oxidation-reduction in *Helicoverpa zea* as a scavenging system against dietary oxidants. *Arch. Insect Biochem. Physiol.* 19, 27–37.

Flohé, L. and Günzler, W.A. (1973) Glutathione peroxidase. In *Glutathione* (L. Flohé, ed.), Academic Press, New York, pp. 132–145.

Fridovich, I. (1988) Superoxide radical: an endogenous toxicant. *Ann. Rev. Pharmacol. Toxicol.* 23, 239–257.

Getzoff, E.D., Tainer, J.A., Weiner, P.K., Kollman, P.A., Richardson, J.S. and Richardson, D.C. (1983) Electrostatic recognition between superoxide and copper, zinc superoxide dismutase. *Nature* 306, 284–287.

Giorgi, F. and Deri, P. (1976) Cytochemistry of late ovarian chambers of *Drosophila melanogaster*. *Histochemistry* 48 , 325–334.

Groden, D. and Beck, E. (1979) H_2O_2 destruction by ascorbate-dependent systems from chloroplasts. *Biochim. Biophys. Acta* 546, 426–435.

Gutteridge, J.M.C. and Quinlan, G.J. (1993) Antioxidant protection against organic and inorganic oxygen radicals by normal human plasma: the important primary role for iron-binding and iron-oxidizing proteins. *Biochim. Biophys. Acta* 1156, 144–150.

Habig, W.H. Pabst, M.H. and Jakoby, W.B. (1974) Glutathione-*S*-transferase. The first enzyme step in mercapturic acid formation. *J. Biol. Chem.* 249, 7130–7139.

Halliwell, B. (1982) Ascorbic acid and the illuminated chloroplast. in *Ascorbic Acid: Chemistry, Metabolism, and Uses* (P.A. Seib and B.M. Tolbert, eds.), *Adv. Chem. Ser.* 200, Am. Chem. Soc., Washington, D.C., pp. 263–274.

Halliwell, B. and Gutteridge, J.M.C. (1990a) The antioxidants of human extracellular fluids. *Arch. Biochem. Biophys.* 280, 1–8.

Halliwell, B. and Gutteridge, J.M.C. (1990b) Role of free radicals and catalytic metal ions in human disease. *Methods Enzymol.* 186, 1–85.

Harris, E.D. (1993) Regulation of antioxidant enzymes. *FASEB J.* 6, 2675–2683.

Hasson, C. and Sugumaran, M. (1987) Protein cross-linking by peroxidase: possible mechanism for sclerotization of insect cuticle. *Arch. Insect Biochem. Physiol.* 5, 13–28.

Heimberger, A. and Eisenstark, A. (1988) Compartmentalization of catalases in *Escherichia coli*. *Biochem. Biophys. Res. Commun.* 154, 392–397.

Hochman, A. and Shemesh, A. (1987) Purification and characterization of a catalase-peroxidase from the photosynthetic bacterium *Rhodopseudomonas capsulata*. *J. Biol. Chem.* 262, 6871–6976.

Jacobson, F.S., Morgan, R.W., Christman, M.F. and Ames, B.N. (1989) An alkyl hydroperoxidase reductase from *Salmonella typhimurium* involved in the defense of DNA against oxidative damage. *J. Biol. Chem.* **264**, 1488–1496.

Jakoby, W.B. (1985) Glutathione transferases: an overview. *Method Enzymol.* **113**, 495–499.

Jones, D.P., Eklow, L., Thor, H. and Orrenius, S. (1981) Metabolism of hydrogen peroxide in isolated hepatocytes: Relative contributions of catalase and glutathione peroxidase in decomposition of endogenously generated H_2O_2. *Arch. Biochem. Biophys.* **210**, 505–516.

Ketterer, B. and Coles, B. (1991) Glutathione transferases and products of reactive oxygen. In *Oxidative Stress Oxidants and Antioxidants* (H. Sies ed.), Academic Press, London, pp. 171–194.

Ketterer, B. and Meyer, D.J. (1989) Glutathione transferases: a possible role in the detoxication and repair of DNA and lipid hydroperoxides. *Mutat. Res.* **21**, 33–40.

Klapper, I., Hagstron, R., Fine, R. and Hnig, B. (1986) Focusing of isoelectric fields in the active site of Cu-Zn superoxide dismutase: effects of ionic strength and amino-acid modifications. *Proteins* **1**, 47–59.

Krinsky, N.I. (1992) Mechanism of action of biological antioxidants (43429). *P.S.E.B.M.* **200**, 248–254.

Law, J.H., Ribeiro, J.M.C. and Wells, M.A. (1992) Biochemical insight derived from insect diversity. *Ann. Rev. Biochem.* **61**, 87–111.

Lee, S.H. and Kim, C.S. (1992) Induction of stress proteins in cultured cells by heat and mercury. *J. Catholic Med. Coll.* **45**, 21–30.

Mannervik, B. (1985) Glutathione peroxidase. *Methods Enzymol.* **113**, 490–495.

Mannervik, B., Guttenberg, C., Jensson, H., Tahir, M.K., Warholm, M. and Jorvall, H. (1985) Identification of three classes of cytosolic glutathione transferases common to several mammalian species: correlation between structural data and enzymatic properties. *Proc. Natl. Acad. Sci. USA* **82**, 7202–7206.

Martin Jr., J.P. and Fridovich, I. (1981) Evidence for a natural gene transfer from the ponyfish to its bioluminescent bacterial symbiont *Photobacter leigonathi*: the close relationship between bacteriocuprein and the copper-zinc superoxide dismutase of teleost fishes. *J. Biol. Chem.* **256**, 6080–6089.

Masuda, A., Longo, D.L., Kobayashi, Y., Appella, E., Oppenheim, J.J. and Matsushima, K. (1988) Induction of mitochondrial manganese superoxide dismutase by interleukin 1. *FASEB J.* **2**, 3087–3091.

McCord, J.M. (1985) Oxygen-derived free radicals in postischemic injury. *New Engl. J. Med.* **312**, 159–163.

Mehlhorn, R.J., Fuchs, J., Sumida, S. and Packer, L. (1990) Preparation of tocopheroxyl radicals for detection by electron spin resonance. *Methods Enzymol.* **186**, 197–205.

Meister, A. and Anderson, M.E. (1993) Glutathione. *Ann. Rev. Biochem.* **52**, 711–760.

Meyer, D.J., Coles, B., Pemble, S.E., Gilmore, K.S., Fraser, G.M. and Ketterer, B. (1991) Theta, a new class of glutathione transferases purified from rat and man. *Biochem. J.* **274**, 409–414.

Meyers, D.M., Ahmad, S. and Pardini, R.S. (1992) Protective role of glutathione-S-transferase against lipid peroxidation in an insect species. *FASEB J.* **6**, 3942 (abstr.).

Michiels, C. and Ramacle, J. (1988) Use of the inhibition of enzymatic antioxidant systems in order to evaluate their physiological importance. *Eur. J. Biochem.* **177**, 435–441.

Misra, H.P. (1979) Reaction of copper-zinc superoxide dismutase with diethyldithiocarbamate. *J. Biol. Chem.* **254**, 11623–11628.

Mitchell, J.M., Ahmad, S. and Pardini, R.S. (1991) Purification of a highly active catalase from cabbage loopers, *Trichoplusia ni*. *Insect Biochem.* **21**, 641–646.

Morgenstern, R. and DePierre, J.W. (1988) Membrane-bound glutathione transferases. In *Glutathione Conjugation. Mechanisms and Biological Significance* (H. Sies and B. Ketterer, eds.), Academic Press, London, pp. 157–174.

Nahmias, J.A. and Bewley, G.C. (1984) Characterization of catalase purified from *Drosophila melanogaster* by hydrophobic interaction chromatography. *Comp. Biochem. Physiol.* **77B**, 355–364.

Nickla, H., Anderson, J. and Palzkill, T. (1983) Enzymes involved in oxygen detoxification during development of *Drosophila melanogaster*. *Experientia* **39**, 610–617.

Nohl, H. and Jordan, W. (1980) The metabolic fate of mitochondrial hydrogen peroxide. *Eur. J. Biochem.* **111**, 203–210.

Pacifici, R.E., Salo, D.C. and Davies, K.J.A. (1989) A 670 kDa proteinase complex that degrades oxidatively denatured proteins in red blood cells. *Free Rad. Biol. Med.* **7**, 521–536.

Packer, L. (1987) Antioxidant responses of the glutathione system, vitamin E, ubiquinones, and a free radical reductase. *J. Proc. 4th Int. Cong. on Oxygen Radicals*, University of California at San Diego, La Jolla, pp. 32–33 (abstr.).

Paoletti, F. and Mocali, A. (1990) Determination of superoxide dismutase activity by purely chemical system based on NAD(P)H oxidation. *Meth. Enzymol.* **186**, 209–220.

Pardini, R.S. and Ahmad, S. (1991) Effects of quinones on antioxidant enzymes of insects. *Proc. ASPET Meeting, The Pharmacologist* **33**, 345.

Pemble, S.E. and Taylor J.B. (1992) An evolutionary perspective on glutathione transferases inferred from class-Theta glutathione cDNA sequences. *Biochem. J.* **287**, 957–963.

Pritsos, C.A., Ahmad, S., Elliott, A.J. and Pardini, R.S. (1990) Antioxidant enzyme level response to prooxidant allelochemicals in larvae of the southern armyworm moth *Spodoptera eridania*. *Free Rad. Res. Commun.* **9**, 127–133.

Prohaska, J.R. (1980) The glutathione peroxidase activity of glutathione-S-transferases. *Biochim. Biophys. Acta* **611**, 87–98.

Puget, K. and Michelson, A.M. (1974) Isolation of a new copper-containing superoxide dismutase bacteriocuprein. *Biochem. Biophys. Res. Commun.* **58**, 830–838.

Puntarulo, S., Galleano, M., Sanchez, R.A. and Boveris, A. (1991) Superoxide anion and hydrogen peroxide metabolism in soybean embryonic axes during germination. *Biochim. Biophys. Acta* **1074**, 277–283.

Rose, R.C. and Bode, A.M. (1992) Tissue-mediated regeneration of ascorbic acid: is the process enzymatic? *Enzyme* **46**, 196–203.

Ross, M.J. and Reith, E.J. (1985) *Histology: A Textbook and Atlas*. Harper and Row, New York.

Salin, M.L. (1987) Preparation of iron-containing superoxide dismutases from eukaryotic organisms. In *CRC Handbook of Methods for Oxygen Radical Research* (R.A. Greenwald, ed.), CRC Press, Boca Raton, FL, pp. 9–13.

Samokyszyn, V.M., Miller, D.M., Reif, D.W. and Aust, S.D. (1989) Inhibition of superoxide and ferritin-dependent lipid peroxidation by Ceruloplasmin. *J. Biol. Chem.* **264**, 21–26.

Schlesinger, M.J. (1992) Heat shock proteins. *J. Biol. Chem.* **265**, 12111–12114.

Schirmer, R.H., Krauth-Siegel, R.L. and Schulz, G.E. (1989) Glutathione reductase. In *Glutathione: Biochemical & Medical Aspects, Part A* (D. Dolphin, R. Poulson and O. Avramovic, eds.), Wiley, New York, pp. 554–596.

Sharp, K., Fine, R. and Honig, B. (1987) Computer simulations of the diffusion of a substrate to an active site of an enzyme. *Science* **236**, 1460–1463.

Sies, H. and Akerboom, T.P.M. (1984) Glutathione disufide (GSSG) efflux from cells and tissues. *Methods Enzymol.* **105**, 445–451.

Singh, A. (1989) Chemical and biochemical aspects of activated oxygen: singlet oxygen, superoxide anion, and related species. In *CRC Handbook of Free Radicals and Antioxidants in Biomedicine* (J. Miquel, A.T. Quintanilha and H. Weber, eds.), CRC Press, Boca Raton, FL pp. 17–28.

Singh, S.V., Leal, T., Ansari, G.A.S. and Awasthi, Y.C. (1987) Purification and characterization of glutathione S-transferases of human kidney. *Biochem. J.* **246**, 179–186.

Stadtman, E.R. (1992) Protein oxidation and aging. *Science* **257**, 1220–1224.

Stallings, W.C., Pattridge, K., Strong, R.K., et al. (1984) Manganese and iron superoxide dismutase are structural homologues. *J. Biol. Chem.* **259**, 10695–10699.

Steinman, H.M. (1982) Copper-zinc superoxide dismutase from *Caulobacter crecentus*. *J. Biol. Chem.* **257**, 10283–10293.

Summers, C.B. and Felton, G.W. (1993) Antioxidant role of dehydroascorbic acid reductase in insects. *Biochim. Biophys. Acta* **1156**, 235–238.

Tainer, J.A., Getzoff, E.D., Richardson, J.S. and Richardson, D.C. (1983) Structure and Mechanism of copper, zinc-superoxide dismutase. *Nature* **306**, 284–287.

Takahashi, K., Avissar, N., Whitin, J. and Cohen, H. (1987) Purification and characterization of human plasma glutathione peroxidase: A selenoglycoprotein distinct from the known cellular enzyme. *Arch. Biochem. Biophys.* **256**, 677–686.

Tan, K.H., Meyer, D.J., Coles, B. and Ketterer, B. (1986) Thymine hydroperoxide, a substrate for rat Se-dependent glutathione peroxidase and glutathione transferase isoenzymes. *FEBS Lett.* **207**, 231–233.

Tartaglia, L.A., Storz, G., Brodsky, M.H., Lai, A. and Ames, B.N. (1990) Alkyl hydroperoxidase reductase from *Salmonella typhimurium*. Sequence and homology to thioredoxin reductase and other flavoprotein disulfide oxidoreductase. *J. Biol. Chem.* **265**, 10535–10540.

Ursini, F., Maiorino, M., Valente, M., Ferri, L. and Gregolin, C. (1982) Purification from pig liver of a protein which protects liposomes and biomembranes from peroxidative degradation and exhibits glutathione peroxidase activities on phosphatidylcholine hydroperoxides. *Biochim. Biophys. Acta* **839**, 197–211.

van Kuijk, F.J.G.M., Sevanian, A., Handelman, G.J. and Dratz, E.A. (1987) A new role for phospholipase A_2: protection of membranes from lipid peroxidation damage. *TIBS* **12**, 31–34.

Weinhold, L.C., Ahmad, S. and Pardini, R.S. (1990) Insect glutathione-*S*-transferase: a predictor of allelochemical and oxidative stress. *Comp. Biochem. Physiol.* **95B**, 355–363.

Weiss, L. (1983) The cell. In *Histology: Cell and Tissue Biology* (L. Weiss, ed.), 5th edition, Elsevier, New York.

Weitzel, F., Ursini, F. and Wendel, A. (1990) Phospholipid hydroperoxide glutathione peroxidase in various mouse organs during selenium deficiency and depletion. *Biochim. Biophys. Acta* **1036**, 88–94.

Williams, C.H. (1992) Lipomide dehydrogenase, glutathione reductase, thioredoxin reductase and mercuric ion reductase—a family of flavoenzyme transhydrogenases. In *Chemistry and Biochemistry of Flavoenzymes*, Vol. III (F. Muller, ed.), CRC Press, Boca Raton, FL, pp. 121–211.

Yoshimura, S., Watanabe, K., Suemuzu, H., Onozawa, T., Mizoguchi, J., Tsuda, K., Hatta, H. and Moriuchi, T. (1991) Tissue specific expression of the plasma glutathione peroxidase gene in rat kidney. *J. Biochem. (Tokyo)* **109**, 918–923.

CHAPTER 8

Antioxidant Defenses of *Escherichia coli* and *Salmonella typhimurium*

Richard P. Cunningham and Holly Ahern

8.1 INTRODUCTION

Single celled organisms like bacteria encounter the same kinds of complex problems that their multicellular counterparts must deal with. Conditions that cause stress to bacterial cells take many forms—fluctuations in environmental temperature or pH, or exposure to toxic chemicals or radiation. The use of oxygen during aerobic respiration creates special problems for bacterial cells. A buildup in reactive oxygen species (ROS) within cells or in a cell's environment may lead to genotoxic or cytotoxic alterations in cellular DNA, RNA, lipids or protein.

During aerobic growth, electrons from nutrients enter the electron transport chain and are eventually transferred to oxygen, the final electron acceptor in the chain. This transfer, with the subsequent reduction of oxygen to water, may result in the creation of oxygen species with reactive properties. These species include superoxide anion radicals (O_2^-), hydroxyl radicals ($^\cdot OH$), and also hydrogen peroxide (H_2O_2). ROS may also result when cells are exposed to oxidizing compounds in their environment, ionizing radiation, and also by oxidation-reduction active drugs, including paraquat and plumbagin. For anaerobic enteric bacteria like *Escherichia coli* and *Salmonella typhimurium*, respiratory bursts of O_2^- and H_2O_2 produced by macrophages and neutrophils during immune responses are life-threatening challenges.

In bacteria, DNA, RNA, lipids and proteins are all targets for ROS. Oxidation of lipids may result in alterations to the cell membrane which can effect energy production or nutrient uptake by the cell. Oxidative damage to amino acids can effect the overall structure and function of both intracellular and extracellular proteins. For a unicellular organism, damage to DNA is particularly critical. ROS can directly attack DNA and cause numerous lesions, including the alteration or removal of bases and strand breakage. In bacteria and other organisms, oxidized bases or their derivatives can be mutagenic or block DNA replication.

The responses of bacteria to oxidative stress and their defensive strategies have been extensively studied in *E. coli* and *S. typhimurium*. These organisms have evolved complex mechanisms to protect themselves from ROS. The first line of defense includes the avoidance or elimination of these species before oxidative damage can occur. Damage control is provided by antioxidants that protect the cell from further damage by preventing the initiation or propagation of free-radical chain reactions. Secondarily, bacteria possess repair pathways for the reversal of oxidation products in DNA lipids, and proteins. Together, these mechanisms protect cells from the damaging effects of oxidative stress.

8.2   DIRECT DEFENSES AGAINST OXIDATIVE STRESS

Oxygen free radicals occur in cells as the result of a succession of one electron reductions of molecular oxygen. This four electron reduction of molecular oxygen to water is catalyzed by a number of membrane-associated respiratory chain enzymes in the following reaction (reaction 8.1):

$$O_2 \xrightarrow{e^-} O_2^- \xrightarrow{e^-,2H} H_2O_2 \xrightarrow{e^-,H+} {}^\cdot OH \xrightarrow{e^-,H+} H_2O \tag{8.1}$$

The generation of oxygen free radicals in biological systems is more fully described in Chapter 1. It suffices to say that there are numerous sources of ROS in bacterial cells. The reactivity of the oxygen radicals increase as the number of electrons increases.

Superoxide anion O_2^- is a relatively weak oxidant that can also act as a reducing agent for transition metals and metal complexes. Thus proteins containing Fe-S clusters are especially sensitive to attacks from O_2^- (Gardner and Fridovich, 1991). The most important reaction of O_2^- involves the spontaneous or enzymatic dismutation of O_2^- to H_2O_2; it can also be protonated to form the hydroperoxyl radical (HOO'). Both of these intermediates are more reactive than O_2^-. The H_2O_2 can then react with reduced iron or copper cations to generate hydroxyl radicals ('OH). These types of reactions can initiate free radical chain reactions that substantially amplify the effects of the initiating oxidant.

As the cellular levels of O_2^- increase, so do the levels of H_2O_2. Studies indicate that the toxicity associated with O_2^- and H_2O_2 is probably due to their conversion to 'OH (Imlay and Linn, 1988; Imlay et al., 1988). Hydroxyl radicals are highly reactive and have been shown to directly oxidize critical targets within cells, including DNA, lipids, and proteins. The intermediates formed from these types of reactions can also have reactive properties.

In all types of cells, the best defense against ROS is avoidance. In aerobic cells, cytochrome oxidases and other metallo-oxidases utilize associated metal groups to directly reduce oxygen to water without an associated release of active intermediates. Cytochrome oxidase activity is responsible for most of the cellular reduction of O_2 and thus greatly decreases the number of ROS in the cell (Fridovich, 1989).

Despite this effective first line of defense, free radicals still arise in cells as intermediates in respiratory pathways, or following exposure to oxidizing agents or other redox active compounds. Macrophages and neutrophils produce respiratory bursts of free radicals as part of their microbiocidal response. The reactive species produced as a result of these events are removed by antioxidant enzymes that dismutate or reduce O_2^- and H_2O_2 into water and molecular oxygen (see chapter 7 for details). Other antioxidants react directly with free radicals, preventing them from attacking cellular components or initiating free radical chain reactions (see chapters 1 and 6). Metal chelators are antioxidants that prevent metal cations present in the cell from participating in oxidation reactions (cf. chapters 6 and 7).

A number of bacterial genes and their gene products that participate in the defense against active oxygen have been identified in

Table 8.1
Antioxidant Enzymes in *E. coli*

Activity		Gene
Defense enzymes:		
Superoxide dismutase	MnSOD	*sodA*
	FeSOD	*sodB*
Catalase	HPI	*katG*
	HPII	*katE*
Alkyl hydroperoxide reductase		*ahpC, ahpF*
Glutathione reductase		*gorA*
Metabolic enzymes:		
NAD(P)H dehydrogenase		*ndh*
Glucose-6-phosphate dehydrogenase		*zwf*

E. coli and *S. typhimurium*. These are shown in Table 8.1. Antioxidant enzymes include two forms of superoxide dismutases that catalyze the dismutation of O_2^- to H_2O_2 and molecular oxygen. *E. coli* also has two types of catalases that convert H_2O_2 to water and oxygen. Both organisms possess a glutathione-independent alkyl hydroperoxide reductase that catalyzes the reduction of hydroperoxides such as thymine hydroperoxide to its corresponding alcohol. Glutathione peroxidase activity, which is found in higher organisms, is apparently absent from bacteria, but *E. coli* does possess a glutathione reductase activity that may function to maintain the pool of reduced glutathione in cells (see Chapter 7).

The best characterized of the antioxidant enzymes is superoxide dismutase (SOD). The SODs dismutate O_2^- into H_2O_2 and O_2. Hydrogen peroxide, also a toxic intermediate, is then reduced by catalases (CATs) into H_2O and O_2, thus eliminating the threat to the cell.

SODs are metalloenzymes that occur ubiquitously in aerobic organisms. They can be categorized into three distinct classes belonging to two evolutionary groups, according to amino acid sequence data (Stallings et al., 1984). The manganese-containing SODs (MnSOD) are found in procaryotes and also in the mitochondria of eucaryotic cells. The procaryotic and mitochondrial MnSODs show a high degree of homology, which supports the theory of an endosymbiotic origin of mitochondria in eucaryotic cells. The iron-containing SODs (FeSOD) occur primarily in procaryotes and also in some plants. The copper and zinc SODs (CuZnSOD), which are evolutionarily unrelated to the other two classes are found in the cytosol of eucaryotic cells and in chloroplasts. They have also been found in a few species of bacteria. All three classes of SODs catalyze the dismutation of O_2^- to H_2O_2 and oxygen in the following reaction:

$$O_2^- + O_2^- + 2H^+ \rightarrow H_2O_2 + O_2 \qquad (8.2)$$

Three isozymic forms of SOD have been identified in *E. coli*; a MnSOD (SodA), an FeSOD (SodB) and a heterodimer consisting of both MnSOD and FeSOD. In *E. coli*, both MnSOD and FeSOD function only with the native metal at the enzyme's active site. However, SODs from other bacterial species catalyze the dismutation reaction with either manganese or iron at their active site.

The catalytic mechanism of FeSOD and MnSOD is not as well characterized as the CuZnSODs, but it is believed to be similar. The catalytic mechanism of CuZnSOD involves the alternate reduction and reoxidation of Cu(II) at the enzyme's active site during successive interactions with O_2^- (Tainer et al., 1983). The Cu (II) group is held in place by the imidazole rings of three histidine residues. In the CuZnSOD, two additional histidine residues and the carboxylate group of an aspartic acid residue bind the Zn(II). The Cu(II) is joined to the Zn(II) by a bridging histidyl imidazole group. Recent evidence implies that substrate binding by the enzyme is guided electrostatically (Cudd and Fridovich, 1982; Getzoff et al., 1983; Klapper et al., 1986; Sharp et al., 1987).

In the CuZnSOD, the imidizolate of the bridging histidine becomes reduced by O_2^-, which releases it from the copper center. The basic imidazolate is then protonated. The second O_2^- to reach the active site simultaneously receives an electron from Cu(I) and a proton from the imidazole group to reestablish the bridge. The resulting HO_2 is protonated in free solution to generate H_2O_2.

It appears that FeSOD and MnSOD have a similar catalytic mechanism, and there is also evidence to implicate electrostatic facilitation of substrate binding in these enzymes as well. The ligands that bind the metal groups to the enzymes' active sites are known to be three histidine residues and an aspartic acid. The active site metals [Mn(III) or Fe(III)] also undergo alternate cycles of reduction and reoxidation during catalysis by the enzymes. This implies that the reaction mechanisms for all of the SODs are similar.

It is becoming increasingly clear that the SODs play an essential role in the defense against oxygen toxicity. Exposing *E. coli* to O_2 results in an increase in SOD biosynthesis. *E. coli sodA, sodB* double mutants do not grow in minimal media under aerobic conditions (Carlioz and Touati, 1986). These cells also show a greatly increased sensitivity to the redox drug paraquat and an increase in spontaneous mutagenesis. The SOD defect can be complemented by human CuZnSOD introduced into mutant cells on a multicopy plasmid (Natvig et al., 1987; Farr et al., 1986). A suggested role for SODs is in the prevention of secondary or free radical chain reactions associated with oxygen metabolism.

Thus it is likely that SODs play an important role in the reduction of oxygen toxicity in bacteria. It also supports the conclusion that the relatively weak oxidant O_2^- is toxic to cells, either directly or via the production of toxic by-products of stronger reactivity.

Alkyl hydroperoxide reductase is an antioxidant enzyme that is found in both *E. coli* and *S. typhimurium* (Farr and Kogoma, 1991). It is a peroxidase capable of reducing lipid and other alkyl hydroperoxides to their corresponding alcohols, using NADH or NADPH as the reducing agent. The physiological substrate for alkyl hydroperoxide reductase has not been determined, but studies have shown that the enzyme will protect against oxidative damage to DNA since its presence prevents a high rate of mutation.

In *S. typhimurium*, the enzyme consists of two subunits with molecular weights of 57,000 and 22,000 Da. The larger subunit has a bound FAD cofactor, and alone it can reduce a number of electron acceptors including dioxygen, 2,6-dichloroindophenol, methylene blue and cytochrome *c*. Thus this larger subunit is both a NAD(P)H oxidase and a NAD(P)H diaphorase. When the two subunits are combined, alkyl hydroperoxides such as thymine hydroperoxide and linoleic acid hydroperoxide are reduced (Jacobson et al., 1989).

The sequence of the larger subunit has been determined and this protein shows sequence similarity to thioredoxin reductase, which suggests that the large subunit may have disulfide reductase activity (Tartaglia et al., 1990). The small subunit is apparently only required for the reduction of organic hydroperoxides and probably plays a role in substrate binding.

The exact physiological role of alkyl hydroperoxide reductase is not clear. Like the SOD, a likely function is to block the propagation of free radical chain reactions.

Catalase (CAT) activity in *E. coli* is believed to defend cells against hydrogen peroxide toxicity. Cells in dilute culture that are deficient in CAT, however, show the same level of sensitivity to H_2O_2 as wild-type cells. It appears that a major function of *E. coli*'s CAT activity is to protect multicellular groups of cells against H_2O_2 challenges.

Colonies or concentrated suspensions of CAT-positive *E. coli* can survive and multiply in the presence of H_2O_2, whereas cells in dilute suspensions are H_2O_2-sensitive. Multicellular groups can also cross-protect groups of CAT-deficient organisms (Muchou and Eaton, 1992). This type of group-defense mechanism in a unicellular organism may have been a driving force in the evolution of multicellularity in bacteria. Another example of this is seen in the complex patterns formed by motile cells of *E. coli* grown on semi-solid media containing mixtures of amino acids or sugars. The cells aggregate in response to a chemoattractant that is excreted in response to oxidative stress (Budrene and Berg, 1991).

The two CATs in *E. coli*, HPI and HPII, are under the control of two distinct regulatory systems (regulation of bacterial oxidative defense systems is discussed in Chapter 11). When *E. coli* and *S. typhimurium* are treated with low doses of H_2O_2, the synthesis of approximately 30 proteins is induced as part of what has been termed the peroxide stress response (Morgan et al., 1986; VanBogelen et al., 1987). Three of the induced proteins are the HPI CAT, alkyl hydroperoxide reductase, and glutathione reductase (Richter and Loewen, 1981; Christman et al., 1985). Most of the other proteins remain uncharacterized. HPI CAT is the product of the *katG* gene in *E. coli* and it degrades H_2O_2 to water and molecular oxygen. Its enzymatic activity is analogous to the dismutation of O_2^- by SOD, where a second H_2O_2 serves as an electron donor to reduce the first molecule, as shown in reaction 8.3.

$$H_2O_2 + H_2O_2 \rightarrow O_2 + 2H_2O \tag{8.3}$$

The apparent role of HPI CAT is to reduce the concentration of H_2O_2 in the cell, thus eliminating a potential source of highly reactive ˙OH radicals.

Although glutathione (GSSG) reductase (GR) is induced along with HPI CAT as part of the peroxide stress response, its exact role is unclear. Glutathione (GSH) is produced by GSH synthetase in *E. coli* and *S. typhimurium*. GR catalyzes the reduction of the oxidized glutathione, GSSG. In its reduced state, GSH can reduce disulfide bridges in oxidized proteins (Farr and Kogoma, 1991). Thus, GR may serve to protect cellular proteins from oxidation by maintaining a pool of reduced GSH in the cell. Under some conditions, GSH can function directly as an antioxidant. Glutathione can react with H_2O_2, O_2^-, or the perhydroxyl radical (HOO˙), forming stable radicals that can be reduced by GR to once again yield reduced glutathione (Meister and Anderson, 1983; Sies, 1986).

Glutathione is also required for the synthesis of glutaredoxin, which is a source of hydrogen during the ribonucleotide diphosphate reductase-catalyzed conversion of ribonucleoside diphosphates into precursors for DNA synthesis (Farr and Kogoma, 1991). This suggests another possible role for GR, which is to maintain the level of GSH in the cell to prevent a decline in the glutaredoxin level during oxidative stress. This is speculative because mutant *E. coli* strains that completely lack GSH are not overly sensitive to H_2O_2 (Greenberg and Demple, 1986).

HPII CAT is a growth related enzyme that is encoded by the *katE* gene in *E. coli*. The expression of HPII CAT appears to be under the control of the *katF* gene along with a group of largely uncharacterized proteins involved in the transition of cells from logarithmic growth phase to the stationary phase. This group also includes the DNA repair enzyme exonuclease III.

HPII CAT like HPI CAT catalyzes the degradation of H_2O_2 to H_2O and O_2, and thus it functions along with HPI to decrease the cellular levels of H_2O_2. Why HPII is produced in response to conditions of stationary growth or starvation, and its relation to other oxidative stress proteins is still unknown.

During O_2^- induced stress, approximately 40 proteins are induced, including MnSOD, the DNA repair enzyme endonuclease IV, the metabolic enzymes glucose-6-phosphate dehydrogenase and NAD(P)H dehydrogenase, and a paraquat-reducing activity. The function of glucose-6-phosphate dehydrogenase in response to oxidation may be to supply NADPH for the reduction of GSH by the enzyme GR to maintain a reduced state in the cell. Glucose-6-phosphate dehydrogenase has also been shown to confer resistance to oxidizing agents in *E. coli* and thus may act directly as an antioxidant (Beutler and Yoshida, 1988). The paraquat-reducing activity that is induced in this response is not well characterized, but it is known that NAD(P)H dehydrogenase is a diaphorase.

In addition to the enzymatic antioxidants, there are a number of substances that have a direct antioxidant effect, although the impact of these compounds on bacteria may not be profound (simple antioxidants are discussed in Chapters 6 and 10). They act as radical scavengers as part of a damage control response to limit the initiation of free radicals in the cell or the propagation of free radical chain reactions. Vitamin E (α-tocopherol) and β-carotene prevent chain reactions in membrane lipids. Vitamin C (ascorbate), a water soluble radical scavenger, is found mostly in the cytoplasm. As mentioned previously, in its reduced state, GSH can act as an antioxidant (although it may have prooxidant tendencies as well). Other low molecular weight antioxidants include thioredoxin and glutaredoxin, which may compensate when there are decreased levels of GSH in the cell. These substances serve to limit the number of free radical initiation or propagation events in the cell.

Thus the direct response of bacterial cells to oxidative stress involves a number of antioxidant enzymes and compounds that can reduce oxidants in order to prevent mutagenesis, loss of essential membrane functions, or the degradation of essential enzymes. Cells rely on SODs, CATs, alkyl hydroperoxide reductase and other enzymes to reduce the number of active oxygen species. Enzymes like GR and glucose-6-phosphate dehydrogenase function to maintain an antioxidant environment within the cell. Antioxidant compounds and metal-chelators prevent free radical chain reactions by scavenging and eliminating ROS. These antioxidants effectively defend cells from the ravages of oxidative stress.

8.3 INDIRECT DEFENSES AGAINST OXIDATIVE STRESS

Antioxidant enzymes and other compounds scavenge and remove many of the toxic ROS generated in bacterial cells. Despite the presence of these complex systems, oxygen radicals remain a cause of damage to macromolecules. The direct repair of these lesions by re-reduction of oxidized components has been shown to occur in some biological macromolecules. However in most cases, oxidative lesions are indirectly repaired by enzymes (proteases, lipases, and nucleases) that degrade the oxidized element and release the damage for excretion while conserving the undamaged constituents for reutilization. Removal of the damage may in fact diminish or prevent the potential cytotoxicity of oxidized macromolecules.

The types of oxidative damage that are lethal to bacteria have not been clearly defined. It is known that treating cells with redox-active drugs damages both DNA and cell membranes, either of which can be lethal. Damage to RNA and proteins by oxygen radicals is most probably not a cause of cell death, but it is a drain on the cells' resources since the damaged molecule must be repaired or replaced.

Oxidative damage to DNA represents a significant threat to survival. Exposing bacterial cells to H_2O_2 or O_2^--generating compounds, ionizing radiation, organic hydroperoxides, singlet oxygen or ozone results in the formation of numerous lesions in DNA. Oxidative attacks on bases can result in the formation of 8-hydroxyguanine, hydroxymethyl urea, urea, thymine glycol, or adenine ring-opened and ring-saturated residues (von Sontag, 1987).

Attacks on deoxyribose by hydroxyl radicals cause strand breaks with 3'-phosphate or 3'-phosphoglycolate ends (Henner et al., 1982; Henner and Rodriguez et al., 1983). Thymine residues in the DNA can react with ·OH to produce 5-hydroxymethyluracil, thymine glycol or urea (Teebor et al., 1988). Additionally, reaction of guanine with ·OH produces 8-hydroxydeoxyguanosine (Dizdaroglu, 1985).

Intermediates generated during lipid peroxidation can also react with DNA, causing site specific strand cleavage (Kappus, 1985). The formation of lipid peroxides can also result in a number of stable end products containing functional oxygen groups. These groups include aldehydes, epoxides, hydroxyl and peroxyl groups, along with alkanes and alkenes. These have been shown to react directly with the DNA, creating alkylated bases or crosslinked stands (Segerback, 1983; Summerfield and Tappel, 1983).

Of these numerous lesions, those that block replication (strand breaks, abasic sites, bulky adducts) are the most serious. Oxidized base residues such as 8-hydroxyguanine may contribute significantly to mutagenesis, but probably do not lead to cell death. To prevent the propagation of mutations in DNA, it is imperative to the cell that such lesions be repaired.

There is evidence to suggest that bacteria are also susceptible to oxidative lipid damage. Oxidative stress causes peroxidation of lipids, demonstrated both in vivo and in vitro (Kappus, 1985). Peroxidation of membrane lipids results in an increase in the membrane's fluidity, which causes a loss of structural integrity for the cell (McElhaney, 1985; Mead, 1976; Dills et al., 1980; Kashket, 1985; Larsen et al., 1974). Structural integrity is necessary for critical membrane functions such as nutrient transport, energy production, motility and maintaining the cell's osmotic balance. It appears that in *E. coli*, there are repair mechanisms that respond to this type of membrane damage (Farr et al., 1988).

Several *E. coli* enzymes, including glutamine synthetase, aconitase, 6-phosphogluconate dehydratase, and ribosomal protein L12 have been shown to be sensitive to oxidative stress. The attack of oxygen free radicals on amino acids results in a number of oxidized derivatives that generally inactivate the enzymes and target them for destruction by proteinases. Many of these enzymes contain [Fe-S] centers, and it is likely that metal-binding sites in enzymes are especially sensitive to oxidation. It appears that *E. coli* can repair protein damage to a limited extent by reducing or degrading the

Table 8.2
DNA Repair Enzymes in *E. coli* Involved in the Repair of Oxidative Lesions

Activity	Protein	Gene
AP endonuclease/repair diesterase	Exonuclease III	*xth*
	Endonuclease IV	*nfo*
	Endonuclease V?	—
Exonuclease/repair diesterase	Exonuclease I	*xon*
AP lyase/glycosylase	Endonuclease III	*nth*
	MutM protein	*mutM*
	MutY protein	*mutY*
Excision repair endonuclease	UvrABC protein	*uvrA*
		uvrB
		uvrC
8-oxo-dGTPase	MutT protein	*mutT*
Repair polymerase	DNA polymerase I	*polA*
DNA ligase	DNA ligase	*lig*

oxidized proteins (Davies, 1986; Gonzales-Porque et al., 1970; Lunn and Pigiet, 1987).

8.3.1 DNA Repair

The ways in which *E. coli* and *S. typhimurium* repair oxidative damage to DNA is the best characterized of the various repair systems. A number of *E. coli* enzymes that participate in the repair of DNA damage have been identified. These are shown in Table 8.2. Although all four bases in DNA can be subject to attack from ROS, considerable evidence points to the purine base guanine and the pyrimidine base thymine as the major targets of oxygen free radicals, although oxidized forms of adenine and cytosine have been found.

The oxidized form of guanine (7,8-dihydro-8-oxoguanine, also known as 8-hydroxyguanine or a GO lesion) and a number of different oxidized thymine products, including thymine glycol, arise following exposure to free radicals or ionizing radiation. There is also evidence to suggest that endogenous cell processes, such as electron transport and lipid peroxidation, can also lead to the creation of reactive oxygen species that attack the bases in DNA.

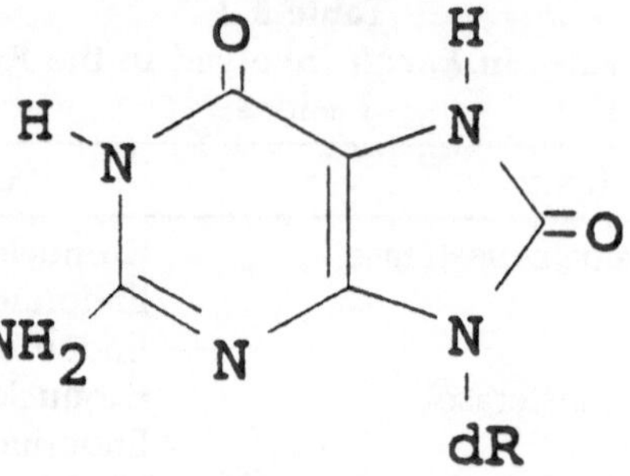

Fig. 8.1 7,8-Dihydro-8-oxoguanidine (8-hydroxyguanidine or GO). The predominant tautomeric form of 8-hydroxyguanidine is depicted.

The GO lesion (Figure 8.1) is a stable product of oxidation, and like the oxidized forms of thymine, has the potential to block replication in vitro and also in vivo (Kuchino et al., 1987). GO lesions have also been shown to form stable mispairs with adenine (Kouchakdjian et al., 1991). Since both types of oxidized bases, or their breakdown products, can be mutagenic and potentially lethal, the damaged bases must be removed from the DNA.

In *E. coli*, the GO system is an intricate system devoted to the repair of oxidized guanines. The repair mechanism appears to involve three levels—direct removal of GO lesions from the DNA, removal of mispaired adenines after a round of replication, and elimination of oxidized guanines from the nucleotide pool before they can be incorporated into the DNA. Mutations that effect the GO system are as serious as mutations disrupting the polymerase III proofreading subunit (*mutD*) and mismatch repair systems (*dam*, *mutHLS*) in *E. coli*. Thus the GO system represents an important error avoidance pathway for bacteria.

The GO system in *E. coli* is composed of at least three enzymes, MutM, MutT, and MutY. MutM and MutY function at the level of the DNA, while MutT is active at the nucleotide level by scavenging the nucleotide pool for oxidized dGTPs. The MutM protein is a glycosylase that was originally identified as formamidopyrimidine DNA glycosylase (FPG), an enzyme active on a variety of modified ring-open purines with associated AP endonuclease and 5'-terminal deoxyribose phosphatase activities (Boiteux et al., 1990; Graves et al., 1992; Boiteux et al., 1992). Mutants deficient in MutM show an increase in specific $G \cdot C \rightarrow T \cdot A$ transversions, which is accounted for by the specificity of misincorporation of adenine opposite GO lesions in DNA (Moriya et al., 1991; Wood et al., 1990). In the GO system, MutM is responsible for the removal of GO lesions from the DNA (Tchou et al., 1991).

MutT protein performs a function similar to MutM at the level of the nucleotide pool. MutT removes potentially mutagenic nucleotides by hydrolyzing 8-oxo-dGTP to 8-oxo-dGMP (Maki and Sekiguchi, 1992). 8-oxo-dGTP has been shown to be inserted by *E. coli* DNA polymerase opposite either cytosines or adenines with equal efficiency, with a 3–4% relative efficiency when compared to dGTP or dTTP (Maki and Sekiguchi, 1992). The elimination of the oxidized form of this triphosphate decreases the opportunity for the misincorporation of this mutagenic precursor opposite template adenines. *E. coli* strains deficient in MutT show a specific increase in A · T → C · G transversions which are mediated by A · G, and not C · T, mispairs (Shaaper and Dunn, 1987). It has also been shown that DNA replication is required for the expression of the *mutT* phenotype (Cox, 1970).

The MutY protein provides a third line of defense against the mutagenic potential of GO lesions in the DNA. MutY is a glycosylase that had previously been shown to correct A · G mispairs by removing the mispaired adenine (Au et al., 1988, 1989). Recently, however, the role of MutY in the removal of undamaged adenines from A · 8-oxoG mispairs has been shown, and this may be the enzyme's primary function (Michaels et al., 1992). The results of complementation assays using MutY and MutM have indicated that although MutY is active on both A · G mispairs and A · GO mispairs, the major substrate in vivo is the A-GO mispair. Like *mutM* strains, *mutY* strains also show an increase in G · C → T · A transversions. Double mutants deficient in both enzymes show a 20-fold increase in the rate of transversions. There is also evidence to suggest that MutY may have an associated AP endonuclease activity (Tsai-Wu et al., 1992).

In *E. coli*, the mutation rate associated with GO lesions is minimized by a multi-level defense plan. The cell relies on MutM to remove oxidized guanine residues from the chromosomal DNA. If a GO lesion escapes detection by MutM and is not removed before the DNA is replicated, misincorporation of adenine may occur. Thus, MutY functions after replication to specifically remove misincorporated adenines from A · GO mispairs, leaving an abasic site. The enzyme remains bound to the site following the removal of the adenine residue to prevent MutM from attacking the GO lesion opposite the AP site. Insertion of a cytosine residue by polymerase

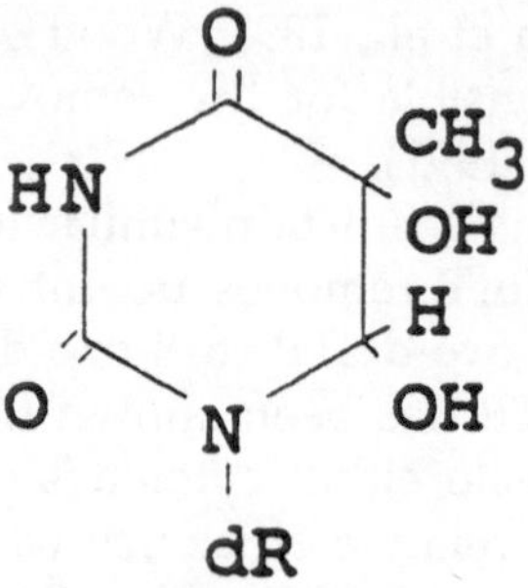

Fig. 8.2   5,6,-Dihydroxy-5,6-dihydrothymidine (thymidine glycol or TG). Thymidine glycol exists as a diastereomeric pair of *cis* glycols in equilibrium with a pair of corresponding C(6) *trans* epimers.

opposite the GO lesion sends the DNA back to MutM for a second chance at correcting the error. MutT assists in this error avoidance pathway by hydrolyzing potentially mutagenic precursors in the nucleotide pool. Thus with this elaborate system, *E. coli* protects itself from the potentially mutagenic effects of oxidized guanines. There are indications that similar proteins exist in other procaryotes and in higher eucaryotes as well.

Thymine residues are a principal target of oxygen free-radicals. A number of oxidized forms of thymine may arise, including thymine glycol and methyltartronylurea, which may be further degraded to other forms (Breimer and Lindahl, 1985). Of these oxidation products, thymine glycol is the major lesion (Figure 8.2).

A mutator phenotype has not been identified for thymine glycol repair-deficient *E. coli* strains. While thymine glycol residues are lethal, 10–12 residues per *E. coli* bacteriophage φX174 double stranded genome are required for a lethal hit (Laspia and Wallace, 1988). This implies that some lesions must be bypassed in vivo. Since thymine glycol lesions are only mildly mutagenic, thymine glycol must base-pair readily with adenine during bypass synthesis. However, enzymes that remove thymine glycol residues from DNA have been isolated from a number of diverse organisms, including both gram positive and gram negative bacteria, and higher organisms (Jorgensen et al., 1987; Gosset et al., 1988; Doetsch et al., 1986; Higgins et al., 1987). This evolutionary conservation implies that thymine glycol residues may be important premutagenic lesions to the cell, and that the repair of these lesions is necessary.

Endonuclease III from *E. coli* is a thymine glycol-DNA glycosylase with an associated AP lyase activity that facilitates the repair of thymine glycol lesions and other thymine ring saturation products. In addition to thymine glycol and 5,6-dihydrothymine, the enzyme recognizes various ring-rearranged and ring-fragmented thymines and also recognizes cytosine hydrates (Breimer and Lindahl, 1984; Boorstein et al., 1989). The reaction mechanism appears to procede via a Schiff base intermediate that leads to cleavage of the phosphodiester bond 3' to AP sites by a β-elimination reaction (Kow and Wallace, 1987).

The protein has been recently crystallized and its three-dimensional structure was solved (Kuo et al., 1992). Endonuclease III contains an [4Fe-4S] cluster, an unusual feature among the other known DNA repair enzymes. It appears that the [4Fe-4S] cluster does not participate directly in the catalysis of either of the two enzymatic activities associated with the protein, but instead appears to function in orienting critical amino acid residues for interaction with the DNA backbone.

The adenine glycosylase MutY has sequence homology to endonuclease III (Michaels et al., 1990), and it appears that MutY may also contain an iron-sulfur cluster. Thus the catalytic mechanism of both enzymes may be similar. There is speculation that endonuclease III may represent a prototype for DNA repair enzymes that recognize and repair damaged DNA in specific ways.

The UvrABC repair system in *E. coli* has also been shown to recognize thymine glycol residues in DNA (Lin and Sancar, 1989; Snowden et al., 1990). This system is well characterized in its ability to repair a number of bulky adducts induced by exposure to UV light and various chemicals. UvrABC was recently shown to cleave DNA on either side of thymine glycol residues in DNA, resulting in excision of the damaged base (Lin and Sancar, 1989).

Oxidative damage to DNA also can produce strand breaks that are not repaired by either DNA ligase alone or by DNA polymerase I and DNA ligase in combination. The termini surrounding these breaks must be processed in order for repair to occur. For strand breakage events, the 3' termini have been identified as 3'-phosphates and 3'-phosphoglycolates (Figure 8.3). Both of these lesions can block the repair of the break by ligases and polymerases.

Exonuclease III was the first enzyme involved in the repair of blocked termini to be identified (Metzel-Landbeck et al., 1976). It was then shown that exonuclease III could remove 3'-phosphates and 3'-phosphoglycolates from ionizing radiation-induced DNA

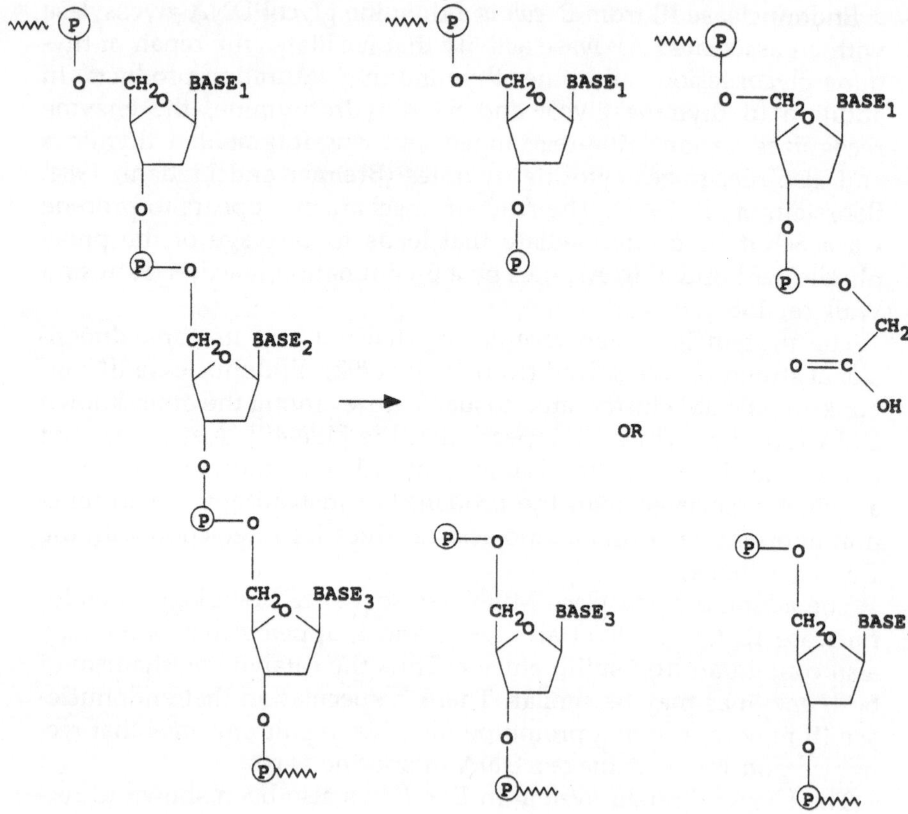

Fig. 8.3   3'-Phosphate (left) and 3'-phosphoglycolate (right) residues remaining at the ends of radical induced single-strand breaks in DNA.

strand breaks (Henner and Gronberg et al., 1983). Exonuclease III mutants are not remarkably sensitive to ionizing radiation (Milcarek and Weiss, 1972), which suggests that other enzymes may also be involved in the repair of these lesions. Mutants of E. coli that are hypersensitive to H_2O_2 were used to show that exonuclease III could remove blocked 3' termini from DNA treated with H_2O_2 (Demple et al., 1986). DNA extracted from xth cells was found to be a good template for DNA polymerase I only after treatment with purified exonuclease III.

Endonuclease IV was also found to reactivate DNA from xth mutants suggesting that it could also repair blocked 3'-termini (Demple et al., 1986). Endonuclease IV was then shown by direct enzyme assay to remove 3' terminal damage from DNA (Levin et al., 1988).

The release of 3'-phosphoglycolate residues by endonuclease IV confirms that endonuclease IV has a role in the repair of strand breaks, although the role must be minor (Siwek et al., 1988) since the level of phosphoglycolatase activity of endonuclease IV is small compared to that of exonuclease III.

E. *coli* mutants deficient in both exonuclease III and endonuclease IV were found to maintain a residual level of activity directed against damaged 3' termini (Bernelot-Moens and Demple, 1989). When extracts of this mutant strain were fractionated, three new repair diesterases were found. Two of these activities may correspond to previously identified enzymes. One has properties similar to deoxyribophosphodiesterase (Franklin and Lindahl, 1988) which has recently been shown to be exonuclease I (Sandigursky and Franklin, 1992). Another is similar to endonuclease V (Gates and Linn, 1977; Demple and Linn, 1982).

It appears that E. *coli* has a multitude of activities to repair the 3' blocked termini resulting from various forms of oxidative stress. The sensitivity of E. *coli xth* mutants to H_2O_2 and their insensitivity to ionizing radiation suggests that different lesions must be formed that are processed more or less efficiently by the remaining repair diesterases.

E. *coli* mutants deficient in endonuclease IV are sensitive to the oxidative antibiotic bleomycin, while exonuclease III mutants are not (Cunningham et al., 1986). An activity which cleaves bleomycin-treated DNA endonucleolytically has been found in extracts of *xth* mutants of E. *coli* (Hagensee and Moses, 1990). This activity was identified as endonuclease IV. Neocarzinostatin-induced AP sites were also reported to be cleaved only by endonuclease IV. It now appears likely that it is the presence of a strand break in the complementary strand, rather than the chemical nature of the AP site, that determines the sensitivity or resistance of DNA to a nuclease attack (Povirk et al., 1989).

H_2O_2-induced damage to DNA is also repaired by other means that are not yet understood at the enzymological level. It is clear that *recA*, *recB*, *polA*, *xth*, *and polC* mutants are all sensitive to H_2O_2 (Carlsson and Carpenter, 1980; Imlay and Linn, 1986; Hagensee and Moses, 1989). There are numerous interpretations of these data and also significant disagreements about the quantitative aspects of these experiments. Therefore, it seems fruitless at this time to attempt to describe a detailed molecular model for the repair of DNA damage caused by H_2O_2.

The repair of oxidatively damaged DNA is a highly sophisticated process. There are multiple activities in E. *coli* that are capable of repairing oxidative lesions in DNA—the same lesions in many instances. This redundancy strongly suggests that DNA repair is an important mechanism to protect cells against oxidative stress.

8.3.2   Protein Repair

The combined effects of the numerous antioxidant enzymes and compounds protects proteins to a great extent from the damaging effects of oxidative stress. Research has shown however that oxidative damage to proteins still occurs. When E. *coli* is exposed to oxygen free radicals or to H_2O_2, a decrease in the rate of protein synthesis along with a dose-dependent increase in protein degradation is observed (Davies and Lin, 1988a). At least one bacterial enzyme has been shown to be susceptible to damage by oxygen free radicals, and the oxidatively modified form of this enzyme is degraded by proteolysis (Levine et al., 1981). Other in vitro studies have confirmed that oxidized proteins are rapidly and preferentially degraded by an activity existing in E. *coli* extracts. This implies the existance of a soluble proteolytic system in E. *coli* that can selectively degrade proteins with oxidative damage (Davies and Lin, 1988a,b). Similar proteolytic systems have been noted in tissues from higher organisms (Davies, 1986).

Studies on both procaryotic and eucaryotic cells suggest that proteolytic susceptibility may be related to the denaturation and hydrophobicity of oxidized proteins. Susceptibility increases with increasing levels of both denaturation and hydrophobicity brought about by an oxidative modification of the protein. Thus it is likely that oxidative denaturation leading to high levels of hydrophobicity may target the protein for selective degradation.

A number of cytoplasmic and periplasmic proteases have been identified in E. *coli* (Swamy and Golberg, 1981). Some are dependent on ATP while others are ATP-independent. While it appears that a number of proteins are degraded by ATP-dependent proteases, the degradation of oxidized proteins in E. *coli* is the function of ATP-independent proteolytic system. As shown in cell extracts, ATP may actually inhibit destruction of oxidatively-damaged proteins. Mutants that lack the ATP-dependent protease La (the product of the *lon* gene in E. *coli*) degrade oxidatively modified proteins at the same rate as wild type cells. The selective destruction of oxidized proteins is thus separate from the ATP-dependent proteolytic pathways.

Proteolytic degradation of oxidatively modified proteins may be secondary to direct repair mechanisms, although this type of repair appears to be limited to the re-reduction of disulfide bonds and the repair of methionine sulfoxides (Davies and Lin, 1988b). The ability to repair oxidative damage in proteins is an interesting possibility, however little is known about this activity in bacteria.

The degradation of oxidatively damaged proteins prevents them from participating in activities that could cause further damage to the cell, such as cross-linking or aggregation reactions. Although it may seem costly to the cell, the resulting pool of amino acids, which contains only a few that are oxidatively damaged, can be recycled for use in the synthesis of new proteins.

8.3.3 Lipid and Membrane Repair

Oxidative stress has been shown to cause lipid peroxidation both in vitro and in vivo (Gonzales-Porque et al., 1970; Lunn and Pigiet, 1987). Lipid peroxidation in bacteria can lead to the formation of numerous reactive intermediates and modified end products, including chain-shortened fatty acids, alkanes, ketones, epoxides and aldehydes. Shortening the chain of fatty acids in lipids results in a loss of membrane fluidity, which can effect a number of essential functions. Many of the products of lipid peroxidation can react with DNA, resulting in alkylated bases or crosslinks between strands. They have also been shown to react with and inactivate enzymes.

In bacteria, the defense against lipid peroxidation occurs both directly and via more complex routes. Given the nature of bacterial cells, lipid peroxidation has less of an effect than in eucaryotic cells. In general, *E. coli* and *S. typhimurium* have mostly saturated or monounsaturated fatty acids in their membranes. This helps to keep the level of lipid peroxidation to a minimum, since the rate of peroxidation increase with the number of unsaturated bonds.

At least in *E. coli*, there may be an inducible repair system to correct oxidative damage to membranes (Farr et al., 1988; Yatvin et al., 1972). It appears that this pathway is under the control of the peroxide stress response, although to what extent is not known. HPI CAT and alkyl hydroperoxide reductase may also play a role in this system by directly reducing fatty acid hydroperoxides. There are also indications that this system is induced by ionizing radiation.

8.4 SUMMARY

The bacterial response to oxidative stress is a complex and well-orchestrated series of events that begins with exposure of the cell to reactive oxygen species (ROS), and ends with the restoration of cellular functions and the repair of damaged DNA. The ways in which *E. coli* and *S. typhimurium* respond to oxidative challenges is apparently dependent upon the oxidizing agent. These organisms have at their disposal an arsenal of antioxidant enzymes and other compounds that protect the cell from both the first and second rounds of oxidative assault. Antioxidant enzymes scavenge ROS and either dismutate or reduce them to less toxic substances, while other antioxidant compounds act to prevent the initiation or propagation of free radical chain reactions.

Oxidants that escape this first line of defense can oxidize critical cellular components. Bacteria can repair oxidative damage that may result from interactions between oxygen free radicals and major cellular macromolecules. *E. coli* possesses repair enzymes that specifically recognize and remove oxidized guanines and thymines from DNA. It is likely that other such DNA repair mechanisms exist. Bacteria also have pathways to repair oxidized proteins and lipids to reverse the damage caused by oxidative stress.

Given the vast array of antioxidant defense mechanisms and their apparent conservation in nature, it would appear that oxidative stress has played a significant role in evolution. Studies on the bacterial responses to oxidative stress will accelerate similar studies in higher organism, where such information can be put to good use in the search for new ways to treat debilitating human conditions associated with oxidative stress.

References

Au, K.G., Cabrera, M., Miller, J.H. and Modrich, P. (1988) *Escheric coli mutY* gene product is required for specific AG—CG mismatch correction. *Proc. Natl. Acad. Sci. USA* **85**, 9163–9166.

Au, K.G., Clark, S., Miller, J.H. and Modrich, P. (1989) *Escherichia coli mutY* gene encodes an adenine glycosylase active on GA mispairs. *Proc. Natl. Acad. Sci. USA* **86**, 8877–8881.

Bernelot-Moens, C. and Demple, B. (1989) Multiple DNA repair activities for 3'-deoxyribose fragments in *Escherichia coli*. *Nucleic Acids Res.* **17**, 587–600.

Beutler, E. and Yoshida, A. (1988) Genetic variation of glucose-6-phosphate dehydrogenase: a catalog and future prospects. *Medicine* **67**, 311–334.

Boiteux, S., Gajewski, E., Laval, J. and Dizdaroglu, M. (1992) Substrate specificity of the *Escherichia coli* Fpg protein (formamidopyrimidine-DNA glycosylase): excision of purine lesions in DNA produced by ionizing radiation or photosensitization. *Biochemistry* **31**, 106–110.

Boiteux, S., O'Connor, T.R., Lederer, F., Gouyette, A. and Laval, J. Homogeneous *Escherichia coli* FPG protein: a DNA glycosylase which excises imidazole ring-open purines and nicks DNA at apurinic sites. *J. Biol. Chem.* **265**, 3916–3922.

Boorstein, R.J., Hilbert, T.P., Cadet, J., Cunningham, R.P. and Teebor, G.W. (1989) UV-induced pyrimidine hydrates in DNA are repaired by bacterial and mammalian DNA glycosylase activities. *Biochemistry* **28**, 6164–6170.

Breimer, L.H. and Lindahl, T. (1984) Excision of oxidized thymine from DNA. *J. Biol. Chem.* **259**, 5543–5548.

Breimer, L.H. and Lindahl, T. (1985) Thymine lesions produced by ionizing radiation in double-stranded DNA. *Biochemistry* **24**, 4018–4022 (1985).

Budrene, E.O. and Berg, H.C. (1991) Complex patterns formed by motile cells of *Escherichia coli. Nature* **349**, 630–633.

Carlioz, A. and Touati, D. (1986) Isolation of superoxide dismutase mutants in *Escherichia coli*: is superoxide dismutase necessary for aerobic life? *EMBO J.* **5**, 623–630.

Carlsson, J. and Carpenter, V.S. (1980) The *recA⁺* gene product is more important than catalase and superoxide dismutase in protecting *Escherichia coli* against hydrogen peroxide toxicity. *J. Bacteriol.* **142**, 319–321.

Christman, M.F., Morgan, R.W., Jacobson, F.S. and Ames, B.N. (1985) Positive control of a regulon for defenses against oxidative stress and some heat shock proteins in *Salmonella typhimurium. Cell* **41**, 753–762.

Cox, E.C. (1970) Mutator gene action and the replication of bacteriophage λ DNA. *J. Mol. Biol.* **50**, 129–135.

Cudd, A. and Fridovich, I. (1982) Electrostatic interactions in the reaction mechanism of bovine erythrocyte superoxide dismutase. *J. Biol. Chem.* **257**, 11443–11447.

Cunningham, R.P., Saporito, S.M., Spitzer, S.G. and Weiss, B. (1986) Endonuclease IV (*nfo*) mutant of *Escherichia coli. J. Bacteriol.* **168**, 1120–1127.

Davies, K.J.A. (1986) Intracellular proteolytic systems may function as secondary antioxidant defenses: an hypothesis. *J. Free Rad. Biol. Med.* **2**, 155–173.

Davies, K.J.A. and Lin, S.W. (1988a) Degradation of oxidatively denatured proteins in *Escherichia coli. Free Rad. Biol. Med.* **5**, 215–223.

Davies, K.J.A. and Lin, S.W. (1988b) Oxidatively denatured proteins are degraded by an ATP-independent proteolytic pathway in *Escherichia coli. Free Rad. Biol. Med.* **5**, 225–236.

Demple, B., Johnson, A. and Fung, D. (1986) Exonuclease III and endonuclease IV remove 3' blocks from DNA synthesis primers in H_2O_2-damaged *Escherichia coli. Proc. Natl. Acad. Sci. USA* **83**, 7731–7735.

Demple, B. and Linn, S. (1982) On the recognition and cleavage mechanism of *Escherichia coli* endodeoxyribonuclease V, a possible DNA repair enzyme. *J. Biol. Chem.* **257**, 2848–2855.

Dills, S.S., Apperson, A., Schmidt, M.R. and Saier, M.H. (1988) Carbohydrate transport in bacteria. *Microbiol. Rev.* **44**, 385–418.

Dizdaroglu, M. (1985) Formation of an 8-hydroxyguanine moiety in deoxyribonucleic acid on gamma-irradiation in aqueous solution. *Biochemistry* 24, 4476–4481.

Doetsch, P.W., Helland, D.E. and Haseltine, W.A. (1986) Mechanism of action of a mammalian DNA repair enzyme. *Biochemistry* 25, 2212–2220.

Farr, S.B., D'Ari, R. and Touati, D. (1986) Oxygen-dependent mutagenesis in *Escherichia coli* lacking superoxide dismutase. *Proc. Natl. Acad. Sci. USA* 83, 8268–8272.

Farr, S.B. and Kogoma, T. (1991) Oxidative stress responses in *Escherichia coli* and *Salmonella typhimurium. Microbiol. Rev.* 55, 561–585.

Farr, S.B., Touati, D. and Kogoma, T. (1988) Effects of oxygen stress on membrane function in *Escherichia coli. J. Bacteriol* 170, 1837–1842.

Franklin, W.A. and Lindahl, T. (1988) DNA deoxyribophosphodiesterase. *EMBO J.* 7, 3617–3622.

Fridovich, I. (1989) Superoxide dismutases. *J. Biol. Chem.* 264, 7761–7764.

Gardner, P. and Fridovich, I. (1991) Superoxide sensitivity of the *Escherichia coli* 6-phosphogluconate dehydratase. *J. Biol. Chem.* 266, 1478–1483.

Gates, F.T. and Linn, S. (1977) Endonuclease V of *Escherichia coli. J. Biol. Chem.* 252, 1647–1653.

Getzoff, E.D., Tainer, J.A., Weiner, P.K., Kollman, P.A., Richardson, J.S., and Richardson, D.C. (1983) Electrostatic recognition between superoxide and copper, zinc superoxide dismutase. *Nature* 306, 284–287.

Gonzales-Porque, P., Baldensten, A. and Reichard, P. (1970) The involvement of the thioredoxin system in the reduction of methionine sulfoxide and sulfate. *J. Biol. Chem.* 254, 2371–2374.

Gosset, J., Lee, K., Cunningham, R.P. and Doetsch, P.W. (1988) Yeast redoxyendonuclease, a DNA repair enzyme similar to *Escherichia coli* endonuclease III. *Biochemistry* 27, 2629–2634.

Graves, R.J., Felzenszwalb, I., Laval, J. and O'Connor, T.R. (1992) Excision of 5'-terminal deoxyribose phosphate from damaged DNA is catalyzed by the Fpg protein of *Escherichia coli. J. Biol. Chem.* 267, 14429–14435.

Greenberg, J.T. and Demple, B. (1986) Glutathione in *Escherichia coli* is dispensible for resistance to H_2O_2 and gamma radiation. *J. Bacteriol.* 168, 1026–1029.

Hagensee, M.E. and Moses, R.E. (1989) Multiple pathways for repair of hydrogen peroxide-induced DNA damage in *Escherichia coli. J. Bacteriol.* 171, 991–995.

Hagensee, M.E. and Moses, R.E. (1990) Bleomycin-treated DNA is specifically cleaved only by endonuclease IV in *E. coli. Biochim. Biophys. Acta* 1048, 19–23.

Henner, W.D., Greenberg, S.M., and Haseltine, W.A. (1982) Sites and structure of γ radiation-induced DNA strand breaks. *J. Biol. Chem.* 257, 11750–11754.

Henner, W.D., Grunberg, S.M. and Haseltine, W.A. (1983) Enzyme action at the 3' termini of ionizing radiation-induced DNA-strand breaks. *J. Biol. Chem.* 258, 15198–15205.

Henner, W.D., Rodriguez, L.O., Hecht, S.M. and Haseltine, W.A. (1983) γ ray induced deoxyribonucleic acid strand breaks. *J. Biol. Chem.* 258, 711–713.

Higgins, S.A., Fenkel, K., Cummings, A. and Teebor, G.W. (1987) Definitive characterization of human thymine glycol N-glycosylase activity. *Biochemistry* 26, 1683–1688.

Imlay, J.A., Chin, S.M. and Linn, S. (1988) Toxic DNA damage by hydrogen peroxide through the Fenton reaction in vivo and in vitro. *Science* **240**, 640–642.

Imlay, J.A. and Linn, S. (1986) Bimodal pattern of killing of DNA-repair-defective or anoxically grown *Escherichia coli* by hydrogen peroxide. *J. Bacteriol.* **166**, 519–527.

Imlay, J.A. and Linn, S. (1988) DNA damage and oxygen radical toxicity. *Science* **240**, 1302–1309.

Jacobson, F.S., Morgan, R.W., Christman, M.F. and Ames, B.N. (1989) An alkyl hydroperoxide reductase from *Salmonella typhimurium* involved in the defense of DNA against oxidative damage. *J. Biol. Chem.* **264**, 1488–1496.

Jorgensen, T.J., Kow, Y-W., Wallace, S.S. and Henner, W.D. (1987) Mechanism of action of *Micrococcus luteus* γ-endonuclease. *Biochemistry* **26**, 6436–6443.

Kappus, H. (1985) Lipid peroxidation: mechanisms, analysis, enzymology and biological relevence. In *Oxidative stress* (H. Sies, ed.), Academic Press, New York, pp. 273–310.

Kashket, E.R. (1985) The proton motive force in bacteria: a critical assessment of methods. *Ann. Rev. Microbiol.* **39**, 219–242.

Klapper, I., Hagstron, R., Fine, R. and Hnig, B. (1986) Focusing of electric fields in the active site of Cu-Zn superoxide dismutase: effects of ionic strength and amino-acid modification. *Proteins* **1**, 47–89.

Kouchakdjian, M., Bodepudi, V., Shibutani, S., Eisenberg, M., Johnson, F., Grollman, A.P. and Patel, D.J. (1991) NMR structural studies of the ionizing radiation adduct 7-hydro-8-oxodeoxyguanosine (8-oxo-7H-dG) opposite deoxyadenosine in a DNA duplex. 8-oxo-7H-dG (syn). dA (anti) alignment at lesion site. *Biochemistry* **30**, 1403–1412.

Kow, Y-W. and Wallace, S.S. (1987) Mechanism of action of Escherichia coli endonuclease III. *Biochemistry* **26**, 8200–8206.

Kuchino, Y., Mori, F., Kasai, H., Inoue, H., Iwai, S., Miura, K., Ohtsuka, E. and Nishimura, S. (1987) Misreading of DNA templates containing 8-hydroxydeoxyguanosine at the modified base and at aduacent residues. *Nature* **327**, 77–79.

Kuo, C.-F., McRee, D.E., Fisher, C.L., O'Handley, S.F., Cunningham, R.P. and Tainer, J.A. (1992) Atomic structure of the DNA repair [4Fe-4S] enzyme endonuclease III. *Science* **258**, 434–440.

Larsen, S.H., Adler, J., Gargus, J.J. and Hogg, R.W. (1974) Chemomechanical coupling without ATP: the source of energy for motility and chemotaxis in bacteria. *Proc. Natl. Acad. Sci. USA* **71**, 1239–1243.

Laspia, M.F. and Wallace, S.S. (1988). Excision repair of thymine glycols, urea residues and apurinic sites in *Escherichia coli*. *J. Bacteriol.* **170**, 3359–3366.

Levin, J.D., Johnson, A.W. and Demple, B. (1988) Homogeneous *Escherichia coli* endonuclease IV. Characterization of an enzyme that recognizes oxidative damage in DNA. *J. Biol. Chem.* **263**, 8066–8071.

Levine, R.L., Oliver, C.N., Fulks, R.M. and Stadtman, E.R. (1981) Turnover of bacterial glutamine synthetase: oxidative inactivation precedes proteolysis. *Proc. Natl. Acad. Sci. USA* **78**, 2120–2124.

Lin, J. and Sancar, A. (1989) A new mechanism for repairing oxidative damage to DNA: (A)BC excinuclease removes AP sites and thymine glycols from DNA. *Biochemistry* **28**, 7979–7984.

Lunn, C.A. and Pigiet, V.P. (1987) The effects of thioredoxin on the radiosensitivity of bacteria. *Int. J. Radiat. Biol.* **51**, 29–38.

Maki, H. and Sekiguchi, M. (1992) MutT protein specifically hydrolyzes a potent mutagenic substrate for DNA synthesis. *Nature* **355**, 273–275.

McElhaney, R. (1985) The effects of membrane lipids on permeability and transport in prokaryotes. In *Structure and Properties of G. Membranes* (G. Benga, ed.), CRC Press, Boca Raton, FL, pp. 75–91.

Mead, J. (1976) Free radical mechanisms of lipid damage and consequences for cellular membranes. In *Free radicals in Biology* (W.A. Pryor, ed.), Academic Press, New York, pp. 51–68.

Meister, A. and Anderson, M.E. (1983) Glutathione. *Ann. Rev. Biochem.* **52**, 711–760.

Metzel-Landbeck, L., Schultz, G. and Hagen, U. (1976) In vitro repair of radiation-induced strand breaks in DNA. *Biochem. Biophys. Acta* **434**, 145–153.

Michaels, M.L., Cruz, C., Grollman, A.P. and Miller, J.H. (1992) Evidence that MutY and MutM combine to prevent mutations by an oxidatively damaged form of guanine in DNA. *Proc. Natl. Acad. Sci. USA* **89**, 7022–7025.

Michaels, M.L., Pham, L., Nghiem, Y., Cruz, C. and Miller, J.H. (1990) MutY, an adenine glycosylase active on GA mispairs, has homology to endonuclease III. *Nucleic Acids Res.* **18**, 3843–3845.

Milcarek, C. and Weiss, B. (1972) Mutants of *Escherichia coli* with altered deoxyribonucleases. *J. Mol. Biol.* **68**, 303–318.

Morgan, R.W., Christman, M.F., Jacobson, F.S. Storz, G. and Ames, B.N. (1986) Hydrogen peroxide-inducible proteins in *Salmonella typhimurium* overlap with heat shock and other stress proteins. *Proc. Natl. Acad. Sci. USA* **83**, 8059–8063.

Moriya, M., Ou, C., Bodepudi, V., Johnson, F., Takeshita, M. and Grollman, A.P. (1991) Site-specific mutagenesis using a gapped duplex vector: a study of translesion synthesis past 8-oxodeoxyguanosine in E. coli. *Mutat. Res.* **254**, 281–288.

Muchou, M. and Eaton, J.W. (1992) Multicellular oxidant defense in unicellular organisms. *Proc. Natl. Acad. Sci. USA* **89**, 7924–7928.

Natvig, D.O., Imlay, K., Touati, D. and Hallewell, R.A. (1987) Human CuZn-superoxide dismutase complements superoxide dismutase-deficient *Escherichia coli* mutants. *J. Biol. Chem.* **262**, 14697–14701.

Povirk, L.F., Han, Y. and Steighner, R.J. (1989) Structure of bleomycin-induced DNA double-strand breaks: predominance of blunt ends and single-base 5′ extensions. *Biochemistry* **28**, 5808–5814.

Richter, H.E. and Loewen, P.C. (1981) Induction of catalase in *Escherichia coli* by ascorbic acid involves hydrogen peroxide. *Biochem. Biophys. Res. Commun.* **100**, 1039–1046.

Sandigursky, M. and Franklin, W.A. (1992) DNA deoxyribophosphodiesterase of *Escherichia coli* is associated with exonuclease I. *Nucleic Acids Res.* **20**, 4699–4703.

Segerback, D. (1983) Alkylation of DNA and hemoglobin in the mouse following exposure to ethene and ethene oxide. *Chem.-Biol. Interact.* **45**, 135–151.

Shaaper, R.M. and Dunn, R.L. (1987) *Escherichia coli mutT* mutator effect during in vitro DNA synthesis. *J. Biol. Chem.* **262**, 16267–16270.

Sharp, K., Fine, R. and Honig, B. (1987) Computer simulations of the diffusion of a substrate to an active site of an enzyme. *Science* **236**, 1460–1463.

Sies, H. (1986) Biochemistry of oxidative stress. *Angew. Chem. Int. Ed. Engl.* **25**, 1058–1071.

Siwek, B., Bricteux-Gregoire, S., Bailly, V. and Verly, W.G. (1988) The relative importance of *Escherichia coli* exonuclease III and endonuclease IV for the hydrolysis of 3'-phosphoglycolate ends in polynucleotides. *Nucleic Acids Res.* **16**, 5031–5038.

Snowden, A., Kah, Y.W. and Van Houten, B. (1990) Damage repertoire of the *Escherichia coli* UvrABC nuclease complex includes basic sites, base damage analogues, and lesions containing adjacent 5' or 3' nicks. *Biochemistry* **29**, 7251–7259.

Stallings, W.C., Pattridge, K.A., Strong, R.K., et al. (1984) Manganese and iron superoxide dismutases are structural homologs. *J. Biol. Chem.* **259**, 10695–10699.

Summerfield, F.W. and Tappel, A.L. (1983) Determination by fluorescence quenching of the environment of DNA crosslinks made by malondialdehyde. *Biochem. Biophys. Acta* **740**, 185–189.

Swamy, K.H.S. and Golberg, A.L. (1981) *E. coli* contains eight soluble proteolytic activities, one being ATP dependent. *Nature* **292**, 652–654.

Tainer, J.A., Getzoff, E.D., Richardson, J.S. and Richardson, D. C. (1983) Structure and mechanism of copper, zinc-superoxide dismutase. *Nature* **306**, 284–287.

Tartaglia, L.A., Storz, G., Brodsky, M.H., Lai, A. and Ames, B.N. (1990) Alkyl hydroperoxide reductase from *Salmonella typhimurium*. Sequence and homology to thioredoxin reductase and other flavoprotein disulfide oxidoreductases. *J. Biol. Chem.* **265**, 10535–10540.

Tchou, J., Kasai, H., Shibutani, S., Chung, M.-H., Laval, J., Grollman, A.P. and Nishimura, S. (1991) 8-oxoguanine (8-hydroxyguanine) DNA glycosylase and its substrate specificity. *Proc. Natl. Acad. Sci. USA* **88**, 4690–4694.

Teebor, G.W., Boorstein, R.J. and Cadet, J. (1988) The reparability of oxidative free radical mediated damage to DNA: a review. *Int. J. Radiat. Biol.* **54**, 131–131.

Tsai-Wu, J-J., Liu, H-F. and Lu, A-L. (1992) *Escherichia coli* MutY protein has both N-glycosylase and apurinic/apyrimidinic endonuclease activities on A·C and A·G mispairs. *Proc. Natl. Acad. Sci. USA* **89**, 8779–8783.

VanBogelen, R.A., Kelley, P.M. and Neidhardt, F.C. (1987) Differential induction of heat shock, SOS, and oxidative stress regulons and accumulation of nucleotides in *Escherichia coli*. *J. Bacteriol.* **169**, 26–32.

von Sontag, C. (1987). *The Chemical Basis of Radiation Biology*. Taylor & Francis, London.

Wood, M.L., Dizdaroglu, M., Gajewski, E. and Essigman, J.M. (1990) Mechanistic studies of ionizing radiation and oxidative mutagenesis: genetic effects of a single 8-hydroxyguanine (7-hydro-8-oxoguanine) residue inserted at a unique site in a viral genome. *Biochemistry* **29**, 7024–7032.

Yatvin, M.B., Wood, P.G. and Brown, S.M. (1972) Repair of plasma membrane injury and DNA single strand breaks in gamma-irradiated *Escherichia coli* B/r and Bs-1. *Biochem. Biophys. Acta* **287**, 390–403.

CHAPTER 9

Antioxidant Defenses of Plants and Fungi

David A. Dalton

9.1 INTRODUCTION

Plants face a common problem shared by all aerobic organisms. Molecular O_2 is required to meet the energy demands of metabolism, but the necessary presence of O_2 leads to the multifarious risks of oxidative damage due to the formation of activated forms of oxygen such as superoxide free radicals (O_2^-), H_2O_2, and hydroxyl radicals ($\cdot OH$). The situation is particularly critical in plants because not only do plants have to tolerate atmospheric O_2 but they have the additional burden of being oxygenic, thus insuring that the internal O_2 concentration remains high in photosynthetic tissues. Indeed photosynthesis can be viewed as a very risky undertaking since large amounts of O_2 are produced in the immediate vicinity (i.e., within chloroplasts) of a powerful oxidation-reduction system that can readily reduce O_2 to O_2^-. Since atmospheric O_2 is derived from photosynthesis, it is perhaps poetic justice that the very organisms responsible for "polluting" the atmosphere with O_2 should also be at high risk from the consequences.

Plants also conduct two other major types of reduction processes that can lead to formation of activated forms of oxygen, namely respiration (reduction of O_2 to H_2O) and nitrogen fixation (reduction of N_2 to NH_3). Plant respiration is not especially unique in this regard. Mitochondria from plants (as well as other organisms) contain flavoprotein dehydrogenases as part of the electron transport chain. As electrons pass through the electron transport chain, some errant electrons from flavoprotein dehydrogenases may reduce O_2 to O_2^- (Massey et al., 1969). Interestingly, cytochrome oxidase is capable of the complete reduction of O_2 to H_2O without the release of partially-reduced intermediates such as O_2^-, H_2O_2 or $\cdot OH$ (Rich and Bonner, 1978).

Although plants per se can not fix N_2, their microbial partners frequently do. The two most notable cases include plants in the legume family that form a symbiosis with bacteria in the genus *Rhizobium* (or *Bradyrhizobium*) and the diverse woody plants such as alder (*Alnus*), *Ceanothus* or *Myrica* that form a symbiosis with actinomycetes of the genus *Frankia*. The conversion of N_2 to NH_3 requires powerful reducing conditions and this leads to very serious oxygen-related problems.

Fortunately plants possess an impressive array of defenses against oxidative damage. These defenses include numerous enzymes as well as small molecules some of whose role is just now being appreciated. Among enzymes, the key players are superoxide dismutase (SOD) which scavenges O_2^- and ascorbate peroxidase which scavenges H_2O_2. The primary non-enzymatic defenses include antioxidants such as ascorbate and glutathione in hydrophilic regions of cells and vitamin E (α-tocopherol) and β-carotene in hydrophobic areas, particularly membrane interiors.

Many other enzymes and antioxidants provide additional defense in plants. Constructing and maintaining these defenses has substantial metabolic costs. The general defense strategy is to further reduce the activated intermediates of oxygen and so render them innocuous. This requires the consumption of reducing power (i.e., energy) which often ultimately (or directly) comes from NADH or NADPH. In addition to the direct utilization of reducing power, cells must invest in numerous secondary processes to support these defenses. For instance five enzymes may be involved in the synthesis of ascorbate and another three enzymes in the regeneration of ascorbate after it becomes oxidized in peroxide scavenging. Considering the cell resources that are required to synthesize and maintain these defenses, it is clear that protection against activated forms of oxygen is one of the fundamentally most important processes in plant cells.

9.2 SOURCES OF ACTIVATED FORMS OF OXYGEN

Several extensive reviews on the production of active oxygen in plants are available (Asada and Takahashi, 1987; Elstner, 1982; Rabinowitch and Fridovich, 1983). The extent of the problem is illustrated by Asada and Takahashi's estimate that about 1% of total O_2 consumed by plants is "leaked to active oxygen" (Asada and Takahashi, 1987). Although plants obviously require O_2 to support respiration, supra-ambient levels of O_2 are deleterious and can even be fatal (Asada and Takahashi, 1987; Halliwell, 1982a).

9.2.1   *Oxygen Toxicity Related to Photosynthesis*

In plants, part of the causes of O_2 toxicity are related to the inhibition of photosynthesis. The inhibition of photosynthesis by O_2 is called the Warburg effect after the pioneering German biochemist who first described this phenomenon in the 1920's. The reasons for this inhibition were unclear until Ogren discovered in 1971 that ribulose-1,5-bisphosphate (RUBP) carboxylase, the chief enzyme responsible for fixing CO_2 into organic form in photosynthesis, had a tragic flaw (Ogren, 1984). In addition to functioning as a carboxylase, this enzyme also acted as an oxygenase by catalyzing the following reaction (eq. 9.1):

Ribulose-1,5-bisphosphate + O_2

$$\rightarrow \text{3-phosphoglycerate} + \text{2-phosphoglycollate} \quad (9.1)$$

One of these products, 3-phosphoglycerate, is a normal metabolite of CO_2 fixation in chloroplasts and so enters easily into the Calvin cycle—the complex series of reactions in chloroplasts that result in the production of numerous sugars and starch. However, 2-phosphoglycollate is not so easily metabolized. Less sophisticated organisms, such as many algae, simply excrete the glycollate resulting in the loss of hard-earned carbon resources. Most higher plants enter into a complex process in which the glycollate is transported into an adjacent organelle, the peroxisome, and then oxidized to glyoxylate in a reaction catalyzed by glycollate oxidase (eq. 9.2).

$$\text{glycollate} + O_2 \rightarrow H_2O_2 + \text{glyoxylate} \quad (9.2)$$

This reaction is noteworthy because it produces substantial amounts of toxic H_2O_2. Consequently, peroxisomes contain large amounts of catalase that catalyzes the removal of the H_2O_2. The peroxisome thus functions to contain the site of H_2O_2 production in the immediate vicinity of the enzyme that will rapidly remove it. The processing of carbon intermediates from glyoxylate involves a complicated shuttling of compounds first into mitochondria, then again into peroxisomes and ultimately back to the chloroplasts where some of the carbon enters the Calvin cycle in the form of 3-phosphoglycerate. Along the way, one carbon atom out of every four is lost as CO_2. This process resembles respiration in the sense that CO_2 is released and O_2 is taken up (in the initial reaction with RUBP as well as in the glycollate reaction). Light stimulates these reactions because light energy is required to produce NADPH to fuel the Calvin cycle and thus regenerate RUBP to act as the initial O_2 acceptor. Consequently, the whole process described above is known as photorespiration. It is disadvantageous since it results in the loss of energy and carbon resources. Many plants have developed elaborate mechanisms of altered photosynthesis, such as C4 pathways and Crassulacean Acid Metabolism (CAM), which minimize the problems of photorespiration. C4 photosynthesis involves a spatial separation of the RUBP carboxylase reaction from O_2, whereas CAM involves a temporal separation. Details of these pathways are provided in numerous excellent reviews (Black, 1973; Edwards and Walker, 1983; Lorimer, 1981; Ting, 1985).

Oxygen in chloroplasts can also lead to serious problems unrelated to photorespiration. The light reactions of photosynthesis involve the generation of strong reducing conditions in order to generate NADPH. Light energy is used to split water molecules in the Hill reaction as follows:

$$H_2O \rightarrow 1/2O_2 + 2H^+ + 2e^- \tag{9.3}$$

The liberated electrons, after being elevated to a higher energy status by photochemical events involving proteins and pigments in a photosystem reaction center, are passed through a series of carriers that are often known as the Z scheme because of the characteristic pattern when the carriers are arranged in a diagram with the lowest (most negative) midpoint redox potentials at the top (Fig. 9.1). One of the most powerful reducing components in this pathway is ferredoxin, a small membrane-associated iron-sulfur protein that undergoes reversible reduction-oxidation reactions as electrons pass through the system. Ferredoxin-NADP reductase, a soluble fla-

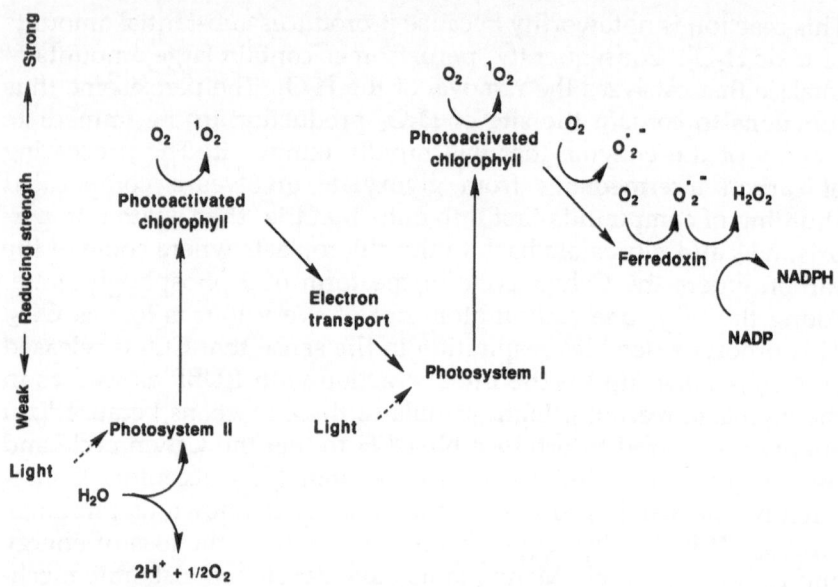

Fig. 9.1 Pathways for generation of activated forms of oxygen in illuminated chloroplasts.

voprotein, catalyzes the reaction in which electrons are passed from ferredoxin to NADP, thus producing NADPH. The midpoint redox potential of ferredoxin (-0.43 V) is nearly identical to that required to reduce O_2 to O_2^- (-0.45 V) so it is not surprising that reduced ferredoxin can univalently reduce O_2 and, in so doing, generate O_2^- via the following reaction (eq. 9.4) (Misra and Fridovich, 1971):

$$\text{Ferredoxin}_{\text{reduced}} + O_2 \rightarrow \text{Ferredoxin}_{\text{oxidized}} + O_2^- \qquad (9.4)$$

O_2^- can also be produced directly by electrons from photosystem I without the intervention of ferredoxin (Elstner, 1982). O_2^- can be further reduced by ferredoxin to form H_2O_2. This ability of chloroplasts to consume O_2 and produce O_2^- is called the Mehler reaction after its original discoverer (Mehler, 1951) and is largely responsible for the high levels of defense activity that are required in chloroplasts.

In addition to O_2^-, chloroplasts are also prone to production of another activated form of oxygen—singlet oxygen ($^1O_2$). Singlet oxygen is produced when an illuminated chlorophyll molecule transfers energy from an excited electron to ground state oxygen instead of passing that excited electron to the next carrier in the Z scheme.

$^1O_2$ auto-destructs quickly in an aqueous environment, but persists much longer in a hydrophobic environment. Unfortunately, the latter is frequently the case since the photosystem reaction centers are embedded in thylakoid membranes. $^1O_2$ can then initiate the process of lipid peroxidation that seriously damages chloroplast membranes (Fig. 9.2). Thylakoid membranes are predisposed to peroxidation since they are enriched in polyunsaturated fatty acids which as a class are especially vulnerable to peroxidation (Halliwell and Gutteridge, 1985). Polyunsaturated fatty acids such as linolenic acid account for as much as 90% of the esterified fatty acids in thylakoids of some plant species (Leech and Murphy, 1976).

9.2.2 Oxygen Toxicity Related to Respiration

Activated forms of oxygen are also a problem in mitochondria, whether of plant, fungal, or animal origin. However, the situation with plant mitochondria is more complicated since the electron transport pathway is often branched and more than one terminal oxidase is present. Regardless of which terminal oxidase is present, four electrons are required to reduce O_2 to water in the final step in the electron transport system:

$$O_2 + 4e^- + 4H^+ \rightarrow 2H_2O$$

Usually, this reaction is catalyzed by cytochrome oxidase, a remarkably efficient enzyme which accomplishes the tetravalent reduction of O_2 without the release of partially reduced intermediates such as O_2^- or H_2O_2. Many plants contain a poorly understood alternative, cyanide-insensitive terminal oxidase. This oxidase also deserves commendation for its efficiency because it produces H_2O but not O_2^- or H_2O_2 (Rich and Bonner, 1978; Siedow, 1982). However, there are other components of the electron transport system in plant mitochondria that are not so efficient. In particular, flavoprotein NADH dehydrogenases can directly reduce O_2 to O_2^- (Rich and Bonner, 1978).

9.2.3 Oxygen Toxicity Related to Nitrogen Fixation

i. Nitrogenase sensitivity to oxygen

Those plants capable of fixing nitrogen face further difficulties related to oxygen toxicity. Although relatively few plants possess the

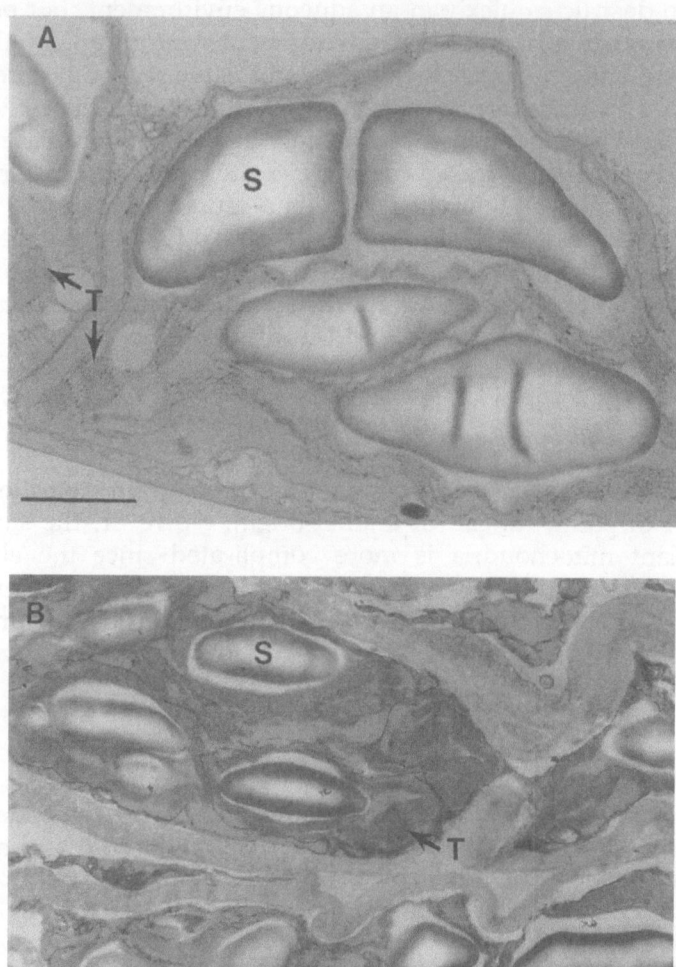

Fig. 9.2   Electron micrographs of cucumber chloroplasts. A. Control chloroplasts showing normal, intact thylakoids (T). B. Chloroplasts containing compressed and barely detectable thylakoids due to lipid peroxidation following treatment with chilling and strong light. S, starch grains. Bar = 1.0 μm for both panels. From Wise and Naylor, 1987. Photographs provided by Dr. Robert R. Wise, Univ. of Wisconsin. Reproduction by permission of American Society of Plant Physiologists.

ability to fix nitrogen, those plants are of great economic and ecological importance. The most notable nitrogen- fixing plants are those in the legume family such as pea, soybean, alfalfa, and clover (Sprent and Sprent, 1990). However, this ability is also present in a scattering of 24 other genera in 8 plant families (Baker and Mullin, 1992). In legumes, nitrogen fixation takes place in root nodules that are infected with bacteria in the genera *Rhizobium* or *Bradyrhizobium*. The bacteria, called bacteroids once they have entered symbiosis, proliferate inside host plant cells (Fig. 9.3). Host plant cells provide energy and carbon resources that are utilized within the bacteroids to support nitrogen fixation.

All nitrogen-fixing organisms face tremendous adversity in dealing with O_2, in large part because nitrogenase, the enzyme that catalyzes the reduction of N_2 to NH_3 (eq. 9.5), is notoriously sensitive to O_2.

$$N_2 + 8e^- + 8H^+ + 16ATP \rightarrow 2NH_3 + H_2 + 16ADP + 16Pi \quad (9.5)$$

The half-life of nitrogenase in air is less than one minute. Consequently, nitrogen-fixing organisms have developed a wide array of special mechanisms to minimize the exposure of nitrogenase to O_2 (reviewed in Gallon, 1992; Robson and Postgate, 1980). Although there is considerable diversity in these mechanisms, they all function to insure that nitrogenase operates in an environment in which the free O_2 concentration is exceedingly low.

The reasons for the O_2 sensitivity of nitrogenase are not clear, but it is probable that nitrogenase reduces O_2 and in so doing produces O_2^-, H_2O_2 or $\cdot OH$ which then rapidly destroys the enzyme. The evidence for this mechanism is indirect. Nitrogenase activity in in vitro preparations can be partially protected from inactivation in air by inclusion of superoxide dismutase and catalase (Mortensen et al., 1974; Robson and Postgate, 1980). While this evidence is perhaps equivocal, the putative substrate status of O_2 is consistent with the well-known lack of substrate specificity of nitrogenase. In fact, nitrogenase can reduce a number of substrates other than N_2 (Table 9.1). Since most of these substrates have a small diatomic core with a bond length of 1.1–1.2 Å, it is logical to extend the list to include O_2. Although most substrates of nitrogenase have a triple bond, the double bond of O_2 is not without precedence since both allene ($CH_2{=}C{=}CH_2$) and cyclopropene ($C{=}C$) are also substrates.

ii. Role of leghemoglobin in active oxygen production

In leguminous plants, nitrogen fixation takes place in root nodules that contain substantial levels of leghemoglobin, a monomeric,

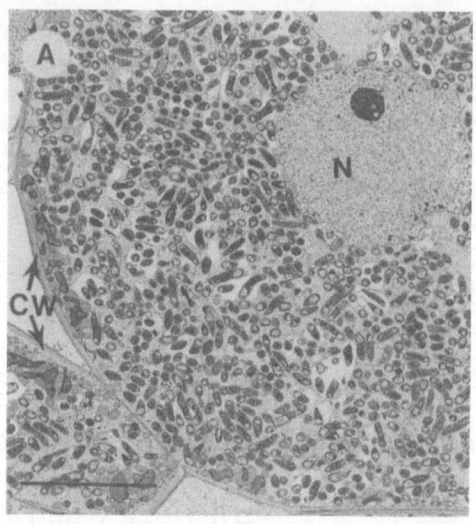

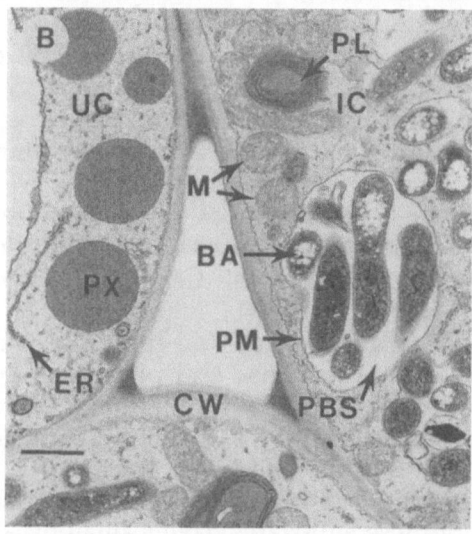

Fig. 9.3 Electron micrographs of nitrogen-fixing root nodules from cowpea (*Vigna unguiculata*). A. A large infected host cell containing numerous bacteroids and a nucleus (N). Bar = 10 μm. B. Detailed view of infected cell (IC) and adjacent uninfected cell (UC). The bacteroids (BA) contain numerous clear granules of poly-β-hydroxybutyrate, a carbon storage compound. The peribacteroid membrane (PM) enclosing the bacteroids and the peribacteroid space (PBS) are clearly visible. Other features include cell wall (CW), endoplasmic reticulum (ER), mitochondria (M), peroxisomes (PX), and plastids (PL). Bar = 1 μm. Photographs provided by Dr. Mary Alice Webb, Purdue Univ.

Table 9.1

A Partial List of Substrates and Products in Reactions Catalyzed by Nitrogenase

Name	Formula	Bond length (Å)	Product(s)	Remarks
Dinitrogen	$N\equiv N$	1.098	$NH_3 + H_2$	The normal reaction
Cyanide	$[C\equiv N]^-$	1.16	$CH_4 + NH_3$	Used for assaying activity
Acetylene	$HC\equiv CH$	1.204	$H_2C{=}CH_2$	
Azide	$[N\equiv N{-}N]^-$	1.18	$N_2 + NH_3 + N_2H_4$	
Nitrous oxide	$N\equiv N{-}O$	1.126	$N_2 + H_2O$	
Oxygen?	$O{=}O$	1.208	$O_2^-?, H_2O_2?,$ and/or $\cdot OH?$	Rapid inactivation of nitrogenase

modified from Postgate, 1982.

O_2-binding protein remarkably similar to mammalian hemoglobins or myoglobins (Appleby, 1984). Rates of nitrogen fixation are highly correlated with leghemoglobin content of nodules (Bergersen, 1982). Leghemoglobin binds free O_2 thus facilitating the flux of O_2 to the rapidly respiring microbial symbiont while maintaining a concentration of free O_2 at a sufficiently low level (about 10 nM) so as to not inhibit nitrogenase. Hemoglobins are also found in nodules of nitrogen-fixing actinorhizal plants such as alder (*Alnus*) and bog myrtle (*Myrica*; Tjepkema et al., 1986).

Leghemoglobin is almost certainly the major source of activated forms of oxygen in nitrogen-fixing root nodules. The oxygenated, reduced form of hemoglobin (and leghemoglobin) is subject to autoxidation in which O_2^- is released (Fridovich, 1979; Puppo et al., 1981). The O_2^- can disproportionate to H_2O_2. While O_2^- and H_2O_2 may have direct deleterious effects, the most danger lies in the subsequent production of highly reactive $\cdot OH$ radicals by the metal (usually Fe or, to a lesser extent, Cu) catalyzed Haber-Weiss reaction (eq. 9.6–9.9) (Fridovich, 1986; Halliwell and Gutteridge, 1985, 1986):

$$2O_2^- + 2H^+ \rightarrow H_2O_2 + O_2 \tag{9.6}$$

$$H_2O_2 + Fe(II) \rightarrow \cdot OH + {}^-OH + Fe(III) \tag{9.7}$$

$$O_2^- + Fe(III) \rightarrow O_2 + Fe(II) \tag{9.8}$$

$$3O_2^- + 2H^+ \rightarrow 2\,O_2 + \cdot OH + {}^-OH \tag{9.9}$$

The second reaction (eq. 9.7) is greatly accelerated in the nodules because H_2O_2 attacks the heme group of hemoglobin, releasing free iron ions which then react to create $\cdot OH$ (Puppo and Halliwell, 1988). H_2O_2 also directly reacts with both the ferrous and ferric forms of leghemoglobin and oxidizes them to the inactive ferryl (Fe^{4+}) form (Aviram et al., 1978; Puppo et al., 1993). Thus at least three major forms of active oxygen (O_2^-, H_2O_2 and $\cdot OH$) are generated by leghemoglobin in a self-perpetuating cycle (Fig. 9.4). The oxidized leghemoglobin that results from the release of O_2^- is unable to bind O_2, thus requiring the presence of a NADH-dependent ferric leghemoglobin reductase to restore the leghemoglobin to the physiologically active form (Ji et al., 1991, 1992).

Although the concentration of free O_2 in nodules is held at low levels by a variable diffusion barrier (Sheehy, 1987), enough O_2 enters to support respiration and to maintain 20% of the leghemoglobin in the oxygenated form (Appleby, 1984). Since the concentration of leghemoglobin in soybean nodules is as high as 3 mM (Appleby, 1984), it is evident that a substantial amount of leghemoglobin-bound

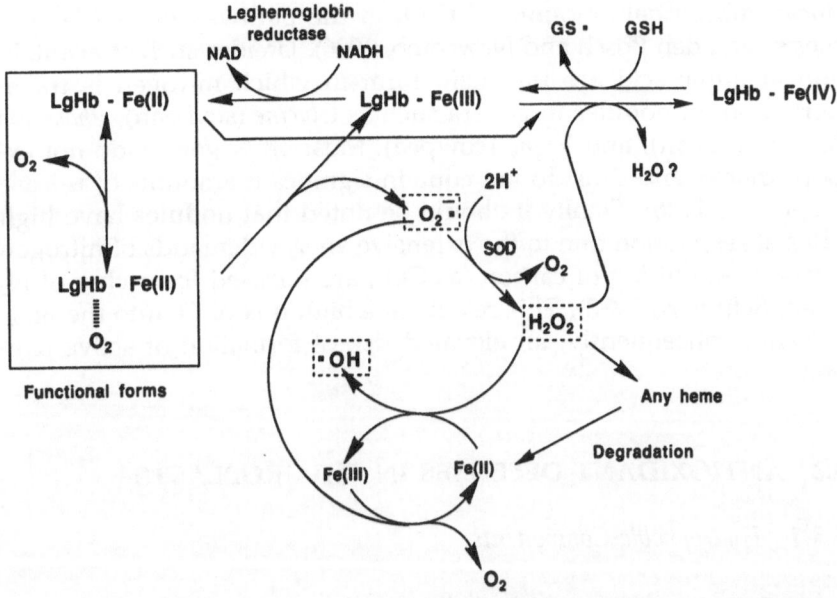

Fig. 9.4 Metabolic pathways for interaction of leghemoglobin with activated forms of oxygen in nitrogen-fixing root nodules. LgHb, leghemoglobin.

O_2 is present in nodules. This oxygenated leghemoglobin represents a major source of active oxygen.

iii. Other sources of active oxygen in nodules

Several other potential sources of active oxygen also exist in nodules. Ferredoxin, the proximal electron donor to nitrogenase, is likely to produce O_2^- just as it does in chloroplasts as discussed earlier. Nodules have relatively high levels of fatty acids (about 10% of dry weight) compared to most plant material (Johnson et al., 1966). Just as is the case in chloroplasts, a high proportion (about 72%) of these fatty acids are unsaturated, thus creating a substantial risk for lipid peroxidation through oxygen free radicals. Most species of *Rhizobium* and *Frankia* contain a H_2-uptake hydrogenase to reclaim energy that would otherwise be lost due to the production of H_2 by nitrogenase (Arp, 1992). This enzyme is a source of O_2^- (Schneider and Schlegel, 1981). Some root nodules contain large amounts of uricase. This enzyme is located in peroxisomes (Fig. 9.3) where it produces substantial amounts of H_2O_2 in the pathway of ureide synthesis (Van den Bosch and Newcomb, 1986). Ureides such as allantoin and allantoic acid are the main form in which nitrogen is transported out of nodules in genera such as *Glycine* (soybean), *Phaseolus* (common bean), and *Vigna* (cowpea). Most other genera do not export ureides and thus do not contain significant amounts of uricase (Schubert, 1986). Finally it should be noted that nodules have high rates of respiration due to the extensive energy demands of nitrogen fixation. About 5 g of carbon (as CO_2) are released for each g of N_2 fixed (Schubert, 1982). This results in a high flux of O_2 into the nodule and, consequently, an elevated risk of formation of active oxygen.

9.3 ANTIOXIDANT DEFENSES IN CHLOROPLASTS

9.3.1 *Hydrophobic Components*

Since chloroplasts are especially prone to production of active oxygen, much attention has been focused on their antioxidant defense systems which are indeed extensive. As discussed earlier, thylakoid membranes are particularly at high risk due to the high levels of unsaturated fatty acids which are especially vulnerable to peroxidation. Lipid peroxidation is a self-perpetuating chain reaction with disastrous consequences for membrane integrity. Lipid peroxidation is a universal problem in biology, with wide-ranging implications in medicine as well in plant processes (Yagi, 1982). The process is initiated by $^1O_2$ attacking the polyunsaturated fatty acids with the subsequent formation of lipid peroxides (Halliwell and Gutteridge, 1985). In chloroplasts, $^1O_2$ originates from photo-excited chlorophyll in the photosystems which are embedded in the thylakoid membranes.

Lipid peroxidation leads to various degradative products including most notably aldehydes, especially malondialdehyde, and volatile hydrocarbons such as ethane and pentane. These products are readily detectable and are commonly used as indicators of peroxidative damage (Janero, 1990; Gutteridge and Halliwell, 1990). Malondialdehyde is measured spectrophotometrically by its reaction with thiobarbituric acid and has been used to assess peroxidative damage in a wide-range of plant tissues including leaves (Heath and Packer, 1968; Dhindsa et al., 1981) and legume root nodules (Puppo et al., 1991; Becana et al., 1986). Ethane production, as measured by gas chromatography, has been used to verify chilling-enhanced photooxidation in chloroplasts (Wise and Naylor, 1987). Lipid peroxidation is greatly accelerated by transition metal ions or by heme. As lipid peroxidation progresses, membrane integrity is lost and the products of decomposition damage many enzymes.

i. Carotenoids

Protection against lipid peroxidation is provided by several hydrophobic compounds located in membranes. The isoprene-based pigment β-carotene serves several roles in this regard (Pallett and Young, 1993). In addition to acting as an accessory pigment in photosynthesis, β-carotene also functions as a scavenger of $^1O_2$ and a quencher of triplet state chlorophyll molecules (Young, 1991).

Recently, it has been suggested that additional photoprotection may be provided by the carotenoid zeaxanthin (Demmig-Adams, 1990; Demmig-Adams and Adams, 1993). Exposure of plants to high light levels leads to rapid enzymic de-epoxidation of violaxanthin to form zeaxanthin. The process is reversible in subdued light or darkness. This cyclic interconversion is known as the xanthophyll cycle and appears to provide photoprotection by dissipating excess excitation energy. This cycle is especially active in sun-tolerant species and may be a key factor in determining ecological success in stressful environments, especially those with high solar irradiance (Thayer and Björkman, 1990).

Carotenoids are probably indispensable as a photoprotective agent since they are present in virtually all photosynthetic organisms. There are exceptions such as some carotenoid-lacking mutants of maize and photosynthetic bacteria, however, these organisms do not occur in the wild and can be maintained only in laboratory conditions where light and O_2 levels are kept low (Siefermann-Harms, 1985; Anderson and Robertson, 1960).

ii. Tocopherols

Additional protection against active oxygen in membranes is provided by α-tocopherol (vitamin E, Fig. 9.5), a ubiquitous component of chloroplasts in all higher plants and algae (Fryer, 1992; Hess, 1993). The structurally related forms of β-, δ- and γ-tocopherol are also widespread, but the α form is the most abundant by far and also the most potent as an antioxidant. α-tocopherol is strongly hydrophobic and thus is restricted to membranes. It is abundant in thylakoid membranes, particularly in association with photosystem I

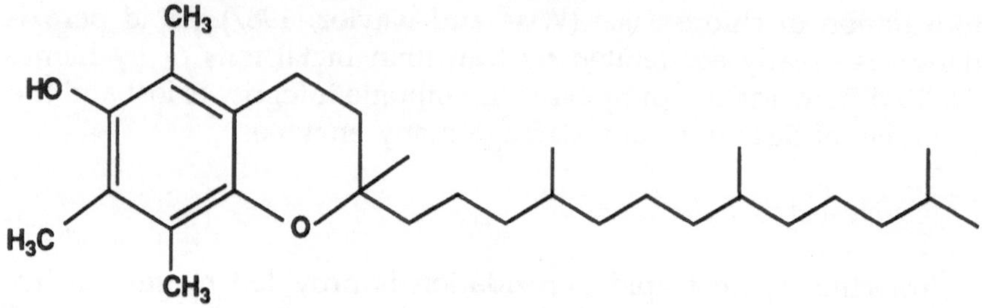

Fig. 9.5   The structure of α-tocopherol (vitamin E).

(Baszynski, 1974). The importance of this compound is reflected in the observation that about one molecule of α-tocopherol is present for each 10 molecules of chlorophyll *a* in spinach chloroplasts (Halliwell, 1982a).

Tocopherols have a diverse range of antioxidant activities. $^1O_2$ is controlled by two distinct processes. First, tocopherols can scavenge $^1O_2$ in a sacrificial process in which the tocopherol molecule is irreversibly oxidized to a mixture of tocopheryl quinones and quinone epoxides (Fryer, 1992; Neeley et al., 1988). Tocopherol also acts as a physical deactivator or quencher of $^1O_2$ in a process similar to that described above for β-carotene. Quenching has the advantage of being relatively non-destructive, although repeated transfers of resonance energy will eventually degrade the tocopherol molecule (Fahrenholtz et al., 1974).

Although $^1O_2$ removal is undoubtedly a beneficial activity of tocopherol, the principal function of tocopherol is its action in terminating the extremely damaging process of lipid peroxidation. Since lipid peroxidation is a self-propagating chain reaction, suppression is a necessity if serious membrane disruption is to be avoided. The mechanism of lipid peroxidation termination by α-tocopherol involves the formation of chromanoxyl free radical which is reduced back to α-tocopherol by ascorbate (Fig. 9.6; Asada and Takahashi, 1987; Fryer, 1992). This produces the ascorbyl free radical (A⁻) which in turn is reduced back to ascorbate by a NADH-dependent A⁻ reductase (see sec. 9.3.4). The chromanoxyl radical may also be reduced by GSH, thus forming a thienyl free radical (GS·; Niki et al., 1982), although this role for GSH is not clearly established. Some authors (e.g., Cadenas, 1989) assert that GSH can not reduce the chromanoxyl radical, while others feel that it can, at least in the presence of some poorly understood membrane-bound heat-labile

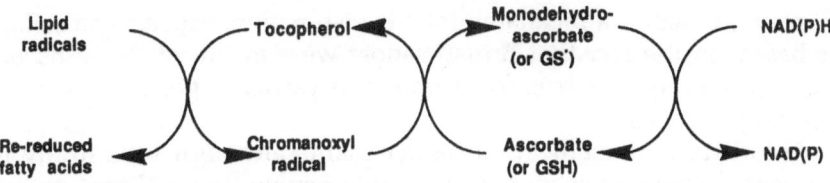

Fig. 9.6 Coupled reactions associated with termination of lipid peroxidation in chloroplasts.

factor (McCay and Powell, 1989). The hydrophilic nature of ascorbate and GSH restricts their presence in the chloroplast to the stroma so the regeneration of α-tocopherol requires an interaction of ascorbate (or perhaps GSH) and the chromanoxyl radical at the membrane surface (Leung et al., 1981; Packer et al., 1979).

Tocopherols have also been implicated in other antioxidant activities in plants. In addition to scavenging $^1O_2$, tocopherols can very efficiently scavenge O_2^-, $HO_2\cdot$ and $\cdot OH$. Although scavenging of these forms of active oxygen has been demonstrated in vitro (Nishikimi et al., 1980; Fukuzawa and Gebicki, 1983), there is some uncertainty about whether this occurs in vivo. Nevertheless, the case for $^1O_2$ scavenging is unequivocal (Fryer, 1992).

Another likely function of tocopherols is the repair of free radical forms of hydrophobic amino acids, especially tryptophan, tyrosine, methionine and histidine. These radicals arise from action of lipid free radicals on amino acid residues of intrinsic membrane proteins (Bisby et al., 1984). This process is not well understood, but apparently proceeds via the chromanoxyl radical which allows for the regeneration of tocopherol by ascorbate and/or GSH (Fryer, 1992; Fig. 9.6).

9.3.2 Small Hydrophilic Molecules

i. Ascorbic acid

Ascorbic acid (vitamin C) is one of the most fundamentally important molecules in biology. The name is derived from the term "antiscorbutic" which means "preventing scurvy." Few compounds have such a long, rich history of both medicinal importance and lore. This history includes centuries of use, notably on sailing explorations or long expeditions away from civilization. The slang word "limey", which is a nickname for a British native, especially a sailor, is based on the fact that British sailors were aware of the antiscorbutic properties of citrus fruits and thus carried a generous supply on long voyages.

Ascorbate is present in all higher plants although little if any is present in dormant seeds (Loewus and Loewus, 1987). Cryptograms and algae also contain ascorbic acid, but there is insufficient information to state definitively that ascorbic acid is ubiquitous in these groups. Generally, prokaryotes lack ascorbic acid, with the notable exception of cyanobacteria (see sec. 9.4). Plant processes dealing with metabolism of ascorbic acid differ considerably from those associated with animals. For details of the plant processes, several excellent reviews on the synthesis and metabolism are available (Loewus, 1980; Loewus and Loewus, 1987).

Ascorbate is recognized chiefly for its properties as an antioxidant, but this is a rather generalized statement than can lead to the false assumptions that the functions are clearly understood. Ascorbate clearly has numerous roles in plants, but defining them precisely remains an "ever-present" challenge that has yet to be overcome (Loewus, 1980). Ascorbate has the potential to react with a variety of forms of active oxygen including $^1O_2$, O_2^-, H_2O_2, and $\cdot OH$ (Bendich et al., 1986; Halliwell, 1982b). The mechanism of action involves the ability of the ascorbate ion to donate an electron to the active oxygen species. The immediate product of this reaction is monodehydroascorbate ($A^{\cdot -}$), a free radical also known as the ascorbyl radical (Fig. 9.7). A further oxidation step involving the removal of another electron results in the formation of dehydroascorbic acid. Monodehydroascorbate may spontaneously disproportionate to ascorbate and dehydroascorbate in the following reaction (eq. 9.10):

$$2 \text{ monodehydroascorbate} \rightarrow \text{ascorbate} + \text{dehydroascorbate} \quad (9.10)$$

Alternatively, monodehydroascorbate may be reduced enzymatically as discussed in section 9.3.4. Since the pK_a of ascorbic acid is 4.25, the unprotonated form (ascorbate) is dominant under physiological conditions. Unfortunately, the nomenclature of dehydroascorbic acid is very misleading since this compound is not an acid at all. Dehydroascorbic acid occurs as a mixture of lactone and hemiketal forms in aqueous solution (Tolbert and Ward, 1982). Neither of these forms has a readily dissociable proton.

Chloroplasts contain concentrations of ascorbate generally in the range of 10–50 mM (Foyer et al., 1983; Halliwell, 1982b; Foyer, 1993)

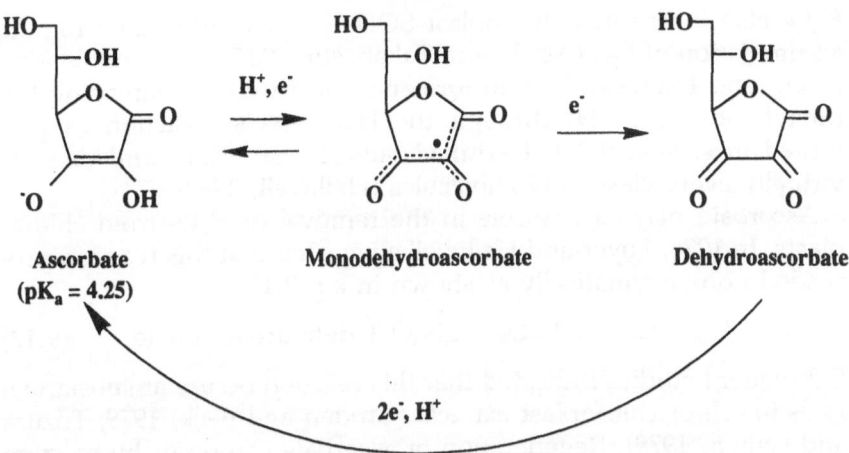

Fig. 9.7 The structure of ascorbate and its oxidation products.

with the majority in the reduced form. This high concentration provides a key protective mechanism against several forms of active oxygen. Ascorbate can react directly with O_2^- as follows:

$$2O_2^- + 2H^+ + \text{ascorbate} \rightarrow 2H_2O_2 + \text{dehydroascorbate} \quad (9.11)$$

This equation is somewhat of an oversimplification since both O_2^- and HO_2 will react with ascorbate and the formation of dehydroascorbate may proceed via the intermediate formation of A^{-} (Bendich et al., 1986). The above reaction has a rate constant of 2.7×10^5 $M^{-1}s^{-1}$ at pH 7.4 (Nishikimi, 1975). This constant is very low when compared to the enzyme-catalyzed dismutation of O_2^- (2×10^9, see section 9.3.3), but this is more than compensated for by the concentration of ascorbate which is much higher than the concentration of SOD. Ascorbate quenches $^1O_2$ rapidly, but carotenoids and α-tocopherol, as discussed above, probably provide the main mechanism for $^1O_2$ quenching in illuminated chloroplasts (Halliwell, 1982b). Ascorbate also reacts with ·OH at a fast rate ($k = 7 \times 10^9$, Halliwell, 1982b).

Regardless of the mechanism of O_2^- removal (direct reaction with ascorbate, non-enzymatic dismutation, or catalysis by SOD), the consequences are the same—the production of H_2O_2. If unchecked, this process would have devastating effects in chloroplasts. H_2O_2 is a powerful inhibitor of the Calvin cycle. Concentrations as low as 10 μM can result in a 50% reduction of CO_2 fixation (Halliwell, 1982b).

H_2O_2 also inactivates chloroplast SOD which would allow for the accumulation of O_2^- (Asada and Takahashi, 1987). The combination of O_2^- and H_2O_2 can lead to formation of extremely damaging hydroxyl radicals ($\cdot OH$) through the Haber-Weiss reaction as presented in section 9.2.3. Hydroxyl radicals can attack and damage virtually every class of biomolecules (Halliwell, 1982b).

Ascorbate plays a key role in the removal of H_2O_2 from chloroplasts. In 1976, Foyer and Halliwell suggested that this reaction proceeded non-enzymatically as shown in eq. 9.12:

$$\text{ascorbate} + H_2O_2 \rightarrow 2H_2O + \text{dehydroascorbate} \qquad (9.12)$$

Subsequent studies indicated that this reaction occurs at substantial rates in cell or chloroplast extracts (Groden and Beck, 1979; Tözüm and Gallon, 1979). Regeneration of ascorbate can occur by reaction with GSH which in turn is regenerated by utilization of the reducing power of NADPH. The seminal proposal by Foyer and Halliwell led eventually to the important concept of the ascorbate-glutathione cycle, although it is now clear that the original scavenging of H_2O_2 involves an ascorbate-specific peroxidase (see section 9.3.4).

ii. Glutathione and its homologues

Glutathione is widely distributed in plants, animals and microorganisms often in intracellular concentrations of 0.1 to 10 mM (Meister, 1983, 1988; Dolphin et al., 1989; Hausladen and Alscher, 1993). In plants, glutathione amounts are especially high in chloroplasts where concentrations are generally in the range of 1.0 to 3.5 mM (Halliwell, 1982a). In the form found in animals and most plants, reduced glutathione (GSH) is a tripeptide (γGlu-Cys-Gly, Fig. 9.8). The oxidized form (GSSG) is a dimer with a central disulfide bridge. Some plants in the Fabaceae (legume family) contain large amounts of homoglutathione (γGlu-Cys-Ala = hGSH) in addition to GSH (Klapheck, 1988). The hGSH/GSH ratio varies considerably within this family, with an especially strong predominance of hGSH in the economically important genera of *Phaseolus* and *Glycine*. Furthermore, many plants in the family Poaceae (grasses) contain a unique glutathione homologue—γGlu-Cys-Ser (Klapheck et al., 1992). Presumably both of the glutathione homologues have antioxidant functions similar to GSH, but this has not been conclusively established.

GSH functions in plants as the main form of transport and storage of sulfur and as an antioxidant (Rennenberg, 1982; Alscher, 1989).

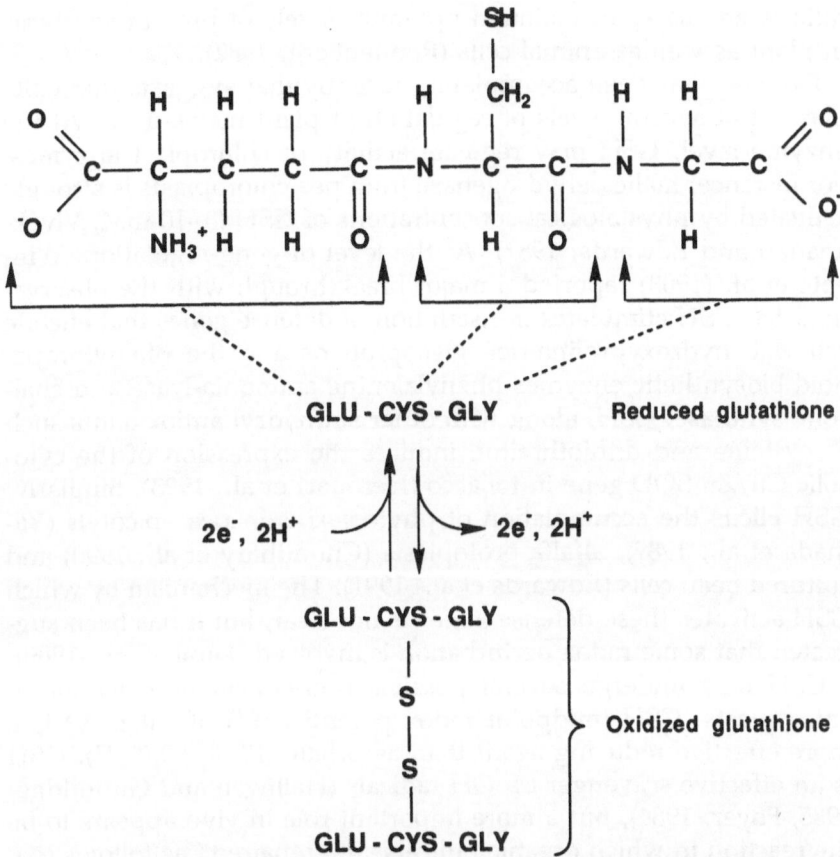

Fig. 9.8 The structure of the reduced and oxidized forms of glutathione.

Both the oxidized and reduced forms can usually be detected in plant tissues, but as with other organisms, the reduced form is usually strongly predominant (Halliwell, 1982a). The free thiol group is essential for GSH's primary role as a reductant. In general, this role involves maintaining proteins, cysteine and homocysteine in the reduced (i.e. physiologically active) form. If the GSH/GSSG ratio drops too low, many enzymes may be inactivated as follows (eq. 9.13) (Halliwell, 1982a):

$$\text{Enzyme-SH} + \text{GSSG} \rightarrow \text{Enzyme-S-SG} + \text{GSH} \qquad (9.13)$$
$$\text{(active)} \hspace{4cm} \text{(inactive)}$$

In addition to the role of a general reductant, a high GSH/GSSG ratio is necessary to maintain optimum levels of protein synthesis in plant as well as animal cells (Rennenberg, 1982).

Evidence has been accumulating recently that suggests that GSH may act at several levels of regulation of plant metabolism. At the enzyme level, GSH may regulate activity of chloroplast enzymes. For instance, malic dehydrogenase from pea chloroplasts is strongly activated by physiological concentrations of GSH (3–10 mM, Vivek-anadan and Edwards, 1987). At the level of gene regulation, Wingate et al. (1988) reported a major breakthrough with the observation that GSH stimulates transcription of defense genes that encode cell wall hydroxyproline-rich glycoproteins and the phenylpropanoid biosynthetic enzymes phenylalanine ammonia-lyase and chalcone synthase. GSH, along with other sulfhydryl antioxidants such as cysteine and dithiothreitol, induces the expression of the cytosolic Cu/Zn SOD gene in tobacco (Hérouart et al., 1993). Similarly, GSH elicits the accumulation of phytoalexins in pea epicotyls (Yamada et al., 1989), alfalfa protoplasts (Choudhary et al., 1990) and cultured bean cells (Edwards et al., 1991). The mechanism by which GSH activates these defense genes is not clear, but it has been suggested that some redox perturbation is involved (Lamb et al., 1989).

GSH may undergo several possible non-enzymatic reactions in chloroplasts. GSH (midpoint redox potential (E_0') of -0.23 V) is a more effective reducing agent than ascorbate ($E_0' = +0.08$ V). GSH is an effective scavenger of $\cdot$OH radicals (Halliwell and Gutteridge, 1985; Foyer, 1984), but a more important role in vivo appears to be the reaction in which organic radicals are "repaired" as follows (eq. 9.14 and 9.15) (Halliwell and Gutteridge, 1985):

$$R\cdot + GSH \rightarrow RH + GS\cdot \qquad (9.14)$$
$$2GS\cdot \rightarrow GSSG \qquad (9.15)$$

GSH also reacts with O_2^- but is less effective than ascorbate in this regard. GSH may react directly with O_2 as follows:

$$2GSH + 1/2O_2 \rightarrow GSSG + H_2O$$

Although the rate of this reaction is slow at neutral pH values, the rate in stroma of chloroplasts may be substantial due to alkaline conditions (Halliwell, 1982a). These high pH values are caused by the light-driven translocation of protons into thylakoid interiors that is necessary to generate the proton motive force to drive ATP synthesis. These alkaline conditions also promote another non-enzymatic reaction of GSH, namely the direct reduction of dehydroascorbate to ascorbate (eq. 9.16) (Foyer, 1984):

$$2GSH + dehydroascorbate \rightarrow GSSG + ascorbate \qquad (9.16)$$

Although this may be a significant reaction at the high pH values (around pH 8, Foyer, 1984) in the stroma of illuminated chloroplasts, it is now clear that chloroplasts contain an enzyme, dehydroascorbate reductase (see section 9.3.4), that greatly facilitates the reaction. As discussed in the following sections, the antioxidant role of GSH is heavily dependent on this enzyme along with glutathione reductase that is required to regenerate GSH in the ascorbate-glutathione pathway.

9.3.3 Superoxide Dismutase

The removal of O_2^- in plants may proceed via a dismutation reaction (according to eq. 9.17):

$$2O_2^- + 2H^+ \rightarrow H_2O_2 + O_2 \qquad (9.17)$$

This is a disproportionation reaction which means that one molecule of reactant is oxidized and another identical molecule is reduced. The rate constant for the uncatalyzed dismutation reaction varies considerably with pH and has been reported to be in a range from <100 to 5×10^5 M^{-1} s^{-1} (Halliwell and Gutteridge, 1985; Fridovich, 1976). The rate of the dismutation reaction is thus quite low which means that measures must be taken to accelerate the process. As with virtually all aerobic organisms, the dismutation reaction in plants is greatly enhanced by the presence of an enzyme, superoxide dismutase (SOD; E.C. 1.15.1.1), that catalyzes the reaction. The catalyzed reaction has a rate constant of around 1.6×10^9 M^{-1} s^{-1}. SOD's are classified into three major types based on their metal cofactor, which is either copper plus zinc (CuZnSOD), manganese (MnSOD), or iron (FeSOD) (Steinman, 1982; Bannister et al., 1987; Salin, 1988; Bowler et al., 1992).

Plants contain all three types of SOD's. CuZnSOD is present in the cytosol and has been purified from numerous plant species including *Pisum*, *Triticum*, *Spinacia*, *Phaseolus*, *Lycopersicon*, *Zea*, and *Cucumis* (see references in Bowler et al., 1992; Bannister et al., 1987). It is a dimer with a molecular weight of 31–33 kD. MnSOD is present in the matrix of mitochondria. In *Zea*, the mitochondrial MnSOD is a tetramer with a molecular weight of 90 kD (Baum and Scandalios, 1981). Chloroplasts contain FeSOD and/or CuZnSOD in the stroma, although there is evidence that a MnSOD may be present in chloroplasts of at least some plants (Bannister, 1987). All types of SOD's are coded for in the nucleus. For organellar SOD's, N-terminal targeting sequences direct the transport (Bowler et al., 1992). SOD's are easily detected and identified as to cofactor type by electrophoresis followed by activity staining in the presence of inhibitors such as H_2O_2 or cyanide (Beauchamp and Fridovich, 1971). Studies of this type have revealed that plants contain several different isozymes and that the number and abundance of different isozymes varies greatly from species to species.

The CuZn form of SOD is the major form in plants including angiosperms, gymnosperms, ferns and mosses (Kanematsu and Asada, 1990). CuZnSOD is absent in cyanobacteria and most green algae (except in the class Chlorophyceae), suggesting that photosynthetic organisms acquired the CuZn form in the later stages of evolution of green algae (Kanematsu and Asada, 1990; Steinman, 1982). Within higher plants, the various isozymes of CuZnSOD can be grouped into two major types—one in the chloroplast and one in the cytosol. While these two types have similar molecular weights, subunit structures and metal contents, they can be distinguished by their amino acid compositions, visible absorption spectra, CD (circular dichroism) spectra, sensitivities to inhibition by H_2O_2, reactivity to antibodies and amino acid sequence homologies (Kanematsu and Asada, 1990). There is less interspecific variation within the chloroplast form, suggesting that the rate of mutation of this form is lower because of the need for optimized scavenging of O_2^-.

In a typical leaf cell, SOD is more abundant in chloroplasts than in the cytosol (Jackson et al., 1978; Tsang et al. 1991; Bowler, 1992). The in situ removal of O_2^- generated in chloroplasts is necessary because O_2^- is a charged molecule and thus does not easily traverse membranes.

Progress in understanding the molecular biology of SOD genes in plants has lagged behind the work in animal or bacterial systems, but this topic is currently under considerable scrutiny. cDNAs for three forms of CuZnSOD and one of MnSOD from maize have been characterized (Cannon et al., 1987; White and Scandalios, 1988; Cannon and Scandalios, 1989). All forms bear significant sequence homologies to the corresponding CuZnSODs and MnSODs of other organisms at both the nucleotide and amino acid level. The amino acid sequence of CuZnSOD from *Brassica* (cabbage) has a 49–56% homology to a wide range of other organisms including mammals, fish, insects, and fungi (Bannister et al., 1987). Recently, cDNA expression libraries from *Nicotiana* and *Arabidopsis* have been used to shotgun clone FeSOD into a SOD-deficient mutant of *E. coli*. (Van Camp et al., 1990). These and similar studies should lead to rapid advances in understanding of the regulation and molecular biology of SOD genes in plants.

Since SOD is critical in the ability of plants to respond to diverse types of stress, it has been suggested that genetic engineering of SOD into plants may be a way to substantially improve their stress tolerance. Several such manipulations have been attempted, with mixed results. Transgenic tobacco and tomato plants that overproduce a chloroplastic CuZnSOD were not more resistant to O_2^- toxicity (Pitcher et al., 1991; Tepperman and Dunsmuir, 1990). Such attempts may seem overly simplistic in view of the observations that some SOD-rich bacteria actually have a paradoxical increase in oxygen sensitivity (Scott et al., 1987). However, overproduction of MnSOD in chloroplasts of transgenic tobacco did lead to enhanced resistance to O_2^- (Bowler et al., 1991). Similarly, overexpression of chloroplastic CuZnSOD in transgenic tobacco plants also results in increased resistance to oxidative stress as measured by paraquat resistance or rates of photosynthesis following chilling-enhanced photooxidation (Gupta et al., 1993). Ultimately, success in achieving consistent and dramatic increases in resistance to oxygen toxicity in plants may require genetic engineering not only of SOD but also for all the components of the ascorbate-glutathione pathway (Bowler et al., 1992; Shaaltiel and Gressel, 1986).

The converse experiment—namely engineering of SOD deficiency into plants—has not been successfully undertaken, although the outcome would certainly be enlightening. SOD deficiency has, however, been successfully generated in the cyanobacterium *Synechococcus* (Herbert et al., 1992). Introduction of a spectinomycin-resistance gene cartridge into the coding region of the *sodB* gene resulted in a mutant lacking FeSOD activity, although MnSOD activity remained unaltered. Under conditions of low light and high CO_2, both the wild-type and mutant strains exhibited similar photosynthetic activity and viability. In contrast, the SOD-deficient mutant exhibited much greater damage to its photosynthetic system (especially photosystem I) than the wild-type strain when grown under high oxygen tension or with methyl viologen. Although similar experiments with higher plants would be a demanding undertaking, the recent application of anti-sense techniques to plants at least makes such a goal feasible.

9.3.4   Ascorbate-Glutathione Pathway

The key antioxidant role of ascorbate and glutathione in chloro-plasts was first thought to be limited to the uncatalyzed reactions discussed above. In particular, ascorbate was thought to reduce H_2O_2 to H_2O. GSH could then regenerate ascorbate by the reduction of dehydroascorbate (Foyer and Halliwell, 1976) in a pathway that has sometimes been referred to as the Halliwell-Asada pathway after two of the key scientists who have been instrumental in elucidating the importance of active oxygen in chloroplasts. In 1979, Groden and Beck demonstrated that the reduction of H_2O_2 by ascorbate was catalyzed by an enzyme, namely ascorbate peroxidase. Earlier at-tempts to detect this enzyme were unsuccessful, perhaps due to low activity at some times of the year (Halliwell, 1982b) or because the enzyme is rapidly inactivated if ascorbate is not included in the ex-traction buffer (Nakano and Asada, 1987).

In the initial reaction of the ascorbate-glutathione pathway, H_2O_2 is scavenged using the reducing power of ascorbate in a reaction catalyzed by ascorbate peroxidase (Fig. 9.9, EC 1.11.1.11). The prod-uct of this reaction is monodehydroascorbate ($A^{\cdot-}$), a free radical that has two possible fates. $A^{\cdot-}$ may be converted directly to ascor-bate by monodehydroascorbate reductase (EC 1.6.5.4, also called ascorbate free radical reductase). Alternatively, $A^{\cdot-}$ may sponta-neously disproportionate to dehydroascorbate and ascorbate. De-hydroascorbate is then converted to ascorbate in a reaction cata-lyzed by dehydroascorbate reductase (EC 1.8.5.1). This reaction converts reduced glutathione (GSH) to the oxidized form (GSSG). GSH is regenerated by glutathione reductase (EC 1.6.4.2). The path-way shares some similarities with the main H_2O_2 scavenging path-way in animals which also involves glutathione. However, in ver-tebrates the initial reduction is accomplished by the selenium-containing enzyme glutathione peroxidase (see chapter 10). This en-zyme is not present in higher plants or invertebrates (Dalton et al., 1986; Smith and Shrift, 1979) although it does appear to be present in *Chlamydomonas* (Yokota et al., 1988).

i.   Ascorbate peroxidase

Since its original discovery in spinach chloroplasts (Groden and Beck, 1979), ascorbate peroxidase has been detected in many angio-sperms (see reviews in Asada, 1992; Dalton, 1990), ferns (Carroll et al., 1988), mosses (Seel et al., 1992) and in eukaryotic algae includ-

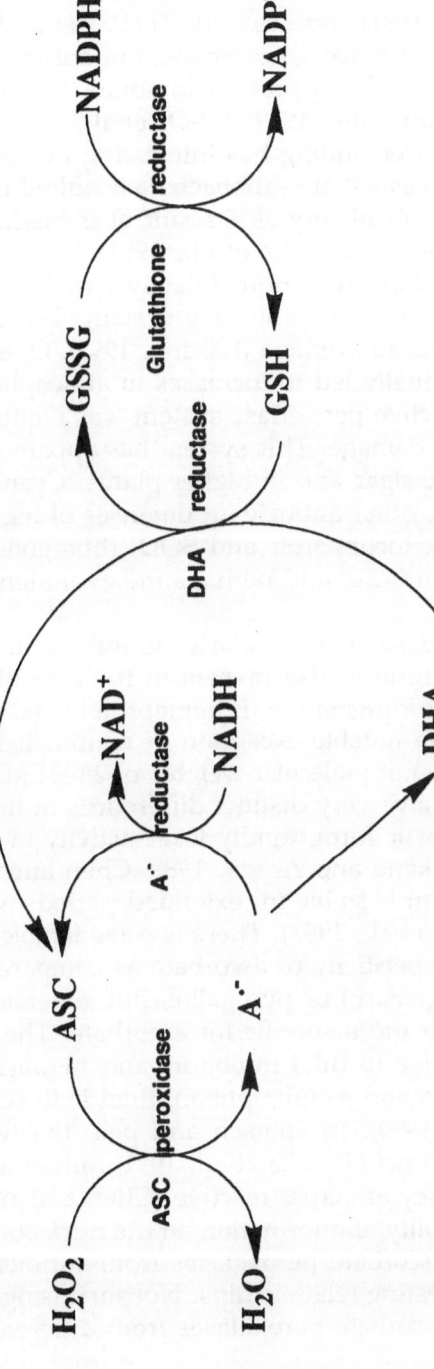

Fig. 9.9 The ascorbate-glutathione pathway by which H_2O_2 is scavenged in plants. A^{-}, ascorbyl free radical (or monodehydroascorbate); DHA, dehydroascorbate.

ing *Euglena* (Shigeoka et al., 1980), *Chlamydomonas* (Yokata et al., 1988; Miyake et al., 1991) and *Zooxanthella* (Lesser and Shick, 1989). This enzyme may be a universal component of eukaryotic plant cells. Ascorbate peroxidase is also present in some, but not all, cyanobacteria (Tözüm and Gallon, 1979; Tel-Or et al., 1985, 1986; Miyake et al., 1991). This latter finding has interesting evolutionary implications since its suggests that cyanobacteria acquired the H_2O_2 scavenging system early in history as a result of increasing concentrations of atmospheric O_2 (Miyake et al., 1991). At the time of first appearance of cyanobacteria, approximately 2 to 3 × 10^9 years ago, there was no need for such protective systems because very little O_2 was present in the atmosphere (Kasting, 1993). O_2 evolution from photosynthesis gradually led to increases in atmospheric O_2 to the point where an effective peroxidase system was required to protect cells from oxidative damage. This system has apparently been conserved in eukaryotic algae and in higher plants. Cyanobacteria also contain many of the other antioxidant defenses of higher plants including carotenoids, tocopherols and SOD, thus giving further indication of the importance, and perhaps the evolutionary origin, of such defenses.

Ascorbate peroxidase is particularly abundant in chloroplasts, however, a related form is also present in the cytosol. The chloroplastic and cytosolic forms are both hemeproteins (as are nearly all peroxidases with the notable exception of mammalian glutathione peroxidase) with subunit molecular weights of 28–34 kDa (Table 9.2), but the two forms have very distinct differences in their characteristics. The chloroplastic form rapidly loses activity in ascorbate-depleted medium (Nakano and Asada, 1987; Chen and Asada, 1989) but the cytosolic form is stable for extended periods in the absence of ascorbate (Dalton et al., 1987). There is considerable variability in the electron donor specificity of ascorbate as compared to artificial substrates such as guaiacol or pyrogallol, but generally the chloroplastic form is much more specific for ascorbate. The chloroplastic form is more sensitive to thiol inhibitors and to suicide inhibitors such as hydroxyurea and *p*-aminophenol than is the cytosolic form (Chen and Asada, 1990). In spinach and pea, the two forms are immunologically distinct (Tanaka et al., 1991; Mittler and Zilinskas, 1991a), but in tea they are cross-reactive (Chen and Asada, 1989).

The recent availability of information on the nucleotide and amino acid sequences of ascorbate peroxidases from various sources has revealed some interesting relationships. Not surprisingly, the amino acid sequences of ascorbate peroxidases from different sources are

Table 9.2

Comparison of Properties of Ascorbate Peroxidases from Various Plant Sources

Source	Subunit mol. wt. (kDa)	Sequence information	K_m (μM) H₂O₂	K_m (μM) Ascorbate	Substrate specificity[a]	Reference
Arabidopsis leaf	28.0	complete cDNA and genomic	—	—	—	Kubo et al., 1992 and 1993
Euglena	—	—	56	410	73.1	Shigeoka, 1980
Pea (*Pisum*) leaf	29.5	complete cDNA and genomic	20	325	28	Mittler and Zilinskas, 1991a, 1991b, 1992
Soybean (*Glycine*) nodule	27.5	complete cDNA	3	70	3700	Dalton et al., 1987; Chatfield and Dalton, 1993
Spinach (*Spinacea*)[b] leaf	30.0–31.0	N-term amino acid	30	300–400	0.7–2.0	Groden and Beck, 1979; Nakano and Asada, 1987; Tanaka et al., 1991
Tea (*Camellia*)[b] leaf	34,000	64% of amino acids	30–80	220–416	41–497	Chen and Asada, 1989; Chen et al., 1992

[a] rate with pyrogallol or pyrocatechol as % of rate with ascorbate
[b] two isozymes described

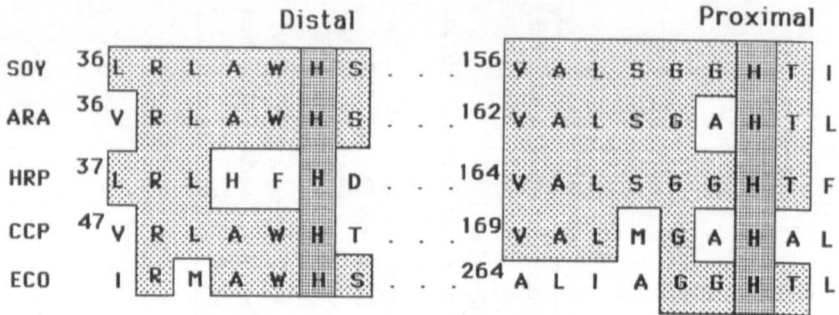

Fig. 9.10 Comparison of amino acid sequences around the conserved histidine residues in various peroxidases. SOY, ascorbate peroxidase from soybean nodules; ARA, ascorbate peroxidase from *Arabidopsis* leaves; HRP, horseradish peroxidase; CCP, yeast cytochrome *c* peroxidase; ECO, NADH peroxidase from *E. coli* (from Chatfield and Dalton, 1993, Chen et al., 1992; Kubo et al., 1992; Mitler and Zilinskas, 1991b; Fujiyama et al., 1990).

quite similar, e.g. the amino acid sequence of cytosolic ascorbate peroxidase from soybean nodules is 90% homologous to the cytosolic peroxidase from pea leaves and 78% homologous to the peroxidase from *Arabidopsis* (Kubo et al., 1992; Mittler and Zilinskas, 1992; Chatfield and Dalton, 1993). With the exception of the heme binding sites, ascorbate peroxidases show very little homology to classical plant guaiacol peroxidases such as horseradish peroxidase. These two classes of peroxidases are also immunologically distinct (Chen and Asada, 1989). Most importantly, ascorbate peroxidases have homology (33% in the case of pea ascorbate peroxidase) to yeast cytochrome *c* peroxidase, an enzyme that also functions to scavenge H_2O_2 (Chen et al., 1992; Mittler and Zinlinskas, 1991b,1992). Homology is especially high around two regions that crystallographic studies of cytochrome *c* peroxidase have shown to be part of the active site (Fig. 9.10; Finzel et al., 1984). These regions include the distal histidine region (including Arg[48] and His[52] of cytochrome *c* peroxidase) and the proximal histidine region (His[164]). The heme moiety is bound to the globin by coordination bonds between the central iron atom and a nitrogen atom in the imidazole ring of the histidines. The terms "distal" and "proximal" reflect the relative distances between the heme and histidine groups. The homology in these regions also extends to the guaiacol peroxidases and even to bacterial NADH peroxidases, although these enzymes do not show much homology in regions outside of these two His sites. These comparisons as well as the shared enzymatic, molecular and functional properties of ascorbate and cytochrome *c* peroxidases, have led to the suggestion that both types of peroxidases originated from the same ancestral protein and that they diverged from guaiacol peroxidases at an early stage in their molecular evolution (Chen et al., 1992).

The genomic DNA structures of the genes for AP have been determined for both pea (Mittler and Zilinskas, 1992) and *Arabidopsis* (Kubo et al., 1993). In both cases, an intron is present in the 5'-untranslated region, an uncommon trait for a plant gene. The significance of this is not clear, but it has been suggested that this could result in increased gene expression (Mittler and Zilinskas, 1992). This might also explain why the amount of AP in soybean nodules is so high—about 1% of total nodule protein (Dalton et al., 1987). The position of the introns in the AP gene (8 total in `Arabidopsis* and 9 in pea) is very different from the positions and numbers (3 or less) of introns in genes for guaiacol peroxidase, further suggesting a long evolutionary distance between the genes for these two types of plant peroxidases.

ii. *Monodehydroascorbate reductase*

Monodehydroascorbate reductase has not been as thoroughly studied as ascorbate peroxidase, but it has been purified from spinach chloroplasts (Hossain et al., 1984), cucumber fruit (Hossain and Asada, 1985), soybean root nodules (Dalton et al., 1992), potato tubers (Borraccino et al., 1986), *Euglena* cytosol (Shigeoka, Yasumoto, et al., 1987) and *Neurospora* (Schulze et al., 1972). Monodehydroascorbate reductase is virtually ubiquitous in plants, algae and animals (Arrigoni et al., 1981). The ubiquity of this enzyme applies not only to its phylogenetic distribution, but also to its subcellular location within a given species. For example, in cucumber fruit this enzyme is present in chloroplasts, mitochondria, microsomes, and cytosol (Yamauchi et al., 1984). In plants, this enzyme is a FAD-containing monomer, but monodehydroascorbate reductase from *Neurospora* contains neither flavin nor heme (Schulze et al., 1972). The molecular weight ranges from 52 kDa in *Euglena* (Shigeoka, Yasumoto, et al., 1987) to 39 kDa in soybean nodules (Dalton et al., 1992). There is a strong preference for NADH as opposed to NADPH in cucumber and soybean, but NADPH is more effective in the *Euglena* form (Hossain and Asada, 1985; Dalton et al., 1992; Shigeoka, Yasumoto, et al., 1987). Alternate electron acceptors include ferricyanide and 2,6-dichloroindophenol, but not dehydroascorbate. The ability to reduce 2,6-dichloroindophenol has been utilized to stain for activity of this enzyme in non-denaturing gels (Dalton et al., 1992). Although sequence analysis of monodehydroascorbate reductase is very limited, a partial analysis of the amino acid sequence has shown a high degree of homology to other flavin-containing oxidoreductases especially in the FAD binding domain (Sano and Asada, 1992).

iii.   Dehydroascorbate reductase

Dehydroascorbate reductase is also widespread in plants and provides an alternate means of ascorbate regeneration in addition to the reaction catalyzed by monodehydroascorbate reductase. It has been purified and characterized from spinach leaves (Foyer and Halliwell, 1977; Hossain and Asada, 1984), potato tubers (Dipierro and Borraccino, 1991) and *Euglena* cytosol (Shigeoka, Yasumoto, et al., 1987). This enzyme is also present in at least some cyanobacteria (Tel-Or et al., 1985, 1986). In higher plants, it is present in both cytosol and chloroplasts (Arrigoni et al., 1981; Hossain and Asada, 1984), but the possible differences between these forms has not been examined. There appear to be at least five isozymes in spinach leaves (Hossain and Asada, 1984). The enzyme is a monomer with a molecular weight of 23 kDa in spinach and potato and 28 kDa in *Euglena*. The K_m values are 2.5 mM and 0.07 mM for GSH and dehydroascorbate, respectively, in spinach (Hossain and Asada, 1984).

The relative importance of dehydroascorbate reductase vis-à-vis monodehydroascorbate is not clear. Both enzymes are present in the cytosol and in chloroplasts, at least in spinach which is the only angiosperm that has been examined in this regard. The ultimate source of reducing power for dehydroascorbate reductase is NADPH which is required to regenerate GSH by glutathione reductase. This is in contrast to the NADH-dependency of most forms of monodehydroascorbate reductase. Hence the presence of alternate means of regenerating ascorbate provides physiological flexibility such that either NADH or NADPH may be utilized.

iv.   Glutathione reductase

Glutathione reductase is by far the most studied and widely distributed of the enzymes of the ascorbate-glutathione cycle. Several recent comprehensive reviews are available (Schirmer et al., 1989; Williams, 1992; Smith et al., 1989). Glutathione reductase is present in bacteria, fungi, plants, protozoa and animals. Indeed there are only a few aerobic organisms that do not contain it and these are often cases where a close substitute (e.g., trypanothione reductase) fulfills a similar role (Krauth-Siegel et al., 1987). A high GSH to GSSG ratio is extremely important for maintaining the proper redox conditions in virtually all cells. Thus the role of glutathione reductase extends beyond participation in the ascorbate-glutathione cycle and this accounts for the wide distribution of this enzyme.

Glutathione reductase has been purified from a wide range of higher plants including maize (Mahan and Burke, 1987), pea (Kalt-Torres et al., 1984; Connell and Mullet, 1986; Bielawski and Joy, 1986; Madamanchi et al., 1992; Edwards et al., 1990), pine (Anderson et al., 1990; Wingsle, 1989) and alfalfa (Kidambi et al., 1990). It has also been purified from *Euglena* (Shigeoka, Onishi et al., 1987) and *Anabaena* (Serrano et al., 1984). Multiple forms of glutathione reductase have been identified in pea leaves with approximately 77% of the activity in chloroplasts, 20% in cytosol and 3% in mitochondria (Edwards et al., 1990; Madamanchi et al., 1992). The native molecular weight of glutathione reductase from various sources ranges from 135 to 190 kD (Smith et al., 1989), with most reports indicating a heterotetramer construction. However, a plastidic homodimer with a native molecular weight of 114 kDa has also been described in pea (Madamanchi et al., 1992). The K_m for GSSG and NADPH has been reported in the ranges of 10-60 and 2-10 μM, respectively. Oxidized homoglutathione (hGSSG, see section 9.3.2) is also a substrate for glutathione reductase. The very limited information that is available indicates that K_m values of plant glutathione reductases are usually slightly higher for hGSSG than for GSSG. For example, with glutathione reductase from bean (*Phaseolus coccineus*), the K_m values are 76 μM for hGSSG and 52 μM for GSSG (S. Klapheck, unpublished data). There is a strong specificity for NADPH as opposed to NADH in all cases of plant glutathione reductases, including the mitochondrial isozyme (Edwards et al., 1990).

Recent sequence information from a cDNA for pea glutathione reductase shows a high degree of amino acid sequence homology to sequences from humans (59.2% similarity), *E. coli* (61.4%) and *Pseudomonas aeruginosa* (70%; Creissen et al., 1991). The degree of conservation is especially high around two cysteine residues (nos. 131 and 136) that form a redox-active disulfide bridge, and around two arginine residues (nos. 287 and 293) that are required for binding of the 2'-phosphate group of NADPH.

The molecular biology of glutathione reductase in plants is an area of very active research at present and may lead to fundamental advances in understanding of how plants respond to oxidative stress, including stress derived from free-radical generating herbicides and from air pollutants such as ozone and SO_2. Some success in engineering improved resistance to active oxygen has already been achieved in transgenic tobacco plants that expressed the gene for bacterial glutathione reductase (Aono et al., 1991). These plants were more resistant to the free radical-generating herbicide paraquat than was the wild type, but ozone resistance was not enhanced. However, similar studies, also with transgenic tobacco plants containing the gene for bacterial glutathione reductase, found that an elevated level of glutathione reductase had only a marginal effect on plant metabolism (Foyer et al., 1991). As with attempts to genetically engineer SOD, it may be that ultimate success could depend on the ability to transfer genes for entire pathways. The problems associated with proper targeting of various plastidic and cytosolic isozymes are also formidable.

9.4   ANTIOXIDANT DEFENSES AND NITROGEN FIXATION

Antioxidant defenses in nitrogen-fixing root nodules are very similar to those present in chloroplasts as discussed earlier. The high capacity for production of active oxygen species, due in large part to the very high concentration of leghemoglobin, necessitates a very active defense system (reviewed by Becana and Rodríguez-Barrueco, 1989).

The chief enzymatic defense against superoxide-SOD- was first described in *Rhizobium* in 1981 (Stowers and Elkan). SOD from plant nodule cells was described a year later (Puppo et al., 1982). The nodule SODs have characteristics similar to other SODs including a Mn-type in bacteroids and a cytosolic CuZn-type (Becana and Salin, 1989; Becana et al., 1989). Puppo and Rigaud (1986) have cataloged a number of cases where SOD activity and nitrogen fixation activity are positively correlated, thus strongly suggesting that SOD is essential for the protection of nitrogen fixation.

Nodules contain high levels of all the enzymes of the ascorbate-glutathione cycle (Dalton et al., 1986). These enzymes have been detected in nodule extracts from nine species of legumes and from red alder (*Alnus rubra*; Dalton et al., 1987) and should probably be considered ubiquitous. Nodules also contain substantial levels of ascorbate (0.8 mM) and glutathione (1.1 mM; Dalton and Langeberg et al., 1993). The first unequivocal evidence for a peroxidase in nodules was provided by Puppo et al. (1980) who demonstrated that soybean nodules contain a peroxidase that could not be confused with leghemoglobin. This is an important distinction because leghemoglobin displays "pseudoperoxidase" activity with many artificial electron donors such as pyrogallol or guaiacol.

In 1986, Dalton et al. demonstrated that soybean nodules contain high peroxidase activity when measured with ascorbate as the electron donor. During the course of early nodule development, ascorbate peroxidase and dehydroascorbate reductase activities and total glutathione content of nodule extracts increase strikingly and positively correlate with acetylene reduction (i.e., nitrogenase) activity and leghemoglobin content (Dalton et al., 1986). The parallel development of peroxide scavenging and nitrogen fixation capacity is strong evidence that these two processes are related. Further evidence of the important relationship between the ascorbate-glutathione pathway and nitrogen fixation was provided by a detailed comparison of effective and ineffective nodules (i.e. nodules unable to fix nitrogen, Dalton and Langeberg, et al., 1993). Effective nodules have higher activities of all 4 enzymes of the ascorbate-glutathione pathway. The concentration of thiol tripeptides (homoglutathione + glutathione) is about 3–4 fold higher in effective nodules. Effective nodules also contain higher levels of NAD^+, $NADP^+$ and NADPH, but not NADH or ascorbate.

Experiments in which nodulated roots of hydroponically-grown soybean plants were exposed to atmospheres containing various concentrations of O_2 have revealed that the activities of all 4 enzymes of the ascorbate-glutathione pathway are higher in nodules exposed to high pO_2 (Dalton et al., 1991). The contents of ascorbate and GSH are also greater in nodules exposed to high O_2. This suggests that the genes of this pathway can be regulated by oxidative stress. This has been confirmed at the molecular level with the gene for cytosolic ascorbate peroxidase from pea (Mittler and Zilinskas, 1992). In this case, transcript levels for ascorbate peroxidase increased in response to several stresses including drought, heat, paraquat, abscisic acid and ethephon.

These studies strongly indicate that an active ascorbate-glutathione pathway is an essential component of oxygen metabolism in nodules. A second major defense in nodules consists of a variable diffusion barrier which regulates the supply of O_2 entering the nodule (Sheehy, 1987; Sinclair et al., 1985; Layzell and Hunt, 1990). As a consequence of this limited permeability to O_2, the pO_2 inside the nodules remains low as measured by microelectrode (Tjepkema and Yocum, 1974) or in situ spectroscopy (Layzell et al., 1990). Mathematical models of the diffusion properties of nodules have been valuable in interpreting responses to drought, acetylene treatment and elevated O_2 concentrations (Denison et al., 1983; Weisz et al., 1985; and Sheehy, 1987). Further investigations are required to elucidate the relative importance of the diffusion barrier vis-à-vis the ascorbate-glutathione pathway.

Ascorbate peroxidase from soybean nodules has been purified to near homogeneity and found to be a hemeprotein (Dalton et al., 1987). Non-physiological reductants such as guaiacol, dianisidine, and pyrogallol function as substrates for the enzyme, but no natural reductants other than ascorbate have been identified. The relative efficiencies of these different reductants distinguish nodule ascorbate peroxidase from both chloroplast ascorbate peroxidase and horseradish peroxidase. Ascorbate peroxidase accounts for almost 1 percent of total cytosol protein—an indication of the importance of this enzyme in nodule metabolism. The deduced amino acid sequence from a cDNA for nodule ascorbate peroxidase shows a high degree of homology to both pea leaf and *Arabidopsis* ascorbate peroxidase (Chatfield and Dalton, 1993).

Monodehydroascorbate reductase from nodules is a flavoprotein that occurs as 2 isozymes with molecular weights of 39 and 40 kDa (Dalton et al., 1992). The K_m values are 5.6, 150, and 7 µM for NADH, NADPH, and monodehydroascorbate, respectively.

Glutathione reductase from nodules has a subunit molecular weight of about 56 kDa and K_m values of 23, 24, and 150 µM for GSSG, NADPH, and NADH, respectively (Dalton, 1993). Glutathione reductase from soybean nodules shares substantial homology with sequences for glutathione reductases from pea, bacteria, and humans. Southern blot analysis of the soybean genome suggests a multigene family for glutathione reductase (Tang and Webb, 1993).

The enzymes of the nodule ascorbate-glutathione pathway are clearly of plant origin. Bacteroids do contain glutathione reductase, but not the other enzymes of the pathway (Dalton et al., 1986). Immunogold electron microscopy of soybean nodules indicates that ascorbate peroxidase is located in the cytoplasm of infected and uninfected cells (Dalton and Baird et al., 1993). Very little ascorbate peroxidase is present in the cell wall. In contrast, similar techniques indicate that monodehydroascorbate reductase is especially abundant in the cell wall. This suggests that ascorbate regeneration is important in the cell wall region and that cytosolic regeneration relies on dehydroascorbate reductase, however more information is required to clarify this. Subcellular fractionation studies have shown some activity of all four enzymes of this pathway in nodule mitochondria, but the majority of activity is in the soluble fraction. Peroxisomes do not contain any of these enzymes. Western blots with purified fractions have confirmed that glutathione reductase is present in both the crude (cytosolic) fraction and in mitochondria (Dalton, 1993). The presence of a putative targeting sequence on a cDNA clone coding for glutathione reductase suggests that glutathione reductase may also be present in nodule plastids (Tang and Webb, 1993).

The ascorbate-glutathione pathway also operates in another important nitrogen-fixing system-cyanobacteria (Tözüm and Gallon, 1979; Tel-Or et al., 1985,1986). These organisms present an interesting opportunity to evaluate the relative importance of this pathway in nitrogen fixation vs. the importance in photosynthesis. Most nitrogen-fixing cyanobacteria separate these two processes into two different cell types with photosynthesis taking place in vegetative cells and nitrogen fixation in thick-walled heterocysts. In *Nostoc muscorum*, the two cell types contain equal ascorbate peroxidase activity (Tel-Or et al., 1986), thus justifying arguments that both photosynthesis and nitrogen fixation are highly dependent on the protective benefits of the ascorbate-glutathione pathway. Similarly, heterocysts of *Anabaena variabilis* also contain the enzymes of the ascorbate-glutathione pathway (and also glutathione peroxidase) at activity levels roughly equal or even slightly higher than levels in vegetative cells (Bagchi et al., 1991). Even more striking is the 14-fold higher level of SOD activity in heterocysts of this cyanobacterium. Increases in nitrogenase activity in heterocysts are accompanied by parallel increases in SOD activity, further suggesting a role of SOD in protection of nitrogenase (Caiola et al., 1991).

9.5 THE ROLE OF CATALASE

Catalase (E.C. 1.11.1.6) is an extremely abundant and widespread enzyme in plants. It catalyzes the following reaction (eq. 9.18):

$$2H_2O_2 \rightarrow 2H_2O + O_2 \qquad (9.18)$$

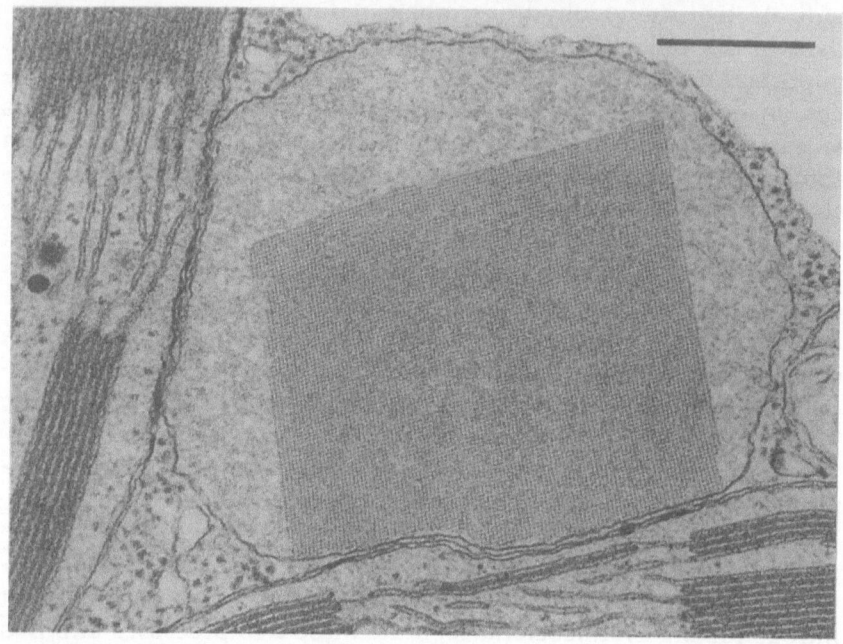

Fig. 9.11   Electron micrograph of a tobacco leaf peroxisome tightly appressed to two chloroplasts (left and below). Crystalline catalase fills the majority of the peroxisome. Bar = 0.5 μm. Photo provided by E.H. Newcomb (Univ. of Wisconsin-Madison). Reproduced from Frederick, G.E. and E.H. Newcomb, J. Cell Biol. 43, 343–353 (1969) by copyright permission of the Rockefeller University Press.

This reaction has the apparent advantage of not requiring an additional source of reducing power (such as ascorbate) as all peroxidases do. However, this apparent advantage is more than offset by an extremely low affinity (high K_m) for H_2O_2 (Halliwell, 1982a). Consequently, catalase is very ineffective in scavenging low concentrations of H_2O_2. Most, if not all, catalase in plants is located in microbodies such as peroxisomes and glyoxysomes. Concentrations of catalase in peroxisomes are often so high that the enzyme can be seen as a nearly pure crystalline lattice in electron micrographs (Fig. 9.11). Concentrations of H_2O_2 are high in microbodies due to the close proximity of enzymes such as glycollate oxidase that produce substantial amounts of H_2O_2. Under these conditions, catalase is effective in removing H_2O_2. Catalase is generally not present in detectable levels in the cytosol or in chloroplasts, where the main task of peroxide scavenging falls to ascorbate peroxidase.

Catalase can take on a cytosolic defense role in unicellular organisms such as green algae and cyanobacteria (Bowler et al., 1992; Tel-Or et al., 1986). However, at least some of these organisms also contain the ascorbate-glutathione pathway and catalase-deficient species appear to do quite well even in the case of *Gloeocapsa* (now called *Gloeothece*), a non-heterocystous cyanobacterium that has the unusual burden of conducting photosynthesis and nitrogen fixation within the same cell (Tözüm and Gallon, 1979).

9.6 STRESS AND ANTIOXIDANT DEFENSES

Antioxidant defenses are a ubiquitous attribute of plant cells and can be regarded as part of the routine housekeeping duties required of an aerobic existence. It is becoming increasingly apparent that diverse kinds of stress, such as heat, chilling, drought, water-logging, intense light, paraquat, and air pollutants, etc., can all lead to oxidative stress and to increased levels of antioxidant defenses (Table 9.3). This has been most extensively examined in the case of SOD and several excellent reviews are available on the stress response of SOD (Scandalios, 1990; Bowler et al., 1992). These responses can be grouped into two classes: 1) short-term physiological responses involving up-regulation of gene expression presumably through some unknown signal transduction pathway and 2) genetic selection of ecotypes that have developed increased defenses in response to some long-term environmental challenge. Numerous examples of the first type of response are given in Table 9.3. Examples of the second category include a high light-resistant biotype of *Conyza bonariensis* with elevated levels of SOD, glutathione reductase and ascorbate peroxidase (Jansen et al., 1989) and four paraquat-resistant lines of ryegrass (*Lolium perenne*) with elevated constitutive levels of SOD, catalase and peroxidase (Harper and Harvey, 1978). Work is in progress in a number of laboratories to address the molecular biology of these responses.

9.7 MISCELLANEOUS ANTIOXIDANTS IN PLANTS

In addition to the compounds and enzymes that have already been discussed, plants contain additional diverse antioxidants. Many of

Table 9.3.
Plant Antioxidant Defenses that Respond Positively to Stress

Defense parameter	Stress factor	Organisms
Ascorbate concentration	light + chilling	spinach
	high O_2	soybean
Carotenoid concentration	light + chilling	spinach
	light + heat	tomato, cucumber, pepper
SOD	drought	tomato, *Tortula* (moss)
	heat	tobacco
	high pO_2	maize
	light + chilling	spinach
	ozone	spruce, pine, spinach, bean
	paraquat	pea, soybean, *Chlorella*, bean, tobacco, maize, duckweed
	SO_2	*Chlorella*, poplar, pine, spruce
	waterlogging	*Iris*
Ascorbate peroxidase	drought	barley, pea, soybean[b]
	high O_2	soybean
	light + chilling	spinach
	light	*Synechococcus* (cyanobacterium)
	ozone	*Sedum*, spruce, spinach
	heat	pea
	ethephone	pea
	abscisic acid	pea
	paraquat	pea
Monodehydroascorbate reductase	high O_2	soybean
	light + chilling	spinach
	ozone	spinach
	paraquat	soybean
Glutathione reductase	drought	wheat, cotton, barley
	high O_2	soybean
	ozone + SO_2 + NO_2	pea
Catalase	cercosporin (toxin)	maize
	drought	*Tortula* (moss)
	light	*Synechococcus* (cyanobacterium)
	ozone	spinach, bean
	paraquat	maize, *Dunaliella* (green alga)

[a] compiled from Mittler and Tel-Or, 1991; Dalton et al., 1991; and Mittler and Zilinskas, 1992, and from sources listed in Scandalios, 1990; Dalton, 1990; and Bowler et al., 1992).
[b] unpublished observations of S. Moe and D. Dalton

these, such as the flavonoids, alkaloids and related secondary natural products are discussed elsewhere in this book (chapters 6 and 10; also see Larson, 1988; Lewis, 1993).

Since free transition metals, primarily iron, strongly promote free radical formation via the Haber-Weiss reaction (Halliwell and Gutteridge, 1986) a very important component of plant antioxidant defenses is based on mechanisms to remove metals from solution. Many plants contain an iron storage protein, phytoferritin, analogous to the ferritin present in mammalian cells (Bienfait and der Mark, 1983; Seckbeck, 1982). These are very large, multi-subunit proteins (20 subunits, 600 kDa in soybean) with a large central core wherein the vast majority of iron in plants is stored. Phytoferritin is especially abundant in nitrogen-fixing nodules where it presumably plays a key role in minimizing free radical damage and in regulating the availability of iron for the biosynthesis of leghemoglobin (Ko et al., 1987).

Phytic acid (*myo*-inositol hexaphosphoric acid) is a powerful antioxidant that comprises 1–5% by weight of most cereals, nuts, legumes, oil seeds, pollen and spores (Graf et al., 1987; Graf and Eaton, 1990). Phytic acid forms a monoferric chelate in which the iron is unavailable for participation in the iron-catalyzed formation of ·OH. This is a rare trait for iron chelators since most chelators actually enhance the catalytic properties of iron. Thus phytic acid is a powerful inhibitor of lipid peroxidation and other oxidative reactions, a property that is especially beneficial in extending seed viability. Phytic acid may be one of the most abundant and fundamentally important antioxidants in plants, but it has been strangely neglected. More attention is clearly merited.

9.8 BENEFICIAL USES OF ACTIVE OXYGEN IN PLANTS

The problem of O_2 toxicity is so universal and demands such strenuous defensive measures that it is frequently overlooked that plants can use potentially damaging oxidants to their advantage. Parallels exist in animals where the "respiratory burst" is used by neutrophils to generate active forms of oxygen and thus kill phagocytosed microbes (Gabig and Babior, 1982; Halliwell and Gutteridge, 1985).

Many plants contain various phenolic compounds that have a strong tendency to oxidize and cause inactivation and "phenolic browning" of proteins. In intact cells, these compounds are usually sequestered away safely in the vacuole, but physical attack of the plant tears open the vacuole and releases the phenolic compounds at which time the would-be predator's digestive enzymes are subject to inactivation (Hrazdina and Wagner, 1985; Loomis, 1974; Pierpoint, 1985). The process requires O_2 and is greatly accelerated by plant enzymes called polyphenol oxidases.

Oxygen free radicals are involved in the induction of plant defenses against microbial attack through some poorly understood signal transduction pathway involving redox perturbation attributable to active oxygen (Devlin and Gustine, 1992; Epperlein et al., 1986; Lamb et al., 1989; Vera-Estrella et al., 1992).

Some photoactivated plant secondary compounds can produce active oxygen with resulting high toxicity to many organisms, especially insects. Ecological interactions of this sort are rapidly emerging as an important aspect of the allelochemic-based defenses of plants (Aucoin et al., 1990; Ahmad, 1992). Insects compensate by altered behavior, such as avoidance of light, and by increased levels of antioxidants and antioxidant enzymes.

Many non-green plant tissues contain lipoxygenase, an enzyme that catalyzes the peroxidation of fatty acids to form unsaturated fatty acid hydroperoxides (Siedow, 1991). Although the physiological functions of this enzyme are poorly understood, numerous beneficial roles have been proposed including various roles in growth and development, stimulation of senescence, response to wounding and other stresses and the synthesis of regulatory molecules (e.g. jasmonic acid).

In addition to the H_2O_2-scavenging role of ascorbate peroxidase, peroxidases in plants have other diverse functions including the formation of lignin and oxidation of indole-3-acetic acid (Gross, 1979; Stonier et al., 1979; Campa, 1991). Some fungi, such as the wood-decaying basidiomycete *Phanerochaete chrysosporium*, produce extracellular peroxidases that degrade lignin (Glenn et al., 1986). These fungi also produce extracellular H_2O_2 to facilitate the process. The function of other peroxidases, particularly the "guaiacol" peroxidases such as horseradish peroxidase, is not clear. These peroxidases are named after their ability to oxidize the artificial substrate guaiacol thus providing a convenient and easy assay for activity in vitro. Apparently, guaiacol peroxidases function not to scavenge H_2O_2, but to catalyze reactions where the products of the electron donors have physiological roles. It may be that peroxidases and H_2O_2 have other beneficial roles in plants that are not understood.

9.9 ANTIOXIDANT DEFENSES OF FUNGI

Fungi are at somewhat less risk of oxidative damage than green plants since fungi do not engage in either photosynthesis or nitrogen fixation. Nevertheless, fungi are aerobic organisms and as such require antioxidant defenses. Fungi share many common defense mechanisms with plants, including SOD, catalase, glutathione and glutathione reductase. Perhaps the main difference between green plants and fungi lies in the chief peroxide scavenging enzyme which in fungi is cytochrome c peroxidase and not ascorbate peroxidase. The main source of active oxygen in fungi is probably mitochondria. Yeast mitochondria produce both O_2^- and H_2O_2 as a consequence of electron transport associated with respiration (Boveris, 1978). As a mitochondrial enzyme, cytochrome c peroxidase is well situated to dispose of this H_2O_2.

Although ascorbic acid has occasionally been reported in yeast, recent work by Nick et al. (1986) has refuted these claims and established instead the presence of erythroascorbic acid in *Saccharomyces* and *Lypomyces*. Erythroascorbic acid has also been detected from *Candida* (Murakawa et al., 1977). Erythorbic acid has been isolated from *Penicillium* (Takahashi et al., 1976). The physiological role of these ascorbate analogs is not clear.

Fungi contain substantial levels of glutathione and glutathione reductase. Considerable attention has been paid to optimizing culture conditions for yeast which are used as commercial sources of glutathione and glutathione reductase (Alfafara et al., 1992; Tsai et al., 1991). Glutathione reductase from yeast has been widely used in biochemical laboratories for the enzymatic recycling assay used to determine glutathione content in biological samples although this method has been largely replaced by HPLC techniques (Griffith, 1980). Glutathione reductase has been purified and thoroughly studied from several fungal sources (see Schirmer et al., 1989; Williams, 1992). It is a dimer with a subunit molecular weight of 56.5 kDa (Massey and Williams, 1965).

CuZnSOD from yeast has also been extensively purified and studied. There is a high degree of amino acid sequence homology between CuZnSOD's from yeast and other sources, although among the eukaryotes, the yeast form has the highest sequence divergence from the bovine enzyme (55% amino acid identities, Bannister et al., 1987).

SOD levels in yeast are regulated by at least two environmental factors—O_2 and Cu. Yeast contain 6.5 times more SOD (and 2.3

times more catalase) when grown under 100% O_2 than when grown anaerobically (Gregory et al., 1974). Cu is, of course, required for activity of CuZnSOD's, but this element also plays another fundamental role in determining SOD activity in yeast. Expression of *SOD1*, the CuZnSOD gene in yeast, is regulated through a transcription activating regulatory protein called ACE1 (Carri et al., 1991; Gralla et al., 1991). The ACE1 protein will bind Cu and then the ACE1-Cu complex can bind to a *cis*-acting region −184 to −206 bases upstream from the promoter region of *SOD1*. This leads to an increase in the transcription of *SOD1*. Yeast strains lacking the ACE1 gene have an atypical inability to increase CuZnSOD mRNA in response to Cu. The *CUP1* gene for metallothionein is also regulated by the same *trans*-acting factor.

The most distinctive feature of antioxidant defenses in fungi is cytochrome *c* peroxidase (E.C. 1.11.1.5), a heme protein with strong amino acid sequence homologies to plant ascorbate peroxidase (see earlier discussion in section 9.3.4). Cytochrome *c* peroxidase (reviewed in Yonetani, 1976 and Bosshard et al., 1991) catalyzes the following reaction:

$$H_2O_2 + 2 \text{ ferrocytochrome } c + 2H^+ \rightarrow \qquad (9.19)$$
$$2 \text{ ferricytochrome } c + 2H_2O$$

Cytochrome *c* peroxidase is located entirely within the mitochondria, with most of the activity in the intracristal space, although some activity is probably also present in the peripheral intermembrane space (Williams and Stewart, 1976). Cytochrome *c* peroxidase may serve as an alternate terminal electron carrier with H_2O_2 as the electron acceptor (Verduyn et al., 1991). The physiological significance of this is unclear since measurable respiration via this route depends on anaerobic conditions and exogenous H_2O_2. Such conditions are unlikely to be encountered in nature.

In recent years, cytochrome *c* peroxidase has become the focus of a large community of researchers interested in interprotein electron transfers. As a result, this enzyme is "possibly the best-characterized heme protein" particularly when one considers the wealth of information available on the catalytic mechanism and the structure of the cytochrome *c* peroxidase-cytochrome *c* complex (Bosshard et al., 1991). The enzyme has been sequenced (Takio et al., 1980) and the structure of the active site determined from crystallographic analyses (Poulos et al., 1980; Finzel et al., 1984). The gene for cytochrome *c* peroxidase has been cloned and sequenced (Goltz et al., 1982; Kaput et al., 1982).

9.10 SUMMARY

Plants are particularly vulnerable to oxidative damage. This risk arises because plants conduct various reductive processes such as photosynthesis, respiration, and nitrogen fixation all of which involve the potential for generation of partially reduced active forms of oxygen such as O_2^-, H_2O_2, and $\cdot OH$. Consequently, plants contain ubiquitous and multi-layered antioxidant defenses that function to remove reactive forms of oxygen. These defenses include small antioxidant molecules as well as several enzymes. Within membranes, hydrophobic scavengers including the carotenoids and tocopherols (vitamin E) that protect against the especially damaging process of lipid peroxidation. Hydrophilic regions of cells are protected by strong antioxidants such as ascorbic acid and glutathione. Plants also contain a wide range of miscellaneous antioxidant defense compounds including flavonoids, alkaloids, iron storage proteins (phytoferritin) and iron chelators (phytic acid).

Antioxidant enzymes include numerous forms of superoxide dismutase (SOD), a O_2^- destroying enzyme that is ubiquitous in aerobic organisms. SOD in the cytosol is of the CuZn type as is typical for eukaryotes. Chloroplasts, which are particularly vulnerable to O_2^- production, contain very high levels of CuZnSOD as well as either Fe- or MnSOD. SOD also plays a critical role in protection of nitrogen fixation. Enzymatic scavenging of H_2O_2 in plants proceeds via ascorbate peroxidase, a heme protein that is perhaps as ubiquitous in plants as SOD is. Ascorbate peroxidase is especially abundant in chloroplasts and in the cytosol of cells from nitrogen-fixing root nodules. This enzyme catalyzes the first step in the ascorbate-glutathione pathway, a series of coupled reduction-oxidation reactions that also involves reactions catalyzed by monodehydroascorbate reductase, dehydroascorbate reductase, and glutathione reductase. Understanding of the molecular biology of antioxidant defenses in plants is progressing rapidly, especially with regards to environmental factors such as drought, heat shock, high light, and xenobiotics that appear to regulate gene expression.

Fungi do not engage in the high risk activities of photosynthesis and nitrogen fixation, but nevertheless require an active antioxidant defense system. As with plants, a key antioxidant defense in fungi is provided by SOD. In yeasts, the CuZnSOD gene is regulated through a transcription activating regulatory protein called ACE1. This model may provide leads towards understanding how SOD is regulated in higher plants. Fungi do not contain the ascorbate-glutathione pathway, but they do possess substantial amounts of glutathione and glutathione reductase. The most distinctive feature of antioxidant defenses in fungi is cytochrome *c* peroxidase, a heme protein with strong amino acid sequence homologies to plant ascorbate peroxidase. Cytochrome *c* peroxidase is a mitochondrial enzyme that scavenges H_2O_2 produced during electron transport associated with respiration. Cytochrome *c* peroxidase has been extensively studied with regards to the details of the catalytic mechanism and the structure of the cytochrome *c* peroxidase-cytochrome *c* complex.

References

Ahmad, S. (1992) Biochemical defence of pro-oxidant plant allelochemicals by herbivorous insects. *Biochem. System. Ecol.* **20**, 269–296.

Alfafara, C.G., Kanda, A., Shioi, T., Shimzu, H., Shioya, S. and Susa, K. (1992) Effect of amino acids on glutathione production by *Saccharomyces cerevisiae*. *Appl. Microbiol. Biotechnol.* **36**, 538–540.

Alscher, R.G. (1989) Biosynthesis and anitoxidant function of glutathione in plants. *Physiol. Plant.* **77**, 457–464.

Anderson, I.C. and Robertson, D.S. (1960) Role of carotenoids in protecting chlorophyll from photodestruction. *Plant Physiol.* **35**, 531–534.

Anderson, J.V., Hess, J.L. and Chevone, B.I. (1990) Purification, characterization, and immunological properties for two isoforms of glutathione reductase from eastern white pine needles. *Plant Physiol.* **94**, 1402–1409.

Aono, M., Kubo, A, Saji, H., Natori, T., Tanaka, K. and Kondo, N. (1991) Resistance to active oxygen toxicity of transgenic *Nicotiana tabacum* that expresses the gene for glutathione reductase from *Escherichia coli*. *Plant Cell Physiol.* **32**, 691–697.

Appleby, C.A. (1984) Leghemoglobin and *Rhizobium* respiration. *Ann. Rev. Plant Physiol.* **35**, 443–478.

Arp, D.J. (1992) Hydrogen cycling in symbiotic bacteria. In *Biological Nitrogen Fixation*, (G. Stacey, R.H. Burris and H.J. Evans, eds.), Chapman & Hall, New York, pp. 432–460.

Arrigoni, O., Dipierro, S. and Borraccino, G. (1981) Ascorbate free radical reductase, a key enzyme of the ascorbic acid system. *FEBS Lett.* **125**, 242–245.

Asada, K. (1992) Ascorbate peroxidase-a hydrogen peroxide-scavenging enzyme in plants. *Physiol. Plant.* **85**, 235–241.

Asada, K. and Takahashi, M. (1987) Production and scavenging of active oxygen in photosynthesis. In *Photoinhibition* (D.J. Kyle, C.B. Osmond and C.J. Arntzen, eds.), Elsevier, New York, pp. 227–287.

Aucoin, R.R., Fields, P., Lewis, M.A. Philogène, B.J.R. and Arnason, J.T. (1990) The protective effect of antioxidants to a phototoxin-sensitive insect herbivore. *J. Chem. Ecol.* **16**, 2913–2924.

Aviram, I., Wittenberg, B.A. and Wittenberg, J.B. (1978) The reaction of ferrous leghemoglobin with hydrogen peroxide to form leghemoglobin (IV). *J. Biol. Chem.* **253**, 5685–5689.

Bagchi, S.N., Ernst, A. and Böger, P. (1991) The effect of activated oxygen species on nitrogenase of *Anabaena variabilis*. *Z. Naturforsch.* **46c**, 407–415.

Baker, D.D. and Mullin, B.C. (1992) Actinorhizal plants. In *Biological Nitrogen Fixation*, (G. Stacey, R.H. Burris and H.J. Evans, eds.), Chapman & Hall, New York, pp. 259–292.

Bannister, J.V., Bannister, W.H. and Rotilio, G. (1987) Aspects of the structure, function, and applications of superoxide dismutase. *CRC Crit. Rev. Biochem.* **22**, 111–180.

Baszynski, T. (1974) The effect of α-tocopherol on reconstitution of photosystem I in heptane-extracted spinach chloroplasts. *Biochim. Biophys. Acta* **347**, 31–35.

Baum, J.A. and Scandalios, J.G. (1981) Isolation and characterization of the cytosolic and mitochondrial superoxide dismutases of maize. *Arch. Biochem. Biophys.* **206**, 249–264.

Beauchamp, C.O. and Fridovich, I. (1971) Superoxide dismutase: improved assays and assay applicable to acrylamide gels. *Anal. Biochem.* **44**, 276–287.

Becana, M., Aparicio-Tejo, P., Irigoyen, J.J. and Sanchez-Diaz, M. (1986) Some enzymes of hydrogen peroxide metabolism in leaves and root nodules of *Medicago sativa*. *Plant Physiol.* **82**, 1169–1171.

Becana, M., Paris, F.J., Sandalio, L.M. and Del Río, L.A. (1989) Isozymes of superoxide dismutase in nodules of *Phaseolus vulgaris* L., *Pisum sativum* L., and *Vigna unguiculata* (L.) Walp. *Plant Physiol.* **90**, 1286–1291.

Becana, M. and Rodríguez-Barrueco, C. (1989) Protective mechanisms of nitrogenase against oxygen excess and partially-reduced oxygen intermediates. *Physiol. Plant.* **75**, 429–438.

Becana, M. and Salin, M.L. (1989) Superoxide dismutases in nodules of leguminous plants. *Can. J. Bot.* **67**, 415–421.

Bendich, A., Machlin, L.J. and Scandurra, O. (1986) The antioxidant role of vitamin C. *Adv. Free Rad. Biol. Med.* **2**, 419–444.

Bergersen, F.J. (1982) *Root Nodules of Legumes: Structure and Functions*. Research Studies Press, Chichester.

Bielawski, W. and Joy, K.W. (1986) Properties of glutathione reductase from chloroplasts and roots of pea. *Phytochemistry*. **25**, 2261–2265.

Bienfait, H.F. and der Mark, V. (1983) Phytoferritin and its role in iron metabolism. In *Metals and Micronutrients: Uptake and Utilization* (D.A. Robb and W.S. Pierpoint, eds.), Academic Press, London, pp. 111–123.

Bisby, R.H., Ahmed, S. and Cundall, R.B. (1984) Repair of amino acid radicals by a vitamin E analogue. *Biochem. Biophys. Res. Commun.* **119**, 245–251.

Black, C.C. (1973) Photosynthetic carbon fixation in relation to net CO_2 uptake. *Ann. Rev. Plant Physiol.* **24**, 253–286.

Borraccino, G., Dipierro, S. and Arrigoni, O. (1986) Purification and properties of ascorbate free-radical reductase from potato tubers. *Planta* **167**, 521–526.

Bosshard, H.R., Anni, H. and Yonetani, T. (1991) Yeast cytochrome c peroxidase. In *Peroxidases in Chemistry and Biology, Vol. II* (J. Everse, K.E. Everse and M.B. Grisham, eds.), CRC Press, Boca Raton, FL, pp. 51–84.

Boveris, A. (1978) Production of superoxide anion and hydrogen peroxide in yeast mitochondria. In *Biochemistry and Genetics of Yeasts*, (M. Bacila, B.L. Horecker and A.O.M. Stoppani, eds.), Academic Press, New York, pp. 65–80.

Bowler, C., Slooten, L., Vandenbranden, S., De Rycke, R., Botterman, J., Sybesma, C., Van Montagu, M. and Inzé, D. (1991) Manganese superoxide dismutase can reduce cellular damage mediated by oxygen radicals in transgenic plants. *EMBO J.* **10**, 1723–1732.

Bowler, C., Van Montagu, M. and Inzé, D. (1992) Superoxide dismutase and stress tolerance. *Ann. Rev. Plant Physiol. Plant Mol. Biol.* **43**, 83–116.

Cadenas, E. (1989) Biochemistry of oxygen toxicity. *Ann. Rev. Biochem.* **58**, 79–110.

Caiola, M.G., Canini, A., Galiazzo, F. and Rotilio, G. (1991) Superoxide dismutase in vegetative cells, heterocysts and akinetes of *Anabaena cylindrica* Lemm. FEMS Microbiol Lett. **80**, 161–166.

Campa, A. (1991) Biological roles of plant peroxidases: known and potential function. In *Peroxidases in Chemistry and Biology, Vol. II* (J. Everse, K.E. Everse and M.B. Grisham, eds.), CRC Press, Boca Raton, FL, pp. 25–50.

Cannon, R.E. and Scandalios, J.G. (1989) Two cDNAs encode two nearly identical Cu/Zn superoxide dismutases in maize. *Mol. Gen. Genet.* **219**, 1–8.

Cannon, R.E., White, J.A. and Scandalios, J.G. (1987) Cloning of cDNA for maize superoxide dismutase 2 (SOD-2). *Proc. Natl. Acad. Sci. USA* **84**, 179–183.

Carri, M.T., Galiazzo, F., Ciriolo, M.R. and Rotilio, G. (1991) Evidence for co-regulation of Cu,Zn superoxide dismutase and metallothionein gene expression in yeast through transcriptional control by copper via the ACE 1 factor. *FEBS Lett.* **278**, 263–266.

Carroll, E.W., Schwarz, O.J. and Hickok, L.G. (1988) Biochemical studies of paraquat-tolerant mutants of the fern *Ceratopteris richardii*. *Plant Physiol.* **87**, 651–654.

Chatfield, J.M.C. and Dalton, D.A. (1993) Ascorbate peroxidase from soybean root nodules. *Plant Physiol.* **103**, 661–662.

Chen, G.-X. and Asada, K. (1989) Ascorbate peroxidase in tea leaves: occurrence of two isozymes and the differences in their enzymatic and molecular properties. *Plant Cell Physiol.* **30**, 987–998.

Chen, G.-X. and Asada, K. (1990) Hydroxyurea and p-aminophenol are the suicide inhibitors of ascorbate peroxidase. *J. Biol. Chem.* **265**, 2775–2781.

Chen, G.-X., Sano, S. and Asada, K. (1992) The amino acid sequence of ascorbate peroxidase from tea has a high degree of homology to that of cytochrome c peroxidase from yeast. *Plant Cell Physiol.* **33**, 109–116.

Choudhary, A.D., Lamb, C.J. and Dixon, R.A. (1990) Stress response in alfalfa (*Medicago sativa* L.) VI. Differential responsiveness of chalcone synthase induction to fungal elicitor or glutathione in electroporated protoplasts. *Plant Physiol.* **94**, 1802–1807.

Connell, J.P. and Mullet, J.E. (1986) Pea chloroplast glutathione reductase: purification and characterization. *Plant Physiol.* **82**, 351–356.

Creissen, G., Edwards, E.A., Enard, C., Wellburn, A. and Mullineaux, P. (1991) Molecular characterization of glutathione reductase cDNAs from pea (*Pisum sativum* L.). *Plant J.* **2**, 129–131.

Dalton, D.A. (1990) Ascorbate peroxidase. In *Peroxidases in Chemistry and Biology*, *Vol. II* (J. Everse, K. Everse and M.B. Grisham, eds.), CRC Press, Boca Raton, FL, pp. 139–153.

Dalton, D.A. (1993) Glutathione reductase from soybean root nodules. *Plant Physiol.* supplement **105**, 839.

Dalton, D.A., Baird, L., Langeberg, L., Taugher, C.Y., Anyan, W.R., Vance, C.P. and Sarath, G. (1993) Subcellular localization of oxygen defense enzymes in soybean (*Glycine max* [L.] Merr.) root nodules. *Plant Physiol.* **102**, 481–489.

Dalton, D.A., Hanus, F.J., Russell, S.A. and Evans, H.J. (1987) Purification, properties and distribution of ascorbate peroxidase in legume root nodules. *Plant Physiol.* **83**, 789–794.

Dalton, D.A., Langeberg, L. and Robbins, M. (1992) Purification and characterization of monodehydroascorbate reductase from soybean root nodules. *Arch. Biochem. Biophys.* **292**, 281–286.

Dalton, D.A., Langeberg, L. and Treneman, N. (1993) Correlations between the ascorbate-glutathione pathway and effectiveness in legume root nodules. *Physiol. Plant.* **87**, 365–370.

Dalton, D.A., Post, C.J. and Langeberg, L. (1991) Effects of ambient oxygen and of fixed nitrogen on concentrations of glutathione, ascorbate and associated enzymes in soybean root nodules. *Plant Physiol.* **96**, 812–818.

Dalton, D.A., Russell, S.A., Hanus, F.J., Pascoe, G.A. and Evans, H.J. (1986) Enzymatic reactions of ascorbate and glutathione that prevent peroxide damage in soybean root nodules. *Proc. Natl. Acad. Sci. USA* **83**, 3811–3815.

Demmig-Adams, B. (1990) Carotenoids and photoprotection in plants: A role for the xanthophyll zeaxanthin. *Biochim. Biophys. Acta* **1020**, 1–24.

Demmig-Adams, B. and Adams, W.W. (1993) The xanthophyll cycle. In *Antioxidants in Higher Plants* (R.G. Alscher and J.L. Hess, eds.), CRC Press, Boca Raton, FL. pp. 91–110.

Denison, R.F., Weisz, P.R. and Sinclair, T.R. (1983) Analysis of acetylene reduction rates of soybean nodules at low acetylene concentrations. *Plant Physiol.* **73**, 648–651.

Devlin, W.S. and Gustine, D.L. (1992) Involvement of the oxidative burst in phytoalexin accumulation and the hypersensitive reaction. *Plant Physiol.* **100**, 1189–1195.

Dhindsa, R.S., Plumb-Dhindsa, P. and Thorpe, T.A. (1981) Leaf senescence: correlated with increased levels of membrane permeability and lipid peroxidation, and decreased levels of superoxide dismutase and catalase. *J. Exp. Bot.* **32**, 93–101.

Dipierro, S. and Borraccino, G. (1991) Dehydroascorbate reductase from potato tubers. *Phytochem.* **30**, 427–429.

Dolphin, D., Poulson, R. and Avramovic, O. (1989) *Glutathione, Coenzymes and Cofactors.* John Wiley, New York.

Edwards, E.A., Rawsthorne, S. and Mullineaux, P.M. (1990) Subcellular distribution of multiple forms of glutathione reductase in leaves of pea (*Pisum sativum* L.). *Planta* **180**, 278–284.

Edwards, G. and Walker, D. (1983) *C3, C4: Mechanisms, and cellular and environmental regulation, of photosynthesis*. University of California Press, Berkeley.

Edwards, R., Blount, J.W. and Dixon, R.A. (1991) Glutathione and elicitation of the phytoalexin response in legume cell cultures. *Planta* **184**, 403–409.

Elstner, E.F. (1982) Oxygen activation and oxygen toxicity. *Ann. Rev. Plant Physiol.* **33**, 73–96.

Epperlein, M.M., Noronha-Dutra, A.A. and Strange, R.N. (1986) Involvement of the hydroxyl radical in the abiotic elicitation of phytoalexins in legumes. *Physiol. Mol. Plant Pathol.* **28**, 67–77.

Fahrenholtz, S.R., Doleiden, F.H., Trozzolo, A.M. and Lamola, A.A. (1974) Quenching of singlet oxygen by α-tocopherol. *Photochem. Photobiol.* **20**, 505–509.

Finzel, B.C., Poulos, T.L. and Kraut, J. (1984) Crystal structure of cytochrome *c* peroxidase refined at 1.7 Ångström resolution. *J. Biol. Chem.* **259**, 13027–13036.

Foyer, C. (1984) *Photosynthesis*. John Wiley, New York.

Foyer, C. (1993) Ascorbic acid. In *Antioxidants in Higher Plants* (R.G. Alscher and J.L. Hess, eds.), CRC Press, Boca Raton, FL, pp. 31–58.

Foyer, C.H. and Halliwell, B. (1976) The presence of glutathione and glutathione reductase in chloroplasts: a proposed role in ascorbic acid metabolism. *Planta* **133**, 21–25.

Foyer, C.H. and Halliwell, B. (1977) Purification and properties of dehydroascorbate reductase from spinach leaves. *Phytochem.* **16**, 1347–1350.

Foyer, C.H., Lelandais, M., Galap, C. and Kunert, K.J. (1991) Effects of elevated cytosolic glutathione reductase activity on the cellular glutathione pool and photosynthesis in leaves under normal and stress conditions. *Plant Physiol.* **97**, 863–872.

Foyer, C., Rowell, J. and Walker, D. (1983) Measurement of the ascorbate content of spinach leaf protoplasts and chloroplasts during illumination. *Planta* **157**, 239–244.

Fridovich, I. (1976) Oxygen radicals, hydrogen peroxide, and oxygen toxicity. In *Free Radicals in Biology, Vol. I.* (W.A. Pryor, ed.), Academic Press, New York, pp. 239–277.

Fridovich, I. (1979) Superoxide and superoxide dismutases. In *Advances in Inorganic Biochemistry* (G.L. Eichhorn and L.G. Marzilli, eds.), Elsevier, New York, pp. 27–90.

Fridovich, I. (1986) Biological effects of the superoxide radical. *Arch. Biochem. Biophys.* **247**, 1–11.

Fryer, M.J. (1992) The antioxidant effects of thylakoid vitamin E (α-tocopherol). *Plant Cell & Environ.* **15**, 381–392.

Fujiyama, K., Takemura, H., Shinmyo, A., Okada, H. and Takano, M. (1990) Genomic DNA structure of two new horseradish-peroxidase-encoding genes. *Gene* **89**, 163–169.

Fukuzawa, K. and Gebicki, J.M. (1983) Oxidation of α-tocopherol in micelles and liposomes by the hydroxyl, perhydroxyl and superoxide free radicals. *Arch. Biochem. Biophys.* **226**, 242–251.

Gabig, T.G. and Babior, B.M. (1982) Oxygen-dependent microbial killing by neutrophils. In *Superoxide Dismutase, Vol. II.* (L.W. Oberley, ed.), CRC Press, Boca Raton, FL, pp. 1–14.

Gallon, J.R. (1992) Reconciling the incompatible: N_2 fixation and O_2. *New Phytol.* **122**, 571–609.

Glenn, J.K., Akileswaran, L and Gold, M.H. (1986) Mn(II) oxidation is the principal function of the extracellular Mn-peroxidase from *Phanerochaete chrysosporium*. *Arch. Biochem. Biophys.* **251**, 688–696.

Goltz, S., Kaput, J. and Blobel, G. (1982) Isolation of the yeast nuclear gene encoding the mitochondrial protein, cytochrome *c* peroxidase. *J. Biol. Chem.* **257**, 11186–11190.

Graf, E. and Eaton, J.W. (1990) Antioxidant functions of phytic acid. *Free Rad. Biol. Med.* **8**, 61–69.

Graf, E., Empson, K.L. and Eaton, J.W. (1987) Phytic acid-a natural antioxidant. *J. Biol. Chem.* **262**, 11647–11650.

Gralla, E.B., Thiele, D.J., Silar, P. and Valentine, J.S. (1991) ACE 1, a copper-dependent transcription factor, activates expression of the yeast copper, zinc superoxide dismutase gene. *Proc. Natl. Acad. Sci. USA* **88**, 8558–8562.

Gregory, E.M., Goscin, S.A. and Fridovich, I. (1974) Superoxide dismutase and oxygen toxicity in a eukaryote. *J. Bact.* **117**, 456–460.

Griffith, O.W. (1980) Determination of glutathione and glutathione disulfide using glutathione reductase and 2-vinylpyridine. *Anal. Biochem.* **106**, 207–212.

Groden, D. and Beck, E. (1979) H_2O_2 destruction by ascorbate-dependent systems from chloroplasts. *Biochim. Biophys. Acta* **546**, 426–435.

Gross, G.G. (1979) Recent advances in the chemistry and biochemistry of lignin. *Rec. Adv. Phytochem.* **12**, 177–220.

Gupta, A.S., Heinen, J.L., Holaday, A.S., Burke, J.J. and Allen, R.D. (1993) Increased resistance to oxidative stress in transgenic plants that overexpress chloroplastic Cu/Zn superoxide dismutase. *Proc. Natl. Acad. Sci. USA* **90**, 1629–1633.

Gutteridge, J.M.C. and Halliwell, B. (1990) The measurement and mechanism of lipid peroxidation in biological systems. *Trends Biochem. Sci.* **15**, 129–135.

Halliwell, B. (1982a) Ascorbic acid and the illuminated chloroplast. In *Ascorbic Acid: Chemistry, Metabolism, and Uses* (P.A. Seib and B.M. Tolbert, eds.), *Adv. in Chemistry Series 200*, Am. Chem. Soc., Washington, DC, pp. 263–274.

Halliwell, B. (1982b) The toxic effects of oxygen on plant tissues. In *Superoxide Dismutase, Vol. I.* (L.W. Oberley, ed.), CRC Press, Boca Raton, FL, pp. 89–123.

Halliwell, B. and Gutteridge, J.M.C. (1985) *Free Radicals in Biology and Medicine*. Clarendon Press, Oxford.

Halliwell, B. and Gutteridge, J.M.C. (1986) Oxygen free radicals and iron in relation to biology and medicine: some problems and concepts. *Arch. Biochem. Biophys.* **246**, 501–514.

Harper, D.B. and Harvey, B.M.R. (1978) Mechanism of paraquat tolerance in perennial ryegrass. II. role of superoxide dismutase, catalase and peroxidase. *Plant Cell Environ.* **1**, 211–215.

Hausladen, A. and Alscher, R.G. (1993) Glutathione. In *Antioxidants in Higher Plants* (R.G. Alscher and J.L. Hess, eds.), CRC Press, Boca Raton, FL, pp. 1–30.

Heath, R.L. and Packer, L. (1968) Photoperoxidation in isolated chloroplasts. I. Kinetics and stoichiometry of fatty acid peroxidation. *Arch. Biochem. Biophys.* **125**, 189–198.

Herbert, S.K., Samson, G., Fork, D.C. and Laudenbach, D.E. (1992) Characterization of damage to photosystems I and II in a cyanobacterium lacking detectable iron superoxide dismutase activity. *Proc. Natl. Acad. Sci. USA* **89**, 8716–8720.

Hérouart, D., Van Montagu, M. and Inzé, D. (1993) Redox-activated expression of the cytosolic copper/zinc superoxide dismutase gene in *Nicotiana. Proc. Natl. Acad. Sci. USA* **90**, 3108–3112.

Hess, J.L. (1993) Vitamin E, α-tocopherol. In *Antioxidants in Higher Plants*, (R.G. Alscher and J.L. Hess, eds.), CRC Press, Boca Raton, FL, pp. 111–134.

Hossain, M.A. and Asada, K. (1984) Purification of dehydroascorbate reductase from spinach and its characterization as a thiol enzyme. *Plant Cell Physiol.* **25**, 85–92.

Hossain, M.A. and Asada, K. (1985) Monodehydroascorbate reductase from cucumber fruit is a flavin adenine dinucleotide enzyme. *J. Biol. Chem.* **260**, 12920–12926.

Hossain, M.A., Nakano, Y. and Asada, K. (1984) Monodehydroascorbate reductase in spinach chloroplasts and its participation in regeneration of ascorbate for scavenging hydrogen peroxide. *Plant Cell Physiol.* **25**, 385–395.

Hrazdina G. and Wagner, G.J. (1985) Compartmentation of plant phenolic compounds; sites of synthesis and accumulation. *Ann. Proc. Phytochem. Soc. Eur.* **25**, 119–133.

Jackson, C., Dench, J., Moore, A.L., Halliwell, B., Foyer, C.H. and Hall, D.O. (1978) Subcellular localisation and identification of superoxide dismutase in the leaves of higher plants. *Eur. J. Biochem.* **91**, 339–344.

Janero, D.R. (1990) Malondialdehyde and thiobarbituric acid-reactivity as diagnostic indices of lipid peroxidation and peroxidative tissue damage. *Free Rad. Biol. Med.* **9**, 515–540.

Jansen, M.A.K., Shaaltiel, Y., Kazzes, D., Canaani, O., Malkin, S. and Gressel, J. (1989) Increased tolerance to photoinhibitory light in paraquat-resistance *Conyza bonariensis* measured by photoacoustic spectroscopy and $^{14}CO_2$-fixation. *Plant Physiol.* **91**, 1174–1178.

Ji, L., Becana, M. and Klucas, R.V. (1992) Involvement of molecular oxygen in the enzyme-catalyzed NADH oxidation and ferric leghemoglobin reduction. *Plant Physiol.* **100**, 33–39.

Ji, L., Wood, S., Becana, M. and Klucas, R.V. (1991) Purification and characterization of soybean root nodule ferric leghemoglobin reductase. *Plant Physiol.* **96**, 32–37.

Johnson, G.V., Evans, H.J. and Temay, C. (1966) Enzymes of the glyoxylate cycle in rhizobia and nodules of legumes. *Plant Physiol.* **41**, 1330–1336.

Kalt-Torres, W., Burke, J.J. and Anderson, J.M. (1984) Chloroplast glutathione reductase: purification and properties. *Physiol. Plant.* **61**, 271–278.

Kanematsu, S. and Asada, K. (1990) Characteristic amino acid sequences of chloroplast and cytosol isozymes of CuZn-superoxide dismutase in spinach, rice and horsetail. *Plant Cell Physiol.* **31**, 99–112.

Kaput, J., Goltz, S. and Blobel, G. (1982) Nucleotide sequence of the yeast nuclear gene for cytochrome c peroxidase precursor. *J. Biol. Chem.* **257**, 15054–15058.

Kasting, J.F. (1993) Earth's early atmosphere. *Science* **259**, 920–926.

Kidambi, S.P., Mahan, J.R. and Matches, A.G. (1990) Purification and thermal dependence of glutathione reductase from two forage legume species. *Plant Physiol.* **92**, 363–367.

Klapheck, S. (1988) Homoglutathione: isolation, quantification and occurrence in legumes. *Physiol. Plant.* **74**, 727–732.

Klapheck, S., Chrost, B., Starke, J. and Zimmermann, H. (1992) γ-glutamylcysteinylserine—a new homologue of glutathione in plants of the family Poaceae. *Bot. Acta.* **105**, 174–179.

Ko, M.P., Huang, P.Y., Huang, J.S. and Barker, K.R. (1987) The occurrence of phytoferritin and its relationship to effectiveness of soybean nodules. *Plant Physiol.* **83**, 299–305.

Krauth-Siegel, R.L., Enders, B., Henderson, G.B., Fairlamb, A.H. and Schirmer, R.H. (1987) Trypanothione reductase from *Trypanosoma cruzi*. Purification and characterization of the crystalline enzyme. *Eur. J. Biochem.* **164**, 123–128.

Kubo, A., Saji, H., Tanaka, K., Tanaka, K. and Kondo, N. (1992) Cloning and sequencing of a cDNA encoding ascorbate peroxidase from *Arabidopsis thaliana*. *Plant Mol. Biol.* **18**, 691–701.

Kubo, A., Saji, H., Tanaka, K. and Kondo, N. (1993) Genomic DNA structure of a gene encoding cytosolic ascorbate peroxidase from *Arabidopsis thaliana*. *FEBS Lett.* **315**, 313–317.

Lamb, C.J., Lawton, M.A., Dron, M. and Dixon, R.A. (1989) Signals and transduction mechanisms for activation of plant defenses against microbial attack. *Cell* **56**, 215–224.

Larson, R.A. (1988) The antioxidants of higher plants. *Phytochemistry* **27**, 969–978.

Layzell, D.B. and Hunt, S. (1990) Oxygen and the regulation of nitrogen fixation in legume nodules. *Physiol. Plant.* **80**, 322–327.

Layzell, D.B., Hunt, S. and Palmer, G.R. (1990) The mechanism of nitrogenase inhibition in soybean nodules. Pulse-modulated spectroscopy indicates that nitrogenase activity is limited by oxygen. *Plant Physiol.* **87**, 296–299.

Leech, R.M. and Murphy, D.J. (1976) The cooperative function of chloroplasts in the biosynthesis of small molecules. In *The Intact Chloroplast* (J. Barber, ed.), Elsevier/North Holland, Amsterdam, pp. 365–401.

Lesser, M.P. and Shick, J.M. (1989) Effects of irradiance and ultraviolet radiation on photoadaptation in the *Zooxanthellae* of *Aiptasia pallida*: primary production, photoinhibition and enzymatic defenses against oxygen toxicity. *Marine Biol.* **102**, 243–255.

Leung, H.-W., Vang, M.J. and Mavis, R.D. (1981) The co-operative interaction between vitamin E and vitamin C in suppression of peroxidation of membrane phospholipids. *Biochim. Biophys. Acta* **664**, 266–272.

Lewis, N.G. (1993) Plant phenolics. In *Antioxidants in Higher Plants* (R.G. Alscher and J.L. Hess, eds.), CRC Press, Boca Raton, FL, pp. 135–169.

Loewus, F. (1980) L-ascorbic acid: metabolism, biosynthesis, function. In *The Biochemistry of Plants*, Vol. 3 (P.K. Stumpf and E.E. Conn, eds.), Academic Press, New York, pp. 77–99.

Loewus, F.A. and Loewus, M.W. (1987) Biosynthesis and metabolism of ascorbic acid in plants. *CRC Crit. Rev. Plant Sci.* **5**, 101–119.

Loomis, W.D. (1974) Overcoming problems of phenolics and quinones in the isolation of plant enzymes and organelles. *Methods Enzymol.* **31**, 528–544.

Lorimer, G.H. (1981) The carboxylation and oxygenation of ribulose-1,5-bisphosphate: the primary events in photosynthesis and photorespiration. *Ann. Rev. Plant Physiol.* **32**, 349–383.

Madamanchi, N.R., Anderson, J.V., Alscher, R.G., Cramer, C.L. and Hess, J.L. (1992) Purification of multiple forms of glutathione reductase from pea (*Pisum sativum* L.) seedlings and enzyme levels in ozone-fumigated pea leaves. *Plant Physiol.* **100**, 138–145.

Mahan, J.R. and Burke, J.J. (1987) Purification and characterization of glutathione reductase from corn mesophyll chloroplasts. *Physiol. Plant.* **71**, 352–358.

Massey, V., Strickland, S., Mayhew, S.G., Howell, L.G., Engel, P.C., Matthews, R.G., Schuman, M. and Sullivan, P.A. (1969) The production of superoxide anion radicals in the reaction of reduced flavins and flavoproteins with molecular oxygen. *Biochem. Biophys. Res. Commun.* **36**, 891–897.

Massey, V. and Wiliams C.H. (1965) On the reaction mechanism of yeast glutathione reductase. *J. Biol. Chem.* **240**, 4470–4480.

McCay, P.B. and Powell, S.R. (1989) Relationship between glutathione and chemically induced lipid peroxidation. In *Glutathione, Coenzymes and Cofactors* (D. Dolphin, R. Poulson and O. Avramovic, eds.), John Wiley, New York, pp. 111–151.

Mehler, A.H. (1951) Studies on reactions of illuminated chloroplasts. I. Mechanism of the reduction of oxygen and other Hill reagents. *Arch. Biochem. Biophys.* **33**, 65–77.

Meister, A. (1983) Glutathione. *Ann. Rev. Biochem.* **52**, 711–760.

Meister, A. (1983) Glutathione metabolism and its selective modification. *J. Biol. Chem.* **263**, 17205–17208.

Misra, H.P. and Fridovich, I. (1971) The generation of superoxide radical during the autoxidation of ferredoxin. *J. Biol. Chem.* **246**, 6886–6890.

Mittler, R. and Tel-Or, E. (1991) Oxidative stress responses and shock proteins in the unicellular cyanobacterium *Synechococcus* R2 (PCC-7942). *Arch. Microbiol.* **155**, 125–130.

Mittler, R. and Zilinskas, B.A. (1991a) Purification and characterization of pea cytosolic ascorbate peroxidase. *Plant Physiol.* **97**, 962–968.

Mittler, R. and Zilinskas, B.A. (1991b) Molecular cloning and nucleotide sequence analysis of a cDNA encoding pea cytosolic ascorbate peroxidase. *FEBS Lett.* **289**, 257–259.

Mittler, R. and Zilinskas, B.A. (1992) Molecular cloning and characterization of a gene encoding pea cytosolic ascorbate peroxidase. *J. Biol. Chem.* **267**, 21802–21807.

Miyake, C., Michihata, F. and Asada, K. (1991) Scavenging of hydrogen peroxide in prokaryotic and eukaryotic algae: acquisition of ascorbate peroxidase during the evolution of cyanobacteria. *Plant Cell Physiol.* **32**, 33–43.

Mortensen, L.E., Walker, M.N. and Walker, G.A. (1974) Effect of magnesium di- and triphosphates on the structure and electron transport function of the components of clostridial nitrogenase. In *Proceedings of the first International Symposium on Nitrogen Fixation* (W.E. Newton and C.J. Nyman, eds.), Washington State Univ. Press, Pullman, pp. 117–149.

Murakawa, S., Sano, S., Yamashita, H. and Takahashi, T.(1977) Biosynthesis of D-erythroascorbic acid by *Candida*. *Agric. Biol. Chem.* **41**, 1799–1806.

Nakano, Y. and Asada, K. (1987) Purification of ascorbate peroxidase in spinach chloroplasts; its inactivation in ascorbate-depleted medium and reactivation by monodehydroascorbate radical. *Plant Cell Physiol.* **28**, 131–140.

Neeley, W.C., Martin, J.M. and Barker, S.A. (1988) Products and relative reaction rates of the oxidation of tocopherols with singlet molecular oxygen. *Photochem. Photobiol.* **48**, 423–428.

Nick, J.A., Leung, C.T. and Loewus, F.A. (1986) Isolation and identification of erythroascorbic acid in *Saccharomyces cerevisiae* and *Lypomyces starkeyi*. *Plant Sci.* **46**, 181–187.

Niki, E., Tsuchiya, J., Tanimura, R. and Kamiya, Y. (1982) Regeneration of vitamin E from α-chromanoxyl radical by glutathione and vitamin C. *Chem. Lett.* **6**, 789–792.

Nishikimi, M. (1975) Oxidation of ascorbic acid with superoxide ion generated by the xanthine-xanthine oxidase system. *Biochem. Biophys. Res. Commun.* **63**, 463–468.

Nishikimi, M., Yamada, H. and Yagi, K. (1980) Oxidation by superoxide of tocopherols dispersed in aqueous media with deoxycholate. *Biochim. Biophys. Acta* **627**, 101–108.

Ogren, W.L. (1984) Photorespiration; pathways, regulation, and modification. *Ann. Rev. Plant Physiol.* **35**, 415–442.

Packer, J.E., Slater, T.F. and Wilson, R.L. (1979) Direct observation of a free radical interaction between vitamin E and vitamin C. *Nature* **278**, 737–738.

Pallett, K.E. and Young, A.J. (1993) Carotenoids. In *Antioxidants in Higher Plants* (R.G. Alscher and J.L. Hess, eds.), CRC Press, Boca Raton, FL, pp. 59–89.

Pierpoint, W.S. (1985) Phenolics in food and feedstuffs: the pleasures and perils of vegetarianism. *Ann. Proc. Phytochem. Soc. Eur.* **25**, 427–451.

Pitcher, L.H., Brennan, E., Hurley, A., Dunsmuir, P., Tepperman, J.M. and Zilinskas, B.A. (1991) Overproduction of petunia chloroplastic copper/zinc superoxide dismutase does not confer ozone tolerance in transgenic tobacco. *Plant Physiol.* **97**, 452–455.

Postgate, J.R. (1982) *The Fundamentals of Nitrogen Fixation*. Cambridge Univ. Press, Cambridge.

Poulos, T.L., Freer, S.T., Alden, R.A., Edwards, S.L., Skogland, U., Takio, K., Eriksson, B., Xuong, N.H., Yonetani, T. and Kraut, J. (1980) The crystal structure of cytochrome *c* peroxidase. *J. Biol. Chem.* **255**, 575–580.

Puppo, A., Dimitrijevic, L. and Rigaud, J. (1982) Possible involvement of nodule superoxide dismutase and catalase in leghemoglobin protection. *Planta* **156**, 374–379.

Puppo, A. and Halliwell, B. (1988) Generation of hydroxyl radicals by soybean nodule leghaemoglobin. *Planta* **173**, 405–410.

Puppo, A., Herrada, G. and Rigaud, J. (1991) Lipid peroxidation in peribacteroid membranes from french-bean nodules. *Plant Physiol.* **96**, 826–830.

Puppo, A., Monny, C. and Davies, M.J. (1993) Glutathione-dependent conversion of ferryl leghaemoglobin into the ferric form: a potential protective process in soybean (*Glycine max*) root nodules. *Biochem. J.* **289**, 435–438.

Puppo, A. and Rigaud, J. (1986) Superoxide dismutase: an essential role in the protection of the nitrogen fixation process? *FEBS Letters* **210**, 187–189.

Puppo, A., Rigaud, J., Job, D., Ricard, J. and Zeba, B. (1980) Peroxidase content of soybean root nodules. *Biochim. Biophys. Acta* **614**, 303–312.

Puppo, A., Rigaud, J. and Job, D. (1981) Role of superoxide anion in leghemoglobin autoxidation. *Plant Sci. Lett.* **22**, 353–360.

Rabinowitch, H.D. and Fridovich, I. (1983) Superoxide radicals, superoxide dismutases and oxygen toxicity in plants. *Photochem. Photobiol.* **37**, 679–690.

Rennenberg, H. (1982) Glutathione metabolism and possible biological roles in higher plants. *Phytochemistry* **21**, 2771–2781.

Rich, P.R. and Bonner, W.D. (1978) The sites of superoxide anion generation in higher plant mitochondria. *Arch. Biochem. Biophys.* **188**, 206–213.

Robson, R.L. and Postgate, J.R. (1980) Oxygen and hydrogen in biological nitrogen fixation. *Ann. Rev. Microbiol.* **34**, 183–207.

Salin, M.L. (1988) Plant superoxide dismutases: a means of coping with oxygen radicals. *Curr. Top. Plant Biochem. Physiol.* **7**, 188–200.

Sano, S. and Asada, K. (1992) Molecular properties of monodehydroascorbate reductase. In *Research in Photosynthesis*, Vol IV (N. Murata, ed.), Kluwer, Dordrecht, Netherlands, pp. 533–536.

Scandalios, J.G. (1990) Response of plant antioxidant defense genes to environmental stress. *Adv. Gene.* **28**, 1–41.

Schirmer, R.H., Krauth-Siegel, R.L. and Schulz, G.E. (1989) Glutathione reductase. In *Glutathione: Biochemical & Medical Aspects, Part A* (D. Dolphin, R. Poulson and O. Avramovic, eds.), John Wiley, New York, pp. 554–596.

Schneider, K. and Schlegel, H.G. (1981) Production of superoxide radicals by soluble hydrogenase from *Alcaligenes eutrophus* H16. *Biochem. J.* **193**, 99–107.

Schubert, K.R. (1982) *The Energetics of Biological Nitrogen Fixation.* Workshop Summary. Amer. Soc. Plant Physiol., Rockville, MD.

Schubert, K.R. (1986) Products of biological nitrogen fixation in higher plants: synthesis, transport, and metabolism. *Ann. Rev. Plant Physiol.* **37**, 539–574.

Schulze, A.U., Schott, H.M. and Staudinger, H. (1972) Isolierung und charakterisierung einer NADH:semidehydroascorbinsäure-oxidoreduktase aus *Neurospora crassa* (E.C. 1.6.5.4). *Hoppe-Seyler's Z. Physiol. Chem.* **353**, 1931–1942.

Scott, M.D., Meshnick, S.R. and Eaton, J.W. (1987) Superoxide dismutase-rich bacteria. Paradoxical increase in oxidant toxicity. *J. Biol. Chem.* **262**, 3640–3645.

Seckback, J. (1982) Ferreting out the secrets of plant ferritin. A review. *J. Plant Nutr.* **5**, 369–394.

Seel, W.E., Hendry, G.A.F. and Lee, J.A. (1992) Effects of desiccation on some activated oxygen processing enzymes and anti-oxidants in mosses. *J. Exp. Bot.* **43**, 1031–1037.

Serrano, A., Rivas, J. and Losada, M. (1984) Purification and properties of glutathione reductase from the cyanobacterium *Anabaena sp.* strain 7119. *J. Bacteriol.* **158**, 317–324.

Shaaltiel, Y. and Gressel, J. (1986) Multienzyme oxygen radical detoxifying system correlated with paraquat resistance in *Conya bonariensis. Pesticide Biochem. Physiol.* **26**, 22–28.

Sheehy, J.E. (1987) Photosynthesis and nitrogen fixation in legume plants. *CRC Crit. Rev. Plant Sci.* **5**, 121–159.

Shigeoka, S., Nakano, Y. and Kitaoka, S. (1980) Purification and some properties of L-ascorbic acid-specific peroxidase in *Euglena gracilis* z. Arch. *Biochem. Biophys.* **201**, 121–127.

Shigeoka, S., Onishi, T., Nakano, Y. and Kitaoka, S. (1987) Characterization and physiological function of glutathione reductase in *Euglena gracilis* z. *Biochem J.* **242**, 511–515.

Shigeoka, S., Yasumoto, R., Onishi, T., Nakano, Y. and Kitaoka, S. (1987) Properties of monodehydroascorbate reductase and dehydroascorbate reductase and their participation in the regeneration of ascorbate in *Euglena gracilis* z. *J. Gen. Microbiol.* **133**, 227–232.

Siedow, J.N. (1982) The nature of the cyanide-resistant pathway in plant mitochondria. In *Recent Advances in Phytochemistry* (L.L. Creasy and G. Hrazdina, eds.), Plenum Press, New York, pp. 47–83.

Siedow, J.N. (1991) Plant lipoxygenase: structure and function. *Ann. Rev. Plant Physiol. Plant Mol. Biol.* **42** , 145–188.

Siefermann-Harms, D. (1985) Carotenoids in photosynthesis. I. Location in photosynthetic membranes and light-harvesting function. *Biochim. Biophys. Acta* **811**, 325–335.

Sinclair, T.R., Weisz, P.R. and Denison, R.F. (1985) Oxygen limitation to nitrogen fixation in soybean nodules. In *World Soybean Research Conference III: Proceedings* (R. Shibles, ed.), Westview Press, Boulder, CO, pp. 797–805.

Smith, I.K., Vierheller, T.L. and Thorne, C.A. (1989) Properties and functions of glutathione reductase in plants. *Physiol. Plant.* **77**, 449–456.

Smith, J. and Shrift, A. (1979) Phylogenetic distribution of glutathione peroxidase. *Comp. Biochem. Physiol.* **63B**, 39–44.

Sprent, J.I. and Sprent, P. (1990) *Nitrogen Fixing Organisms: Pure and Applied Aspects.* Chapman and Hall, London.

Steinman, H.M. (1982) Superoxide dismutases: protein chemistry and structure-function relationships. In *Superoxide Dismutase Vol. I* (L.W. Oberley, ed.), CRC Press, Boca Raton, FL, pp. 11–68.

Stonier, T., Stasions, S. and Reddy, K.B.S.M. (1979) The masking of peroxidase-catalyzed oxidation of IAA in *Vigna. Phytochem.* **18**, 25–28.

354   Oxidative Stress and Antioxidant Defenses in Biology

Stowers, M.D. and Elkan, G.H. (1981) An inducible iron-containing superoxide dismutase in *Rhizobium japonicum*. *Can. J. Microbiol.* **27**, 1202–1208.

Takahashi, T., Yamashita, H., Kato, E., Mitsumoto, M. and Murakawa, S. (1976) Purification and some properties of D-glucono-γ-lactone dehydrogenase. D-erythorbic acid producing enzyme of *Penicillium cyaneo-fulvum*. *Agric. Biol. Chem.* **40**, 121–129.

Takio, K., Titani, K., Ericsson, L.H. and Yonetani, T. (1980) Primary structure of yeast cytochrome *c* peroxidase. II. The complete amino acid sequence. *Arch. Biochem. Biophys.* **203**, 615–629.

Tanaka, K., Takeuchi, E., Kubo, A., Sakaki, T., Haraguchi, K. and Kawamura, Y. (1991) Two immunologically different isozymes of ascorbate peroxidase from spinach leaves. *Arch. Biochem. Biophys.* **286**, 371–375.

Tang, X. and Webb, M. A. (1993) Cloning and characterization of cDNAs encoding glutathione reductase from soybean root nodules. In *New Horizons in Nitrogen Fixation* (R. Palacios, J. Mora and W. E. Newton, eds.), Kluwer Academic ,Dordrecht, p. 374.

Tel-Or, E., Huflejt, M. and Packer, L., (1985) The role of glutathione and ascorbate in hydroperoxide removal in cyanobacteria. *Biochem. Biophys. Res. Comm.* 132, 533–539.

Tel-Or, E., Huflejt, M. E. and Packer, L., (1986) Hydroperoxide metabolism in cyanobacteria. *Arch. Biochem. Biophys.* 246, 396–402.

Tepperman, J.M. and Duinsmuir, P. (1990) Transformed plants with elevated levels of chloroplastic SOD are not more resistant to superoxide toxicity. *Plant Mol. Biol.* **14**, 501–511.

Thayer, S.S. and Björkman, O. (1990) Leaf xanthophyll content and composition in sun and shade determined by HPLC. *Photosyn. Res.* **23**, 331–343.

Ting, I.P. (1985) Crassulacean acid metabolism. *Ann. Rev. Plant Physiol.* **36**, 595–622.

Tjepkema, J.D., Schwintzer, C.R. and Benson, D.R. (1986) Physiology of actinorhizal nodules. *Ann. Rev. Plant Physiol.* **37**, 209–232.

Tjepkema, J.C. and Yocum, C.W. (1974) Measurement of oxygen partial pressure within soybean nodules by oxygen micro-electrodes. *Planta* **119**, 351–360.

Tolbert, B.M. and Ward, J.B. (1982) Dehydroascorbic acid. In *Ascorbic Acid: Chemistry, Metabolism, and Uses* (P.A. Seib and B.M. Tolbert, eds.), Adv. in Chem. Ser. 200. Am. Chem. Soc. Washington, DC, pp. 81–100.

Tözüm, S.R.D. and Gallon, J.R. (1979) The effects of methyl viologen on *Gleocapsa sp.* LB795 and their relationship to the inhibition of acetylene reduction (nitrogen fixation) by oxygen. *J. Gen. Microbiol.* **111**, 313–326.

Tsai, Y.-C., Yang, T.-Y., Cheng, S.-W., Li, S.-N. and Wang, Y.-J. (1991) High yield extraction and purification of glutathione reductase from baker's yeast. *Prep. Biochem.* **21**, 175–185.

Tsang, E.W.T., Bowler, C., Hérouart, D., Van Camp, W., Villarroel, R., Genetello, C., Van Montagu, M. and Irké, D. (1991) Differential regulation of superoxide dismutases in plants exposed to environmental stress. *Plant Cell* **3**, 783–792.

Van Camp, W., Bowler, C., Villarroel, R., Tsang, E.W.T., Van Montagu, M. and Inzé, D. (1990) Characterization of iron superoxide dismutase cDNAs from plants obtained by genetic complementation in *Escherichia coli. Proc. Natl. Acad. Sci. USA* **87**, 9903–9907.

Van den Bosch, K.A. and Newcomb, E.H. (1986) Immunogold localization of nodule-specific uricase in developing soybean root nodules. *Planta* **167**, 425–436.

Vera-Estrella, R., Blumwald, E. and Higgins, V.J. (1992) Effect of specific elicitors of *Cladosporium fulvum* on tomato suspension cells. *Plant Physiol.* **99**, 1208–1215.

Verduyn, C., van Wijngaarden, C.J., Scheffers, W.A. and van Dijken, J.P. (1991) Hydrogen peroxide as an electron acceptor for mitochondrial respiration in the yeast *Hansenula polymorpha. Yeast* **7**, 137–146.

Vivekanadan, M. and Edwards, G.E. (1987) Activation of NADP-malate dehydrogenase in C3 plants by reduced glutathione. *Photosyn. Res.* **14**, 113–124.

Weisz, P.R., Denison, R.F. and Sinclair, T.R. (1985) Response to drought stress of nitrogen fixation (acetylene reduction) rates by field-grown soybeans. *Plant Physiol.* **78**, 525–530.

White, J.A. and Scandalios, J.G. (1988) Isolation and characterization of a cDNA for mitochondrial manganese superoxide dismutase (SOD-3) of maize and its relation to other manganese superoxide dismutases. *Biochim. Biophys. Acta* **951**, 61–70.

Williams, C.H. (1992) Lipoamide dehydrogenase, glutathione reductase, thioredoxin reductase, and mecuric ion reductase-—a family of flavoenzyme transhydrogenases. In *Chemistry and Biochemistry of Flavoenzymes, Vol. III* (F. Müller, ed.), CRC Press, Boca Raton, FL, pp. 121–211.

Williams, P.G. and Stewart, P.R. (1976) The intramitochondrial location of cytochrome *c* peroxidase in wild-type and petite *Saccharomyces cerevisiae. Arch. Microbiol.* **107**, 63–70.

Wingate, V.P.M., Lawton, M.A. and Lamb, C.J. (1988) Glutathione causes a massive and selective induction of plant defense genes. *Plant Physiol.* **87**, 206–210.

Wingsle, G. (1989) Purification and characterization of glutathione reductase from Scots pine needles. *Physiol. Plant.* **76**, 24–30.

Wise, R.R. and Naylor, A.W. (1987) Chilling-enhanced photooxidation. The peroxidative destruction of lipids during chilling injury to photosynthesis and ultrastructure. *Plant Physiol.* **83**, 272–277.

Yagi, K. (1982) *Lipid Peroxides in Biology and Medicine.* Academic Press, New York.

Yamada, T., Hashimoto, H., Shiraishi, T. and Oku, H. (1989) Suppression of pisatin, phenylalanine ammonia-lyase mRNA, and chalcone synthase mRNA accumulation by a putative pathogenicity factor from the fungus *Mycosphaerella pinodes. Mol. Plant. Microb. Interact.* **2**, 256–261.

Yamauchi, N., Yamawaki, K., Ueda, Y. and Chachin, K. (1984) Subcellular localization of redox enzymes involving ascorbic acid in cucumber fruit. *J. Jpn. Soc. Hort. Sci.* **53**, 347–353.

Yokota, A., Shigeoka, S., Onishi, T. and Kitaoka, S. (1988) Selenium as inducer of glutathione peroxidase in low-CO_2-grown *Chlamydomonas reinhardtii. Plant Physiol.* **86**, 649–651.

Yonetani, T. (1976) Cytochrome c peroxidase. In *The Enzymes, Vol. XIII* (P.D. Boyer, ed.), Academic Press, New York, pp. 345–361.

Young, J.A. (1991) The protective role of carotenoids in higher plants. *Physiol. Plant.* **83**, 702–708.

CHAPTER 10

Oxidative Stress of Vertebrates and Invertebrates

Gary W. Felton

10.1 INTRODUCTION

"Thanks to green plants, oxidation is the fate that awaits us all."
George Hendry (1992)

Animals are at the mercy of green plants. Plants are releasing O_2 at rates faster than can be consumed by animal respiration. Approximately 1.3 billion years ago, the percentage of atmospheric O_2 was 1% of its current level of 22% (Hendry, 1992; Harman 1992). Living in a world of increased O_2 is costly—almost any chemical process that requires O_2 may produce injurious reactive oxygen species (ROS). Protection from ROS is of fundamental biological importance regardless of whether an animal's life span is measured in days or decades.

I will discuss the antioxidant defenses of animals using the categories described by Fridovich (1989). The first and most effective defense is *avoidance* which may be achieved via behavioral or biochemical means. The second strategy is *enzymatic removal* of ROS provided by enzymes such as superoxide dismutase (SOD) and catalase (CAT). The third line of defense is *prevention* and involves antioxidants which prevent initiation and/or propagation of free radical chain reactions. The final defense is to *repair* oxidative damage by removing damaged lipid, protein or DNA molecules via phospholipases, proteases, etc.

Due to the rapid emergence of this field (e.g., Halliwell and Gutteridge, 1989; Czapski, 1991; Arouma, 1993), it is impossible to make this a comprehensive review. While a virtual explosion of biomedical research on oxidative stress has occurred, an appreciation of oxidative stress in lower vertebrates and invertebrates has been hampered by a lack of clearly defined diseases or pathologies associated with the phenomenon. Additionally, there has been a perception that oxidative assault is a chronic process that may be primarily important for relatively long-lived organisms. I will highlight a few examples of animals from diverse taxa that demonstrate the importance of antioxidant systems to the biology of these organisms.

10.2 AVOIDANCE OF OXIDATIVE STRESS

10.2.1 Behavioral

Animals may employ behavioral defenses to avoid dietary prooxidants. Several insect species avoid the toxicity of photoactivated prooxidants by rolling up the leaves of plants rich in coumarins (e.g., family Apiaceae) and feeding within them so that activation by sunlight is avoided (Berenbaum, 1978). Larvae of the chrysomelid beetle, *Chrysolina hyperici*, which feed on the phototoxic *Hypericum perforatum*, a plant containing hypericin, display light avoidance behavior (Fields et al., 1990). Larvae belonging to the moth family Tortricidae use leaf-tying to avoid light when feeding upon this plant (Sandberg and Berenbaum, 1989). The cabbage looper, *Trichoplusia ni*, feeds on the underside of celery leaves thus avoiding the photoactivation of furanocoumarins (Ahmad, 1992).

10.2.2 Biochemical

Biochemical avoidance occurs via enzymes such as cytochrome oxidase that can carry out the tetravalent reduction of molecular oxygen without releasing reactive intermediates (Fridovich, 1989). Cytochrome oxidase is responsible for most of the reduction of molecular oxygen in respiring cells and thus eliminates much of the production of ROS. Inhibition of cytochrome C oxidase leads to the stimulation of mitochondrial H_2O_2 production (Sohal, 1993). Cytochrome oxidase activity declines during aging in many organisms (e.g., insects and mammals) and may be responsible for the increased rate of ROS formation in mitochondria (Sohal, 1993).

10.3   ENZYMATIC REMOVAL OF ROS

The second line of antioxidant defense includes the enzymes superoxide dismutase (SOD), catalase (CAT), glutathione peroxidase (GPOX), and glutathione transferase (GST) (see chapter 7). These enzymes may directly remove intermediates of dioxygen reduction or remove damaging oxidants.

10.3.1   Superoxide Dismutase

SOD (EC 1.15.1.1) protects against oxidative damage by catalyzing dismutation of O_2^- to O_2 and H_2O_2 (Fridovich, 1989; Scandalios, 1993).

$$2O_2^- + H^+ \longrightarrow O_2 + H_2O_2 \tag{10.1}$$

The rate is diffusion limited and enhanced by electrostatic guidance, making SOD one of the fastest enzymes known (V_{max} of $\sim 2 \times 10^9$ M^{-1} s^{-1}; Getzoff et al., 1992). Reviews on the biochemistry, regulation, and genetics of SOD are available (Fridovich, 1989; Harris, 1992a; Harris, 1992b; Scandalios, 1993). My focus is recent papers showing the importance of SOD to invertebrate and vertebrate biology.

Three distinct SODs are known and may be derived from two evolutionary families (Fridovich, 1989). The manganese-containing superoxide dismutases (MnSOD) occur in prokaryotes and in eukaryotic mitochondria (Fridovich, 1989). The related iron-containing superoxide dismutases (FeSOD) occur in prokaryotes and in a few plant families. Amino acid sequence data have affirmed a close relationship between MnSOD and FeSOD (Fridovich, 1989). The unrelated family of SODs are the copper and zinc enzymes (CuZnSOD) found in the cytosol of eukaryotes, in plant chloroplasts, and in a few bacteria (Fridovich, 1989; Liou et al., 1993). In most cases, CuZnSOD occurs in the cytosol, nucleus, and lysosomes (Liou et al., 1993). All SODs characterized to date are multimeric: cytosolic CuZnSOD, prokaryotic MnSOD and FeSOD are homodimers and the mitochondrial MnSOD is a homotetramer (Fridovich, 1989). The glycosylated CuZnSOD from mammalian extracellular fluids (e.g., plasma, lymph, and synovial fluids) is tetrameric (Halliwell and Gutteridge, 1990b). The extracellular SOD (EC-SOD) is the major SOD isozyme in the extracellular space and possesses heterogeneous affinity for heparin (Adachi and Ohta, et al., 1992; Adachi and Kodera, et al., 1992). The enzyme affinity for heparin is divided into fraction A without affinity, fraction B with intermediate affinity, and fraction C with high affinity (Adachi and Kodera, et al., 1992).

i. Phylogenetic distribution

The MnSOD and CuZnSOD predominate in animals. FeSOD is usually restricted to prokaryotes, but has also been found in some algae and higher plants. Blum and Fridovich (Blum and Fridovich, 1984) discovered FeSOD in tissues of a deep-sea hydrothermal vent animal, the tube worm *Riftia pachyptila*. This was the first report of an FeSOD in animals and it was found that the enzyme was associated with symbiotic bacteria in the trophosomes of the tubeworm. Due to the large number of symbiotic relationships between animals and bacteria, it would not be surprising if FeSOD occurs in many other animal tissues. The importance of this enzyme to the survival of the host organism is unknown.

SOD activity varies widely in invertebrates, but is generally lower than in vertebrates with a few notable exceptions (e.g., symbiotic relationships; Livingstone et al., 1992). Invertebrate CuZnSODs are similar to the mammalian enzymes. The adult helminth *Trichinella spiralis* possesses two molecular forms of CuZnSOD, each comprised of two 17 kDa units (Rhoads, 1983). Rhoads (1983) found that the SOD was excreted in media and may exist as an extracellular SOD in vivo. Two isoforms of CuZnSOD occur in the worm *Ascaris suum* with molecular weights of 39.8 and 42.6 kDa (Sanchez-Moreno et al., 1989). A MnSOD (73 kDa) was also characterized from the mitochondria (Sanchez-Moreno et al., 1989). Several CuZnSOD isoforms from the worm *Caeonorhabditis elegans* exist as homodimers of 37.5 to 40 kDa molecular mass (Vanfleteren, 1992).

Insect hemocytes are similar to vertebrate erythrocytes and leukocytes in possessing primarily CuZnSOD (Ahmad et al., 1991). In the larval cabbage looper moth *Trichoplusia ni*, MnSOD predominates in all tissues except hemocytes (Ahmad, 1991). This is in contrast to vertebrates where cytosolic CuZnSOD prevails over the MnSOD of mitochondria in most tissues (Ahmad et al., 1988a). Conversely, in the larval southern armyworm, *Spodoptera eridania*, the cytosolic CuZnSOD exceeds mitochondrial MnSOD levels (Ahmad et al., 1988b). The CuZnSOD makes up >86% of total SOD in whole body homogenates of the Mediterranean fruit fly *Ceratitis capita* throughout development (Fernandez-Sousa and Michelson, 1976). However, MnSOD activity increases significantly during development, peaking at the adult stage when utilization of energy peaks during flight (Fernandez-Sousa and Michelson, 1976). Thus it is difficult to generalize about the relative importance of different SOD forms among insects due to the wide variations in the distribution of these enzymes.

The SODs of lower vertebrates have been studied (Van Balgooy and Roberts, 1979; Abe et al., 1984; Montesano et al., 1989; Perez-Campo et al., 1993). SODs from amphibians show the highest activity in the brain and the lowest in the lung, in contrast to higher vertebrates where liver activity is generally highest (Perez-Campo et al., 1993). The CuZnSOD of the bull frog *Rana catesbeiana* shows exceptional amino acid homology with human, bovine, and yeast SODs (Abe et al., 1984). Montesano et al. (1989) were the first to report the complete nucleotide sequence of an amphibian CuZnSOD gene in *Xenopus laevis*. The CuZnSOD gene from *Xenopus* showed a 66% similarity with the human SOD gene (Montesano et al., 1989). CuZnSOD was first characterized from bovine erythrocytes (McCord and Fridovich, 1969) and the CuZnSOD gene has now been characterized in several mammalian species including humans, rodents, pigs, horses, sheep, dogs, cats, donkeys, and others (Marklund, 1984; Halliwell and Gutteridge, 1989; Hsu et al., 1992). These studies on SOD gene structure reveal that the CuZnSOD gene is highly conserved, suggesting a common ancestral gene (Hsu et al., 1992).

Due to different methodologies often used for SOD assay, it is difficult to compare enzyme activities among studies. Marklund (1984) found that the average SOD activity of sheep was the highest of nine mammal species examined including rodents, pig, cow, cat, dog, and man. SOD activities range between approximately 15 to 40 units/mg protein in the brain and liver of primates (Tolmasoff et al., 1980). SOD activity in the heart and liver of humans was highest among the primates tested including tree shrews, squirrel monkeys,

Table 10.1
Comparison of SOD in Invertebrates and Vertebrates

SOD Forms	Invertebrates	Vertebrates
FeSOD	present (from symbiotic organism)	absent
MnSOD	present	present
CuZnSOD	present	present
Subcellular Locations		
mitochondria	present	present
cytosol	present	present
peroxisomes	unknown	present
extracellular	?	present
Range of Specific Activity (units/min/mg protein)	~1–405	~13–100

Sources of data: (Van Balgooy and Roberts, 1979; Marklund, 1984; Keller et al., 1991 Ahmad, 1992; Livingstone et al., 1992; Vanfleteren, 1992; Scandalios, 1993; Perez-Campo et al., 1993).
? indicates existence is questionable until purification and characterization has been made.

lemurs, baboons, gorillas, chimpanzees, orangutans, etc. (Tolmasoff et al., 1980).

For most mammals, CuZnSOD activity is highest in liver tissues followed by kidney, heart, lung, and brain (Marklund, 1984; Halliwell and Gutteridge, 1989; Perez-Camp et al., 1993). MnSOD is also highest in liver and kidneys (Marklund, 1984). The overall content of EC-SOD is normally lower than MnSOD or CuZnSOD (Marklund, 1984). EC-SOD is relatively high in human, pig, sheep, cow, rabbit, and mouse but much less in dog, cat, and rat tissues (Marklund, 1984). EC-SOD is the dominant SOD in plasma and other extracellular fluids, but is also present at significant levels in the lungs and kidneys of some animals (Marklund, 1984). A comparison of SOD between invertebrates and vertebrates is summarized in Table 10.1.

ii. Biological importance

The significance of SOD in the aging process, in the etiology of certain diseases, and in ecology has been demonstrated in several systems. SOD levels have been manipulated by molecular techniques, by the use of selective inhibitors, by hormones, and by nutritional means to study its functional role (e.g., Petrovic et al., 1982; Ahmad, 1992; Bartoli et al., 1992; Ceballos-Picot, 1992; Ito et al., 1992; White, 1992).

There is ample evidence that SOD is important in protection against oxidative stress and lipid peroxidation (Yoshioka et al., 1977; Bartoli et al., 1992; Tsan et al., 1992; McNamara and Fridovich, 1993; Tsan, 1993). In studies on cultured rat hepatocytes, selective inhibition of SOD with diethylthiocarbamate sensitized cells to ROS-induced cytotoxicity (Ito et al., 1992). Rats fed a copper deficient diet show reduced CuZnSOD activity by greater than 70% (Bartoli et al., 1992). Blood cells from the rats with depleted SOD were more sensitive to lipid peroxidation; had lower vitamin E levels, and higher (Na^+, K^+) and Mg^{2+} ATPase activities compared to cells from rats fed on copper-sufficient diets (Bartoli et al., 1992). In other studies, rats fed copper-deficient diets showed significantly depressed aortic SOD activity and increased lipid peroxidation in the aortae (Nelson et al., 1992).

The free radical theory of aging has stimulated considerable research on antioxidant mechanisms, and in particular, the relationship of SOD to life span (Tolmasoff et al., 1980; Epstein et al., 1987; Seto et al., 1990; White et al., 1991; Bondy, 1992; LeBel and Bondy, 1992; Warner, 1992; Ames et al., 1993). Tolmasoff et al. (1980) assayed SOD in the liver, brain, and heart of two rodent and 12 primate species and found that SOD did not correlate with maximum life span potentials. However, the ratio of SOD activity to the specific metabolic rate of the tissue in these animals was highly correlated with increasing life span potentials (Tolmasoff et al., 1980). Selective alteration of gene expression provides a more definitive method to assess the relationship of SOD to life span than previous correlational studies. The life span of transgenic mice with moderately elevated human CuZnSOD is two-three times greater than control mice when breathing 99% O_2 (White et al., 1991). *Drosophila* strains with a null mutation for the CuZnSOD gene show a reduced life span and increased hypersensitivity to pro-oxidants, e.g., transition metals, paraquat (Phillips et al., 1989). Seto et al. (1990) reported that the overexpression of CuZnSOD in *Drosophila* did not affect lifespan. However, it is unknown why the investigators failed to report statistical analyses of their data, despite the fact that the maximum life span of the transgenics was up to six days longer than the control strain. Reveillaud et al. (1992) used two transgenic strains of *Drosophila* that express SOD at levels up to 35% higher total SOD activity. The strains were developed by microinjection with p-elements containing the bovine CuZnSOD cDNA (Reveillaud et al., 1992; Fleming et al., 1992). Lifespan was significantly increased in the transgenic flies when exposed to oxidative stresses (e.g., 100% O_2, paraquat) but not to non-oxidative stress (e.g., starvation). However, flies overexpressing SOD by more than 35% typically die during the pupal stage, presumably due an overload of H_2O_2 (Fleming et al., 1992).

Transgenic mice were constructed with a human CuZnSOD gene to specifically investigate the role of SOD during aging in the brain (Ceballos-Picot et al., 1992). Transgenic mice that overexpressed CuZnSOD (1.9 fold) in the brain, showed enhanced rates of lipid peroxidation and increased MnSOD activity in these tissues (Ceballos-Picot et al., 1992). Apparently, there was no concomitant increase in GPOX activity to scavenge the excess H_2O_2 generated by SOD. These results are analogous to data from Down's syndrome patients in whom a 50% increase in cytoplasmic SOD is observed that is associated with free radical-induced pathogenesis (Fleming et al., 1992). Investigations on the role of SOD in protecting against ischemia-reperfusion injury have also demonstrated a similar phenomenon (Mao et al., 1993). A bell-shaped dose-response curve with SOD has been frequently observed, whereby SOD at higher levels loses its effectiveness and may even enhance the extent of reperfusion injury (Mao et al., 1993).

Several factors may explain the deleterious effects of high SOD activity. First, high SOD activity, along with physiological levels of Fe^{2+} greatly increased the production of $\cdot OH$ (Mao et al., 1993). Second, results with ESR spin trapping indicate that $\cdot OH$ is produced by the reaction of H_2O_2 when free copper is released from oxidatively damaged CuZnSOD (Sato et al., 1992). Finally, the deleterious effects of high SOD may be due to a peroxidative activity of the enzyme (Yim et al., 1993). In the presence of small anionic radical scavengers (e.g., phosphate, cyanide, halides, azide), SOD maintains a peroxidative activity, and free radicals derived from the scavengers are produced (Yim et al., 1993). Thus SOD can catalyze the formation of free radicals using anionic scavengers and H_2O_2 as substrates (Yim et al., 1993). The conflicting data regarding the protective properties of SOD point out that antioxidant protection is not merely provided by increases in single components, but must be considered in the context of functionally coupled systems (i.e., SOD with CAT and/or GPOX). Thus a deregulation (e.g., overexpression) of an antioxidant system may disturb the steady-state levels of ROS within cells resulting in oxidative stress.

There is substantive evidence that deregulation of SOD expression is involved in several neuronal pathologies such as Lou Gehrig's disease or familial amyotrophic lateral sclerosis (ALS), Parkinson's disease, and brain ischemia (Ceballos-Picot et al., 1992; McNamara and Fridovich, 1993; Ames et al., 1993). The role of CuZnSOD in the etiology of ALS has been demonstrated (Rosen et al., 1993; Deng, 1993). Rosen et al. (1993) found that a mutation in the CuZnSOD(1) gene was present in individuals with familial ALS. Deng et al. (1993) reported that the structural defect in the CuZnSOD(1) gene resulted in more than 50% reduction in SOD levels in red blood cells. Thus it was concluded free radical damage in this neurodegenerative disease results from a decreased ability to scavenge O_2^-. Rosen et al. (1993) are currently investigating whether mutations in the genes for MnSOD or EC-SOD may be responsible for cases of ALS not associated with the CuZnSOD(1) gene or that the mutations may be involved in other neurodegenerative disorders.

O_2^- may be a prime toxic molecule associated with the pathologies of several viral diseases, particularly influenza virus (Oda et al., 1989; Maeda and Akaike, 1991). When mice infected with a lethal influenza virus strain were injected intravenously with bovine CuZnSOD, the SOD was rapidly cleared from the blood and little pharmacological benefit was observed. Oda et al. (1989) prepared a higher molecular weight SOD by conjugating the native enzyme with divinylether maleic acid/anhydride copolymer (pyran copolymer) to increase the biostability of the enzyme. When the conjugated SOD was injected in infected mice once daily, the survival rate of infected mice greatly improved (Oda et al., 1989).

ROS have been implicated in tumor promotion and progression (Kensler and Guyton, 1992). Church et al. (1993) examined the effect of transfecting sense and antisense human MnSOD cDNAs into melanoma cell lines. Cell lines expressing high levels of sense MnSOD cDNA lost their abilities to form tumors in mice (Church et al., 1993). In contrast, the introduction of antisense MnSOD cDNA had no effect on tumorigenicity (Church et al., 1993).

Considerable biomedical research has focused on enhancing O_2 tolerance by increasing levels of pulmonary antioxidant enzymes (Tsan, 1993). Administration of liposome encapsulated or polyethylene conjugated CuZnSOD to tracheal tissue has been successfully used to protect rats against O_2 toxicity (Padmanabhan et al., 1985;

Tang et al., 1993). However, intraperitoneal or intravenous administration of the encapsulated CuZnSOD was ineffective unless a conjugated CAT was co-administered (White et al., 1989; Jacobson et al., 1990). The importance of MnSOD in defense against O_2 toxicity was elegantly demonstrated by Wispe et al. (1992). Transgenic mice that express human MnSOD mRNA in alveolar and bronchiolar epithelial cells were shown to be protected against pulmonary O_2 toxicity (Wispe et al., 1992).

In some instances, SOD levels may show dramatic differences during development that may be associated with changes in habitat and/or O_2 metabolism (Allen, 1991). SOD levels in the anal gill of the mosquito, *Aedes aegypti*, increase with successive larval molts which may be related to the geometric increase in O_2 uptake and consumption associated with each molt (Nivsarkar et al., 1991). SOD (as well as POD and CAT) increases markedly during development in *Drosophila* and reaches a peak at the adult stage when O_2 consumption is maximal (Nickla et al., 1983). Similar results were obtained with the Mediterranean fruit fly, *Ceratitis capitis*, in which MnSOD peaked in the active adult stage (Fernandez-Sousa and Michelson, 1976). Barja de Quiroga et al. (1989a) found significant increases in SOD during the metamorphic climax of *Discoglossus pictus* tadpoles to the aerial environment. The predominantly aquatic frog, *Rana ridibunda*, had liver and brain SOD levels ca. one order of magnitude greater than the largely terrestrial toad *Discoglossus pictus* (Barja de Quiroga et al., 1984). However, in the case of the bullfrog *Rana catesbeiana*, SOD levels were not different in the aquatic tadpole stage compared to the more terrestrial adult stage (Abe et al., 1984).

SOD in the lung and other tissues of animals undergoes postnatal increases (Allen 1991; Munim et al., 1992). This may reflect a low potential for lipid peroxidation during intrauterine life (Yoshioka et al., 1980). Neonatal rats, mice, and rabbits survived hyperoxia and showed rapid increases in SOD, but neonatal guinea pigs and hamsters had no SOD response to hyperoxia and soon died (Frank et al., 1978). Immunostaining for CuZnSOD and MnSOD in rats indicates that both SODs increase in lung, kidney, and other tissues during the late gestational period (Munim et al., 1992). However, SOD in rats breathing atmospheric O_2 during early extrauterine life was not affected (Munim et al., 1992), suggesting that the increases in SOD during late gestation are a general adaptive mechanism for the increase in oxidative stress during extrauterine life. Allen (1991) reviewed investigations on developmental increases in SOD and concluded that only the CuZnSOD isozyme was sensitive to changes in ambient O_2.

SOD may play a role in symbiotic associations (Allen, 1991). Cnidarians that contain symbiotic algae face toxicity resulting from their interaction with photosynthetically produced O_2. In a series of elegant and extensive studies, J.M. Shick and coworkers have demonstrated the importance of SOD in algal-invertebrate symbioses. Sea anemones, *Anthopleura elegantissima* containing zooxanthellae (symbiotic dinoflagellate *Symbiodinium microadriaticum*) possessed SOD activities in their gastrodermal tissues nearly two orders of magnitude greater than individuals lacking zooxanthellae (Dykens and Shick, 1982). A later examination of 34 species of symbiotic invertebrates in four phyla (i.e., Cnidaria, Porifera, Mollusca, Chordata) from the Great Barrier Reef showed a direct relationship between chlorophyll content (from algal symbiosis) and SOD and CAT activities (Shick and Dykens, 1985). Differences in levels of these antioxidant enzymes also depend on localization of the algal symbionts (intracellular vs. extracellular) and on the extent of solar irradiance experienced by the symbionts. The highest SOD activities (> 200 units/mg protein) occurred in the alcyonarian Cnidarians that possess intracellular symbionts and live in unshaded microhabitats (Shick and Dykens, 1985). The SODs of the symbiotic algae from anemone show 30–40% increases in response to UV irradiation (Lesser and Shick, 1989). However, in the free-living nematodes, *Turbatrix* and *Caeonorhabditis elegans*, hyperoxic conditions did not affect SOD or CAT levels (Blum and Fridovich, 1983). Nonetheless, treatment of the nematodes with the SOD inhibitor, diethylthiocarbamate, results in increased susceptibility to hyperoxia (Blum and Fridovich, 1983). CuZnSOD levels in the frog *Xenopus laevis* appear to be independent of O_2 metabolism (Montesano et al., 1989).

SOD may also play a role in host-parasite associations. Many host organisms, both vertebrate and invertebrate, produce ROS as a response to invading parasites (Anderson et al., 1992). The parasite may manipulate host metabolism to reduce the flux of ROS (Batra et al., 1989) or scavenge host oxidants via antioxidant enzymes (Brophy and Barrett, 1990a; Brophy and Barrett, 1990b; Brophy et al., 1990; Brophy and Barrett, 1990a; Batra et al., 1990; Batra et al., 1992). Data suggest that SOD plays an important role in protecting parasites from host oxidative responses (Callahan et al., 1988; Leid et al., 1989; Callahan et al., 1993). An extracellular CuZnSOD was isolated from the parasitic helminth, *Trichinella spiralis* (Rhoads, 1983).

The enzyme may function as an integral defense mechanism against O_2^- encountered intracellularly as a normal product of biological oxidation, or externally as a component of host defense (Rhoads, 1983).

There is a growing recognition that an important component of plant defense against herbivores may be comprised of plant responses that involve the formation of ROS (Felton and Duffey, 1991a; Felton et al., 1992; Appel, 1993). Several lines of evidence indicate that SOD is important in protection against natural dietary prooxidants in herbivorous insects. First, inhibition of SOD in *Papilio polyxenes* and *Spodoptera eridania* by diethyldithiocarbamate, dramatically increased the toxicity of the prooxidant flavonoid quercetin (Pritsos et al., 1991). Many lepidopteran insects are particularly susceptible to dietary prooxidants such as *ortho*-dihydroxyphenolics (e.g., caffeic acid, quercetin) because of the strong alkalinity of their midguts (e.g., 8.0 to 12.4 pH) which favors the autoxidation of these compounds (Felton and Duffey, 1991b). Perhaps it is not surprising that SOD and other antioxidant enzymes (e.g., CAT, GST, and DHA-reductase) are found at higher levels in the digestive system than in many other tissues (Felton and Duffey 1991a; Ahmad et al., 1991; Ahmad, 1992; Lee and Berenbaum, 1992). Second, SOD activity is induced ~2 to 3 fold when *T. ni* larvae feed on diet containing quercetin or xanthotoxin, a photoactive furanocoumarin (Lee and Berenbaum, 1989; Ahmad and Pardini, 1990). In *T. ni* dietary harmine, a photoactive β-carboline alkaloid, induced SOD by 3.8 fold (Lee and Berenbaum, 1989). Third, in some insects adapted to feed on plants rich in photoactive furanocoumarins (e.g., *Papilio polyxenes and Papilio cresphontes*), considerable levels of SOD are found in the integument that is exposed to sunlight (Lee and Berenbaum, 1992). The fourth line of evidence is based upon studies on model systems used by several laboratories. Interestingly, May Berenbaum's lab using *Papilio* species as a model system (*P. polyxenes, P. cresphontes, P. glaucus glaucus*), Sami Ahmad's group using their model system of lepidopterans (*P. polyxenes, S. eridania*, and *T. ni*) and B.J.R. Philogene and J.T. Arnason's lab with a lepidopteran model system (*Anaitis plagiata, Ostrinia nubilalis*, and *Manduca sexta*), found that the insect species possessing the greatest natural exposure to prooxidants (i.e., *P. polyxenes, P. cresphontes, Anaitis plagiata, S. eridania*) had the highest constitutive levels of SOD and other antioxidant enzymes (Pardini et al., 1989; Aucoin et al., 1991; Lee and Berenbaum, 1992). However, SOD may be less important in other insects. For example, phototoxic furanocoumarins do not increase the levels of SOD in the parsnip webworm *Depressaria pastinacella*, an insect also adapted to feeding on plants rich in furancoumarins (Lee and Berenbaum, 1990). Further studies by Lee and Berenbaum indicate that cytochrome P-450 enzyme acts as the primary detoxication system in this insect against phototoxins, and that antioxidant enzymes act as a backup system (see chapter 5).

Studies with freshwater fish suggest that diet choice may partially explain variations in SOD activities. Radi et al. (1985) studied the antioxidant enzyme levels in freshwater fish and found that activities varied among fish with different feeding behaviors. The SOD levels were higher in two herbivorous fish species compared to those in three omnivorous species (Radi et al., 1985). The CAT and GPOX, however, were lower in the herbivores (Radi et al., 1985). It is difficult to generalize about the relationship between antioxidant enzymes and feeding behavior with so few examples tested, but it is an area for future investigation.

SOD may be particularly important for bioluminescent organisms where O_2 levels may reach dangerously high levels in the light producing tissues. Luminescent and non-luminescent elaterid beetles differ strikingly with respect to their SOD content with a five to 15-fold higher SOD level in luminescent species relative to their non-luminescent counterparts (Neto et al., 1986). The tissues containing the highest number of photocytes and consequently the highest O_2 levels (i.e., prothorax and last two abdominal segments) had the highest SOD activities (Neto et al., 1986).

Interestingly, thiobiotic metazoans that live in sediments beneath the oxygenated surface layer, possess higher levels of SOD and CAT than oxybiotic species (Morrill et al., 1988). The higher activities in the thiobiotic meiofauna (primarily turbellarians) may be related to sulfide concentration and metabolism, because O_2 exposure is an unlikely explanation. Possibly the normal substrates for these enzymes are not ROS but instead are sulfur radicals.

The physiological role of EC-SOD is unclear (Halliwell and Gutteridge, 1990b; Tsan, 1993). Human recombinant EC-SOD significantly reduces myocardial infarct size when retrogradely infused in pigs (Hatori et al., 1992). Human recombinant EC-SOD administered intravenously to rats following reperfusion-induced myocardial damage significantly reduced myocardial damage, whereas bovine CuZnSOD had no effect (Wahlund et al., 1992). These results may relate to the relatively high affinity of EC-SOD for the vascular epithelium that provides for a long vascular half-life.

No specific role for the carbohydrate moiety of EC-SOD is known (Edlund et al., 1992). An expression vector defining non-glycosylated EC-SOD was constructed by mutagenesis and the vector was transfected into a Chinese hamster ovarian cell line (Edlund et al., 1992). The enzymatic activity of the expressed non-glycosylated SOD was retained and the major difference with the native enzyme was a reduction in solubility. A slightly greater turnover of the mutant enzyme was observed in plasma, but the investigators concluded that no specific role for glycosylation was revealed (Edlund et al., 1992).

10.3.2 Catalase

One of the products of O_2^- dismutation is H_2O_2. H_2O_2 is detoxified by peroxidases and/or CAT (Halliwell and Gutteridge, 1989). CAT was one of the earliest identified enzymes and catalyses the following reaction (eq. 10.2):

$$2H_2O_2 \longrightarrow H_2O + O_2 \tag{10.2}$$

CAT (EC 1.11.1.6) is present in virtually all aerobic organisms tested to date (Harris, 1992a). The wide occurrence of CAT in prokaryotes and eukaryotes indicates a very ancient origin for this enzyme in aerobes. The enzyme is generally a tetrameric molecule of ~240 kDa made up of four identical subunits.

i. Phylogenetic distribution

CAT has broad phylogenetic occurrence. There are only a few reports of the absence of this enzyme among aerobic organisms such as parasitic protozoa and helminths (Barrett and Beis, 1982). Barrett and Beis (1982) compared CAT activity in parasitic vs. free-living platyhelminths (cestodes and trematodes) and found that CAT was quite active in the free-living species but could not be detected in the parasitic species. Rhoads (1983) was unable to detect CAT in the parasitic worm, *T. spiralis*.

Apart from the few cases where CAT is absent, invertebrates generally possess greater CAT activity than vertebrates (Livingstone et al., 1992). This may be associated with the diminished role or absence of GPOX in many invertebrates (see GPOX below). CAT often exhibits both CAT activity and POD activity depending upon substrate availability (Sichak and Dounce, 1986). However, in some cases POD activity may be absent (e.g., common mussel *Mytilus edulis*, Livingstone et al., 1992).

CAT has exceptionally high activity in insects and may compensate for the absence of Se-GPOX. CAT has been studied in several insect orders including Diptera, Coleoptera, and Lepidoptera (Aucoin et al., 1991; Ahmad, 1992; Allen et al., 1983). CAT activity in excess of 100 µmol/min/mg protein has frequently been reported (Ahmad et al., 1988b). In the swallowtail butterfly, *P. polyxenes*, CAT activity is nearly 1.1 mmol/min/mg protein in the midgut and fat body (Lee and Berenbaum, 1992). This exceptionally high activity is postulated to be an evolutionary consequence of selection pressure from the insect's prooxidant diet that is composed of furanocoumarin-rich host plants (Lee and Berenbaum, 1992).

CAT has been purified from the dipteran *D. melanogaster* (Nahmias and Bewley, 1984) and the lepidopterans *T. ni* (Mitchell et al., 1991) and *S. eridania* (Ahmad, 1992). In *T. ni*, CAT possessed minimal POD activity (Mitchell et al., 1991), but in *S. eridania*, 40% of its POD activity was due to CAT (Ahmad, 1992). The purified CAT of *T. ni* was tetrameric with an apparent molecular weight of 63 kDa for each subunit (Mitchell et al., 1991). The subunit size for *Drosophila* CAT is 58 kDa (Nahmias and Bewley, 1984) and the enzyme possesses a strong POD activity (Best-Belpomme and Ropp, 1982).

CAT is normally located in peroxisomes where many H_2O_2 producing enzymes are present (del Rio et al., 1992; van den Bosch et al., 1992). Insect CAT is unusual in that the activity is not restricted to the microsomal fraction that contains the peroxisomes, but is also present in the cytosol and mitochondria (Ahmad et al., 1988b; Jimenez and Gilliam, 1988). Ahmad et al. (1988a) suggested that the wide intracellular distribution of CAT in insects is an apparent evolutionary adaptation to the absence of selenium-dependent GPOX. Similarly, in the guinea pig, an animal deficient in GPOX in the liver and kidney, CAT is found predominantly in the soluble cytosol fraction of these tissues (Himeno et al., 1993). In contrast, CAT is hardly detectable in the cytosol of rat liver that possesses high GPOX activity (Himeno et al., 1993). However, due to the ease at which peroxisomes may be ruptured during homogenization, the non-peroxisomal occurrence of CAT in some reports should be considered with some skepticism until more definitive studies are completed.

The tissue distribution of CAT has been examined in only a few insect species. In *T. ni*, CAT is concentrated in tissues of highest metabolic activity (e.g., gut tissues, muscles and fat body) but virtually absent from hemolymph, Malpighian tubules, and salivary glands. We have been unable to detect CAT in the Malpighian tubules of *Helicoverpa zea* (unpublished data). Other antioxidant sys-

Table 10.2
Comparison of Catalase in Invertebrates and Vertebrates

CAT Forms	Invertebrates	Vertebrates
w/peroxidase activity	present	present
Subcellular Locations		
mitochondria	present	present
cytosol	present	present
peroxisomes	present	present
extracellular	unknown	present
Range of Specific Activity (μmoles/min/mg protein)	0–1700	~130–740

Sources of data: (Best-Belpomme and Ropp 1982; Jimenez and Gilliam 1988; Ahmad, 1992; Livingstone et al., 1992).

tems for scavenging H_2O_2 may replace CAT in these tissues (see Ascorbate below).

CAT shows a general pattern of tissue distribution in vertebrates with liver activity⟩kidney⟩heart⟩brain, in fish (Aksnes and Njaa, 1981; Morris and Albright, 1981) in amphibians (Barja de Quiroga et al., 1985), and in mammals (Halliwell and Gutteridge, 1989). CAT is virtually absent in the extracellular fluids (e.g., blood plasma, tissue fluid, spinal fluid, synovial fluid, and seminal plasma) from healthy humans (Halliwell and Gutteridge, 1990b). But, human serum CAT is markedly increased in patients with adult respiratory distress syndrome or advancing human immunodeficiency virus (HIV) infection (Leff et al., 1992). These increases apparently are not due to lysis of erythrocytes. A summary of CAT in invertebrates and vertebrates is presented in Table 10.2

ii. Biological importance

The role of CAT in regulating life span has been investigated in several insect species (Allen et al., 1983, 1984; Bewley and Mackay 1989; Orr et al., 1992; Orr and Sohal 1992; Baker 1993; Mahaffey et al., 1993). Allen et al. (1983) examined the effect of CAT inactivation by the inhibitor, 3-amino-1,2,4-triazole, on adult life span of the housefly *M. domestica*. The loss of CAT activity did not affect life span due to adaptive responses (Allen et al., 1983). In contrast, Bewley and Mackay (1989) reported a significant positive correlation between life span and CAT activity in *Drosophila*.

Orr et al. (1991) obtained two mutant strains of *Drosophila* with one strain containing no CAT activity and the other possessing 14% of the normal parent strain. Their results contrasted with those of Bewley and Mackay (1989) in that the absence of CAT did not affect life span. The null mutant flies exhibited adaptive responses to the loss of CAT including a lowering in the rate of O_2 consumption (Orr et al., 1992). The mutants possessing 14% CAT activity showed increased GSH levels of 28%. GSH may directly react with H_2O_2 rendering it harmless. Orr and Sohal (1992) used a genomic fragment containing the *Drosophila* CAT gene to produce transgenic *Drosophila* lines via P-element-mediated transformation. The transgenic flies overexpressed CAT with up to 80% higher activity and showed decreased sensitivity to H_2O_2. However, the transgenic flies did not show prolongation of life span. These results indicate that normal levels of CAT exceed the minimum physiological requirements for this insect (Orr et al., 1992; Orr and Sohal, 1992). Their results challenge the prevailing assumption that enhanced antioxidant levels have a beneficial effect upon longevity. Sohal and Brunk (1992) have argued strongly that after investigating antioxidant enzymes (CAT, SOD, GPOX and GR) in six different mammalian species (mouse, rat, guinea pig, rabbit, pig, and cow), that no relationship between antioxidant defenses and longevity exists in these species. They argue, instead, that life span is determined by the rate of mitochondrial production of O_2^- and H_2O_2 (Sohal and Brunk, 1992). However, it is important to stress that antioxidant defense is comprised not only of a few antioxidant enzymes, but exists as a network of protection (chemical and enzymatic mechanisms) that varies among tissue and cellular compartments. Orr and Sohal (1994) developed transgenic *Drosophila* lines that overexpressed both CAT and SOD genes, which resulted in significantly greater lifespan.

CAT may play an important role in protection against dietary oxidative stress in insects and other animals. As noted above with SOD, the levels of CAT also tend to be high in insects exposed to naturally high levels of prooxidants (Ahmad, 1992). CAT is inducible by oxidative stress in certain instances. Aucoin et al. (1991) found that CAT was induced in larvae of the geometrid moth, *Anaitis plagiata* when feeding on its primary host, the phototoxic plant *H. perforatum*. In *T. ni* dietary harmine, a photoactive β-carboline alkaloid, induced CAT by 20% (Lee and Berenbaum, 1989). However, other studies may indicate a less important role for CAT in protection against diet-related oxidative stress. Furanocoumarins had no effect on larval CAT activity in the parsnip webworm (Lee and Beren-

baum, 1990) or in *P. polyxenes* (Lee and Berenbaum, 1993). Xanthotoxin and a prooxidant flavonoid, quercetin, had no effect on CAT
in *T. ni* (Ahmad and Pardini, 1990). CAT was not induced when *H.
zea* larvae fed on tomato *Lycopersicon esculentum*, a host plant rich in
prooxidant phenolics such as chlorogenic acid (Felton and Duffey,
1991a). However, in these cases constitutive levels of CAT may be
more than sufficient to accommodate protection against diet-in-
duced oxidative stress. When the CAT inhibitor, amino-1,2,4-tria-
zole, was included in the diet of *T. ni*, larval mortality to quercetin
was greatly increased within 24 hours (Ahmad, 1992).

The importance of CAT in host-parasite relationships is unclear.
Barret and Beis (1982) found that the lack of CAT activity in platyhelminths correlated with the parasitic mode of life. The filarial
parasites, *Dirofilaria immitis* and *Onchocerca cervicalis* possess negligible levels of CAT, whereas GSTpx constitutes a major defense
against H_2O_2 (Callahan et al., 1988). However, other studies indicate
that the ability of a parasite to survive in its host correlates with its
capacity to deal with H_2O_2 via CAT. The parasitic nematode, *Nematospiroides dubius*, which can thrive for months in its host contains
~2 times higher CAT activity than *Nipostrongylus spiralis*, a nematode species quickly expelled from its host (Smith and Bryant 1986).
Adult filarial worms, *Acanthocheilonema vitaee*, detoxify H_2O_2 primarily through CAT (Batra et al., 1990). Appreciable amounts of
CAT are released by this worm into its ambient medium that may
protect the worm from ROS released by its host *Mastomys natalensis*
(Batra et al., 1990). Several antifilarial compounds appear to render
the parasite prone to H_2O_2 toxicity via inhibition of CAT (Batra et
al., 1990).

Ectothermic vertebrates normally are able to tolerate much higher
environmental O_2 tension than endothermic vertebrates. This is especially true for adult amphibians that can survive for long periods
of time in normobaric 100% O_2 (Barja de Quiroga et al., 1989a). Tadpoles of the toad *Discoglossus pictus* more readily adapt to hyperoxia
compared to tadpoles of the frog *Rana ridibunda perezi* which may
be related to the very high constitutive levels of CAT and SOD in
the toad (Barja de Quiroga, 1989a). Barja de Quiroga et al. (1989a)
also found highly significant increases in CAT, SOD, and GPOX
during the metamorphic climax of amphibian tadpoles to the aerial
environment. Tolerance may also be due to lower mitochondrial
production of ROS occurring in animals at lower body temperatures.

The metamorphosis of animals from an aquatic to an aerial environment (e.g., amphibians) is analogous to the mammalian lung at birth. These transitions involve an adaptation to a comparatively O_2-rich environment that may involve changes in the activity of CAT and other antioxidant enzymes. Increases in CAT and other enzymes have been frequently described in mammalian lungs during the transition from fetus to newborn (Yam et al., 1978; Yoshioka et al., 1980; Clerch and Massaro 1992).

Again, the armory of antioxidant defenses may be adequate to cover deficiencies in single components of antioxidant defense. Under conditions of "normal" oxidative stress, the production of intracellular H_2O_2, is handled by GPOX. For even humans that suffer from a genetic defect in the CAT gene (which produces an unstable CAT enzyme), show few life-threatening effects (Halliwell and Gutteridge, 1989). Manifestation of human acatalasemia does result in periodontal disease that may lead to tooth loss (Shaffer et al., 1987). If the intracellular production of H_2O_2 is increased by certain drugs, loss of CAT may then become very crucial (Halliwell and Gutteridge, 1989).

10.3.3   Se-Dependent Glutathione Peroxidase

H_2O_2 may also be destroyed by peroxidases (PODs) that catalyze the general reaction (eq. 10.3):

$$SH_2 + H_2O_2 \longrightarrow S + 2H_2O \tag{10.3}$$

SH_2 is a substrate that becomes oxidized.

Several PODs are known in organisms including "non-specific" PODs, cytochrome *c* POD, NADH POD, ascorbate POD, and GPOX (Halliwell and Gutteridge, 1989). GPOX is a "specific" selenium-dependent POD that catalyzes the destruction of H_2O_2 and organic hydroperoxides (ROOH/LOOH) by GSH (eq. 10.4).

$$\text{ROOH} + 2\,\text{GSH} \xrightarrow{\text{glutathione peroxidase}} \text{GSSG} + \text{ROH} + H_2O \tag{10.4}$$

GPOX activity may be due to the expression of multiple isozymes. The classical cellular glutathione peroxidase (GPOX; EC 1.11.1.9) has been studied in many mammal and bird species and exists as a cytosolic, tetrameric enzyme containing four gram atoms of Se per molecule and ranging in size from 76 to 105 kDa (Stadtman, 1991; Harris, 1992a; Schrauzer 1992). An extracellular, plasma form of GPOX

(PL-GPOX) exists as a tetrameric glycoprotein, but it is antigenically distinct from the cytosolic enzyme, and possesses a lower affinity for GSH (Maddipati and Marnett 1987; Takahashi et al., 1987). The human PL-GPOX shares only 44% amino acid sequence homology with cellular GPOX and has an approximate molecular weight of 23 kDa (Takahashi et al., 1990). Due to the fact that the K_m for GSH for the plasma GPOX is in the millimolar range and that GSH levels in plasma are normally in the micromolar range, the functional role of the plasma enzyme remains an anomaly (Halliwell and Gutteridge, 1990b). Halliwell and Gutteridge (1990b) concluded that extracellular antioxidant enzymes such as SOD and GPOX contribute little to the antioxidant status of extracellular fluids. Another unique Se-containing GPOX found in human milk, more closely resembles the plasma enzyme than the cytosolic enzyme (Avissar et al., 1991). This enzyme apparently accounts for about 10% of the total GPOX activity of milk with the plasma GPOX accounting for the remaining 90% of activity (Avissar et al., 1991). It is not yet known if this enzyme is the product of a separate gene from the plasma GPOX.

Cytosolic GPOX reduces free fatty acid hydroperoxides and thus requires the action of a phospholipase A_2 to liberate free fatty acid hydroperoxides from membranes. The phospholipase and GPOX act in concert to reduce harmful LOOH to less toxic LOH. Ursini et al. (1982) reported another GPOX, phospholipid monomeric GPOX (PH-GPOX), in pig liver that acts directly on phospholipid hydroperoxides, without the prior requirement for hydrolysis of free fatty acids from membranes. Consequently, the liberation of potentially excessive amounts of LOOH by phospholipases is avoided (Krinsky, 1992). The PH-GPOX has broad tissue distribution and is considerably more resistant to Se deficiency than the cellular enzyme (Weitzel et al., 1990). However, the PH-GPOX requires much higher Se levels to maintain full activity (Harris, 1992a).

Chu et al. (1993) reported the DNA sequence of a fourth form of GPOX (GPOX-GI), a tetrameric protein localized in the cytosol that does not cross react with antisera for other GPOX forms (Chu et al., 1993). GPOX-GI catalyzed the reduction of LOOH and H_2O_2 but not phosphatidylcholine hydroperoxide. GPOX-GI was readily detected in human liver and colon, occasionally in breast samples, but not in kidney, heart, lung, placenta, or uterine tissues (Chu et al., 1993). In rodent tissues, Chu et al. (1993) found GPOX-GI only in the gastrointestinal tract. The fact that GPOX-GI appeared to be the major GPOX activity in the gastrointestinal tract suggests that it may have

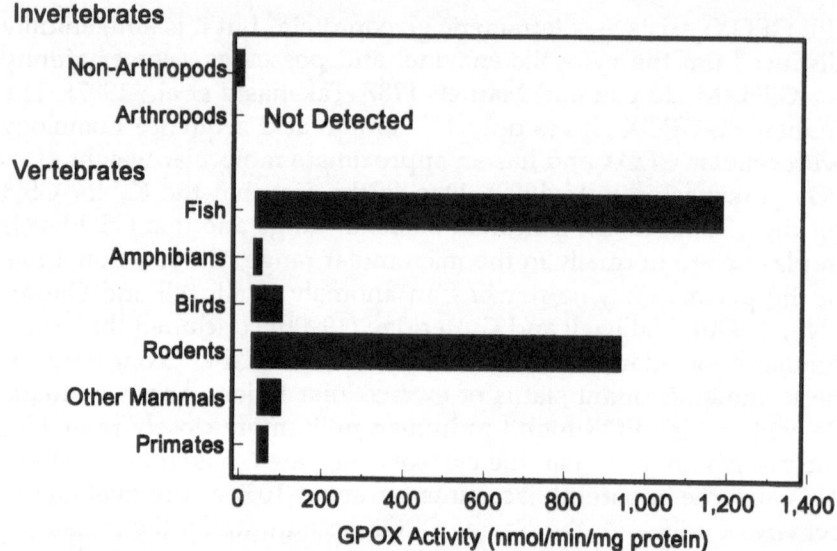

Fig. 10.1 Range of glutathione peroxidase activity in animals. Activity is generally measured with hydrogen peroxide, cumene hydroperoxide or *t*-butylhydroperoxide as substrates. Data from: (Tappel et al., 1982; Smith and Shrift, 1979; Shick and Dykens, 1985; Barja de Quiroga, 1989a; Winston, 1990; Livingstone et al., 1992; Ahmad, 1992; Kurata et al., 1993).

a major role in protecting mammals from the toxicity of ingested LOOH.

On the basis of similarities in tertiary structures, it has been suggested that the genes for GPOX evolved from GST genes or gluta-redoxin/thioltransferase genes (Fahey and Sundquist, 1991). Further analyses of sequence data will be required to better establish these relationships. Fahey and Sundquist (1991) suggested that the utilization of polyunsaturated fatty acids and cholesterol in eukaryotic membranes provided sensitive sites for lipid peroxidation and thus supplied the driving force for the evolution of new GSH dependent peroxidases.

i. Phylogenetic distribution

GPOX has a restricted phylogenetic distribution (Smith and Shrift, 1979; See Fig. 10.1). GPOX was originally purified from mammalian sources and has also been detected in fish, amphibians, reptiles, and birds (Smith and Shrift 1979; Wendel, 1980; Tappel et al., 1982).

The enzyme has not been unequivocally identified in arthropods and many other invertebrates. Relatively low GPOX activity was reported in a crayfish and snail, but was not detectable in the Chinese oak silk moth *Antheraea pernyi*, a fleshfly *Sarcophaga spp.* or the earthworm *Lumbricus terrestris* (Smith and Shrift, 1979). GPOX was not detected in the beetle, *Rhyzopertha dominica* (Chaudhry and Price, 1992) and in three species of herbivorous lepidopteran insects, *Ostrinia nubilalis*, *M. sexta*, or *Anaitis plagiata* (Aucoin et al., 1991). GPOX also was not detectable in the prawn *Pandalus nipponensis* and in the scallop *Patinopecten yessoensis* (Nakano et al., 1992).

A GPOX-like activity towards H_2O_2 was found to range between 2.0 and 12.8 units in three lepidopteran insects *T. ni*, *P. polyxenes*, and *S. eridania* examined by Ahmad's laboratory (Ahmad, 1992). Compared to mammals, these insects possess trivial GPOX activity. Because of the low levels of activity and that the essentially of Se has not been demonstrated for the insect enzyme, it is possible that the GSH-dependent reduction of H_2O_2 is due to another enzyme (see section on Ascorbate). Moreover, GPOX-like activity was not present in the cytosol of the southern armyworm, *Spodoptera eridania*, the normal subcellular location of GPOX in mammals (Ahmad et al., 1988a; Ahmad, et al., 1988b).

Most of the enzymatic activity towards LOOH in insects is due to the peroxidase activity of glutathione transferase (GSTpx) activity (Simmons et al., 1989; Ahmad, 1992). Lee (1991) found that the plant phenolics, quercetin, ellagic acid and juglone, which inhibited GSTpx in vitro also suppressed GPOX. These results support the previous findings in insects that GPOX reduction of ROOH is a Se-independent GSTpx activity (Cochrane, 1987; Ahmad et al., 1989; Simmons et al., 1989). The broad subcellular distribution of GSTpx and GR in the lepidopteran insects examined by Ahmad's laboratory suggests that these enzymes form a team to reduce ROOH and GSSG (Ahmad, 1992). Some researchers declared that CAT provides the sole enzymatic mechanism for removal of H_2O_2 in insects (Orr and Sohal, 1992; Orr et al., 1992). However, PODs are present in some insects and may provide protection against H_2O_2 (Chaudhry and Price, 1992). Considering that over one million insect species are known and that GPOX has been examined in only a minute number of species by few scientists, it may be premature to draw generalizations about the enzymatic reduction of ROOH among insects.

GPOX activity exists in other invertebrates, but in most cases a Se-dependent activity has not been clearly separated from non-selenium dependent GSTpx activity. However, RNA transcript sequences showing homology to human Se-dependent GPOX have been detected in the digestive gland of the mollusk *Mytilus edulis*, the common mussel (Goldfarb et al., 1989). The K_m values for H_2O_2 GPOX in the mussel were in the micromolar range which were similar to the K_m for mammalian GPOX (Livingstone et al., 1992). In this study, the GPOX activity towards H_2O_2 was quite low and barely exceeded the rate of chemical reduction of H_2O_2 by GSH (Livingstone et al., 1992). However, Winston (1990) reported higher GPOX activities in the same tissue and organism with rates approaching 20 nmol/min/mg protein depending upon substrate.

The GSH-dependent enzymatic reduction of the model ROOH, cumene hydroperoxide has also been reported in other invertebrates including mollusks, crustaceans, filarial worms, nematodes, cestodes, digeneans, etc. (Blum and Fridovich, 1984; Batra et al., 1989; Batra et al., 1990; Batra et al., 1992; Brophy and Pritchard, 1992). The GPOX activity measured with cumene hydroperoxide was considerably higher (i.e., up to 11.5 nmol/min/mg protein) in gastrointestinal nematodes than in a majority of cestodes and digeneans examined (Brophy and Pritchard, 1992). Parasitic nematodes are faced with an onslaught of ROS produced by host leucocytes.

GPOX is absent from the eggs of marine invertebrates such as sea urchins (Turner et al., 1988). In these instances an amino acid, ovothiol C (1-methyl-N^α, N^α-dimethyl-4-mercaptohistidine) exists in the sea urchin *Strongylocentrotus purpuratus* that acts as a functional GPOX (Turner et al., 1988). Ovothiol (See Fig. 10.3) occurs at a concentration of 5 mM in the sea urchin and has also been found in eggs from rainbow trout *Salmo gairdneri* and Coho salmon (Turner et al., 1988). The importance of this thiol amino acid in other organisms and tissues is unknown.

GPOX activity occurs widely in fish (Smith and Shrift, 1979; Aksnes and Njaa, 1981; Braddon-Galloway and Balthrop, 1985; Matkovics et al., 1987; Di Giulio et al., 1989; Sakai et al., 1992; Nakano et al., 1992; Gallagher and Di Giulio, 1992). The molecular weights reported for GPOX in fish (e.g., 72 to 100 kDa) are consistent with those reported in mammals (Wendel, 1980; Harris, 1992a). GPOX activity is highest in liver, blood, and ovaries (Smith and Shrift, 1979). Smith and Shrift (1979) in a phylogenetic survey of GPOX, reported that fish, as a group, had the highest GPOX activities (i.e., up to 280 nmol/min/mg protein) compared to GPOX from human, turtle, frog, toad, salamander, insect, or crustacean tissues (e.g., blood, muscle, ovaries, liver). Higher activities have since been reported in many other mammalian tissues and species (Tappel et al., 1982). There has been

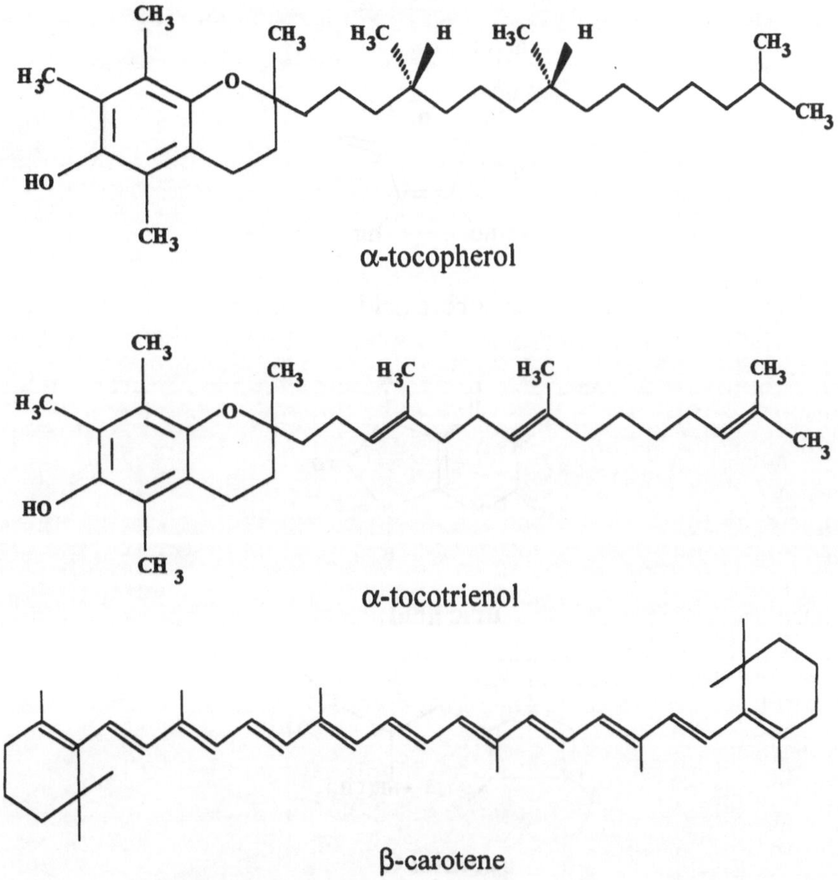

Fig. 10.2 Structures of the lipid soluble antioxidants α-tocopherol, α-Tocotrienol and β-carotene.

a surge of research on fish antioxidant systems because fish contain high levels of unsaturated fatty acids that form LOOH and aldehydes that may adversely affect the taste and smell of seafood (Nakano et al., 1992).

GPOX is present in Amphibia, but the activity is comparatively low in these lower vertebrates (Smith and Shrift, 1979; Tappel et al., 1982; Barja de Quiroga, et al., 1985). Tappel et al. (1982) found very low total glutathione peroxidase activity [GPOX + GSTpx] in the American toad *Bufo terrestris* (1.3 to 3.3 nmol/min/mg protein in various tissues) and in the Western newt *Taricha torosa* (0 to 2.5 nmol/

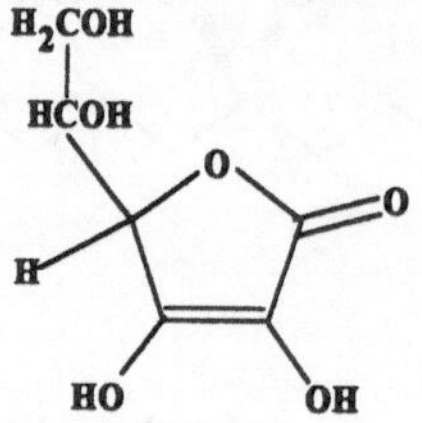

ascorbic acid

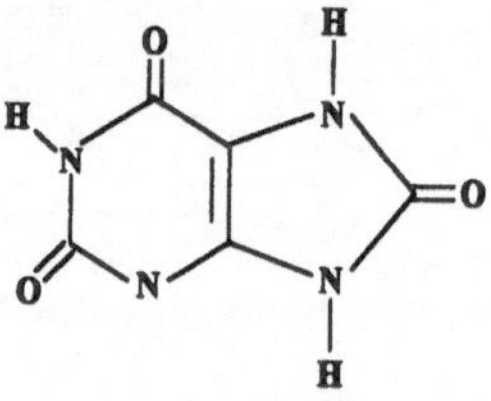

uric acid

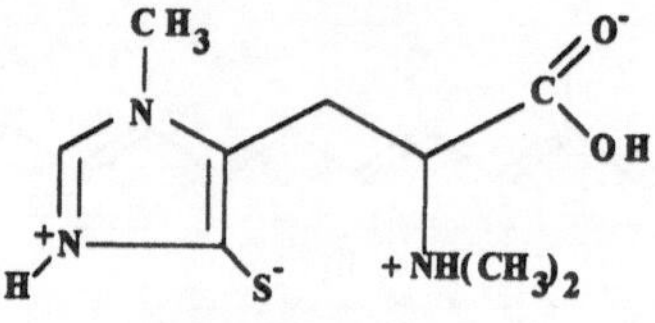

ovothiol C

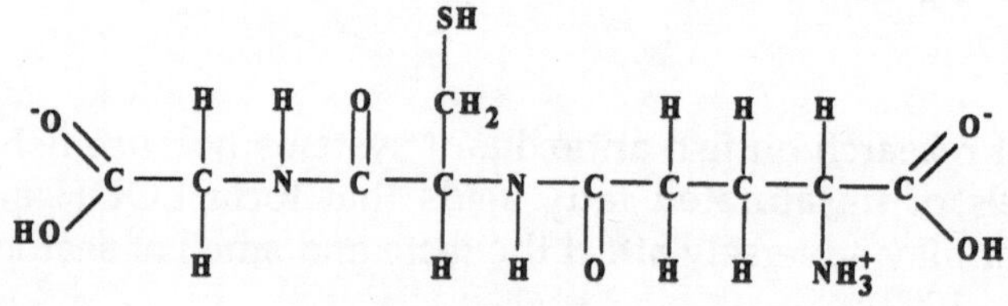

glutathione

Fig. 10.3 Structures of the water soluble antioxidants ascorbic acid, uric acid, ovothiol C and glutathione.

min/mg). GPOX activities in the lung of the tadpole *Discoglossus pictus* were substantially higher and exceeded 25 nmol/min/mg protein at certain larval stages (Barja de Quiroga, 1989a). In the adult frog, *Rana perezi*, GPOX activity was relatively high in the liver and kidneys, 64 and 235 nmol/min/mg protein, respectively. Smith and Shrift (1979) found higher activities in the salamander *Gyrinophilus porphyriticus* (i.e., 66 nmol/min/mg) than in the frog *Rana clamitans* or the toad *Bufo americanus* (5 to 16 nmol/min/mg). The implication from these few studies is that GPOX is relatively less important for protection from oxidative stress in Amphibia than in higher vertebrates. This is surprising due to the amazing tolerance of these organisms to high environmental O_2 levels. Tolerance may be a consequence of CAT and/or GSTpx for protection against ROOH (Barja de Quiroga, et al., 1989a; 1989b).

GPOX is present in Reptilia, but like Amphibia the activity is low compared to endothermic vertebrates (Smith and Shrift, 1979, Tappel et al., 1982). GPOX activity was as high as 22 nmol/min/mg protein in the liver of the western fence lizard *Sceloporus occidentalis* (Tappel et al., 1982) and a relatively low 10 nmol/min in the blood of the turtle (Smith and Shrift, 1979). Again, a precautionary note is required; so few studies have been made in these organisms that it is impossible to formulate sound generalizations about the relative importance of GPOX in endothermic vertebrates.

GPOX seems to be more important for protection in birds than in lower vertebrates and invertebrates. In the broiler chicken, GPOX activity was as high as 40 nmol/min/mg protein in some tissues while activity exceeded 100 nmol/min in the duck (Miyazaki and Motoi, 1992). Dietary trans-stilbene oxide increased the activity of liver GPOX to greater than 300 nmol/min/mg in the chick (Miyazaki, 1991). Birds also possess a second Se-dependent GPOX enzyme, the monomeric GPOX (PH-GPOX) with activity towards phospholipid hydroperoxides (Miyazaki, 1991). The PH-GPOX identified in the chick liver has a molecular weight of 18.5 kDa (Miyazaki, 1991). The activity of this enzyme makes up a significant percentage (i.e., 10–28%) of the total GPOX activity in the livers of chicken, quail, and ducks (Miyazaki and Motoi, 1992).

The vast majority of research on the enzymatic reduction of ROOH has been conducted with mammals (Wendel, 1980; Stadtman, 1991; Harris, 1992b). Among mammals, rodents generally possess the highest activities (Tappel et al., 1982; Godin and Garnett, 1992a, 1992b). The GPOX activity of rodent liver, comprised of only GPOX-GI and PH-GPOX, is normally at least 10 times higher than human liver activity that expresses all four forms of GPOX (Chu et al., 1993). In fact, rat liver activity can exceed 5.0 μmol/min/mg protein (Prohaska et al., 1992; Rikans and Cai, 1992; Chu et al., 1993) and mouse liver 1.8 μmol/min/mg protein (Permadi et al., 1992). Among the many rodent species tested (e.g., rats, mice, rabbits, hamsters, squirrels, gerbils), only guinea pigs lack Se-GPOX in most tissues (Tappel et al., 1982; Himeno et al., 1993). GPOX mRNA is hardly detectable in the liver, kidney, and heart of the guinea pig (Toyoda et al., 1989). The absence of Se-GPOX activity in the guinea pig tissues has apparently been compensated for with an active GSTpx and CAT (Tappel et al., 1982; Himeno et al., 1993). However in erythrocytes, the Se-GPOX activity of guinea pigs is comparable to rats and mice (Himeno et al., 1993). GPOX is apparently the sole enzyme responsible for the removal of LOOH in rodent erythrocytes, where GSTpx is virtually absent (Himeno et al., 1993).

GPOX has been studied in other mammals such as dogs, cats, sheep, and cattle, where its activity is generally lower than in rodents (Tappel et al., 1982). GPOX has been comparatively well studied in primates, and in particular humans. Kurata et al. (1993) compared GPOX in the erythrocytes of several primates and found that GPOX activity in common marmosets was similar to humans, but rhesus monkeys and common tree shrews showed three and two times higher activity, respectively.

Monomeric PH-GPOX that acts directly upon membrane LOOH was originally identified from pig liver by Ursini et al. (Ursini et al., 1982, 1985). The PH-GPOX of birds is very similar to the pig enzyme, but in contrast to birds where PH-GPOX comprises a significant component of total GPOX activity in the liver (i.e., 10–28%), the activity of PH-GPOX in the livers of rat, pig, or bovine makes up less than one percent of the total GPOX (Duan et al., 1988).

In conclusion, a phylogenetic comparison of GPOX, albeit far from extensive, indicates a decreasing role for GPOX in invertebrates compared to vertebrates. Whereas GPOX plays a major role for ROOH reduction in most mammalian tissues, CAT, GSTpx, or other enzymes may be relatively more important in invertebrates, in certain developmental stages of amphibians (i.e., tadpoles), and in some guinea pig tissues (Livingstone et al., 1992; Ahmad, 1992).

iii. Biological importance

In comparison with SOD and CAT, there have been fewer definitive studies on the biological importance of GPOX. Because the genes for SOD and CAT were characterized comparatively early, more definitive studies have been possible with these enzymes compared to GPOX.

GPOX may be less important than CAT for adapting to change from an aquatic to aerial environment in the toad *Discoglossus pictus* (Barja de Quiroga et al., 1989a). GPOX activities did not change during acclimation to hyperoxia (Barja de Quiroga et al., 1989a). Excessively high GPOX activity may also contribute to rheumatoid arthritis by generating excessive amounts of oxidized glutathione, a potent activator of collagenase (Chorazy et al., 1992).

Researchers have studied GPOX function by manipulating dietary selenium levels. GPOX is one of the few mammalian selenoproteins (Stadtman, 1991; Burk and Hill, 1993) and dietary deficiencies in selenium normally diminish GPOX activity (Baliga et al., 1992). A deficiency in GPOX to less than 1% of control levels does not affect the health of rats (Burk and Hill, 1993). However, severe selenium deficiency in rats leads to glomerular diseases and is accompanied by severe decreases in liver and kidney GPOX (Baliga et al., 1992). Selenium deficiency exerts its effect on pre-translational GPOX gene expression (Christensen and Burgener, 1992). Moderate GPOX deficiency is common among human Mediterranean populations, but is generally asymptomatic (Beutler, 1989). More severe genetic deficiencies in erythrocyte GPOX result in severe hemolysis in humans (Halliwell and Gutteridge, 1989). A degenerative heart disease known as Keshan disease is caused by selenium deficiency and thus may be related to low GPOX activity (Halliwell and Gutteridge, 1989). A deficiency in GPOX may, in certain instances, be compensated by CAT for removal of H_2O_2, and by GSTpx for removal of LOOH (Halliwell and Gutteridge, 1989). The compensation for low GPOX by CAT is demonstrated by the fact that the activities of GPOX and CAT often are inversely correlated among various species. Godin and Garnett (1992a) found that the activities of CAT and GPOX in red blood cells of various species were negatively correlated. This was very evident in quail red blood cells that exhibited negligible CAT activity but had the highest GPOX activity of all species tested.

Because GPOX deficiency is often asymptomatic, some researchers have suggested that GPOX serves as a storage molecule for selenium (Burk and Hill, 1993). A great deal remains to be discovered about GPOX function.

10.3.4 Glutathione Transferase

GSTs (EC 2.5.1.18) are a family of multifunctional enzymes involved in detoxication and conjugation of xenobiotics and in pro-

Table 10.3
Summary of Biochemical Features of Cytosolic GST Gene Families

Gene family	GSTpx activity	CDNB activity	pI	Subunit MW
α	moderate	moderate	basic	intermediate
μ	low	high	near-neutral	high
π	very low	moderate	acidic	lowest
θ	very high	none	??	highest

GSTpx = activity towards cumene hydroperoxide; CDNB = activity towards 1-chloro-2,4-dinitrobenzene.
Data from: (Mannervik et al., 1985; Meyer et al., 1991; Pemble and Taylor, 1992).

tecting against peroxidative damage. Extensive study has been made on the role of GSTs in xenobiotic metabolism and conjugation. It is well beyond the scope of this chapter to review this voluminous literature; I refer the reader to several excellent reviews and papers (Mannervik and Danielsen, 1988; Morgenstern and DePierre, 1988; Pickett and Lu, 1989; Waxman, 1990; Brophy and Barrett, 1990b; Fahey and Sundquist, 1991; Sheehan and Casey, 1993a; Sheehan and Casey, 1993b). I will focus on the peroxidase activity, GSTpx, activity of this enzyme (eq. 10.5) (Prohaska 1980), and its evolutionary relationships among animal taxa:

$$2\,GSH + ROOH \longrightarrow GSSG + ROH + H_2O \qquad (10.5)$$

GSTs are found from bacteria to vertebrates and are ubiquitous in the cytosol and microsomes of eukaryotes. Virtually all animal species tested possess GST (Stenersen et al., 1987). Structural and functional studies on vertebrate GSTs have divided the enzyme into four distinct classes of cytosolic proteins designated as α, μ, π (Mannervik et al., 1985) and θ (Meyer, Coles et al., 1991; Pemble and Taylor, 1992). These classes are based upon amino-terminal amino acid sequences, substrate specificities, sensitivity to inhibitors, and pI (Mannervik et al., 1985). The θ class is quite different from other GSTs in that it is not retained by a GSH affinity matrix, has a much higher K_m, and it lacks activity towards 1-chloro-2,4-dinitrobenzene (CDNB), (Meyer, Coles, et al., 1991; Pemble and Taylor, 1992).

On the basis of sequence homologies with plant GSTs, it has been suggested that the θ class is similar to the progenitor GST. Thus, the θ family may have arisen in purple bacteria and cyanobacteria to protect against O_2 based upon its high activity towards LOOH (Pembroke and Taylor, 1992).

Much less is known about GST gene families from invertebrates. It is unclear if the mammalian classification of GST gene families can be extended to invertebrates. GSTs from helminths (digenean and cestode) show sequence and biochemical relationships to vertebrate α, μ, and π GSTs but can not be clearly placed in one gene family (Brophy and Barrett, 1990b). The principal GST from *Schistosoma mansoni* showed α, μ, and π biochemical features but low sequence homology to the π family (Brophy and Barrett, 1990b). Four isoforms of GST have been isolated from the free-living nematode *Panagrellus redivivus* and the major GST form showed no biochemical homology to α, μ, and π families (Papadopoulos et al., 1989). However, a minor GST form showed a strong biochemical relationship to the α family including catalytic activity towards cumene hydroperoxide (Papadopoulos et al., 1989). Four forms of GST from the tapeworm, *Moniezia expansa*, showed a mixture of α- and μ-family characteristics (Brophy et al., 1989). Two of the forms possessed peroxidase activity (Brophy et al., 1989).

The deduced amino acid sequence and exon-intron gene structure of two GSTs from the squid *Ommastrephes sloani pacificus* were consistent with π class of vertebrate GST (Tomarev et al., 1993). An α-class GST is also present in squid (Harris et al., 1991). In contrast a *Drosophila* GST, encoded by an intronless gene, is related to the θ GST of vertebrates (Toung et al., 1990; Toung and Tu, 1992). There is also little information available on the GST gene families present in amphibians or reptiles (Pemble and Taylor, 1992; Aceto et al., 1993). I have summarized the available descriptions of GST gene families among various animal taxa:

Studies demonstrating a specific role for GSTpx in diseases and/or pathologies are sparse (Halliwell and Gutteridge, 1989). Genetic manipulation of GST expression should begin to provide answers.

10.3.5 Tyrosinase

The copper enzyme tyrosinase (EC 1.14.18.1) occurs widely among eukaryotes (Tsukamoto et al., 1992). Tyrosinase functions in melanin biosynthesis by oxidizing L-tyrosine to L-dopa followed by rapid oxidation to dopaquinone (Tsukamoto et al., 1992). O_2^- was shown to greatly enhance the oxidation of L-tyrosine by human tyrosinase (Wood and Schallreuter, 1991). Saturating concentrations of O_2^- en-

Table 10.4.
Phylogenetic Relationships of GST Gene Families

Animal group	Probable gene families present
Platyheliminthes	α, μ, π
Nematodes	α, μ, π
Mollusks (squid)	α-crystallin, π
Arthropods	α, θ
Fish (Salmonids)	π
Amphibians (toad)	α, ?
Reptiles	No reports
Birds (chicken)	α, μ, π, θ
Mammals	α, μ, π, θ

? = gene family: distinct from mammalian GST families.
Data from: (Mannervik et al., 1985; Dominey et al., 1991; Harris et al., 1991; Liu and Tam, 1991; Meyer, Coles et al., 1991; Meyer, Bilmore, et al., 1991; Bogaards et al., 1992; Pemble and Taylor, 1992; Waxman et al., 1992; Toung and Tu, 1992; Aceto et al., 1993; Tomarev et al., 1993).

hanced the rate of dopachrome formation from tyrosine by 40-fold. The discovery that O_2^- is a much preferred substrate for human tyrosinase over O_2 indicates that melanin biosynthesis in human skin represents a free radical trapping that may protect melanocytes against O_2^- generated by UV light and/or by leucocytes and lymphocytes (Wood and Schallreuter, 1991). This overlooked antioxidant role for tyrosinases remains to be established but should stimulate additional research on tyrosinase in many other biological systems.

10.4   PREVENTION OR INTERCEPTION OF FREE RADICAL PROCESSES

ROS escaping the action of antioxidant enzymes, may be scavenged by an array of lipid- (see Fig. 10.2) and water-soluble (see Fig. 10.3) antioxidants. The major nonenzymatic antioxidants are listed in Table 10.5. The antioxidants, vitamins E and C and carotene, form a network of interlinked processes that protect cells from further oxidative damage. The antioxidants may act catalytically through their interactions with other antioxidants or through direct enzymatic systems involved in their regeneration. Thus these antioxidants may act synergistically or additively to inhibit oxidative stress. For instance vitamin E and β-carotene act additively to inhibit peroxida-

Table 10.5
Major Lipid- and Water-Soluble Antioxidants in Animals

Lipid-Soluble	Water-Soluble
tocopherols	ascorbate
tocotrienols	GSH
carotenoids	uric acid
	albumins
	glucose
	metal binding proteins

Information obtained from: (Halliwell and Gutteridge, 1990b; Krinsky, 1992).

tion of rat liver microsomes (Palozza et al., 1992). A combination of β-carotene and α-tocopherol inhibits lipid peroxidation in a membrane model system significantly greater than the sum of their individual reactions (Palozza and Krinsky, 1992). This was the first evidence that β-carotene could act synergistically with α-tocopherol as a radical trapping antioxidant in membranes.

There have been many excellent recent reviews and research articles on the role of chemical antioxidants (particularly carotenoids, vitamin E, and vitamin C) in protection against an array of human diseases and pathologies including arthritis, atheroscelerosis, cancer, Alzheimer's disease, Parkinson's disease, diabetes, ulcerative colitis, neurotoxicity, Crohn's disease, ethanol-induced cellular injuries, and free radical injury associated with exercise. It is beyond the scope of this review to consider the many aspects of antioxidant protection against disease. I refer the reader to Chapter 2 of this book and several excellent recent papers (e.g., Adams et al., 1991; Jaeschke, 1991; Strain, 1991; Babbs, 1992; Bondy, 1992; Byers and Perry, 1992; Hemila, 1992; LeBel and Bondy, 1992; Nordmann et al., 1992; Olson and Kobayashi, 1992; Packer, 1992; Singh, 1992).

10.4.1 Nonprotein Antioxidants

i. Carotenoids

Over 600 naturally occurring carotenoids have been identified (Canfield et al., 1992). The mechanistic basis of their antioxidant activity, β-carotene specifically, is not well understood. It is hypothesized that β-carotene acts as a radical trap, rather than a chain-breaking antioxidant such as ascorbate or tocopherol (Buettner, 1993). Peroxyl radicals add covalently to the conjugated system of β-carotene, thereby breaking the chain reaction of lipid peroxidation (Canfield et al., 1992; Buettner, 1993). The highly conjugated double bond system of carotenoids is responsible for their highly efficient scavenging of $^1O_2$ (Goodwin, 1986). β-Carotene has an extremely high rate constant for quenching $^1O_2$ (i.e., 13-30 × 10^9 M sec)$^{-1}$, while the carotenoid lycopene is the most effective quencher (31 × 10^9 M sec^{-1}; Goodwin, 1986; Ahmad, 1992). Furthermore, the plasma levels of lycopene in human plasma are higher than β-carotene (Di Mascio et al., 1989). β-Carotene is unusual in that it is most effective at low O_2 tensions (Buettner, 1993); however, the peroxyl radical scavenging properties of β-carotene are highly effective within the normal physiological range of O_2 tensions (Kennedy and Liebler, 1992).

The contribution of carotenoids to color pattern in animals and insects, in particular, has been comparatively well studied (Goodwin, 1986; Berenbaum, 1987), but the effectiveness of these compounds in antioxidant protection has received minimal study (Ahmad, 1992). Most studies have focused on their origin and metabolic transformation (Goodwin, 1986). A few studies document antioxidant effects in invertebrates and lower vertebrates. For example, dietary β-carotene decreases the toxicity of the photosensitizing dye erythrosine B in the housefly *Musca domestica* (Robinson and Beatson, 1985). Aucoin et al. (1990) found that dietary β-carotene or vitamin E dramatically reduced mortality of the lepidopteran, *M. sexta*, to the phototoxin α-terthienyl. The protective effects of carotene against $^1O_2$-induced photosensitivity by hematoporphyrin was also demonstrated in the mouse (Matthews-Roth, 1964).

Viarengo et al. (1991) examined seasonal variations in antioxidant defenses in the digestive gland of the common mussel and found that not only the antioxidant enzymes (SOD, CAT, and GPOX) declined during winter months, but so did the chemical antioxidants (carotenoids, vitamin E, and GSH). These declines correlated well with an enhanced susceptibility of mussel tissues to oxidative stress as indicated by high levels of lipid peroxidation (Viarengo et al., 1991). The factors responsible for the winter decrease in antioxidants are unknown, but are presumably linked to changes in metabolic status and/or food availability. In particular, the carotenoid content seems to be related to differential food availability because it resembles the fluctuations of phytoplankton biomass (Viarengo et al., 1991).

The inhibition of lipid peroxidation by carotenoids has been frequently observed in test animals (Kunert and Tappel, 1983). β-Carotene protects against oxidative modification of low density lipoprotein (LDL), and consequently, may play a prominent role in preventing or slowing atheroscelerosis (Jialal et al., 1991; Rice-Evans and Diplock, 1993). β-Carotene is a more effective inhibitor of LDL oxidation than α-tocopherol, the most abundant antioxidant in human LDL (Jialal et al., 1991). It seems likely that a combination of water soluble antioxidants (e.g., ascorbate) and lipid-soluble antioxidants (tocopherols and carotenoids) may afford the most effective defense against LDL oxidation and atheroscelerosis. However, pig LDL is very susceptible to oxidation and lacks associated carotenoids, β-carotene, lycopenezeaxanthin, and crytoxanthin (Knipping et al., 1990).

Insects possess a high density lipoprotein (lipophorin) in their blood (i.e., hemolymph) that acts a carrier for α- or β-carotene (Bergman and Chippendale, 1992). A direct relationship between dietary α-carotene and that associated with lipophorin was demonstrated, but the roles of these carotenoids as antioxidants were not reported. These results are interesting in light of the fact that insect hemolymph possesses very low levels of antioxidant enzymes SOD, CAT, GPOX, and GR (Ahmad, 1992). Ahmad (1992) found that the lepidopteran larvae *S. eridania* and *P. polyxenes* selectively resorb lutein (β-ε-carotene-3,3'-diol) over all other carotenoids. Lutein has a quenching constant for $^1O_2$ similar to β-carotene but is generally more soluble in aqueous media than most other carotenoids. Perhaps lutein plays an important antioxidant role in insect plasma. Birds tend to accumulate the oxygenated carotenoids, xanthophylls, in their plasma and other tissues to the virtual exclusion of carotenes (Goodwin, 1986; Lim et al., 1992). In chick plasma, xanthophylls are effective chain breaking antioxidants against aqueous peroxyl radicals (Lim et al., 1992). These findings indicate that not only water soluble antioxidants (e.g., ascorbate, uric acid) are important protectants in plasma, but that lipid-soluble carotenoids (i.e., xanthophylls) are important components of antioxidant defense.

In humans and other mammals, profound differences in carotenoid patterns exist among various tissues. Generally, liver, adrenal gland, and testes contain much higher levels of carotenoids than kidney, ovaries, fat, and brain stem tissue. β-Carotene is the major carotenoid in liver, adrenal gland, kidney, ovary, and fat, whereas lycopene predominates in the testes (Sies and Stahl et al., 1992). The oxycarotenoids, lutein and zeaxanthin, are the major carotenoids in macula area of the retina where β-carotene is absent (Sies and Murphy et al., 1992). These differences in tissue distribution suggest that the physiological roles of the various carotenoids may be highly specialized.

The chemopreventative properties of carotenoids against human cancers are the subject of considerable recent research (Ziegler, 1989; Meyskens, 1990; Ziegler, 1991; Malone, 1991; Krinsky, 1993). β-Carotene is believed to protect against carcinogenesis by scavenging radicals involved in tumor formation. Dietary β-carotene may decrease risk to several cancers including cervical, lung, squamous cell oral cancer, and several UV- and chemical-induced neoplasias (Ziegler, 1989; 1991; Canfield et al., 1992). Despite extensive studies, little information is currently available on the in vivo molecular mechanisms of anticarcinogenesis of carotenoids. In addition to their ability to scavenge $^1O_2$ and free radicals, the anticarcinogenic properties of carotenoids may be due to their ability to be metabolized to retinoids or apocarotenoids, which also possess anticancer properties (Lippman et al., 1987). Additionally, the protective properties of carotenoids and their metabolites could relate to their positive effects on the immune system or their inhibition of arachidonic metabolism (Goodwin, 1986; Canfield et al., 1992).

ii. *Vitamin E*

Vitamin E is comprised of naturally occurring tocopherols and tocotrienols (Packer, 1992). Vitamin E is the major lipid-soluble antioxidant in protecting cellular membranes from free radical attack. Tocopherols are radical scavengers and quench lipid radicals (R·) thus regenerating RH molecules as well as producing tocopheroxyl radicals (Shahidi et al., 1992). In general, both tocopherol and ascorbate are highly effective chain-breaking antioxidants because: (1) their radicals are relatively harmless, being neither strongly oxidizing or reducing; (2) they are thermodynamically poorly oxidizing radicals; (3) their radicals react poorly with O_2 producing little O_2^- via electron transfer; (4) their kinetic properties require only small amounts to serve as effective antioxidants; and (5) they can be recycled either by chemical means or by enzymatic means (Buettner, 1993). In vitro evidence has shown that Vitamin E reacts directly with peroxyl radicals, O_2^- and $^1O_2$ (Machlin, 1991; van Acker et al., 1993). In vitamin E-deficient animals increased LOOH in fat tissue, increased exhalation of ethane and pentane, and accumulation of lipofuscin has been observed (Machlin, 1991).

Tocopherol is oxidized to tocopheroxyl radical and then to tocopherol quinone during its reactions with free radicals. Packer (1992) hypothesized that vitamin E acts catalytically, by being efficiently reduced from its tocopheroxyl radical form, back to its native form. The recycling may occur via nonenzymatic and/or enzymatic mechanisms (Liebler et al., 1989; Packer et al., 1989; Scholz et al., 1989; Packer 1992). Thiols (e.g., GSH) or ascorbate may react with tocopheroxyl (chromanoxyl) radicals to regenerate tocopherol (Packer et al., 1979; McCay, 1985; Niki, 1987; Sies, Murphy et al., 1992; Sies, Stahl et al., 1992; Ho and Chan, 1992; See Fig. 10.4). Packer et al. (1989) reported that mitochondrial and microsomal membranes possessed a free radical reductase activity that prevents chromanoxyl radical accumulation. However, evidence for enzymatic regeneration of tocopherol is still insufficient (McCay, 1985 Ho et al., 1992; Murphy et al., 1992; Buettner, 1993). A specific "vitamin E free radical reductase" has not been purified or characterized (Sies, Murphy, et al., 1992; Buettner 1993). The recycling of GSH by GSH-reductase or the recycling of ascorbate by an NADH-dependent ascorbate free radical reductase may be involved in the apparent enzymatic reduction of vitamin E. A GSH-dependent, heat labile system in rat liver has been observed that regenerates vitamin E, but the system was not further characterized (Scholz et al., 1989).

Rat liver hepatocytes, mitochondria, microsomes, and cytosol possess a tocopherol quinone reductase activity that is NADPH-dependent (Hayashi et al., 1992). The quinone reductase activity reduces the tocopherol quinone to tocopherolhydroquinone which retains some antioxidant properties (Hayashi et al., 1992). The activity may be attributable to DT-diaphorase rather than a specific tocopherolquinone reductase. In rat polymorphonuclear leukocytes, vitamin E regeneration is predominantly a chemical reaction (Ho and Chan, 1992). Our laboratory has been unable to detect a tocopherol quinone reductase activity in *H. zea* using several different cofactors including GSH, NADPH, and NADH. This insect does possess a fairly active DT-diaphorase (Felton and Duffey, 1992), but is unable to use quinones such as tocopherol quinone as a substrate.

The importance of vitamin E as an antioxidant is being investigated in a variety of systems. α-Tocopherol in combination with ascorbate is an extremely effective antioxidant for human low density lipoprotein (Sato et al., 1990; Ingold et al., 1993). Pigs have been used frequently as a model for human atheroscelerosis. Whereas vitamin E is considered an important antioxidant in human low den-

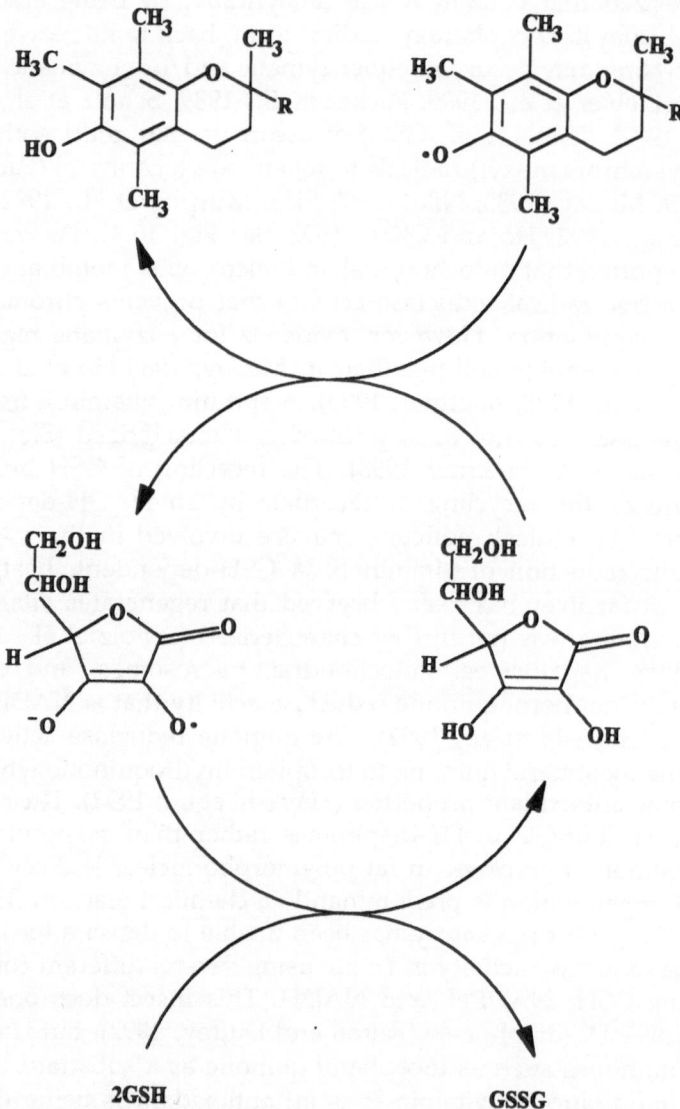

Fig. 10.4   Reduction of tocopheryl radical by ascorbate. Oxidized ascorbate is reduced by glutathione.

sity lipoprotein (LDL), in pigs vitamin E levels are on the average of four times lower (Knipping et al., 1990).

iii.   Ascorbate and auxiliary enzymes

Ascorbate is an essential nutrient for primates, flying mammals, guinea pigs, passeriformes birds, fish such as carp, trout, and Coho salmon, and most insects and invertebrates (Kramer, 1982). It is generally believed that in animals that the ability to synthesize ascorbate arose in amphibians where the key enzyme in ascorbate biosynthesis L-gulanolactone oxidase first appears. Other reports of lower invertebrates not requiring ascorbate may be due to the presence of symbionts.

Ascorbate has many functions in biological systems but an important feature is its ability to serve as a water-soluble antioxidant (Meister, 1992a). Because of the low redox potential of the ascorbate radical/ascorbate couple (i.e., +282 mV), ascorbate will react as an antioxidant with nearly every oxidizing radical that may arise in a biological system (Buettner, 1993). The oxidation of ascorbate may proceed either through single electron transfer with the formation of ascorbate free radical or through two-electron transfer with the direct formation of dehydro-L-ascorbic acid (DHA). The ascorbate free radical and DHA are not stable and spontaneously decay to biologically irreversible products such as diketogulonic acid.

The antioxidant properties of ascorbate are becoming increasingly appreciated (Byers and Perry, 1992). Ascorbate may be the most effective water-soluble antioxidant in plasma (Frei et al., 1989). Ascorbate can react directly with aqueous free radicals such as hydroxyl and peroxyl radicals (Moser, 1991). The antioxidant roles of vitamin C are summarized below:

It is important to note that as is the case with several antioxidants (e.g., GSH), ascorbate may act as a prooxidant and is even used as a ·OH generating system with Fe^{2+} and H_2O_2. In the presence of iron ions ascorbate may react to form ·OH radicals. The prooxidant-antioxidant properties of ascorbate are concentration dependent. At low concentrations ascorbate acts to reduce transition metals and promote free radical formation, whereas at high concentrations ascorbate scavenges O_2^- and ·OH (Halliwell and Gutteridge, 1990b).

Ascorbate possess antioxidant properties for insects. Ascorbate deficiency in larvae of the moth *M. sexta* greatly increased its susceptibility to the $^1O_2$-generating phototoxin α-terthienyl (Aucoin et al., 1990). We have found that the toxicity of the prooxidant phe-

Table 10.6
Antioxidant Roles of Vitamin C With Some Rate Constants (M^{-1} s^{-1})

reduces tocopheryl radicals (1.55×10^6)
scavenges $^1O_2$ (rate constant = 1×10^7)
scavenges $O_2^{\cdot-}$ (2.7×10^5)
scavenges water-soluble peroxyl radicals (2×10^8)
scavenges $\cdot OH$ (>10^9)
scavenges hypochlorite (HOCl)
scavenges thiyl and sulphenyl radicals
reduces nitroxide radicals
reduces nitrosamines to inactive products
protects against oxidative DNA damage
protects against lipid peroxidation
inhibits oxidation of low density lipoprotein
reduces semiquinones and quinones
spares GSH in GSH-deficient animals

Information condensed from (Packer et al., 1979; Halliwell and Gutteridge, 1990b; Sies et al., 1992; Hemila, 1992; Meister, 1992a).

nolic, caffeic acid, to larval *H. zea* is eliminated by increased levels of dietary ascorbate.

Besides quenching aqueous radicals, ascorbate indirectly affects other antioxidant systems (see Fig. 10.4). Vitamin E or tocopherol may be spared by ascorbate (Niki, 1987). During antioxidant reactions with fatty acid oxidation products, tocopherol is oxidized to the tocopheryl free radical. Ascorbate can donate an electron to the tocopheroxyl free radical to regenerate tocopherol (Moser 1991).

The oxidized forms of ascorbate, ascorbate free radical and de-hydro-L-ascorbic acid, may be reduced to regenerate ascorbate. This can be accomplished via direct chemical reactions (e.g., glutathione)

Table 10.7
Effect of Ascorbate on the Toxicity of Caffeic Acid

Treatment	% Survival	RGR
Control	93.3b	0.34b
Caffeic Acid (4 mM)	46.7a	0.27a
Caffeic Acid (4 mM) + Ascorbate (11.7 mg/gm diet)	88.3b	0.34b

RGR = Relative Growth Rate (mg/day/mg larva)

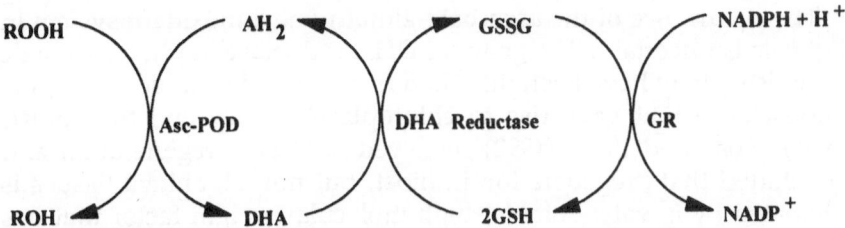

Fig. 10.5 Enzymatic reduction of peroxide by ascorbate peroxidase and associated reducing enzymes, AH_2 = ascorbate, DHA = dehydroascorbic acid.

or via enzymatic reductions. GSH is essential for the physiological function of ascorbate because it is required for the reduction of DHA (Martensson and Meister, 1991; Martensson et al., 1993). The enzymatic means for ascorbate recycling has been characterized in chloroplasts, cytosol and root nodules of plants, and in cyanobacteria (see Chapter 9). This system for removal of H_2O_2 involves an initial POD reaction (eq. 10.6) catalyzed by ascorbate POD (EC 1.11.1.7), which utilizes ascorbate as an antioxidant and generates DHA (Dalton et al., 1986) (See Fig. 10.5).

$$H_2O_2 + \text{ascorbate} \xrightarrow{\text{ascorbate peroxidase}} 2H_2O + \text{dehydroascorbic acid} \tag{10.6}$$

Ascorbate is regenerated in a GSH-dependent reaction catalyzed by DHA reductase (eq. 10.7) (EC 1.8.5.1; Hossain and Asada, 1984; Dalton et al., 1986; Dipierro and Borraccino, 1991).

$$\text{dehydroascorbic acid} + 2\,GSH \xrightarrow{\text{dehydroascorbic acid reductase}} \text{ascorbic acid} + GSSG \tag{10.7}$$

The GSSG is regenerated to GSH by glutathione reductase (Dalton et al., 1986; Hausladen and Kunert, 1990).

An ascorbate free radical reductase (AFR), which directly converts ascorbate free radical back to ascorbate (eq. 10.8), is also present in chloroplasts (Arrigoni et al., 1981; Hossain et al., 1984; Dalton et al., 1986; Borraccino et al., 1989; Bowditch and Donaldson, 1990).

$$\text{ascorbate free radical} + NADPH \xrightarrow{\text{ascorbate free radical reductase}} \text{ascorbate} + NADP \tag{10.8}$$

The occurrence of the ascorbate-glutathione antioxidant system in animals is uncertain. The gene for DHA reductase in photosynthetic organisms may have been inherited from an endosymbiont (i.e., cyanobacteria) that gave rise to chloroplasts (Fahey and Sundquist, 1991). Rose and Bode (1992) reviewed ascorbate regeneration and concluded that "regeneration in most, but not all, animal tissues is mediated by a water-soluble, high molecular weight factor that has several characteristics of an enzyme". However, they noted that no investigator has provided unequivocal evidence of a purified protein that has the unique property of reducing dehydro-L-ascorbic acid.

Alternatively, Winkler (1992) noted that the nonenzymatic reduction of DHA by GSH occurs at a rate close to that of "reductase catalyzed" rates in spleen, liver, kidney, lung, heart, and brain tissues of pigs (Coassin et al., 1991) and in eye tissues of rabbits and rats (Winkler, 1992). At least in these tissues and species, the recycling of ascorbate probably depends upon the nonenzymatic redox coupling between DHA and GSH. Nevertheless, the marked instability of DHA at physiological temperatures (i.e., 37 °C) raises questions about the efficiency of the redox couple between GSH and DHA in maintaining reduced ascorbate in most mammalian systems (Winkler, 1987).

There is evidence supporting an enzymatic role for DHA reduction in animals. Studies by Hughes (1964) and Hughes and Kilpatrick (1964) found that a liver extract fraction that separated between 52 and 66% ammonium sulfate saturation catalyzed the reduction of DHA with GSH as a cofactor. Furthermore, heat treatment stopped DHA reduction. They also found evidence that this "reductase" activity was inducible because erythrocytes from patients with pernicious anemia showed a 250% increase in DHA reductase activity. Grimble and Hughes (1967) studied a DHA reductase factor in guinea pigs and found activity in the stomach, brain, adrenals, lung, and intestines. The activity was trypsin sensitive, had a pH optimum of 6.9, and sulfhydryl compounds were essential for activity. Yamamoto et al. (1977b) reported that a homogenate from the hepatopancreas of carp possessed DHA reductase activity. The specific activity of DHA reductase was increased by 100-fold during purification by ammonium sulfate precipitation and chromatography with Sephadex G-75 and DEAE-Sephadex. The DHA reductase activity was specific for GSH as a hydrogen donor while NADH, NADPH, and cysteine were inactive (Yamamoto et al., 1977b).

Rose (1989) postulated that a cytosolic DHA reductase was present in the renal cortex of kidneys from rats and guinea pigs. The evidence in support of an enzymatic activity for DHA reduction included the fact that the activity was found in a 55–70% ammonium sulfate fraction, was retained by 12 kDa molecular weight tubing and by a 30 kDa molecular weight cut off Centricon filter, was heat and pH sensitive, was inhibited by thiol reagents and was most active in the presence of both NADPH and GSH. The enzyme molecular weight was at least as large as cytochrome c (i.e., 12,384 Da). Choi and Rose (1989) also studied ascorbate regeneration in the rat colon and found a DHA reductase activity that was again precipitated by 55–70% ammonium sulfate, retained by 12,000 MW dialysis tubing, and was larger than cytochrome c. However, the rat colon DHA reductase activity required NADPH rather than GSH as a hydrogen donor that suggests that this activity is distinct from the DHA reductase enzyme EC 1.8.5.1. They also found a small amount of GSH-dependent reductase activity in one of the ammonium sulfate fractions, suggesting that multiple enzymes may participate in the reduction of DHA. Choi and Rose (1989) point out that the colonic mucosa of mammals is particularly subject to the damaging effects of dietary prooxidants and oxidative stresses.

DHA reductase occurs in the liver, kidney, and hepatopancreas of several fish species (Yamamoto et al., 1977a; 1977b; Dabrowski, 1990). The DHA reductase activity in carp intestine was 20-fold higher than in hepatic tissue (Yamamoto et al., 1977b). Dabrowski (1990) also reported very low DHA reductase activity in the hepatopancreas and kidney of the carp where at least 60–80% of the DHA-reduction was due to chemical reduction. Also, the rate of reduction was not significantly modified by different dietary levels of ascorbic acid (Dabrowski, 1990). Similarly, there was no major difference in the DHA reductase activity between scurvy-prone guinea pigs and rats (Tsumimura et al., 1983).

On the basis of these reports of DHA reductase activity, our laboratory began investigating ascorbate recycling in ascorbate-dependent insects. Our initial studies began with the highly polyphagous insect, *H. zea*, that is commonly exposed to numerous dietary prooxidants such as quinones and dihydroxyphenolics. In addition to an AFR activity, we found considerable DHA reductase activity present in the midgut epithelium of the insect (Felton and Duffey, 1992). Evidence supporting an enzymatic role for DHA reduction includes the fact that the activity was retained by dialysis at 12,000 MW cutoff, was heat labile, sensitive to pH and thiol reagents, and was purified nearly 184-fold by ammonium sulfate precipitation, Sephadex G-75 chromatography, and isoelectric focusing with a BioRad

Table 10.8
Effect of Dietary Phenolics on DHA Reductase Activity in Fifth Instar
Helicoverpa Zea

Dietary Component	DHA Reductase activity (nmol/min/mg protein)
control	2.69b
gentisic acid	4.20c
t-cinnamic acid	0.79a
p-coumaric acid	0.69a
caffeic acid	2.12b
salicyclic acid	3.28bc

Activity is from whole body homogenates. Means not followed by the same letter are significantly different at $P < 0.05$. Phenolics are at 4 mM concentration in diet.

Rotofor (Felton and Duffey, 1992; Summers and Felton, 1993). The activity was dependent upon GSH while cysteine, NADH, NADPH, or dithiothreitol were inactive as hydrogen donors (Summers and Felton, 1993). The Km for the DHA reductase of *H. zea* was 0.27 mM with DHA that compares similarly to plant DHA reductases.

Recent studies in our laboratory indicate that the greatest DHA reductase activity is present in the Malpighian tubules (101 nmol/min/mg protein) and the midgut (45 nmol/min/mg protein), organs functionally analogous to mammalian kidneys and colon, respectively. In the cabbage looper *T. ni*, high levels of GR are found in the Malpighian tubules, where GSH may be oxidized during reduction of DHA by DHA reductase (Ahmad, 1992).

We have found DHA reductase activity in virtually every insect species tested including insects from the orders Coleoptera, Lepidoptera, Diptera, Hymenoptera, and Odonata (Summers and Felton, 1993; unpublished data). We began studies on the effect of different host plants and phytochemicals on DHA reductase in *H. zea* and have found that the enzyme may be strongly inhibited by certain phenolic compounds (Table 10.8) or may be strongly inducible by various host plants (Table 10.9).

The findings of DHA reductase in eukaryotes may indicate that the gene for DHA reductase did not arise from endosymbiont cyanobacteria (chloroplast origin) or that there may an independent origin in animals. Protein and nucleic acid sequences are needed in animals to test these hypotheses. The possibility exists that DHA reduction occurs by another currently recognized enzyme, perhaps

Table 10.9
Effect of Host Plant on DHA Reductase Activity in Fifth Instar
Helicoverpa Zea

Host Plant	Relative DHA Reductase Activity
control (artificial diet)	100a
cotton (cv. DPL-50)	108a
soybean (cv. Forrest)	623c
tomato (cv. Castlemart)	221b
red clover (feral)	610c

Activity is from whole body homogenates. Means not followed by the same letter are significantly different at $P < 0.05$.

in a oxido-reductase role (Rose and Bode, 1992). Mammalian thiol-transferase and protein disulfide isomerase possess DHA reductase activity (Wells et al., 1990). Wells et al. (1990) suggested that these two enzymes may be the missing link between GSH and ascorbate metabolism in animal tissues. However, it is improbable that the thioltransferases are the enzymes responsible for DHA reduction in colon and kidney tissues reported by Rose's laboratory (Rose and Bode, 1992), because thioltransferases have molecular weights of about 12 kDa, substantially smaller than the enzyme reported by Rose and colleagues.

Evidence also exists that AFR is present in animal tissues (Rose and Bode, 1992; Minakata et al., 1993). Lumper et al. (1967) discovered a microsomal enzyme system that catalyzed the electron transfer from NADH to AFR. Later reports in the 1970's confirmed the presence of an AFR or "semidehydroascorbate oxidoreductase" (EC 1.6.5.4) in microsomes prepared from rat liver (Hara, 1971; Green, 1973). A primary subcellular localization for AFR activity was the outer mitochondrial membrane in the bovine adrenal medulla (Diliberto et al., 1982). Diliberto et al. (1982) suggested that this activity played an important role in protecting membranes from lipid peroxidation. Ito et al. (1981) also identified an NADH-dependent AFR activity in the outer mitochondrial membrane of the rat liver. A significant portion of the AFR activity in the outer mitochondrial membrane may be due to the participation of a cytochrome b_5-like hemoprotein (Ito et al., 1981). Nishino and Ito (1986) later reported the presence of high AFR activity in the liver, kidney, and adrenal gland of rats with lesser activity in the heart, brain, lung, and spleen. Again the outer mitochondrial is the primary site of AFR activity.

Coassin et al. (1991) concluded that virtually all ascorbate enzymatic recycling in pig tissues is carried out by AFR. Significant AFR occurs in the liver, kidney, and heart but low activity exists in the spleen, brain, and lung. In the heart, AFR was present primarily in the mitochondrial and microsomal membranes (Coassin et al., 1991). In the central nervous system, ascorbate free radical is recycled by an AFR, but DHA is apparently reduced by only nonenzymatic means (Pietronigro et al., 1985). Khatami et al. (1986) found significant AFR activity in the retina and pigment epithelium of bovine ocular tissues but little activity in the lens and aqueous humor. Navas et al. (1988) found an NADH-dependent AFR in the plasma membrane of rat liver that was inhibited by preincubation with trypsin, neuraminidase, or the lectins concanavalin A, wheat germ agglutinin, and *Limax flavus* lectin. These results demonstrate a role of cell surface glycoproteins in the AFR activity of rat liver plasma membranes.

Rose (1990) obtained a semipurified preparation of AFR from mammalian colonic mucosa following ammonium sulfate fractionation and dialysis. The AFR activity was heat sensitive, acid precipitable, and contained within 1,000 MW cutoff dialysis. Unlike the standard AFR (EC 1.6.5.4), this reductase uses GSH and cysteine to reduce the free radical instead of NADH or NADPH as cofactors. It was hypothesized that the AFR protects against free radicals formed in the colon (Rose, 1990). Results from these studies suggest that several different enzyme systems may exist for ascorbate free radical reduction in various subcellular localities.

Our laboratory began investigating AFR activity in *H. zea* (Felton and Duffey, 1992). We have found considerable reductase activity in the midgut epithelium (e.g., ~168 nmol/min/mg protein). The activity is heat sensitive, precipitable by ammonium sulfate and utilizes NADPH or NADH, but not GSH. The AFR is not due to DHA reductase because we have separated the two activities by isoelectric focusing and gel filtration. We have found that AFR is also strongly influenced by the insect's diet (Table 10.10).

It is possible that animal tissues also contain a specific ascorbate POD (EC 1.11.1.7) (Meister, 1992a). The only published report of an ascorbate POD in mammals that I am aware of is in bovine eye tissue, but the enzyme differs in enzymatic properties from the plant enzyme (Kaul et al., 1988). The enzyme has not been reported from any other nonphotosynthetic organisms except *Trypanosomo cruzi* (Boveris et al., 1980; Asada, 1992). Our laboratory has been unable to detect any ascorbate POD activity or guaiacol POD activities in

Table 10.10
Effect of Phytochemicals on Midgut AFR

Diet Treatment	Relative AFR Activity
control	100a
4.0 mM caffeic acid	111.7a
4.0 mM chlorogenic acid	180.2a
4.0 mM t-cinnamic acid	138.6a
4.0 mM p-coumaric acid	122.6a
4.0 mM ferulic acid	442.7bc
4.0 mM salicylic acid	334.9b
4.0 mM gentisic acid	475.2c

Means not followed by the same letter are significantly different at $P < 0.05$.

the midgut of several lepidopteran insects including *M. sexta, Spodoptera exigua, T. ni,* or *Heliothis virescens* (see also Ahmad, 1992). Surprisingly, we detected substantial ascorbate and guaiacol POD activities in the midgut and other tissues of *H. zea*. At the present time we do not know if the ascorbate POD activity is an "ascorbate" specific POD similar to that found in plants. These preliminary data seem to indicate that the enzymatic activities may be due, in part, to separate enzymes because the tissue distributions for each activity do not follow similar patterns. Furthermore, we have ruled out the peroxidative activity of CAT as a source of ascorbate POD activity because the enzyme activities are cleanly separated by gel filtration and isoelectric focusing.

Ascorbate POD activity may compensate for the absence of GPOX in insects. Whereas, the GSTpx of insects possesses activity towards

Table 10.11
Relative Percent POD Activity in *Helicoverpa Zea* Fifth Instar

Tissue	Guaiacol POD	Ascorbate POD
fat body	37.1	55.6
midgut	29.0	20.0
salivary glands	40.3	100
Malphigian tubules	100	31.1

Activity: 100% Guaiacol Activity $= .062 \, \Delta \, 470/min/mg$ protein; 100% Ascorbate POD Activity $= 18.33 \, \mu mol$ ascorbate oxidized/min/mg protein. Sample size = 4 replicates of 15–20 insects/rep.

LOOH, CAT has been assumed to be the sole enzyme for the removal of H_2O_2 in insects. The ascorbate POD activity is substantial (i.e., μmol/min/mg) and is of comparable magnitude as CAT. We are in the process of determining if the ascorbate POD activity may also act upon model LOOH such as cumene hydroperoxide.

iv.   Glutathione (GSH) and the auxiliary enzyme (GR)

The role of GSH as a cofactor for several antioxidant enzymes (i.e., GST, GPOX, DHA reductase) was discussed. GSH is the most abundant nonprotein thiol and is responsible for maintaining an intracellular reducing environment (Meister, 1992a). Intracellular levels of GSH are in millimolar range (e.g., 0.5 to 10 mM; Meister and Anderson, 1983). GSH synthesis occurs by consecutive reactions of γ-glutamylcysteine synthetase and GSH synthetase (Meister and Anderson, 1983; Meister, 1992b).

GSH is a multifunctional peptide involved in many processes including protein and DNA synthesis, transport, enzyme activity, xenobiotic conjugation, and antioxidant protection (Meister and Anderson, 1983). Compounds with electrophilic centers facilely conjugate with GSH in spontaneous reactions or by enzymatic catalysis by GST (Meister and Anderson, 1983). The role of GSH in detoxication reactions of xenobiotic compounds has been extensively studied (Meister and Anderson, 1983).

GSH plays a multifunctional role in antioxidant protection. GSH maintains ascorbate, tocopherols, and other reductants in their reduced state (Meister, 1992a; Martensson et al., 1993). GSH, synthesized in most animal cells, scavenges $^1O_2$ and ·OH, but reacts slowly with O_2^- (Halliwell and Gutteridge, 1989). GSH does provide protection from LOOH and H_2O_2 (Meister, 1988; Hu and Tappel, 1992). Pioneering work on selective modification of GSH synthesis by Alton Meister has provided the means to investigate the in vivo functions of GSH (Meister, 1992b). Ample evidence suggests that GSH in invertebrates and vertebrates is an important protectant against oxidative stress arising from both endogenous (Allen and Sohal, 1986; Lash et al., 1986; Martensson et al., 1991) and dietary/environmental sources (Starrat and Bond, 1990; Christensen, 1990; Coles and Ketterer, 1990). Mitochondrial membrane disintegration is a major consequence of prolonged GSH deficiency (Meister, 1992b). GSH depletion in newborn rats causes decreased tissue levels of ascorbate, damage to lung, hepatic, renal, and brain tissues followed by death (Martensson et al., 1991). Lifespans in adult houseflies also were decreased by GSH depletion, but were not prolonged by GSH supplementation (Allen et al., 1985). GSH synthesis declines with increasing age in primates, rabbits, mice, and other animals (Rathbun and Holleschau, 1992). Loss of GSH is associated with age-related pathologies such as senile cataract (Teramoto et al., 1992). Deficiencies in GSH synthesis in humans lead to brain dysfunction, peripheral neuropathologies, and a tendency towards hemolysis (Meister, 1988).

The intracellular ratio of GSH to GSSG is maintained at approximately 300:1 through the action of the NADPH-dependent enzyme glutathione reductase (GR) that catalyses the reaction (Meister and Anderson 1983; Ziegler 1985):

$$GSSG + NADPH + H^+ \rightarrow 2\,GSH + NADP^+ \qquad (10.9)$$

GR (EC 1.6.4.2) is a disulfide oxidoreductase, an important class of flavoenzymes that includes lipoamide dehydrogenase, thioredoxin reductase, and mercuric reductase (Douglas, 1986; Schirmer, 1989). Studies of their amino acid sequences suggest that GR, mercuric reductase, and lipoamide reductase may have arisen from a common ancestor. GR is one of the most ancient proteins (Fahey and Sundquist 1991). Comparisons of the DNA sequence coding for the active site of the enzyme indicate that the human gene may have been inherited from purple bacteria (related to *E. coli*) via the mitochondria (Fahey and Sundquist, 1991). Overall there is a 55% similarity between aligned sequences for human and *E. coli* GR genes (Fahey and Sundquist, 1991). This is one of the highest degrees of similarity between any enzymes of human and prokaryotic origin. The structure of the human erythrocyte enzyme is known in exceptional detail (Krauth-Siegel, et al., 1982; Karplus and Schulz, 1987; Karplus and Schulz, 1989).

Virtually all organisms that synthesize GSH possess GR (EC 1.6.4.2; Douglas, 1986; Fahey and Sundquist, 1991). GR has been purified from human, rabbit, gerbil, pig, rat, mouse, calf, sea urchin, sheep, and sea mussel sources (Fahey and Sunquist, 1991). In most instances, GR exists as a dimer of ~50 kDa subunits with a preference for NADPH over NADH (Fahey and Sunquist, 1991). GR has not been purified or well characterized in invertebrates (Ahmad, 1992).

In mammals, liver tissue generally possess the greatest activity followed by heart, lung, kidneys, and erythrocytes (Halliwell and Gutteridge, 1989). Studies on the insect, *T. ni* show that gonads and muscle possess high specific activity but GR is virtually absent from the hemolymph. GR possesses broad subcellular distribution in insects (Ahmad, 1992) but is limited to cytosol and mitochondria in mammals (Chance et al., 1979).

GR functions in the production of free SH groups for deoxynucleotide synthesis, detoxication of peroxides (via GST, GPOX, or DHA reductase) (eq. 10.9) and many other processes (Bucheler et al., 1992). In particular, maintenance of tissue reducing power by GR may be a strong determinant of age-related pathologies and mean survival rates. For instance, humans with GR deficiency develop cataracts at an early age (Meister, 1988). Loss of GR induction following CAT inhibition in old frogs results in a sharp decrease in survival rate (Lopez-Torres et al., 1993).

10.4.2   Protein Antioxidants

i.   Metal binding proteins

Halliwell and Gutteridge (1990b) have argued that a major antioxidant defense of human plasma proteins is the binding of transition metals. Iron-dependent free radical formation is inhibited by proteins such as transferrins with high binding capacity for iron ions, whereas albumins are powerful antioxidants against copper-dependent free radical processes due to their high affinity for copper. Transferrin and the iron-oxidizing protein, ceruloplasmin, comprise about 4% of the total human plasma protein, and appear to be the primary antioxidants in protecting plasma against iron-catalyzed free radical damage (Gutteridge and Quinlan, 1993).

Transferrins are a superfamily of single chain ~80 kDa glycoproteins that function in iron transport. Transferrin in human plasma from healthy individuals is ca. 20–30% loaded with iron so that the free iron in plasma is very low. Iron bound to transferrin will not participate in ·OH formation or in lipid peroxidation (Halliwell and Gutteridge, 1990a). Transferrins typically bind two iron molecules per protein molecule.

Iron-binding proteins have received limited attention in invertebrates. Iron binding proteins have been described from a tunicate, a crab, a spider, the hornworm *M. sexta* and the cockroach *Blaberus discoidalis* (Huebers et al., 1988; Jamroz et al., 1993). The transferrin from the hemolymph of the roach is similar to vertebrate transferrins based upon amino acid sequences (i.e., 32–34% amino acid sequence positional identity with vertebrate transferrins), molecular weight (i.e., 78 kDa) and iron binding capacity (i.e., two iron molecules per protein molecule; Jamroz et al., 1993). The transferrin characterized from the hornworm, however, binds only one iron atom (Huebers et al., 1988). The biological significance of these differences in iron-binding capacity is unknown. Jamroz et al. (1993) conclude that transferrins are widely distributed among insects and that vertebrate and insect transferrins descend from a sequence that was present in the last common ancestor of vertebrates and insects, ~600 million years ago. The importance of arthropod transferrins in antioxidant protection in the hemolymph is unknown. The low levels of CAT, POD, and GPOX in insect hemolymph suggest that other antioxidants, perhaps transferrins, may play an important protective role in these tissues.

Ceruloplasmin (CP, ferro-O_2-oxidoreductase, EC 1.16.3.1) is a multifunctional, copper-protein (Samokyszyn et al., 1989). Ceruloplasmin is an extracellular glycoprotein synthesized in the liver and released into the plasma of all vertebrates (Goldstein et al., 1979; Ortel et al., 1984; Sato and Gitlin, 1991). Human CP, 132 kDa, and its gene were sequenced (Ortel et al., 1984; Takahashi et al., 1984). CP has several important functions including transport and mobilization of copper, ferroxidase activity, bactericidal activity, as well as an antioxidant function (Ortel et al., 1984; Klebanoff, 1992). CP catalyzes the oxidation of Fe(II) to Fe(III) (eq. 10.9) and Cu(I) to Cu(II) (eq. 10.10), without the production of partially reduced O_2 species (Samokyszyn et al., 1989; de Silva and Aust, 1992):

$$CP\text{-}[Cu(II)]_4 + 4\,Fe(II) \longrightarrow CP\text{-}[Cu(I)]_4 + 4\,Fe(III) \qquad (10.9)$$

$$CP\text{-}[Cu(I)]_4 + O_2 + 4H^+ \longrightarrow CP\text{-}[Cu(II)]_4 + 2H_2O \qquad (10.10)$$

CP promotes the incorporation of Fe(III) into the iron storage protein, apoferritin (Boyer and Schori, 1983).

CP occurs in mammalian heart, brain, and kidney tissues, macrophages, lymphocytes, and in bronchoalveolar lavage fluid (Linder and Moor, 1977; Krsek-Staples et al., 1992). Ceruloplasmin has been well studied in humans, rats, mice, and pigs (Linder and Moor, 1977; Goldstein et al., 1979; Alcain et al., 1992; Lynch and Klevay, 1992; Yamada et al., 1992), but its presence in invertebrate animals is unknown.

CP in plasma is often elevated in chronic and acute infections including rheumatoid arthritis, carcinomas, leukemia and liver disease (Ravin 1961; DiSilvestro et al., 1992). CP is absent or deficient in Wilson's disease, a genetic disease in which excessive amounts of copper are deposited in many tissues (Ortel et al., 1984). A mutant rat strain (LEC) possesses a deficiency in plasma CP and also suffers from excessive copper accumulation (Yamada et al., 1992).

The bactericidal effect of CP is associated with its ability to oxidize Fe(II) (Klebanoff, 1992). Several mechanisms may account for the antioxidant function of CP. CP inhibits Fe(II)-catalyzed lipid peroxidation in several systems (e.g., Samokyszyn et al., 1989; Krsek-Staples et al., 1992). CP possesses SOD activity (Goldstein et al., 1979). However, on a weight basis the O_2^- scavenging capacity of CP is substantially less than SOD (Goldstein et al., 1979). The reaction of ceruloplasmin with O_2^- and H_2O_2 is essentially stoichiometric and not catalytic (Halliwell and Gutteridge, 1990b). Halliwell and Gutteridge (1990b) point out that the SOD or CAT-like activity of ceruloplasmin may be due to contamination of ceruloplasmin EC-SOD or with copper complexes that can scavenge O_2^-. The O_2^- scavenging ability of CP may also be due to its ability to inhibit O_2^--dependent mobilization of ferritin iron (Samokyszyn et al., 1989).

The ferroxidase activity of CP is no doubt responsible for its ability to inhibit iron-dependent lipid peroxidation and ·OH formation (Samokyszyn et al., 1989; Gutteridge and Quinlan, 1993). Unlike the nonenzymatic oxidation of Fe(II)$^+$, the ceruloplasmin catalyzed oxidation of Fe(II) does not release any ROS because they are kept on the active site of the protein (Halliwell and Gutteridge, 1990b). Also, by maintaining iron in the oxidized state, CP prevents it from participating in the Haber-Weiss reaction that initiates lipid peroxidation (Krsek-Staples et al., 1992). The ferroxidase activity of CP inhibits cellular protein carbonyl formation that may prevent loss of enzyme function (Krsek-Staples and Webster, 1993). Ceruloplasmin also nonspecifically binds copper atoms and can then inhibit copper-stimulated formation of ROS (Halliwell and Gutteridge, 1990b). Moreover, CP transports copper to SOD in the liver (Harris, 1992c).

CP and the iron-binding protein transferrin may be the primary antioxidants in human plasma (Gutteridge and Quinlan, 1993). CP and transferrin offer considerable protection against ROS generated by iron and ascorbate (Gutteridge and Quinlan, 1993). Nevertheless, in studies not employing iron-driven systems to generate ROS, other antioxidants such as ascorbate (Frei et al., 1988; Frei et al., 1989), bilirubin (Stocker et al., 1987a; 1987b), or uric acid (Ames, 1981), may be more important. The relative importance of iron redox cycling to oxidative stress in plasma is in need of further investigation.

Albumin also binds copper ions and may inhibit copper-dependent lipid peroxidation and ·OH formation (Halliwell and Gutteridge, 1990b). Metallothioneins (MTs) are cysteine-rich low molecular weight, metal-binding proteins found in a wide range of organisms from bacteria to humans (Cousins, 1985; Halliwell and Gutteridge, 1989). MTs bind ions such as copper, zinc, cadmium, and mercury may function in the storage of these heavy metals in nontoxic forms. Several isoforms of MTs from yeast and monkeys may functionally substitute for CuZnSOD (Tamai et al., 1993). Purified MTs possess substantial Cu-dependent SOD activity in vitro (Tamai et al., 1993). Furthermore, MT mRNA levels may be dramatically elevated in response to ROS (Tamai et al., 1993).

ii. DT-Diaphorase

Quinones may be toxic via several mechanisms: (1) the quinone or their semiquinones may form addition reactions with -SH groups of essential molecules such as proteins or GSH; (2) direct inhibition of vital processes such as DNA synthesis; and (3) as redox-cycling compounds through autooxidation of the semi- or hydroquinone following the enzymatic reduction of the oxidized counterpart (Halliwell and Gutteridge, 1989; O'Brien, 1991). Through the process of cycling, ROS such as O_2^- are formed (eq. 10.11) (O'Brien, 1991).

$$\text{semiquinone radical} + O_2 \longrightarrow \text{quinone} + O_2^- + H^+ \quad (10.11)$$

The one-electron reduction of the quinone (eq. 10.12) occurs via flavoenzymes such as NADPH-cytochrome P-450 reductase, NADH-cytochrome b_5 reductase, and mitochondrial NADH dehydrogenase (Buffington et al., 1989).

$$\text{quinone} \xrightarrow[\text{1 } e^- \text{ reduction}]{\text{cellular reducing system}} \text{semiquinone radical} \quad (10.12)$$

DT-diaphorase (EC 1.6.99.2), formerly called quinone reductase, catalyses the direct two-electron reduction of quinones to hydroquinones (see chapter 7 for details):

$$\text{quinone} \xrightarrow[\text{2 } e^- \text{ reduction}]{\text{NAD(P)H}} \text{hydroquinone} \quad (10.13)$$

The two-electron reduction by DT-diaphorase decreases O_2^- formation by removing quinones and thus preventing their reduction to semiquinones by competing with the single electron reduction systems (Halliwell and Gutteridge, 1989).

DT-diaphorase, primarily a cytosolic enzyme, is present in most animal tissues (Ernster, 1987). The enzyme has been studied in several herbivorous insects and may protect against redox cycling of phenolics (Yu, 1987; Lindroth, 1989; Felton and Duffey, 1992).

10.5 REPAIR PROCESSES

Despite a concatenation of defenses, protection against ROS is incomplete. Many diseases (e.g., atheroscelerosis, cancer, Alzheimer's disease, arthritis, cataracts, cardiovascular disease) may result from the accumulation of oxidatively damaged macromolecules (Smith et al., 1991; Harman, 1992; Retsky et al., 1993; Ames et al., 1993). Organisms, however, possess several selective processes that can mitigate the accumulation of oxidatively damaged proteins, lipids, and nucleic acids.

10.5.1 Protein Damage

ROS may react with amino acids and can alter protein structure and function (Davies, 1987; Davies and Delsignore, 1987; Davies et al., 1987; Stadtman, 1993). In fact, all amino acid residues of a protein are susceptible to attack by ·OH (Stadtman, 1992). ROS may cause aggregation, fragmentation, amino acid modification and changes in proteolytic susceptibility (Davies and Goldberg, 1987). The accumulation of oxidatively modified protein has been correlated with the aging process (Carney et al., 1991; Smith et al., 1991). The enzyme glutamine synthetase is particularly susceptible to inactivation by ROS and losses in glutamine synthetase activity are strongly associated with Alzheimer disease (Smith et al., 1991).

Both fragmentation and aggregation can induce increased susceptibility of oxidatively damaged proteins to proteases such as trypsin or subtilisin (Wolff et al., 1986; Davies et al., 1987; Starke-Reed and Oliver, 1989). Thus, the higher turnover rate of damaged proteins may prevent their accumulation. The accumulation of oxidized proteins is also inhibited by selective degradation processes. Animal tissues contain a neutral protease that degrades the oxidized forms of proteins or enzymes but has little affinity for their unoxidized counterparts (Rivett, 1985; Stadtman, 1992). Rivett (1985) purified four intracellular proteases from mouse and rat liver that preferentially degrade oxidatively modified (catalytically inactive form) glutamine synthetase. The enzyme was oxidatively modified by ascorbate and ferric chloride that specifically causes a loss of a single histidine residue and the introduction of a carbonyl group (Rivett, 1985). Two of the proteases were calcium-dependent and degraded the oxidized glutamine synthetase at 26 to 32 times the rate of the unoxidized, active enzyme. There is evidence that the in vivo degradation of endogenous proteins in liver, heart, and red blood cells is greatly enhanced by exposure of cells to ROS generating systems or H_2O_2 (Davies and Goldberg, 1987; Stadtman, 1992). Davies et al. (1987) and Davies and Goldberg (1987) reported that oxidatively-damaged bovine serum albumin was preferentially degraded by novel, ATP- and calcium-independent proteases from cell-free rabbit erythrocytes at rates up 50 times faster than untreated albumin. Reticulocytes contain ATP-independent and ATP-dependent proteases that rapidly degrade oxidatively damaged intracellular proteins (Fagan et al., 1986).

Stadtman (1992) and Giulivi and Davies (1993) suggest that perhaps all animal tissues contain an intracellular, multicatalytic protease (in proteasomes) that selectively degrades oxidized proteins. These proteases appear to be present in both prokaryotic (see chapter 8) and eukaryotic cells.

Despite selective removal of damaged proteins, many key enzymes and proteins accumulate as catalytically inactive or denatured forms during oxidative stress and aging processes (Wolff et al., 1986; Starke-Reed and Oliver, 1989). Lipofuscin (a peroxidized lipid-protein aggregate) accumulates during aging or under conditions of oxidative stress (Wolff et al., 1986). This accumulation may occur as a result of declines in antioxidant defenses or loss of repair functions (Ames et al., 1993). Starke-Reed and Oliver (1989) found that in young rats, oxidative stress induces alkaline protease activities in hepatocytes, but in hepatocytes from old animals protease activities are low and not inducible. Accompanying the decline in protease activities is an accumulation of oxidatively modified proteins (Starke-Reed and Oliver, 1989; Stadtman, 1992). The intracellular accumulation of protein "junk" may severely affect cell function (Wolff, Garner et al., 1986). Carney et al. (Carney, Starke-Reed et al., 1991) reported that chronic treatment of aging gerbils with PBN resulted in an increase in neutral proteases of brain tissue with accompanying reversal of several age-related impairments in memory function.

10.5.2 Lipid Damage

LOOH are destroyed by GPOX and/or GSTpx as discussed previously in this chapter. Peroxidized lipids may be directly excised from membranes by a GPOX (PH-GPOX) or first liberated from membranes by a phospholipase and then degraded by PODs.

10.5.3 DNA Damage

The repair of DNA damage before becoming fixed as a mutation is of fundamental importance. There are several steps involved in DNA repair (Halliwell and Gutteridge, 1989). First, an excision enzyme (e.g., DNA glycosylase) removes bases damaged by ROS by cleaving base-sugar bonds to leave an apurinic/apyrimidine site. An endonuclease then nicks the DNA strand at this site to remove the damaged strand. DNA synthesis fills the gap in the strand and a DNA ligase joins the new DNA to the remaining strand (Halliwell and Gutteridge, 1989). This repair mechanism exists in prokaryotic and eukaryotic organisms. However, because the greatest body of literature is available from microbial systems, I refer the reader to Chapter 8 for a more through coverage of this topic.

10.6 SUMMARY AND FINAL COMMENTS

The diversity of scientific publications addressing oxidative stress certainly illustrates its broad importance to biology. An increased interest in invertebrate systems has taken place due to the recognition that reactive oxygen species (ROS) are important in many biological processes that affect growth, development, reproduction, and longevity. While damaging effects on DNA may not show themselves until weeks or even years after assault, damage to compartments within cells is nearly immediate. Therefore, protection against ROS is a necessity regardless if an organism lives but a few days or many decades. Specific disease states arising from oxidative stress in lower organisms have not been widely reported but should provide a basis for intensive study. Also, the need for developing nonmammalian model systems for biomedical research will no doubt encourage the use of new invertebrate systems for study.

Many of the antioxidant enzymes (e.g., CAT, SOD, GST, GR) are of very broad phylogenetic occurrence; however GPOX has more limited occurrence as primarily a vertebrate enzyme. Many of the other antioxidant proteins and enzymes (e.g., ceruloplasmin, AFR, and DHA reductase) have received such limited attention by invertebrate biologists that it is impossible to generalize about their phy-

Table 10.12.
Comparison of the Major Antioxidant Proteins in Humans and Insects

	Humans	Insects
Superoxide Dismutase (SOD)	Moderate Activity (21–38 units)	Low Activity (0.9–10.3 units)
CuZnSOD (cytosol)	50–85% of total	24–76% of total
MnSOD (mitochondrial)	15–50% of total	33–76% of total
Catalase (CAT)	Moderate Activity (50–100 units)	High Activity (100–1000 units)
Glutathione Peroxidase (GPOX)	High Activity	Absent
Glutathione Transferase (GSTpx) (Activity towards cumene hydroperoxide)	Moderate Activity	High Activity
Glutathione Reductase	High Activity	Moderate Activity
Ceruloplasmin	High Activity	No Reports
Transferrins	High Levels	High Levels
DT-Diaphorase	Moderate Activity	Moderate Activity
Ascorbate Peroxidase	No Reports	Possible
Ascorbate Free Reductase	Probable	Probable
Dehydroascorbic Acid Reductase	?	High Activity

Activities reported from humans are from liver, lung, kidneys, or heart; insects are midgut, fat body, muscle or whole body.
? = activity attributed to specific enzyme is questionable.
Source of Data: (Allen et al., 1983; Pritsos et al., 1988a, 1988b; Felton and Duffey, 1991a; Ahmad et al., 1991; Ahmad, 1992; Lee and Berenbaum, 1992; McElroy et al., 1992; Jamroz et al., 1993).

logenetic occurrence. In Table 10.12, I have compared data on the antioxidant systems in humans and insects, examples of what are probably the best studied vertebrate and invertebrate systems.

Two directions in antioxidant research will greatly contribute to this field. First, the ability to selectively alter gene expression (e.g., antisense DNA and RNA methodologies), has accelerated the understanding of individual antioxidant enzymes such as CAT and SOD. Site specific mutants of amino acid side chains implicated in electrostatic guidance of the human SOD have created SOD mutants that possess faster activity (Getzoff et al., 1992). These types of studies will no doubt shed considerable light of the functional, ecological, and evolutionary roles of antioxidant enzymes.

Second, there is a need for a greater understanding of the underlying biochemical basis of entire oxidative processes rather than individual reactions. Each component reaction in vivo takes place within a local environment that can significantly affect the rate and extent of overall oxidative processes. The understanding of environmental effects upon individual chemical reactions within the context of the biological system has been termed "holobiochemistry" (Minton, 1990). Many of the current studies on rates and equilibrium constants of antioxidants are conducted in homogenous and/or dilute solutions where volume occupancy and boundary effects are negligible. However, most physiological fluids contain between 5 and 50% by weight of macromolecules. The cytoplasm of animal cells contains numerous compartments, organelles, and extensively folded membranes that occupy as much as 25% of the cytoplasmic volume. Consequently, antioxidants, ROS, and other chemical components interact differentially with components of their local environments. Many of the oxidation-reduction reactions in vivo are taking place at or immediately adjacent to the surface of membranes, organelles, or other structural elements. The development of physical-chemical techniques for the assessment of these local environmental effects will accelerate the understanding of antioxidant function in vivo.

ACKNOWLEDGMENTS

I sincerely thank Dr. Sami Ahmad for his direction and guidance during the preparation of this chapter. I am especially grateful for his patience during the writing of this chapter, which came during a difficult personal time. I am very thankful for the excellent research by Mr. C. B. Summers that has been invaluable in elucidating antioxidant systems in insects. The grant support of the USDA and the Arkansas Science and Technology Authority has made this work possible.

References

Abe, Y., Okazaki, T., Shukuya, R. and Furuta, H. (1984) Copper and zinc superoxide dismutase from the liver of bullfrog *Rana catesbiana*. *Comp. Biochem. Physiol.* **77B**, 125–130.

Aceto, A., Dragani, B., Bucciarelli, T., Saccheta, P., Martini, F., Angelucci, S., Amicarelli, F., Miranda, M. and Di Ilio, C. (1993) Purification and characterization of the major glutathione transferase from adult toad (*Bufo bufo*) liver. *Biochem. J.* **289**, 417–422.

Adachi, T., Kodera, T., Ohta, H., Hayashi, K. and Hirano, K. (1992) The heparin binding site of human extracellular-superoxide dismutase. *Arch. Biochem. Biophys.* **297**, 155–161.

Adachi, T., Ohta, H., Hayashi, K., Hirano, K. and Marklund, S.L. (1992) The site of nonenzymic glycation of human extracellular-superoxide dismutase in vitro. *Free Rad. Bio. Med.* **13**, 205–210.

Adams, J.D., Klaidman, L.K., Odunze, I.N., Shen, H.C. and Miller, C.A. (1991) Alzeimer's and Parkinson's disease—brain levels of glutathione, glutathione disulfide, and vitamin E. *Mol. Chem. Neuropathol.* **14**, 213–226.

Ahmad, S. (1992) Biochemical defence of pro-oxidant plant allelochemicals by herbivorous insects. *Biochem. Syst. Ecol.* **20**, 269–296.

Ahmad, S., Duval, D.I., Weinhold, L.C. and Pardini, R.S. (1991) Cabbage looper antioxidant enzymes: tissue specificity. *Insect Biochem.* **21**, 563–572.

Ahmad, S. and Pardini, R.S. (1990) Antioxidant defense of the cabbage looper, *Trichoplusia ni*: enzymatic responses to the superoxide-generatin flavonoid, quercetin, and photodynamic furanocoumarin, xanthotoxin. *Photochem. Photobiol.* **51**, 305–311.

Ahmad, S., Pritsos, C.A., Bowen, S.M., Heisler, C.R., Blomquist, G.J. and Pardini, R.S. (1988a) Antioxidant enzymes of larvae of the cabbage looper moth, *Trichoplusia ni*: subcellular distribution and activities of superoxide dismutase, catalase, and glutathione reductase. *Free Rad. Res. Commun.* **4**, 403–408.

Ahmad, S., Pritsos, C.A., Bowen, S.M., Heisler, C.R., Blomquist, G.J. and Pardini, R.S. (1988b) Subcellular distribution and activities of superoxide dismutase, catalase, and glutathione reductase in the southern armyworm, *Spodoptera eridania*. *Arch. Insect Biochem. Physiol.* **7**, 173–186.

Aksnes, A. and Njaa, L.R. (1981) Catalase, glutathione peroxidase and superoxide dismutase in different fish species. *Comp. Biochem. Physiol.* **69B**, 893–896.

Alcain, F.J., Villalba, J.M., Low, H., Crane, F.L. and Navas, P. (1992) Ceruloplasmin stimulates NADH oxidation of pig liver plasma membrane. *Biochem. Biophys. Res. Commun.* **186**, 951–955.

Allen, R.G. (1991) Oxygen-reactive species and antioxidant responses during development: the metabolic paradox of cellular differentiation. *Proc. Soc. Exp. Biol. Med.* **196**, 117–129.

Allen, R.G., Farmer, K.J., Newton, R.K. and Sohal, R.S. (1984) Effects of paraquat administration on longevity, oxygen consumption, superoxide dismutase, catalase, glutathione reductase, inorganic peroxides and glutathione in the adult housefly. *Comp. Biochem. Physiol.* **78C**, 283–288.

Allen, R.G., Farmer, K.J. and Sohal, R.S. (1983) Effect of catalase inactivation on levels of inorganic peroxides, superoxide dismutase, glutathione, oxygen consumption, and life span in adult houseflies (*Musca domestica*). *Biochem. J.* **216**, 503–506.

Allen, R.G. and Sohal, R.S. (1986) Role of glutathione in the aging and development of insects. In *Insect Aging* (K.G. Gillatz and R.S. Schul, eds.), Springer-Verlag, Berlin, pp. 168–181.

Allen, R.G., Toy, P.L., Newton, R.K., Farmer, K.J. and Sohal, R.S. (1985) Effects of experimentally altered glutathione levels on life span, metabolic rate, superoxide dismutase, catalase and inorganic peroxides in the adult housefly, *Musca domestica*. *Comp. Biochem. Physiol.* **82C**, 399–402.

Ames, B.N., Cathcart, R., Schwiers, E. and Hochstein, P. (1981) Uric acid provides an antioxidant defense in humans against oxidant- and radical-caused aging and cancer: a hypothesis. *Proc. Natl. Acad. Sci. USA* **78**, 6858–6862.

Ames, B.N., Shigenaga, M.K. and Hagen, T.M. (1993) Oxidants, antioxidants, and the degenerative diseases of aging. *Proc. Natl. Acad. Sci. USA* **90**, 7915–7922.

Anderson, R.S., Oliver, L.M. and Brubacher, L.L. (1992) Superoxide anion generation by *Crassostrea virginica* hemocytes as measured by nitroblue tetrazolium reduction. *J. Invertebr. Pathol.* **59**, 303–307.

Appel, H.M. (1993) Phenolics in ecological interactions: the importance of oxidation. *J. Chem. Ecol.* **19**, 1521–1552.

Arouma, O.I. (1993) *Free Radicals in Tropical Diseases*. Harwood Academic, New York.

Arrigoni, O., Dipierro, S. and Borracino, G. (1981) Asorbate free radical reductase, a key enzyme of the ascorbic acid system. *FEBS Lett.* **125**, 242–244.

Asada, K. (1992) Ascorbate peroxidase-a hydrogen peroxide-scavenging enzyme in plants. *Physiol. Plant.* **85**, 235–241.

Aucoin, R.R., Fields, P., Lewis, M.A., Philogene, B.J.R. and Arnason, J.T. (1990) The protective effect of antioxidants to a phototoxin-sensitive insect herbivore, *Manduca sexta*. *J. Chem. Ecol.* **16**, 2913–2924.

Aucoin, R.R., Philogene, B.J.R. and Arnason, J.T. (1991) Antioxidant enzymes as biochemical defenses against phototoxin-induced oxidative stress in three species of herbivorous Lepidoptera. *Arch. Insect Biochem. Physiol.* **16**, 139–152.

Avissar, N., Slemmon, J.R., Plamer, I.S. and Cohen, H.J. (1991) Partial sequence of human plasma glutathione peroxidase and immunological identification of milk glutathione peroxidase as the plasma enzyme. *J. Nutr.* **121**, 1243–1249.

Babbs, C.F. (1992) Oxygen radicals in ulcerative colitis. *Free Rad. Biol. Med.* **13**, 169–181.

Baker, G.T. (1993) Effects of various antioxidants on aging in *Drosophila*. *Toxicol. Environ. Hlth.* **9**, 163–186.

Baliga, R., Baliga, M. and Shah, S.V. (1992) Effect of selenium-deficient diet in experimental glomerular disease. *Am. J. Physiol.* **263**, F56–F61.

Barja de Quiroga, G., Gil, P. and Alonso-Bedate, M. (1985) Catalase enzymatic activity and electrophoretic patterns in adult amphibians-a comparative study. *Comp. Biochem. Physiol.* **80B**, 853–858.

Barja de Quiroga, G., Gutierrez, P., Rojo, S. and Alonso-Bedate, M. (1984) A comparative study of superoxide dismutase in amphibian tissues. *Comp. Biochem. Physiol.* **77B**, 589–593.

Barja de Quiroga, G., Lopez-Torres, M. and Gil, P. (1989a) Hyperoxia decreases lung size of amphibian tadpoles without changing GSH-peroxidases or tissue peroxidation. *Comp. Biochem. Physiol.* **92A**, 581–588.

Barja de Quiroga, G., Lopez-Torres, M. and Perez-Campo, R. (1989b) Catalase is needed to avoid tissue peroxidation in *Rana perezi* in normoxia. *Comp. Biochem. Physiol.* **94C**, 391–398.

Barrett, J. and Beis, I. (1982) Catalase in free-living and parasitic platyhelminths. *Experientia* 38, 536.

Bartoli, G.M., Palozza, P. and Piccioni, E. (1992) Enhanced sensitivity to oxidative stress in Cu, ZnSOD depleted rat erythrocytes. *Biochem. Biophys. Acta* 1123, 291–295.

Batra, S., Chatterjee, R.K. and Srivastava, V.M.L. (1990) Antioxidant enzymes in *Acanthocheilonema viteae* and effect of antifilarial agents. *Biochem. Pharmacol.* 40, 2363–2369.

Batra, S., Singh, S.P., Fatma, N., Sharma, S., Chatterjee, R.K. and Srivastava, V.M.L. (1992) Effect of 2,2'-dicarbomethoxylamino-5,5'-dibenzimidazolyl ketone on antioxidant defenses of *Acanthocheilonema viteae* and its laboratory host *Mastomys natalensis*. *Biochem. Pharmacol.* 44, 727–731.

Batra, S., Singh, S.P., Srivastava, V.M.L. and Chatterjee, R.K. (1989) Xanthine oxidase, superoxide dismutase, catalase and lipid peroxidation in *Mastomys natalensis*: effect of *Dipetalonema viteae* infection. *Ind. J. Exp. Biol.* 27, 1067–1070.

Berenbaum, M. (1978) Toxicity of a furanocoumarin to armyworms: A case of biosynthetic escape from insect herbivores. *Science* 201, 532–534.

Berenbaum, M. (1987) Charge of the light brigade. In *Light Activated Pesticides. ACS Symposium Ser. 339* (J.R. Heitz and K.R. Dowrum, eds.), Am. Chem. Soc., Washington, D.C., pp. 206–216.

Bergman, D.K. and Chippendale, G.M. (1992) Carotenoid transport by the larval lipohorin of the southwestern corn borer, *Diatraea grandiosella*. *Entomol. Exp. Appl.* 62, 81–85.

Best-Belpomme, M. and Ropp, M. (1982) Catalase is induced by ecdysterone and ethanol in *Drosophila* cells. *Eur. J. Biochem.* 121, 349–355.

Beutler, E. (1989) Nutritional and metabolic aspects of glutathione. *Ann. Rev. Nutr.* 9, 287–302.

Bewley, G.C. and Mackay, W.J. (1989) Development of a genetic model for acatalasemia: testing the oxygen free radical theory of aging. In *Genetic Effects on Aging* (D.E. Harrison, ed.), Telford Press, Caldwell, New Jersey, 359–378.

Blum, J. and Fridovich, I. (1983) Superoxide, hydrogen peroxide, and oxygen toxicity in two free-living nematode species. *Arch. Biochem. Biophys.* 222, 35–43.

Blum, J. and Fridovich, I. (1984) Enzymatic defenses against oxygen toxicity in the hydrothermal vent animals *Riftia pachyptila* and *Calyptogena magnifica*. *Arch. Biochem. Biophys.* 228, 617–620.

Bogaards, J.J.P., van Ommen, B., and van Bladeren, P.J. (1992) Purification and characterization of eight glutathione S-transferase isoenzymes of hamster. *Biochem. J.* 286, 383–388.

Bondy, S.C. (1992) Reactive oxygen species: relation to aging and neurotoxic damage. *Neurotoxicol.* 13, 87–100.

Borraccino, G., Dipierro, S., and Arrigoni, O. (1989) Interaction of ascorbate free radical reductase with sulphhydryl reagents. *Phytochemistry* 28, 715–717.

Boveris, A.H., Sies, H., Martino, E.E., Decampo, R., Turreus, J.F., and Stoppani, A.O.M. (1980) Deficient metabolic utilization of hydrogen peroxide in *Trypanosoma cruzi*. *Biochem. J.* 188, 643–648.

Bowditch, M.I. and Donaldson, R.P. (1990) Ascorbate free-radical reduction by glyoxysomal membranes. *Plant Physiol.* **94**, 531–537.

Boyer, R.F. and Schori, B.E. (1983) The incorporation of iron into apoferritin as mediated by ceruloplasmin . *Biochem. Biophys. Res. Commun.* **116**, 244–250.

Braddon-Galloway, S. and Balthrop, J.E. (1985) Se-dependent GSH-peroxidase isolated from black sea bass (*Centropristis striata*). *Comp. Biochem. Physiol.* **82C**, 297–300.

Brophy, P.M. and Barrett, J. (1990a) Strategies for detoxication of aldehyde products of lipid peroxidation in helminths. *Mol. Biochem. Parasitol.* **42**: 205–212.

Brophy, P.M. and Barrett, J. (1990b) Glutathione transferase in helminths. *Parasitol.* **100**, 345–349.

Brophy, P.M., Crowley, P. and Barrett, J. (1990) A novel NADPH/NADH-dependent aldehyde reduction enzyme isolated from the tapeworm *Moniezia expansa*. *FEBS Lett.* **263**, 305–307.

Brophy, P.M. and Pritchard, D.I. (1992) Metabolism of lipid peroxidation products by the gastro-intestinal nematodes *Necator americanus*, *Ancylostoma ceylanicum* and *Heligmosomoides polygrus*. *Int. J. Parasitol.* **22**, 1009–1012.

Brophy, P.M., Southan, C. and Barrett, J. (1989) Glutathione transferases in the tapeworm *Moniezia expansa*. *Biochem. J.* **262**, 939–946.

Bucheler, U.S., Werner, D. and Schirmer, R.H. (1992) Generating compatible translation initiation regions for heterologous gene expression in *Escherichia coli* by exhaustive periShine-Dalgarno mutagenesis. Human glutathione reductase cDNA as a model. *Nucleic Acids Res.* **20**(12), 3127–3133.

Buettner, G.R. (1993) The pecking order of free radicals and antioxidants: lipid peroxidation, a-tocopherol, and ascorbate. *Arch. Biochem. Biophys.* **300**, 535–543.

Buffington, G.D., Ollinger, K., Brunmark, A. and Cadenas, E. (1989) DT-diaphorase-catalysed reduction of 1,4-naphthoquinone derivatives and glutathionyl conjugates. *Biochem. J.* **257**, 561–571.

Burk, R.F. and Hill, K.E. (1993) Regulation of selenoproteins. *Ann. Rev. Nutr.* **13**, 65–81.

Byers, T. and Perry, G. (1992) Dietary carotenes, vitamin C, and vitamin E as protective antioxidants in human cancers. *Ann. Rev. Nutr.* **12**, 139–159.

Callahan, H.L., Crouch, R.K. and James, E.R. (1988) Helminth antioxidant enzymes: a protective mechanism against host oxidants? *Parasitol. Today* **4**, 219–229.

Callahan, H.L., Hazen-Martin, D., Crouch, R.K. and James, E.R. (1993) Immunolocalization of superoxide dismutase in *Dirofilaria immitis* adult worms. *Infect. Immun.* **61**, 1157–1163.

Canfield, L.M., Forage, J.W. and Valenzuela, J.G. (1992) Carotenoids as cellular antioxidants. *Proc. Soc. Exp. Biol. Med.* **200**, 260–265.

Carney, J.M., Starke-Reed, P.E., Oliver, C.N., Landum, R.W., Cheng, M.S., Wu, J.F. and Floyd, R.A. (1991) Reversal of age-related increase in brain protein oxidation, decrease in enzyme activity, and loss in temporal and spatial memory by chronic administration of the spin-trapping compound N-tert-butyl-a-phenylnitrone. *Proc. Natl. Acad. Sci. USA* **88**, 3633–3636.

Ceballos-Picot, I., Nicole, A., Clement, M., Bourre, J. and Sinet, P. (1992) Age-related changes in antioxidant enzymes and lipid peroxidation in brains of control and transgenic mice overexpressing copper-zinc superoxide dismutase. *Mut. Res.* **275**, 281–293.

Chance, B., Sies, H. and Boveris, A. (1979) Hydroperoxide metabolism in mammalian organs. *Physiol. Rev.* **59**, 527–605.

Chaudhry, M.Q. and Price, N.R. (1992) Comparison of the oxidant damage induced by phosphine and the uptake and tracheal exchange of ^{32}-P-radiolabelled phosphine in the susceptible and resistant strains of *Rhyzopertha dominica* (F.) (Coleoptera: Bostrychidae). *Pestic. Biochem. Physiol.* **42**, 167–179.

Choi, J.-L. and Rose, R.C. (1989) Regeneration of ascorbic acid by rat colon. *Proc. Soc. Exp. Biol. Med.* **190**, 369–374.

Chorazy, P.A., Schumaker, H.R. and Edlind, T.D. (1992) Role of glutathione peroxidase in rheumatoid arthritis: analysis of enzyme activity and DNA polymorphism. *DNA Cell Biol.* **11**, 221–225.

Christensen, H.N. (1990) Role of amino acid transport and countertransport in nutrition and metabolism. *Physiol. Rev.* **70**, 43–77.

Christensen, M.J. and Burgener, K.W. (1992) Dietary selenium stabilizes glutathione peroxidase mRNA in rat liver. *J. Nutr.* **122**, 1620–1626.

Chu, F.F., Doroshow, J.H. and Esworthy, R.S. (1993) Expression, characterization, and tissue distribution of a new cellular selenium-dependent glutathione peroxidase, GSHPx-GI. *J. Biol. Chem.* **268**, 2571–2576.

Church, S.L., Grant, J.W., Ridnour, L.A., Oberley, L.W., Swanson, P.E., Meltzer, P.S. and Trent, J.M. (1993) Increased manganese superoxide dismutase expression suppresses the malignant phenotype of human melanoma cells. *Proc. Natl. Acad. Sci. USA* **90**, 3113–3117.

Clerch, L.B. and Massaro, D. (1992) Rat lung antioxidant enzymes: differences in perinatal gene expression and regulation. *Am. J. Physiol.* **263**, L466–L470.

Coassin, M., Tomasi, A., Vannini, V. and Ursini, F. (1991) Enzymatic recycling of oxidized ascorbate in pig heart: one-electron vs two-electron pathway. *Arch. Biochem. Biophys.* **290**, 458–462.

Coles, B. and Ketterer, B. (1990) The role of glutathione and transferases in chemical carcinogenesis. *Crit. Rev. Biochem. Mol. Biol.* **25**, 47–30.

Cousins, R.J. (1985) Absorption, transport, and hepatic metabolism of copper and zinc: special reference to metallothionein and ceruloplasmin. *Physiol. Rev.* **65**, 238–309.

Czapski, G. (1991) *Fifth Conference on Superoxide and Superoxide Dismutase* (G. Czapski, guest ed.) Harwood Academic, New York.

Dabrowski, K. (1990). Absorption of ascorbic acid and ascorbic sulfate and ascorbate metabolism in common carp (*Cyprinus carpio* L.). *J. Comp. Physiol. B* **160**, 549–561.

Dalton, D.A., Russell, S.A., Hanus, F.J., Pascoe, G.A. and Evans, H.J. (1986) Enymatic reactions of ascorbate and glutathione that prevent peroxide damage in soybean root nodules. *Proc. Natl. Acad. Sci. USA* **83**, 3811–3815.

Davies, K.J.A. (1987) Protein damage and degradation by oxygen radicals I. General Aspects. *J. Biol. Chem.* **262**, 9895–9901.

Davies, K.J.A. and Delsignore, M.E. (1987) Protein damage and degradation by oxygen radicals III. Modification of secondary and tertiary structures. *J. Biol. Chem.* **262**, 9908–9913.

Davies, K.J.A., Delsignore, M.E. and Lin, S.W. (1987) Protein damage and degradation by oxygen radicals II. Modification of amino acids. *J. Biol. Chem.* **262**, 9902–9907.

Davies, K.J.A. and Goldberg, A.L. (1987) Proteins damaged by oxygen radicals are rapidly degraded in extracts of red blood cells. *J. Biol. Chem.* **262**, 8227–8234.

de Silva, D. and Aust, S.D. (1992) Stoichiometry of Fe(II) oxidation during ceruloplasmin-catalyzed loading of ferritin. *Arch. Biochem. Biophys.* **298**, 259–264.

del Rio, L.A., Sandalio, L.M., M., P.J., Bueno, P. and Corpas, F.J. (1992) Metabolism of oxygen radicals in peroxisomes and cellular implications. *Free Rad. Biol. Med.* **13**, 557–580.

Deng, H., Hentati, A., Tainer, J.A., Iqbal, Z., Cayabyab, A., Hung, W. et al. (1993) Amyotrophic lateral sceloris and structural defects in Cu,Zn superoxide dismutase. *Science* **261**, 1047–1051.

Di Giulio, R.T., Washburn, P.C., Wenning, R.J., Winston, G.W. and Jewell, C.S. (1989) Biochemical responses in aquatic animals: a review of determinants of oxidative stress. *Environ. Toxicol. Chem.* **8**, 1103–1123.

Di Mascio, P., Kaiser, S. and Sies, H. (1989) Lycopene as the most efficient biological carotenoid singlet oxygen quencher. *Arch. Biochem. Biophys.* **274**, 532–538.

Diliberto, E.J., Dean, G., Carter, C. and Allen, P.L. (1982) Tissue, subcellular, and submitochondrial distributions of semidehydroascorbate reductase: possible role of semidehydroascorbate reductase in cofactor regeneration. *J. Neurochem.* **39**, 563–568.

Dipierro, S. and Borraccino, G. (1991) Dehydroascorbate reductase from potato tubers. *Phytochemistry* **30**, 427–429.

DiSilvestro, R.A., Marten, J. and Skehan, M. (1992) Effects of copper supplementation on ceruloplasmin and copper-zinc superoxide dismutase in free-living rheumatoid arthritis patients. *J. Amer. Coll. Nutr.* **11**, 177–180.

Dominey, R.J., Nimmo, I.A., Cronshaw, A.D. and Hayes, J.D. (1991) The major glutathione S-transferase in salmond fish livers is homologous to the mammalian pi-class GST. *Comp. Biochem. Physiol.* **100B**, 93–98.

Douglas, K.T. (1986) Mechanism of action of glutathione-dependent enzymes. *Adv. Enzymol.* **58**, 103–167.

Duan, Y.-J., Komura, S., Fiszer-Szafarz, B., Safarz, D. and Yagi, K. (1988) Purification and characterization of a novel monomeric glutathione peroxidase from rat liver. *J. Biol. Chem.* **263**, 11003–19008.

Dykens, J.A., Shick, J.M., Benoit, C., Buettner, G.R. and Winston, G.W. (1992) Oxygen radical production in the sea anemone *Anthopleura elegantissima* and its endosymbiotic algae. *J. Exp. Biol.* **168**, 219–241.

Edlund, A., Edlund, T., Hjalmarsson, K., Marklund, S.L., Sandstrom, J., Stromquist, M. and Tibell, L. (1992) A non-glycosylated extracellular superoxide dismutase variant. *Biochem. J.* **288**, 451–456.

Englard, S. and Seifter, S. (1986) The biochemical functions of ascorbic acid. *Ann. Rev. Nutr.* **6**, 365–406.

Epstein, C.J., Abraham, K.B., Lovett, M., Smith, S., Elroy-Stein, O., Rotman, G., Bry, C. and Groner, Y. (1987) Transgenic mice with increased Cu/Zn superoxide dismutase activity: animal model of dosage effects in Down syndrome. *Proc. Natl. Acad. Sci. USA* **84**, 8044–8048.

Ernster, L. (1987) DT diaphorase: a historical review. *Chem. Scripta* **27A**, 1–13.

Fagan, J.M., Waxman, L. and Goldberg, A.L. (1986) Red blood cells contain a pathway for the degradation of oxidant-damaged hemoglobin that does not require ATP or ubiquitin. *J. Biol. Chem.* **262**, 5505–5713.

Fahey, R.C. and Sundquist, A.R. (1991) Evolution of glutathione metabolism. In *Advances in Enzymology and Related Areas of Molecular Biology*, Vol. 64 (A. Meister, ed.), John Wiley & Sons, New York, 1–53.

Felton, G.W. and Duffey, S.S. (1991a) Protective action of midgut catalases in lepidopteran larvae against oxidative plant defenses. *J. Chem. Ecol.* **17**, 1715–1732.

Felton, G.W. and Duffey, S.S. (1991b) A reassessment of the role of gut alkalinity and detergency in insect herbivory. *J. Chem. Ecol.* **17**, 1821–1836.

Felton, G.W. and Duffey, S.S. (1992) Ascorbate oxidation-reduction in *Heliocoverpa zea* as a scavenging system against dietary oxidants. *Arch. Insect Biochem. Physiol.* **19**, 27–37.

Felton, G.W., Workman, J. and Duffey, S.S. (1992) Avoidance of an antinutritive plant defense: Role of midgut pH in the Colorado Potato beetle. *J. Chem. Ecol.* **18**, 571–583.

Fernandez-Sousa, J.M. and Michelson, A.M. (1976) Variation of superoxide dismutase during the development of the fruit fly *Ceratitis capitata* (L). *Biochem. Biophys. Res. Commun.* **73**, 217–223.

Fields, P., Arnason, J.T. and Philogene, B.J.R. (1990) The behavioural and physical adaptations of three insects that feed on the phototoxic plant *Hypericum perforatum*. *Can. J. Zool.* **68**, 339–346.

Fleming, J.E., Reveillaud, I. and Niedzwiecki, A. (1992) Role of oxidative stress in Drosphila aging. *Mut. Res.* **275**, 267–279.

Frank, L., Bucher, J.R. and Roberts, R.J. (1978) Oxygen toxicity in neonatal and adult animals of various species. *J. Appl. Physiol.* **45**, 699–704.

Frei, B., Stocker, R. and Ames, B.N. (1988) Antioxidant defenses and lipid peroxidation in human blood plasma. *Proc. Natl. Acad. Sci. USA* **85**, 9748–9752.

Frei, B., England, L. and Ames, B.N. (1989) Ascorbate is an outstanding antioxidant in human blood plasma. *Proc. Natl. Acad. Sci. USA* **86**, 6377–6381.

Fridovich, I. (1989) Superoxide dismutase-an adaptation to a paramagnetic gas. *J. Biol. Chem.* **264**, 7761–7764.

Gallagher, E.P. and Di Giulio, R.T. (1992) A comparison of glutathione-dependent enzymes in liver, gills and posterior kidney of channel catfish (*Ictalurus punctatus*). *Comp. Biochem. Physiol.* **102C**, 543–547.

Geiss, D. and Schulze, H.-U. (1975) Isolation and chemical composition of the NADH: semidehydroascorbate oxidoreductase rich membranes from rat liver. *FEBS Lett.* **60**, 374–379.

Getzoff, E.D., Cabelli, D., Fisher, C.L., Parge, H.E., Viezzoli, M.S., Banci, L. and Hallewell, R.A. (1992) Faster superoxide dismutase mutants designed by enhancing electrostatic guidance. *Nature* **358**, 347–351.

Giuliva, C. and Davies, K.J.A. (1993) Dityrosine and tyrosine oxidation products are endogenous markers for the selected proteolysis of oxidatively modified red blood cell hemoglobin by the (19S) proteasome. *J. Biol. Chem.* **268**, 8752–9759.

Godin, D.V. and Garnett, M.E. (1992a) Species-related variations in tissue antioxidant status—I. Differences in antioxidant enzyme profiles. *Comp. Biochem. Physiol.* **103B**, 737–742.

Godin, D.V. and Garnett, M.E. (1992b) Species-related variation in tissue antioxidant status—II. Differences in susceptibility to oxidative challenge. *Comp. Biochem. Physiol.* **103B**, 743–748.

Goldfarb, P., Spry, J.A., Dunn, D., Livingstone, D.R., Wiseman, A. and Gibson, G.G. (1989) Detection of mRNA sequences homologous to the human glutathione peroxidase and rat cytochrome P-450IVA1 genes in *Mytilus edulis*. *Marine Environ. Res.* **28**, 57–60.

Goldstein, I.M., Kaplan, H.B., Edelson, H.S. and Weissmann, G. (1979) Ceruloplasmin: a scavenger of superoxide anion radicals. *J. Biol. Chem.* **254**, 4040–4045.

Goodwin, T.W. (1986) Metabolism, nutrition, and function of carotenoids. *Ann. Rev. Nutr.* **6**, 273–297.

Green, R.C. and O'Brien, P.J. (1973) The involvement of semidehydroascorbate reductase in the oxidation of NADH by lipid peroxide in mitochondria and microsomes. *Biochim. Biophys. Acta* **293**, 334–342.

Grimble, R.F. and Hughes, R.E. (1967) A "dehydroascorbic acid reductase" factor in guinea-pig tissues. *Experientia* **23**, 362.

Gutteridge, J.M.C. and Quinlan, G.J. (1993) Antioxidant protection against organic and inorganic oxygen radicals by normal human plasma: the important primary role for iron-binding and iron-oxidizing proteins. *Biochim. Biophys. Acta* **1156**, 144–150.

Halliwell, B. and Gutteridge, J.M.C. (1989) *Free Radicals in Biology and Medicine*. Clarendon Press, Oxford.

Halliwell, B. and Gutteridge, J.M.C. (1990a) Role of free radicals and catalytic metal ions in human disease: an overview. *Methods Enzymol.* **186**, 1–85.

Halliwell, B. and Gutteridge, J.M.C. (1990b) The antioxidants of human extracellular fluids. *Arch. Biochem. Biophys.* **280**, 1–8.

Hara, T. and Minakami, S. (1971) On functional role of cytochrome b5. II. NADH-linked ascorbate radical reductase activity in microsomes. *J. Biochem.* **69**, 325–330.

Harman, D. (1992) Free radical theory of aging. *Mut. Res.* **275**, 257–266.

Harris, E.D. (1992a) Regulation of antioxidant enzymes. *FASEB J.* **6**, 2675–2683.

Harris, E.D. (1992b) Regulation of antioxidant enzymes. *J. Nutr.* **122**, 625–626.

Harris, E.D. (1992c) Copper as a cofactor and regulator of copper, zinc superoxide dismutase. *J. Nutr.* **122**, 636–640.

Harris, J., Coles, B., Meyer, D.J. and Ketterer, B. (1991) The isolation and characterization of the major glutathione S-transferase from the squid *Loligo vulgaris*. *Comp. Biochem. Physiol.* **98B**, 511–515.

Hatori, N., Sjoquist, P., Marklund, S.J. and Ryden, L. (1992) Effects of recombinant human extracellular-superoxide dismutase type C on myocardial infarct size in pigs. *Free Rad. Biol. Med.* **13**, 221–230.

Hausladen, A. and Kunert, K.J. (1990) Effects of artificially enhanced levels of ascorbate and glutathione on the enzymes monodehydroascorbate reductase, dehydroascorbate reductase, and glutathione reductase in spinach (*Spinacia oleracea*). *Physiol. Plant.* **79**, 385–388.

Hayashi, T., Kanetoshi, A., Nakamura, M., Tamura, M. and Shirahama, H. (1992) Reduction of α-tocopherolquinone to α-tocopherolhydroquinone in rat hepatocytes. *Biochem. Pharmacol.* **44**, 489–493.

Hemila, H. (1992) Vitamin C and plasma cholesterol. *Crit. Rev. Food Sci. Nutr.* **32**, 33–57.

Hendry, G. (1992) Oxygen, the great destroyer. *Natural History* **8**, 46–52.

Himeno, S., Takekawa, A. and Imura, N. (1993) Species difference in hydroperoxide-scavenging enzymes with special reference to glutathione peroxidase. *Comp. Biochem. Physiol.* **104B**, 27–31.

Ho, C.T. and Chan, A.C. (1992) Regeneration of vitamin E in rat polymorphonuclear leucocytes. *FEBS* **306**, 269–272.

Hossain, M.A. and Asada, K. (1984) Purification of dehydroascorbate reductase from spinach and its characterization as a thiol enzyme. *Plant Cell Physiol.* **25**, 85–92.

Hossain, M.A., Nakano, Y. and Asada, K. (1984) Monodehydroascorbate reductase in spinach chloroplasts and its participation in regeneration of ascorbate for scavenging hydrogen peroxide. *Plant Cell Physiol.* **25**, 385–395.

Hsu, J., Visner, G.A., Burr, I.A. and Nick, H.S. (1992) Rat copper/zinc superoxide dismutase gene: isolation, characterization, and species comparison. *Biochem. Biophys. Res. Comm.* **186**, 936–943.

Hu, M.-L. and Tappel, A.L. (1992) Glutathione and antioxidants protect microsomes against lipid peroxidation and enzyme inactivation. *Lipids* **27**, 42–45.

Huebers, H.A., Huebers, E., Finch, C.A., Webb, B.A., Truman, J.W., Riddiford, L.M., Martin, A.W. and Massover, W.H. (1988) Iron binding proteins and their roles in the tobacco hornworm, *Manduca sexta* (L.). *J. Comp. Physiol. B* **158**, 291–300.

Hughes, R.E. (1964) Reduction of dehydroascorbic acid by animal tissues. *Nature* **203**, 1068–1069.

Hughes, R.E. and Kilpatrick, G.S. (1964) The reduction of dehydroascorbic acid by haemolysates of pernicious anaemia erythrocytes. *Clin. Chim. Acta* **9**, 241–244.

Hwang, C., Sinskey, A.J. and Lodish, H.F. (1992) Oxidized redox state of glutathione in the endoplasmic reticulum. *Science* **257**, 1496–1502.

Ingold, K.U., Bowry, V.W., Stocker, R. and Walling, C. (1993) Autooxidation of lipids and antioxidation by α-tocopherol and ubiquinol in homogeneous solution and in aqueous dispersions: unrecognized consequences of lipid particle size as exemplified by oxidation of human low density lipoprotein. *Proc. Natl. Acad. Sci. USA* **90**, 45–49.

Ito, A., Hayashi, S. and Yoshida, T. (1981) Participation of a cytochrome b5-like hemoprotein of outer mitochondrial membrane (OM cytochrome b) in NADH-semidehydroascorbic acid reductase activity of rat liver. *Biochem. Biophys. Res. Commun.* **101**, 591–598.

Ito, Y., Hiraishi, H., Razandi, M., Terano, A., Harada, T. and Ivey, K. (1992) Role of cellular superoxide dismutase against reactive oxygen metabolite-induced cell damage in cultured rat hepatocytes. *Hepatology* **16**, 247–254.

Jacobson, J.M., Michael, J.R., Jafri, M.A. and Gurtner, G.H. (1990) Antioxidant enzymes protect against oxygen toxicity in the rabbit. *J. Appl. Physiol.* **68**, 1252–1259.

Jaeschke, H. (1991) Reactive oxygen and ischemia/reperfusion injury of the liver. *Chem.-Biol. Interact.* **79**, 115–136.

Jamroz, R.C., Gasdaska, J.R., Bradfield, J.Y. and Law, J.H. (1993) Transferrin in a cockroach: molecular cloning, characterization, and suppression by juvenile hormone. *Proc. Natl. Acad. Sci. USA* **90**, 1320–1324.

Jialal, I., Norkus, E.P., Cristol, L. and Grundy, S.M. (1991) β-Carotene inhibits the oxidative modification of low-density lipoprotein. *Biochim. Biophys. Acta* **1086**, 134–138.

Jimenez, D.R. and Gilliam, M. (1988) Cytochemistry of peroxisomal enzymes in microbodies of the midgut of the honey bee, *Apis mellifera*. *Comp. Biochem. Physiol.* **90B**, 757–766.

Karplus, P.A. and Schulz, G.E. (1987) Refined structure of glutathione reductase at 1.54 angstrom resolution. *J. Mol. Biol.* **195**, 701–729.

Karplus, P.A. and Schulz, G.E. (1989) Substrate binding and catalysis by glutathione reductase as derived from refined enzyme: substrate crystal structures at 2 angstrom resolution. *J. Mol. Biol.* **210** 163–180.

Kaul, K., Lam, K.-W., Fong, D., Lok, C., Berry, M. and Treble, D. (1988) Ascorbate peroxidase in bovine retinal pigment epithelium and chroid. *Curr. Eye. Res.* **7**, 675–679.

Keller, G.A., Warner, T.G., Steimer, K.S. and Hallewell, R.A. (1991) Cu, Zn superoxide dismutase is a peroxisomal enzyme in human fibriblasts and hepatoma cells. *Proc. Natl. Acad. Sci. USA* **88**, 7381–7385.

Kennedy, T.A. and Liebler, D.C. (1992) Peroxyl radical scavenging by β-carotene in lipid bilayers. *J. Biol. Chem.* **267**, 4658–4663.

Kensler, T.W. and Guyton, K.Z. (1992) Modulation of carcinogenesis by antioxidants. In *Biological Consequences of Oxidative Stress-Implications for Cardiovascular Disease and Carcinogenesis* (L. Spatz and Bloom, A.D., eds.), Oxford University Press, New York, 162–186.

Kevers, C., Goldberg, R., Driessche, T.V. and Gaspar, T. (1992) A relationship between ascorbate peroxidase activity and the conversion of 1-aminocyclopropane-1-carboxylic acid into ethylene. *J. Plant Physiol.* **139**, 379–381.

Khatami, M., Roel, L.E., Li, W. and Rockey, J.H. (1986) Ascorbate regeneration in bovine ocular tissues by NADPH-dependent semidehydroascorbate reductase. *Exp. Eye Res.* **43**, 167–175.

Klebanoff, S.J. (1992) Bactericidal effect of Fe^{2+}, ceruloplasmin, and phosphate. *Arch. Biochem. Biophys.* **295**, 302–308.

Knipping, G., Rotheneder, M., Striegl, G. and Esterbauer, H. (1990) Antioxidants and resistance against oxidation of porcine LDL subfractions. *J. Lipid Res.* **31**, 1965–1972.

Kramer, K.J. and Seib, P.A. (1982) Ascorbic acid in the growth and development of insects. In *Ascorbic Acid: Chemistry, Metabolism, and Uses* (P.A. Seib and Tolbert, B.N. eds.) Am. Chem. Soc., Washington D.C., 275–291.

Krauth-Siegel, R.L., Blatterspiel, R., Saleh, M., Schiltz, E., Schirmer, R.H. and Untucht-Grau, R. (1982) Glutathione reductase from human erythrocytes. *Eur. J. Biochem.* 121, 259–267.

Krinsky, N.I. (1992) Mechanism of action of biological antioxidants. *P.S.E.B.M.* 200, 248–254.

Krinsky, N.I. (1993) Actions of carotenoids in biological systems. *Ann. Rev. Nutr.* 13, 561–587.

Krsek-Staples, J.A., Kew, R.R. and Webster, R.O. (1992) Ceruloplasmin and transferrin levels are altered in serum and brocnoalveolar lavage fluid of patients with respiratory distress syndrome. *Am. Rev. Respir. Dis.* 145, 1009–1015.

Krsek-Staples, J.A. and Webster, R.O. (1993) Ceruloplasmin inhibits carbonyl formation in endogenous cell proteins. *Free Rad. Biol. Med.* 14, 115–125.

Kunert, K.J. and Tappel, A.L. (1983) The effect of vitamin C on in vivo lipid peroxidation in guinea pigs as measured by pentane and ethane production. *Lipids* 18, 271–274.

Kurata, M., Suzuki, M. and Takeda, K. (1993) Diferrences in levels of erythrocyte glutathione and its metabolizing enzyme activities among primates. *Comp. Biochem. Physiol.* 104B, 169–171.

Lash, L.H., Hagen, T.M. and Jones, D.P. (1986) Exogenous glutathione protects intestinal epithelial cells from oxidative injury. *Proc. Natl. Acad. Sci. USA* 83, 4641–4645.

LeBel, C.P. and Bondy, S.C. (1992) Oxidative damage and cerebral aging. *Progress Neurobiol*, 38, 601–609.

Lee, H.S., Louriminia, S.S., Clifford, A.J., Whitaker, J.R. and Feeny, R.A. (1978) Effect of reductive alkylation of the ε-amino group of lysyl residues of casein on its nutritive value in rats. *J. Nutrition* 108, 687–697.

Lee, K. (1991) Glutathione S-transferase activities in phytophagous insects: induction and inhibition by plant phototoxins and phenols. *Insect Biochem.* 21, 353–361.

Lee, K. and Berenbaum, M.R. (1989) Action of antioxidant enzymes and cytochrome P-450 monoxygenases in the cabbage looper in response to plant phototoxins. *Arch. Insect Biochem. Physiol.* 10, 151–162.

Lee, K. and Berenbaum, M.R. (1990) Defense of parsnip webworm against phototoxic furanocoumarins: role of antioxidant enzymes. *J. Chem. Ecol.* 16, 2451–2460.

Lee, K. and Berenbaum, M.R. (1992) Ecological aspects of antioxidant enzymes and glutathione S-transferases in three *Papilio* species. *Biochem. Syst. Ecol.* 20 197–207.

Lee, K. and Berenbaum, M.R. (1993) Food utilization and antioxidant enzyme activities of black swallowtail in response to plant phototoxins. *Arch. Insect Biocheh. Physiol.* 23, 79–89.

Leff, J.A., Oppegard, M.A., Curiel, T.J., Brown, K.S., Schooley, R.T. and Repine, J.E. (1992) Progressive increases in serum catalase activity in advancing human immunodeficiency virus infection. *Free Radical Biol. Med.* 13, 143–149.

Leid, R.W., Suquet, C.M. and Tanigoshi, L. (1989) Oxygen detoxifying enzymes in parasites: a review. *Acta Leiden* 57, 107–114.

Lesser, M.P. and Shick, J.M. (1989) Effects of irradiance-and ultraviolet radiation on photoadaptation in the zooxanthellae of *Aiptasia pallida*: primary production, photoinhibition, and enzymic defenses against oxygen toxicity. *Marine Biol.* 102, 243–255.

Liebler, D.C., Kaysen, K.L. and Kennedy, T.A. (1989) Redox cycles of vitamin E: hydrolysis and ascorbic acid dependent reduction of 8α-(alkyldioxy)tocopherones. *Biochemistry* 28, 9772–9777.

Lim, B.P., Nagao, A., Terao, J., Tanaka, K., Suzuki, T. and Takama, K. (1992) Antioxidant activity of xanthophylls on peroxyl radical-mediated phospholipid peroxidation. *Biochim. Biophys. Acta* 1126, 178–184.

Linder, M.C. and Moor, J.R. (1977) Plasma ceruloplasmin-evidence for its presence in and uptake by heart and other organs of the rat. *Biochim. Biophys. Acta* 499, 329–336.

Lindroth, R.L. (1989) Chemical ecology of the luna moth: effects of host plant on detoxication enzyme activity. *J. Chem. Ecol.* 15, 2019–2029.

Liou, W., Chang, L.Y., Geuze, H.J., Strous, G.J., Crapo, J.D. and Slot, J.W. (1993) Distribution of CuZn superoxide dismutase in rat liver. *Free Rad. Biol. Med.* 14, 201–207.

Lippman, S.M., Kessler, J.F. and Meyskens, F.L. (1987) Retinoids as preventive and therapeutic anticancer agents. *Cancer Treat. Rep.* 71, 391–405.

Liu, L. and Tam, M.F. (1991) Nucleotide sequence of a class m glutathione S-transferase from chick liver. *Biochim. Biophys. Acta* 1090, 343–344.

Livingstone, D.R., Lips, F., Martinez, P.G. and Pipe, R.K. (1992) Antioxidant enzymes in the digestive gland of the common mussel *Mytilus edulis*. *Marine Biol.* 112, 265–276.

Lopez-Torres, M., Perez-Campo, R., Rojas, C., Cadenas, S. and Barja, G. (1993) Simultaneous induction of SOD, glutathione reductase, GSH, and ascorbate in liver and kidney correlates with survival during aging. *Free Rad. Biol. Med.* 15, 133–142.

Lumper, V.L., Schneider, W. and Staudinger, H. (1967) Untersuchungen zur Kinetik der mikrosomalen NADH: semidehydroascorbat-oxidoreduktase. *Hoppe Seyler Z. Physiol. Chem.* 348, 323–328.

Lynch, S.M. and Klevay, L.M. (1992) Effects of a dietary copper deficiency on plasma coagulation factor activities in male and female mice. *J. Nutr. Biochem.* 3, 387–391.

Machlin, L.J. (1991) Vitamin E. In *Handbook of Vitamins*, (L.J. Machlin, ed.,) 2nd ed., Marcel Dekker, New York, 99–144.

Maddipati, K.R. and Marnett, L.J. (1987) Characterization of the major hydroperoxide-reducing activity of human plasma. *J. Biol. Chem.* 262 17398–17403.

Maeda, H. and Akaike, T. (1991) Oxygen free radicals as pathogenic molecules in viral diseases. *Proc. Soc. Exp. Biol. Med.* 198, 721–727.

Mahaffey, J.W., Griswold, C.M. and Matthews, A.L. (1993) Molecular genetic mechanisms in oxidative damage and aging. *Toxicol. Indust. Hlth.* 9, 215–222.

Malone, W.F. (1991) Studies evaluating antioxidants and β-carotene as chemopreventatives. *Am. J. Clin. Nutr.* 53, 305S–313S.

Mannervik, B., Alin, P., Guthenberg, C., Jensson, H., Tahir, M. K., Warholm, M. and Jornvall, H. (1985) Identification of three classes of cytosolic glutathione transferase common to several mammalian species: correlation between structural data and enzymatic properties. *Proc. Natl. Acad. Sci. USA* **82**, 7202–7206.

Mannervik, B. and Danielsen, U.H. (1988) Glutathione transferases-structure and catalytic activity. *CRC Crit. Rev. Biochem.* **23**, 283–337.

Mao, G.D., Thomas, P.D., Lopaschuk, G.D. and Poznansky, M.J. (1993) Superoxide dismutase (SOD)-catalase conjugates. *J. Biol. Chem.* **268**, 416–420.

Marklund, S.L. (1984) Extracellular superoxide dismutase and other superoxide dismutase isoenzymes in tissues from nine mammalian species. *Biochem. J.* **222**, 649–655.

Martensson, J., Han, J., Griffith, O.W. and Meister, A. (1993) Glutathione ester delays the onset of scurvy in ascorbate-deficient guinea pigs. *Proc. Natl. Acad. Sci. USA* **90**, 317–321.

Martensson, J. and Meister, A. (1991) Glutathione deficiency decreases tissue ascorbate levels in newborn rats: ascorbate spares glutathione and protects. *Proc. Natl. Acad. Sci. USA* **88**, 9360–9364.

Martensson, J., Jain, A., Stole, E., Frayer, W., Auld, P.A.M. and Meister, A. (1991) Inhibition of glutathione synthesis in the newborn rat: a model for endogenously produced oxidative stress. *Proc. Natl. Acad. Sci. USA* **88**, 9360–9364.

Matkovics, B., Witas, H., Gabrielak, T. and Szabo, L. (1987) Paraquat as an agent affecting antioxidant enzymes of common carp erythrocytes. *Comp. Biochem. Physiol.* **87C**, 217–219.

Matthews-Roth, M.M. (1964) Portective effect of β-carotene against lethal photosensitivation by hematoporyphyrin. *Nature* **203**, 1092.

McCay, P.B. (1985) Vitamin E: interactions with free radicals and ascorbate. *Ann. Rev. Nutr.* **5**, 323–340.

McCord, J.M. and Fridovich, I. (1969) Superoxide dismutase: an enzymic function for erythrocuprein (hemocuprein). *J. Biol. Chem.* **244**, 6049–6055.

McElroy, M.C., Postle, A.D. and Kelly, F.J. (1992) Catalase, superoxide dismutase and glutathione peroxidase activities of lung and liver during human development. *Biochim. Biophys. Acta* **1117**, 153–158.

McNamara, J.O. and Fridovich, I. (1993) Did radicals strike Lou Gehrig? *Nature* **362**, 20–21.

Meister, A. (1982) Selective modification of glutathione metabolism. *Science* **220**, 472–478.

Meister, A. (1988) Glutathione metabolism and its selective modification. *J. Biol. Chem.* **263**, 17205–17208.

Meister, A. (1992a) Commentary-on the antioxidant effects of ascorbic acid and glutathione. *Biochem. Pharmacol.* **44**, 1905–1915.

Meister, A. (1992b) A trail of research: from glutamine synthetase to selective inhibition of glutathione synthesis. *Chemtracts-Biochem. Molec. Biol.* **3**, 75–106.

Meister, A. and Anderson, M.E. (1983) Glutathione. *Ann. Rev. Biochem.* **52**, 711–760.

Meyer, D.J., Coles, B., Pemble, S.E., Gilmore, K.S., Fraser, G.M. and Ketterer, B. (1991). Theta, a new class of glutathione transferases purified from rat and man. *Biochem. J.* **274**, 409–414.

Meyer, D.J., Gilmore, K.S., Coles, B., Dalton, K., Hulbert, P.B. and Ketterer, B. (1991). Structural distinction of rat GSH transferase subunit 10. *Biochem. J.* **274**, 619–623.

Meyskens, F.L. (1990) The chemoprevention of cancer. *New Engl. J. Med.* **323**, 825–828.

Minakata, K.O., Saito, S. and Harada, N. (1993) Ascorbate radical levels in human sera and rat plasma intoxicated with paraquat and diquat. *Arch. Toxicol.* **67**, 126–130.

Minton, A.P. (1990) Holobiochemistry: the effect of local environment upon the equilibria and rates of biochemical reactions. *Int. J. Biochem.* **22**, 1063–1067.

Mitchell, M.J., Ahmad, S. and Pardini, R.S. (1991) Purification and properties of a highly active catalase from cabbage loopers, *Trichoplusia ni*. *Insect. Biochem.* **21**, 641–646.

Miyazaki, S. (1991) Effect of chemicals on glutathione peroxidase of chick liver. *Res. Vet. Sci.* **51**, 120–122.

Miyazaki, S. and Motoi, Y. (1992) Tissue distribution of monomeric glutathione peroxidase in broiler chicks. *Res. Veterinary Sci.* **53**, 47–51.

Montesano, L., Carri, M.T., Mariottini, P., Amaldi, F. and Rotilio, G. (1989) Developmental expression of Cu,Zn superoxide dismutase in *Xenopus*. Constant level of the enzyme in oogenesis. *Eur. J. Biochem.* **186**, 421–426.

Morgenstern, R. and DePierre, J.W. (1988) Membrane-bound glutathione transferases. In *Glutathione Conjugation. Mechanisms and Biological Significance* (H. Sies and B. Ketterer, eds.), Academic Press, New York, pp. 157–174.

Morrill, A.C., Powell, E.N., Bidigare, R.R. and Shick, J.M. (1988) Adaptations to life in the sulfide system: a comparison of oxygen detoxifying enzymes in thiobiotic and oxybiotic meiofauria (and freshwater planarians). *J. Comp. Physiol. B* **158**, 335–344.

Morris, S.M. and Albright, J.T. (1981) Superoxide dismutase, catalase, and glutathione peroxidase in the swim bladder of the physoclistous fish, *Opsanus tau* (L.). *Cell Tissue Res.* **220**, 739–752.

Moser, U. and Bendlich, A. (1991) Vitamin C. In *Handbook of Vitamins* (L.J. Machlin, ed.), 2nd ed., Marcel Dekker, New York, pp. 195–232.

Munim, A., Asayama, K., Dobashi, K., Suzuki, K., Kawaoi, A. and Kato, K. (1992) Immunohistochemical localization of superoxide dismutases in fetal and neonatal rat tissues. *J. Histochem. Cytochem.* **40**, 1705–1713.

Murphy, M.E., Scholich, H. and Sies, H. (1992) Protection by glutathione and other thiol compounds against the loss of protein thiols and tocopherol homologs during muicrosomal peroxidation. *Eur. J. Biochem.* **210**, 139–146.

Nahmias, J.A. and Bewley, G.C. (1984) Characterization of catalase purified from *Drosphila melanogaster* by hydrophobic interaction chromatography. *Comp. Biochem. Physiol.* **77B**, 355–364.

Nakano, T., Sato, M. and Takeuchi, M. (1992) Glutathione peroxidase of fish. *J. Food Sci.* **57**, 1116–1119.

Navas, P., Estevez, A., Buron, M.I., Villalba, J.M. and Crane, F.L. (1988) Cell surface glycoconjugates control the activity of the NADH-ascorbate free radical reductase of rat liver plasma membrane. *Biochem. Biophys. Res. Commun.* **154**, 1029–1033.

Nelson, S.K., Huang, C., Mathias, M.M. and Allen, K.G.D. (1992) Copper-marginal and copper-deficient diets decrease aortic prostacyclin production and copper-dependent superoxide dismutase activity, and increase aortic lipid peroxidation in rats. *J. Nutr.* **122**, 2101–2108.

Neto, P.C., Bechara, E.J.H. and Costa, C. (1986) Oxygen toxicity aspects in luminescent and non-luminescent elaterid larvae. *Insect Biochem.* **16**, 381–385.

Nickla, H., Anderson, J. and Palzkill, T. (1983) Enzymes involved in oxygen detoxification during development of *Drosophila melanogaster*. *Experientia* **39**, 610–612.

Niki, E. (1987) Interaction of ascorbate and α-tocopherol. *Annals N.Y. Acad. Sci.* **498**, 186–198.

Nishino, H. and Ito, A. (1986) Subcellular distribution of OM cytochrome b-mediated NADH-semidehydroascorbate reductase activity in rat liver. *J. Biochem.* **100**, 1523–1531.

Nivsarkar, M., Kumar, G.P., Laloraya, M. and Laloraya, M.M. (1991) Superoxide dismutase in the anal gills of the mosquito larvae of *Aedes aegypti*: its inhibition by *alpha*-terthienyl. *Arch. Insect Biochem. Physiol.* **16**, 249–255.

Nordmann, R., Ribiere, C. and Rouach, H. (1992) Implication of free radical mechanisms in ethanol-induced cellular injury. *Free Rad. Biol. Med.* **12**, 219–240.

O'Brien, P.J. (1991) Molecular mechanisms of quinone cytotoxicity. *Chem.-Biol. Interact.* **80**, 1–41.

Oda, T., Akaike, T., Hamamoto, T., Suzuki, F., Hirano, T. and Maeda, H. (1989) Oxygen radicals in influenza-induced pathogenesis and treatment with pyran polymer-conjugated SOD. *Science* **244**, 974–976.

Olson, J.A. and Kobayashi, S. (1992) Antioxidants in health and disease: overview. *Proc. Soc. Exp. Biol. Med.* **200**, 245–247.

Orr, W.C., Arnold, L.A. and Sohal, R.S. (1992) Relationship between catalase activity, life span and some parameters associated with antioxidant defenses in *Drosophila melanogaster*. *Mechan. Ageing Develop.* **63**, 287–296.

Orr, W.C. and Sohal, R.S. (1992) The effects of catalase gene overexpression on life span and resistance to oxidative stress in transgenic *Drosophila melanogaster*. *Arch. Biochem. Biophysics.* **297**, 35–41.

Ortel, T.L., Takahashi, N. and Putnam, F.W. (1984) Structural model of human ceruloplasmin based on internal triplication, hydrophilic/hydrophobic character, and secondary structure of domains. *Proc. Natl. Acad. Sci. USA* **81**, 4761–4765.

Packer, L. (1992) Interactions among antioxidants in health and disease: vitamin E and its redox cycle. *Proc. Soc. Exp. Biol. Med.* **200**, 271–276.

Packer, L., Maguire, J.J., Mehlhorn, R.J., Serbinova, E. and Kagan, V.E. (1989) Mitochondria and microsomal membranes have a free radical reductase activity that prevents chromanoxyl radical accumulation. *Biochem. Biophys. Res. Commun.* **159**, 229–235.

Packer, J.E., Slater, T.F. and Willson, R.L. (1979) Direct observation of a free radical interaction between vitamin E and vitamin C. *Nature* **278**, 737–738.

Padmanabhan, R.V., Gudapaty, R., Liener, I.E., Schwartz, B.A. and Hoidal, J.R. (1985) Protection against pulmonary oxygen toxicity in rats by the tracheal administration of liposome-encapsulated superoxide dismutase or catalase. *Am. Rev. Respir. Dis.* **132**, 164–167.

Palozza, P. and Krinsky, N.I. (1992) β-carotene and α-tocopherol are synergistic antioxidants. *Arch. Biochem. Biophys.* **297**, 184–187.

Palozza, P., Moualla, S. and Krinsky, N.I. (1992) Effects of β-carotene and α-tocopherol on radical-initiated peroxidation of microsomes. *Free Rad. Biol. Med.* **13**, 127–136.

Papadopoulos, A., Brophy, P.M., Crowley, P., Ferguson, M. and Barrett, J. (1989) Glutathione transferase in the free-living nematode *Panagrellus redivivus*. *FEBS Lett.* **253**, 76–78.

Pardini, R.S., Pritsos, C.A., Bowen, S.M., Ahmad, S. and Blomquist, G.J. (1989). In *Adaptations to Plant Pro-oxidants in a Phytophagous Insect Model: Enzymatic Protection from Oxidative Stress* (M.G. Simic, K.A. Taylor, J.F. Ward and von Sonntag, C., eds.), Plenum, New York, pp. 725–728.

Pemble, S.E. and Taylor, J.B. (1992) An evolutionary perspective on glutathione transferases inferred from class-Theta glutathione cDNA sequences. *Biochem. J.* **287**, 957–963.

Perez-Campo, R., Lopez-Torres, M., Rojas, C., Cadenas, S. and Barja, G. (1993) A comparative study of free radicals in vertebrates—I. Antioxidant enzymes. *Comp. Biochem. Physiol.* **105B**, 749–755.

Permadi, H., Lundgren, B., Andersson, K. and DePierre, J.W. (1992) Effects of perflouro fatty acids on xenobiotic-metabolizing enzymes, enzymes which detoxify reactive forms oxygen and lipid peroxidation in mouse liver. *Biochem. Pharmacol.* **44**, 1183–1191.

Petrovic, V.M., Spasic, M., Saicic, Z., Milic, B. and Radojicic, R. (1982) Increase in superoxide dismutase activity induced by thyroid hormones in the brains of neonate and adult rats. *Experentia* **38**, 1355–1356.

Phillips, J.P., Campbell, S.D., Michauld, D., Charbonneau, M. and Hilliker, A.J. (1989) Null mutation of copper/zinc superoxide dismutase in *Drosophila* confers hypersensitivity to paraquat and reduced longevity. *Proc. Natl. Acad. Sci. USA* **86**, 2761–2765.

Pickett, C.B. and Lu, A.Y.H. (1989) Glutathione S-transferases: gene structure, regulation, and biological function. *Ann. Rev. Biochem.* **58**, 743–764.

Pritsos, C.A., Ahmad, S., Bowen, S.M., Elliot, A.J., Blomquist, G.J. and Pardini, R.S. (1988a) Antioxidant enzymes of the black swallowtail butterfly, *Papilio polyxenes*, and their response to the prooxidant allelochemical, quercetin. *Arch. Insect Biochem. Physiol.* **8**, 101–112.

Pritsos, C.A., Ahmad, S., Bowen, S.M., Blomquist, G.J. and Pardini, R.S. (1988b) Antioxidant enzyme activities in the southern armyworm, *Spodoptera eridania*. *Comp. Biochem. Physiol.* **90C**, 423–427.

Pritsos, C.A., Pastore, J. and Pardini, R.S. (1991) Role of superoxide dismutase in protection and tolerance to the prooxidant allelochemical quercetin in *Papilio polyxenes*, *Spodoptera eridania*, and *Trichoplusia ni*. *Arch. Insect Biochem. Physiol.* **16**, 273–282.

Prohaska, J.R. (1980) The glutathione peroxidase activity of glutathione S-transferases. *Biochim. Biophys. Acta* **611**, 87–98.

Prohaska, J.R., Sunde, R.A. and Zinn, K.R. (1992) Livers from copper-deficient rats have lower glutathione peroxidase activity and mRNA levels but normal liver selenium levels. *J. Nutr. Biochem.* 3, 429–436.

Radi, A.A.R., Hai, D.Q., Matkovics, B. and Gabrielak, T. (1985). Comparative antioxidant enzyme study in freshwater fish with different types of feeding behavior. *Comp. Biochem. Physiol.* 81C, 395–399.

Rathbun, W.B. and Holleschau, A.M. (1992). The effects of age on glutathione synthesis enzymes in lenses of Old World simians and prosimians. *Current Eye Research* 11, 601–607.

Ravin, H.A. (1961) An improved colorimetric enzymatic assay of ceruloplasmin. *J. Lab. Clin. Med.* 58, 161–168.

Retsky, K., Freeman, M.W. and Frei, B. (1993) Ascorbic acid oxidation product(s) protect human low density lipoprotein against atherogenic modification. *J. Biol. Chem.* 268, 1304–1309.

Reveillaud, I., Kongpachith, A., Park, R. and Fleming, J.E. (1992) Stress resistance of *Drosphila* transgenic for bovine CuZn superoxide dismutase. *Free Rad. Res. Commun.* 17, 73–85.

Rhoads, M.L. (1983) *Trichinella spiralis*: identification and purification of superoxide dismutase. *Exp. Parasitol.* 56, 41–54.

Rice-Evans, C.A. and Diplock, A.T. (1993) Current status of antioxidant therapy. *Free Radical Biol. Med.* 15, 77–96.

Rikans, L.E. and Cai, Y. (1992) Age-associated enhancement of diquat-induced lipid peroxidation and cytotoxicity in isolated rat hepatocytes. *J. Pharmacol. Exp. Therap.* 262, 271–278.

Rivett, A.J. (1985) Preferential degradation of the oxidatively modified form of glutamine synthetase by intracellular mammalian proteases. *J. Biol. Chem.* 260, 300–305.

Robinson, J.R. and Beatson, E.P. (1985) Enhancement of dye-sensitized phototoxicity to house fly larvae in vivo by dietary ascorbate, diazobicyclooctane, and other additives. *Pestic. Biochem. Physiol.* 24, 375–383.

Rose, R.C. (1989) Renal metabolism of the oxidized form of ascorbic acid (dehydro-L-ascorbic acid). *Am. J. Physiol.* 256, F52–F56.

Rose, R.C. (1990) Ascorbic acid metabolism in protection against free radicals: a radiation model. *Biochem. Biophys. Res. Commun.* 169, 430–436.

Rose, R.C. and Bode, A.M. (1992) Tissue-mediated regeneration of ascorbic acid: is the process enzymatic? *Enzyme* 46, 196–203.

Rosen, D.R., Siddique, T., et al. (1993) Mutations in Cu/Zn superoxide dismutase gene are associated with familial amyotrophic lateral sclerosis. *Nature* 362, 59–62.

Sakai, T., Murata, H., Yamauchi, K., Sekiya, T. and Ukawa, M. (1992) Effects of dietary lipid peroxides contents on in vivo lipid peroxidation, α-tocopherol contents, and superoxide dismutase and glutathione peroxidase activities in the liver of yellowtail. *Nippon Suisan Gakkaishi* 58, 1483–1486.

Samokyszyn, V.M., Miller, D.M., Reif, D.W. and Aust, S.D. (1989) Inhibition of superoxide and ferritin-dependent lipid peroxidation by ceruloplasmin. *J. Biol. Chem.* 264, 21–26.

Sanchez-Moreno, M., Garcia-Ruiz, M.A., Sanchez-Navas, A. and Monteoliva, M. (1989) Physico-chemical characteristics of superoxide dismutase in *Ascaris suum*. *Comp. Biochem. Physiol.* **92B**, 737–740.

Sandberg, S.L. and Berenbaum, M. (1989) Leaf-tying by tortricid larvae as an adaptation for feeding on phototoxic *Hypericum perforatum*. *J. Chem. Ecol.* **15**, 875–885.

Sato, K., Akaike, T., Kohno, M., Ando, M. and Maeda, H. (1992) Hydroxyl radical production by H_2O_2 plus Cu,Zn-superoxide dismutase reflects the activity of free copper released from the oxidatively damaged enzyme. *J. Biol. Chem.* **267**, 25371–25377.

Sato, K., Niki, E. and Shimasaki, H. (1990) Free radical-mediated chain oxidation of low density lipoprotein and its synergistic inhibition by vitamin E and vitamin C. *Arch. Biochem. Biophys.* **279**, 402–405.

Sato, M. and Gitlin, J.D. (1991) Mechanisms of copper incorporation during the biosynthesis of human ceruloplasmin. *J. Biol. Chem.* **266**, 5128–5134.

Scandalios, J.G. (1993) Oxygen stress and superoxide dismutases. *Plant Physiol.* **101**, 7–12.

Schirmer, R.H. (1989) Glutathione reductase. In *Glutathione, Chemical, Biochemical and Medical Aspects. Part A* (O. Avraminov and R. Poulson, eds.), John Wiley & Sons, New York, pp. 553–596.

Scholz, R.W., Graham, K.S., Gumpricht, E. and Reddy, C.C. (1989) Mechanism of interaction of vitamin E and glutathione in protection against membrane lipid peroxidation. *Annals N.Y. Acad. Sci.* **570**, 514–517.

Schrauzer, G.N. (1992) Selenium. Mechanistic aspects of anticarcinogenic action. *Biol. Trace El. Res.* **33**, 51–62.

Seto, N.O., Hayashi, S. and Tener, G.M. (1990) Overexpression of Cu-Zn superoxide dismutase in *Drosphila* does not affect life-span. *Proc. Natl. Acad. Sci. USA* **87**, 4270–4274.

Shaffer, J.B., Sutton, R.B. and Bewley, G.C. (1987) Isolation of a cDNA clone for murine catalase and analysis of an acatalasemic mutant. *J. Biol. Chem.* **262**, 12908–12911.

Shahidi, F., Janitha, P.K. and Wanasundara, P.D. (1992) Phenolic antioxidants. *Crit. Rev. Food Sci. Nutr.* **32**, 67–103.

Sheehan, D. and Casey, J.P. (1993a) Microbial glutathione S-transferases. *Comp. Biochem. Physiol.* **104B**, 1–6.

Sheehan, D. and Casey, J.P. (1993b) Evidence for alpha and mu class glutathione S-transferases in a number of fungal species. *Comp. Biochem. Physiol.* **104B**, 7–13.

Shick, J.M. and Dykens, J.A. (1985) Oxygen detoxification in algal-invertebrate symbioses from the Great Barrier Reef. *Oecologia* **66**, 33–41.

Sichak, S.P. and Dounce, A.L. (1986) Analysis of the peroxidatic mode of action of catalase. *Arch. Biochem. Biophys.* **249**, 286–295.

Sies, H., Murphy, M.E., Di Mascio, P. and Stahl, W. (1992) Tocopherols, carotenoids and the glutathione system. In *Lipid-Soluble Antioxidants: Biochemistry and Clinical Applications*. Birkhauser Verlag, Basel, pp. 160–165.

Sies, H., Stahl, W. and Sundquist, A.R. (1992) Antioxidant functions of vitamins. *Annals N.Y. Acad. Sci.* **669**, 7–20.

Simmons, T.W., Jamall, I.S. and Lockshin, R.A. (1989) Selenium-independent glutathione peroxidase activity associated with glutathione S-transferase from the housefly, *Musca domestica. Comp. Biochem. Physiol.* 9, 323–327.

Singh, V.N. (1992) A current perspective on nutrition and exercise. *J. Nutr.* 122, 760–765.

Smith, C.D., Carney, J.M., Starke-Reed, P.E., Iliver, C.N., Stadtman, E.R., Floyd, R.A. and Markesberry, W.R. (1991) Excess brain protein oxidation and enzyme dysfunction in normal aging and in Alzheimer disease. *Proc. Natl. Acad. Sci. USA* 88, 10540–10543.

Smith, J. and Shrift, A. (1979) Phylogenetic distribution of glutathione peroxidase. *Comp. Biochem. Physiol.* 63B, 39–44.

Smith, N.C. and Bryant, C. (1986) The role of host generated free radicals in helminth infections: *Nippostrogylus brasiliensis* and *Nematospiroides dubius* compared. *J. Parasitol.* 16, 617–622.

Sohal, R.S. (1993) Aging, cytochrome oxidase activity, and hydrogen peroxide release by mitochondria. *Free Rad. Biol. Med.* 14, 583–588.

Sohal, R.S. and Brunk, U.T. (1992) Mitochondrial production of pro-oxidants and cellular senescence. *Mut. Res.* 275, 295–304.

Stadtman, E.R. (1992) Protein oxidation and aging. *Science* 257, 1220–1224.

Stadtman, E.R. (1993) Oxidation of free amino acids and amino acid residues in proteins by radiolysis and by metal-catalyzed reactions. *Ann. Rev. Biochem.* 62, 797–821.

Stadtman, T.C. (1991) Biosynthesis and function of selenocysteine-containing enzymes. *J. Biol. Chem.* 266, 16257–16260.

Starke-Reed, P.E. and Oliver, C.N. (1989) Protein oxidation and proteolysis during aging and oxidative stress. *Arch. Biochem. Biophys.* 275, 559–567.

Starrat, A.N. and Bond, E.J. (1990) Recovery of glutathione levels in susceptible and resistant strains of *Sitphilus granarius* (L.) (Coleoptera: Curculionidae) following methyl bromide treatment. *J. Stored. Prod. Res.* 26, 39–41.

Stenersen, J., Kobro, S., Bjerke, M. and Arend, U. (1987). Glutathione transferases in aquatic and terrestrial animals from nine phyla. *Comp. Biochem. Physiol.* 86C, 73–82.

Stocker, R., Glazer, A.N. and Ames, B.N. (1987) Antioxidant activity of albumin-bound bilirubin. *Proc. Natl. Acad. Sci. USA* 84, 5918–5922.

Stocker, R., Yamamoto, Y., McDonagh, A.F., Glazer, A.N. and Ames, B.N. (1987a) Bilirubin is an antioxidant of possible physiological importance. *Science* 240, 1043–1046.

Strain, J.J. (1991) Disturbances of micronutrient and antioxidant status in diabetes. *Proc. Nutr. Soc.* 50, 591–604.

Summers, C.B. and Felton, G.W. (1993) Antioxidant role of dehydroascorbic acid reductase in insects. *Biochim. Biophys. Acta* 1156, 235–238.

Takahashi, K., Akasaka, M., Yamamoto, Y., Kobayashi, C., Mizoguchi, J. and Koyama, J. (1990) Primary structure of human plasma glutathione peroxidase deduced from cDNA sequences. *J. Biochem.* 108, 145–148.

Takahashi, K., Avissar, N., Whitin, J. and Cohen, H. (1987). Purification and characterization of human plasma glutathione peroxidase: a selenoglycoprotein distinct from the known cellular enzyme. *Arch. Biochem. Biophys.* **256**, 677–686.

Takahashi, N., Ortel, T.L. and Putnam, F.W. (1984) Single-chain structure of human ceruloplasmin: the complete amino acid sequence of the whole molecule. *Proc. Natl. Acad. Sci. USA* **81**, 390–394.

Tamai, K.T., Gralla, E.B., Ellerby, L.M. and Valentine, J.S. (1993) Yeast and mammalian metallothioneins functionally substitute for yeast copper-zinc superoxide dismutase. *Proc. Natl. Acad. Sci. USA* **90**, 8013–8017.

Tang, G., White, J.E., Gordon, R.J., Lumb, P.D. and Tsan, M.F. (1993) Polyethylene glycol-conjugated superoxide dismutase protects rats against oxygen toxicity. *J. Appl. Physiol.* **74**, 1425–1431.

Tappel, M.E., Chaudiere, J. and Tappel, A.L. (1982) Glutathione peroxidase activities of animal tissues. *Comp. Biochem. Physiol.* **73B**, 945–949.

Teramoto, S., Fukuchi, Y., Uejima, Y., Ito, H. and Orimo, H. (1992) Age-related changes in GSH content of eyes in mice—a comparison of senescence-accelerated mouse (SAM) and C57BL/J mice. *Comp. Biochem. Physiol.* **102A**, 693–696.

Thomsen, B., Drumm-Herrel, H. and Mohr, H. (1992) Control of the appearance of ascorbate peroxidase (EC 1.11.1.11) in mustard seedling cotyledons by phytochrome and photooxidative treatments. *Planta* **186**, 600–608.

Tolmasoff, J.M., Ono, T. and Cutler, R.G. (1980) Superoxide dismutase: correlation with life-span and specific metabolic rate in primate species. *Proc. Natl. Acad. Sci. USA* **77**, 2777–2781.

Tomarev, S.I., Zinovieva, R.D., Guo, K. and Piatigorsky, J. (1993) Squid glutathione S-transferase. *J. Biol. Chem.* **268**, 4534–4542.

Toung, Y.S., Hsieh, T. and Tu, C.D. (1990) *Drosophila* glutathione S-transferase 1-1 shares a region of sequence homology with the maize glutathione S-transferase III. *Proc. Natl. Acad. Sci. USA* **87**, 31–35.

Toung, Y.S. and Tu, C.D. (1992) Drosophila glutathione S-transferases have sequence homology to the stringent starvation protein of *Escherichia coli*. *Biochem. Biophys. Res. Commun.* **182**, 355–360.

Toyoda, H., Himeno, S. and Imura, N. (1989) The regulation of glutathione peroxidase gene expression relevant to species difference and the effects of dietary selenium manipulation. *Biochim. Biophys. Acta* **1008**, 301–308.

Tsan, M. (1993) Superoxide dismutase and pulmonary oxygen toxicity. *P.S.E.B.M.* **203**, 286–290.

Tsan, M., White, J.E., Del Vecchio, P.J. and Schaffer, J.B. (1992) IL-6 Enhances TNF-a- and IL-1-induced increase of Mn superoxide dismutase mRNA and O_2 tolerance. *Am. J. Physiol.* **263**, L22–L26.

Tsukamoto, K., Jimenez, M. and Hearing, V.J. (1992) The nature of tyrosinase isozymes. *Pigment Cell Res. Suppl.* **2**, 84–89.

Tsumimura, M., Fukada, T. and Kasai, T. (1983) On the reduction of dehydroascorbic acid in guinea pig and rat. *Nutr. Rep. Int.* **28**, 881–890.

Turner, E., Hager, L.J. and Shapiro, B.M. (1988) Ovothiol replaces glutathione peroxidase as a hydrogen peroxide scavenger in sea urchin eggs. *Science* **242**, 939–941.

Ursini, F., Maiorino, M., Valente, M., Ferri, L. and Gregolin, C. (1982) Purification from pig liver of a protein which protects liposomes and biomembranes from peroxidative degradation and exhibits glutathione peroxidase activities on phosphatidylcholine hydroperoxides. *Biochim. Biophys. Acta* **710**, 197–211.

Ursini, F., Maiorino, M. and Gregolin, C. (1985) The selenoenzyme phospholipid hydroperoxide glutathione peroxidase. *Biochim. Biophys. Acta* **839**, 62–70.

van Acker, S.A.B.E., Koymans, L.M.H. and Bast, A. (1993) Molecular pharmacology of vitamin E: structural aspects of antioxidant activity. *Free Rad. Biol. Med.* **15**, 311–328.

Van Balgooy, J.N.A. and Roberts, E. (1979) Superoxide dismutase in normal and malignant tissues in different species. *Comp. Biochem. Physiol.* **62B**, 263–268.

van den Bosch, H., Schutgens, R.B.H., Wanders, R.J.A. and Tager, J.M. (1992) Biochemistry of peroxisomes. *Ann. Rev. Biochem.* **61**, 157–197.

Vanfleteren, J.R. (1992) Cu-Zn Superoxide dismutase from *Caenorhabditis elegans*: purification, properties and isoforms. *Comp. Biochem. Physiol.* **102B**, 219–229.

Viarengo, A., Canesi, L., Pertica, M. and Livingstone, D.R. (1991) Seasonal variations in the antioxidant defence systems and lipid peroxidation of the digestive gland of mussels. *Comp. Biochem. Physiol.* **100C**, 187–190.

Wahlund, G., Marklund, S.L. and Sjoquist, P.-O. (1992) Extracellular-superoxide dismutase type C (EC-SOD C) reduces myocardial damage in rats subjected to coronary occlusion and 24 hours of reperfusion. *Free Rad. Res. Commun.* **17**, 41–47.

Warner, H.R. (1992) Overview: mechanisms of antioxidant action on lifespan. In *Antioxidants: Chemical, Physiological, Nutritional and Toxicological Aspects* (H. Sies, J. Erdman, Jr., G. Baker III, C. Henry and G. Williams), Princeton Scientific Publishing, pp. 151–160.

Waxman, D.J. (1990) Glutathione S-transferases: role in alkylating agent resistance and possible target for modulation chemotherapy—a review. *Cancer. Res.* **50**, 6449–6454.

Waxman, D.J., Sundseth, S.S., Srivastava, P.K. and Lapenson, D.P. (1992) Gene-specific oligonucleotide probes for a, m, p, and microsomal rat glutathione S-transferases: analysis of liver transferase expression and its modulation by hepatic enzyme inducers and platinum anticancer drugs. *Cancer Res.* **52**, 5797–5802.

Weitzel, F., Ursini, F. and Wendel, A. (1990) Phospholipid hydroperoxide glutathione peroxidase in various mouse organs during selenium deficiency and repletion. *Biochim. Biophys. Acta* **1036**, 88–94.

Wells, W.W., Xu, D.P., Yang, Y. and Rocque, P.A. (1990) Mammalian thioltransferase (glutaredoxin) and protein disulfide isomerase have dehydroascorbate reductase activity. *J. Biol. Chem.* **265**, 15361–15364.

Wendel, A. (1980) Glutathione peroxidase. In *Enzymatic Basis of Detoxication*, (W.B. Jakoby, ed.), Academic Press, New York, pp. 333–353.

White, C.W., Avraham, K.B., Shanley, P.R. and Groner, Y. (1991) Transgenic mice with expression of elevated levels of copper-zinc superoxide dismutase in the lungs are resistant to pulmonary oxygen toxicity. *J. Clin. Invest.* **87**, 2162–2168.

White, C.W., Jackson, J.H., Abuchowski, A., Kazo, G.M., Mimmack, R.F., Berger, E.M., Freeman, B.A., McCord, J.M. and Repine, J.E. (1989) Polyethylene glycol-attached antioxidant enzymes decrease pulmonary oxygen toxicity in rats. *J. Appl. Physiol.* **66**, 584–590.

Winkler, B.S. (1987) The *in vitro* oxidation of ascorbic acid and its prevention by GSH. *Biochim. Biophys. Acta* **925**, 258–264.

Winkler, B.S. (1992) Unequivocal evidence in support of nonenzymatic redox coupling between glutathione/glutathione disulfide and ascorbic acid/dehydroascorbic acid. *Biochim. Biophys. Acta* **1117**, 287–290.

Winston, G.W. (1990) Oxygen reduction metabolism by the digestive gland of the common marine mussel, *Mytilus edulis* L. *J. Exp. Zool.* **255**, 296–308.

Wispe, J.R., Warner, B.B., Clark, J.C., Dey, C.R., Neuman, J., Glasser, S.W., Crapo, J.D., Chang, L.Y. and Whitsett, J.A. (1992) Human Mn-superoxide dismutase in pulmonary epithelial cells of transgenic mice confers protection from oxygen injury. *J. Biol. Chem.* **267**, 23937–23941.

Wolff, S., Garner, A. and Dean, R.T. (1986) Free radicals, lipids, and protein degradation. *T.I.B.S.* **11**, 27–31.

Wood, J.M. and Schallreuter, K.U. (1991) Studies on the reactions between human tyrosinase, superoxide anion, hydrogen peroxide, and thiols. *Biochim. Biophys. Acta* **1074**, 378–385.

Yam, J., Frank, L. and Roberts, R.J. (1978) Age-related development of pulmonary antioxidant enzymes in the rat (40040). *Proc. Soc. Exp. Biol. Med.* **157**, 293–296.

Yamada, T., Muramatsu, Y., Agui, T. and Matsumoto, K. (1992) A new restriction fragment length polymorphism of the ceruloplasmin gene in rat. *Biochem. Int.* **27**, 243–249.

Yamamoto, Y., Sato, M. and Ikeda, S. (1977a) Biochemical studies on L-ascorbic acid in aquatic animals-VII. Reduction of dehydroascorbic acid in fishes. *Bull. Jap. Soc. Sci. Fish* **43**, 59–67.

Yamamoto, Y., Sato, M. and Ikeda, S. (1977b) Biochemical studies on L-ascorbic acid in aquatic animals. VIII. Purification and properties of dehydro-L-ascorbic acid reductase from carp hepatopancreas. *Bull. Jap. Soc. Sci. Fish* **43**, 59–67.

Yim, M.B., Chock, P.B. and Stadtman, E.R. (1993) Enzyme function of copper, zinc superoxide dismutase as a free radical generator. *J. Biol. Chem.* **268**, 4099–4105.

Yoshioka, T., Shimada, T. and Sekiba, K. (1980) Lipid peroxidation and antioxidants in the rat lung during development. *Biol. Neonate* **38**, 161–168.

Yoshioka, T., Utsumu, K. and Sekiba, K. (1977) Superoxide dismutase activity and lipid peroxidation of the rat liver during development. *Biol. Neonate* **32**, 147–153.

Yu, S.J. (1987) Quinone reductase of phytophagous insects and its induction by allelochemicals. *Comp. Biochem. Physiol.* **87B**, 621–624.

Ziegler, D.M. (1985) Role of reversible oxidation-reduction of enzyme thiols-disulfides in metabolic regulation. *Ann. Rev. Biochem.* **54**, 305–329.

Ziegler, R.G. (1989) A review of epidemiological evidence that carotenoids reduce the risk of cancer. *J. Nutr.* **119**, 116–122.

Ziegler, R.G. (1991) Vegetables, fruits and carotenoids and the risk of cancer. *Am. J. Clin. Nutr.* **53**, 251S–259S.

CHAPTER 11

Genetic Regulation of Antioxidant Defenses in *Escherichia coli* and *Salmonella typhimurium*

Holly Ahern and Richard P. Cunningham

11.1 INTRODUCTION

Escherichia coli and *Salmonella typhimurium* have multiple mechanisms for protection from oxidative stress (Chapter 8). Since a bacterial cell can be exposed to fluctuations in levels and types of reactive oxygen species (ROS), it must be able to respond to these changes in order to insure that the cell maintains an adequate level of protection. This chapter will review the mechanisms by which bacteria regulate their gene expression to protect against damage from ROS. At present only two of the many ROS, the superoxide anion radical (O_2^-) and hydrogen peroxide (H_2O_2), appear to induce cellular responses; however, conditions of starvation can also induce enzymes that are required for resistance to oxidative stress.

11.2 RESPONSES TO OXIDATIVE STRESS

Several criteria have been used to define responses which protect against oxidative stress. The induction of a resistant state by sublethal doses of oxidants (Demple and Halbrook, 1983; Farr et al., 1985) has been taken as evidence for the existence of adaptive responses to oxidative damage. Two dimensional gel electrophoresis of proteins from whole cell extracts made from cultures before and after oxidative stress (Christman et al., 1985; Walkup and Kogoma, 1989; Greenberg and Demple, 1989) have revealed various patterns of protein induction. Specific enzymatic activities thought to play a role in protection against oxidative damage have been monitored after exposure to oxidative stress and have been found to increase in activity (Chan and Weiss, 1987; Hassan and Fridovich, 1977b). Random operon fusions of reporter genes have been used to define genes induced by oxidative stress (Kogoma et al. 1988; Mito et al., 1993). Regulatory genes controlling responses to oxidative stress have been identified either through the isolation of constitutive mutants which offer cells resistance to oxidative stress (Christman et al., 1985; Greenberg et al., 1990) or the use of operon fusions to find "up" regulatory mutations (Tsaneva and Weiss, 1990). Using these approaches, both stimulons and regulons which respond to a stress have been identified. Stimulons consist of all of the responses of a cell to a particular stress. Regulons are the subset of responses controlled by a unique regulatory element.

11.3 RESPONSES TO H_2O_2: THE PEROXIDE STIMULON AND THE oxyR REGULON

It was originally observed that *E. coli* exhibited inducible repair of oxidative DNA damage (Demple and Halbrook, 1983). Cells exposed to micromolar levels of H_2O_2 developed resistance to subsequent challenges with millimolar concentrations of H_2O_2. Chloramphenicol treatment at the same time as H_2O_2 pretreatment blocked subsequent development of resistance. A component of the resistance appeared to be due to DNA repair since phage treated with H_2O_2 showed a higher plating efficiency on adapted cells. In addition, increased levels of catalase (CAT) were induced by H_2O_2, as had been previously noted (Richter and Loewen, 1981).

A similar result was obtained with *S. typhimurium* (Christman et al., 1985) with the exception that damaged phage were not reactivated in adapted *Salmonella*. Two dimensional gel analysis of adapted cells showed that the expression of 30 proteins was induced. There is a temporal control to this induction, with 12 proteins being expressed early (in the first 10 minutes) and 18 proteins being expressed later in the response (from 10 to 30 minutes after treatment). Several H_2O_2 resistant mutants were isolated and were found to be phenotypically similar. One was chosen for study and the mutation it carried was designated as *oxyR1*. This mutant constitutively expresses 9 of the 12 early proteins induced by H_2O_2 treatment. Three enzymes were shown to be expressed constitutively in the mutant; HPI catalase (HPI CAT), glutathione reductase (GR), and alkyl hydroperoxide reductase. A deletion mutant for the *oxyR* gene was shown to be H_2O_2 sensitive. It did not overexpress the 9 early proteins constitutively expressed by *oxyR1* mutants, nor did it induce these proteins after H_2O_2 treatment.

OxyR deletion mutants show higher spontaneous mutation rates than wild type cells (Storz et al., 1987). The mutation rate is greatly enhanced by mucA and mucB suggesting that oxidative lesions are processed to create mutations. Anaerobic growth reduces mutagenesis, but it is still considerably above wild type levels even in cells grown aerobically. On the other hand, the *oxyR1* mutation reduces spontaneous mutagenesis. Overproduction of either HPI CAT or alkyl hydroperoxide reductase reduces spontaneous mutation rates to wild type levels; however, spontaneous mutation rates are not affected in *katG* (HPI CAT) or *ahp* (alkyl hydroperoxide reductase) mutants. The effects of *oxyR* seem primarily to be due to the induction of enzymes which scavenge ROS.

An analysis of *oxyR* mutants of *E. coli* that had acquired suppressors making them H_2O_2 resistant (Greenberg and Demple, 1988) revealed that they overproduced HPI CAT, HPII CAT or alkyl hydroperoxide reductase. In addition, the overproduction of alkyl hydroperoxide reductase could restore resistance to cumene hydroperoxide and overproduction of either CAT or alkyl hydroperoxide reductase could restore resistance to redox-cycling agents. The elevated levels of these three enzymes also suppressed spontaneous mutation rates in *oxyR* cells. Genetic analysis suggests that the suppressor mutations for HPI CAT and alkyl hydroperoxide reductase map to the promoter of the *katG* and *ahp* genes, respectively.

The *oxyR* gene has been cloned and sequenced (Christman et al., 1989; Tao et al., 1989) and the OxyR protein shows homology to a family of bacterial regulatory proteins. The sequence of the *oxyR2* mutation showed that a valine to proline change was responsible for the constitutive phenotype of the mutants. In addition, it was shown that OxyR protein could bind to the oxyR promoter (Christman et al., 1989) and negatively control its own expression. The footprint was quite large, extending from -27 to $+21$ relative to the start of the oxyR promoter.

Studies with purified OxyR protein have provided the best insight into the control of the *oxyR* regulon (Storz et al., 1990). OxyR protein binds to its own promoter and also to the *katG* and *ahpC* promoters. OxyR induced transcription of the *katG* gene even when purified from cells that had not been treated with H_2O_2. When transcription assays were carried out in the presence of 100 mM DTT, OxyR did not induce synthesis of *katG* message. From these results and footprinting studies, the following picture emerges. OxyR responds rapidly and reversibly to O_2, probably through the production of H_2O_2 from dissolved oxygen in buffers. OxyR can exist in two states, an "oxidized" state and a "reduced" state, that reflect the presence or absence of reducing agents. Oxidized OxyR stimulates transcription of *katG* and *ahpC*. In its reduced form, OxyR does not stimulate the transcription of the same genes. OxyR in either state represses transcription of *oxyR*. OxyR can bind to promoters in both oxidized and reduced forms. It can undergo a conformational change in response to H_2O_2. The oxidized form of the protein induces transcription of the genes which it positively controls in vivo, while the protein maintains negative control of its own synthesis. To summarize, OxyR is bound to the promoter of the genes it regulates. When oxidized, presumably by H_2O_2, the protein undergoes a conformational change that induces transcription of a number of genes, including *katG*. When CAT levels increase and H_2O_2 levels are lowered, OxyR returns to its inactive state. The promoters which OxyR recognizes are not very similar. Recent studies (Tartaglia et al., 1992) suggest that there is a multidegenerate recognition code. Whether this degenerate base recognition serves a function in response to H_2O_2 remains to be determined, but may allow for various genes to respond differentially to oxidative stress.

The H_2O_2 stress response consists of OxyR and non-OxyR genes. The nine OxyR genes are expressed early and several code for enzymes which have known protective functions. The *katG* gene product, HPI CAT, reduces concentrations of H_2O_2. The *ahpCF* gene products, the heterodimeric protein alkyl hydroperoxide reductase, can remove organic peroxides which can damage cellular constituents. Additionally, alkyl hydroperoxide reductase can reduce peroxidized cellular components that can initiate new rounds of peroxidation. One group has reported a DNA repair component of the H_2O_2 response (Demple and Halbrook, 1983), while several other laboratories have not found the reactivation of H_2O_2-damaged phage (Christman et al., 1985; Imlay and Linn, 1986; Tyrell, 1985). Thus, at present, the primary mechanism for H_2O_2 resistance must be ascribed to enzymes which scavenge ROS.

11.4 RESPONSE TO O_2^-: THE SUPEROXIDE STIMULON AND THE *soxRS* REGULON

Several enzymes have been found to be induced by O_2^--generating conditions. SOD (Gregory and Fridovich, 1973; Hassan and Fridovich, 1977a), glucose-6-phosphate dehydrogenase and NAD(P)H dehydrogenase (Kao and Hassan, 1985), and endonuclease IV (Chan and Weiss, 1987) all show increased levels after exposure of *E. coli* to various agents that increase the levels of O_2^- in cells. In addition, it was shown that *E. coli* exposed to low levels of plumbagin, a redox-cycling agent that produces O_2^-, became resistant to a subsequent challenge by high levels of plumbagin (Farr et al., 1985). Bacteriophage λ which was damaged by exposure to O_2^- was reactivated by *E. coli* that had been previously treated with low levels of plumbagin. This study suggested that O_2^- could induce a new DNA repair pathway in *E. coli*. Random fusions of *E. coli* genes and operons to *lacZ* were created and tested for inducibility by paraquat, also a redox-cycling agent. Three gene fusions were found which were inducible and were called *soi*::*lacZ* fusions (for superoxide inducible) (Kogoma et al., 1988). These fusions were not regulated by the global regulatory genes *oxyR*, *htpR*, or *recA*. Furthermore, two of the fusions conferred sensitivity to paraquat. When the synthesis of proteins was monitored following treatment of *E. coli* with various O_2^- generating compounds, approximately 30 proteins were expressed at higher levels (Walkup and Kogoma, 1989; Greenberg and Demple, 1989). Among the proteins whose synthesis increase were endonuclease IV, Mn containing SOD (MnSOD), glucose-6-phosphate dehydrogenase, and the products of the genes defined by *soi*::*lacZ* fusions.

The regulation of a portion of these O_2^- inducible proteins was described when the *soxR* locus was isolated. An *nfo*::*lacZ* gene fusion was used to isolate extragenic mutations affecting its own expression (Tsaneva and Weiss, 1990). The mutations defined a new locus, *soxR*. *SoxR^c* mutants overexpress endonuclease IV and MnSOD, while *soxR$^-$* fail to induce these enzymes and are sensitive to O_2^- generating compounds. Mutants of *E. coli* that were selected for resistance to menadione (Greenberg et al., 1990) also were mutant for the *soxR* locus. These *soxR^c* mutants overproduce endonuclease IV, glucose-6-phosphage dehydrogenase, MnSOD and six other proteins that had previously been shown to be O_2^- inducible. *SoxR$^-$* mutants showed no induction of the nine proteins and were sensitive to a number of O_2^- generators. The results of both studies indicate that strains carrying deletions of the *soxR* locus are not inducible, and that the *soxR* regulon is positively regulated.

Subsequent experiments showed that two genes, *soxR* and *soxS*, are involved in regulation and are divergently transcribed. The *SoxS* protein shows homology to C-terminal regions of members of the AraC family of positive regulatory proteins (Wu and Weiss, 1991; Amabile-Cuevas and Demple, 1991) while SoxR shows homology to the MerR family of positive regulatory proteins (Amabile-Cuevas and Demple, 1991). An intriguing model for regulation by SoxR and SoxS has been proposed (Wu and Weiss, 1992; Nunoshiba et al., 1992). The SoxS protein appears to be a positive regulatory protein which can turn on the synthesis of *sox* genes when expressed at a high level. The SoxR protein appears to be a sensor and the activator of the *soxS* gene. A cluster of cysteine residues near the C-terminus may bind a metal that could function as a sensor of oxidative stress. Some proteins containing iron-sulfur clusters are sensitive to redox reactions with O_2^- radicals (Gardner and Fridovich, 1991a,b) and this could be the mechanism by which the SoxR protein senses a high level of O_2^-. SoxR protein can bind to the *soxS* promoter (Nunoshiba et al., 1992) although it is not known if the active or inactive form of the protein, or both forms, bind.

The induction of the *soxRS* regulon is a two stage process in which SoxR protein senses a cellular signal and activates transcription of *SoxS*, which in turn activates the transcription of the remaining genes in the regulon. The nature of the signal that initiates this minicascade is unknown. Superoxide itself could be the signal and could interact with a metal group in the SoxR protein. Another suggestion (Liochev and Fridovich, 1992) is that the protein senses a decrease in the ratio of NADPH to NADP$^+$. Superoxide can deplete NADPH and alter the redox balance of the cell. The recent demonstration that SoxR protein binds the *soxS* promoter (Nunoshiba et al., 1992), suggests that such studies with purified SoxR protein under defined conditions will determine the factors which activate the SoxR protein.

The O_2^- response consists of *soxRS* and non-*soxRS* genes. The nine *SoxRS* genes code for several known enzymes. MnSOD removes O_2^-, endonuclease IV repairs oxidatively damaged DNA and glucose-6-phosphate dehydrogenase maintains a high ratio of NADPH to $NADP^+$. In addition, fumarase C is induced to replace fumarases A and B which are O_2^- sensitive.

Some *soxRS* genes are also regulated by another locus. Amongst a set of menadione resistant mutants, one isolate showed elevated levels of only 6 of the 9 *soxRS* proteins. The mutation was mapped to 34 minutes on the *E. coli* genetic map and was named *soxQ1* (Greenberg et al., 1991). MnSOD and glucose-6-phosphate dehydrogenase are controlled by *soxQ*. The expression of seven other non-soxRS proteins are also controlled by *soxQ*.

11.5 RESPONSE TO STARVATION/STATIONARY PHASE: THE STATIONARY PHASE STIMULON AND THE *katF* REGULON

When cells are starved for a nutrient such as nitrogen, phosphorous or carbon, specific regulons are induced to secure more of the limiting nutrient and to allow growth under conditions of the limiting nutrient. If the induction of these regulons does not provide sufficient nutrients, the cell enters stationary phase. The entrance into stationary phase elicits a response which helps maintain cell viability during starvation and which allows cells to emerge from starvation when nutrients reappear. Protein synthesis is required for this response (Reeve et al., 1984) since protein synthesis inhibitors applied at the time when cells enter stationary phase cause reduced survival. Approximately 30 new proteins are synthesized in *E. coli* (Groat et al., 1986) or *S. typhimurium* (Spector et al., 1986) when cells enter stationary phase. Thus there appears to be a stimulon which responds to the entrance of bacterial cells into stationary phase. One aspect of this response is the resistance it confers upon starved cells to a number of environmental stresses including oxidative stress. Carbon-starved *E. coli* are remarkably resistant to H_2O_2 (Jenkins et al., 1988). The nature of this resistance can be traced to the *katF* regulon. The *katF* gene was originally described as a gene required for HPII CAT expression (Loewen and Triggs, 1984). Subsequently, it was shown that the *nur* mutation was an allele of *katF* (Sammartano et al., 1986). The *nur* mutation results in sensitivity to near-UV irradiation (Tuveson, 1980). Surprisingly, while *katF* is required for H_2O_2 resistance and near-UV resistance, *katE*, the structural gene for HPII CAT, is not (Sammartano et al., 1986). The suggestion that *katF* controlled the expression of more enzymes than just HPII CAT was confirmed when it was discovered that the expression of exonuclease III was also regulated by the *katF* gene product (Sak et al., 1989). The *katF* gene has been cloned (Mulvey et al., 1988) and sequenced (Mulvey and Loewen, 1989) and appears to be an alternative sigma factor. Subsequent studies (McCann et al., 1991; Lange and Hengge-Aronis, 1991) have shown that *katF* plays a central role in the development of the general starvation response in *E. coli* and controls the expression of over 30 proteins after carbon starvation.

Bacteria which are growing have the capacity to respond to oxidative stress by turning on sets of genes such as the *soxRS* and the *oxyR* regulons. Cells in stationary phase do not have the luxury of being able to respond rapidly to environmental stress via the induction of protective enzymes. It appears that a pre-programmed response to oxidative stress is part of a cell's strategy for survival during starvation conditions.

11.6   MULTI-LAYERED REGULATION: THE *sodA* GENE

E. coli responds to oxygen limitation by several adaptive responses (Spiro and Guest, 1991; Iuchi and Lin, 1991). These responses deal primarily with establishing the most favorable metabolic pathways for obtaining energy from fermentation or respiration. The Arc system represses the synthesis of gene products involved in aerobic respiration when anaerobiosis prevails. The Fnr system activates the synthesis of a number of gene products involved in anaerobic metabolism when anaerobiosis prevails. Both systems utilize sensing systems for anoxia or changes in redox potential. Fur protein controls the synthesis of a number of genes required for the biosynthesis of iron siderophores in *E. coli* (Bagg and Neilands, 1987). Thus, Fur controls the level of iron in the cell.

Arc, Fnr, and Fur all regulate the transcription of the *sodA* gene (Tardat and Touati, 1991; Hassan and Sun, 1992). When aerobic metabolism is reduced, ArcA represses the expression of *sodA* since there is little chance of O_2^- flux. When iron is present, Fur represses *sodA* since the iron SOD (FeSOD) should be active and also to prevent excess FeSOD activity which could produce high levels of H_2O_2 leading to the production of ˙OH by excess iron. When cells are growing anaerobically Fnr is a positive activator of many genes required for anaerobic respiration. It acts to repress *sodA*, presumably since no active oxygen species will be produced during anaerobiosis. The role of *soxRS* is completely independent of Arc, Fnr and Fur.

11.7 SUMMARY

E. coli and *S. typhimurium* can respond to oxidative stress in the form of hydrogen peroxide (H_2O_2) and superoxide anion radicals (O_2^-). These responses involve a large number of genes (more than 60 in number) and multiple regulatory circuits. The regulation of a subset of these genes is under the control of the *oxyR* regulon and the *soxRS* regulon. The regulatory mechanisms governing the remainder of the genes in the peroxide stimulon and the O_2^- stimulon are not understood, nor are the functions of the proteins expressed by the responses. In addition, *E. coli* protects itself from a number of stresses when it enters stationary phase as part of a developmental program controlled by *katF*. At least in the case of *sodA*, it is apparent that other regulatory circuits also exercise control over oxidative stress genes.

It seems clear that we are just beginning to understand the genetics and biochemistry of oxidative stress and that major advances will be made in the near future. Several of the more exciting advances to be anticipated are the elucidation of the regulatory circuits controlling the remainder of the genes in the peroxide and O_2^- radical stimulons, and the biochemical characterization of many of the proteins and enzymes induced by oxidative stress.

References

Amábile-Cuevas, C.F. and Demple, B. (1991) Molecular characterization of the *soxRS* genes of *Escherichia coli*: two genes control a superoxide stress regulon. *Nucleic Acids Res.* **19**, 4479–4484.

Bagg, A. and Neilands, J.B. (1987) Molecular mechanism of regulation of siderophore-mediated iron assimilation. *Microbiol. Rev.* **51**, 509–518.

Chan, E. and Weiss, B. (1987) Endonuclease IV of *Escherichia coli* is induced by paraquat. *Proc. Natl. Acad. Sci. USA* **84**, 3189–3193.

Christman, M.F., Morgan, R.W., Jacobson, F.S. and Ames, B.N. Positive control of a regulon for defenses against oxidative stress and some heat shock proteins in *Salmonella typhimurium. Cell* **41**, 753–762.

Christman, M.F., Storz, G. and Ames, B.N. (1989) OxyR, a positive regulator of hydrogen peroxide-inducible genes in *Escherichia coli* and *Salmonella typhimurium*, is homologous to a family of bacterial regulatory proteins. *Proc. Natl. Acad. Sci. USA* **86**, 3484–3488.

Demple, B. and Halbrook, J. (1983) Inducible repair of oxidative DNA damage in *Escherichia coli. Nature* **304**, 466–468.

Farr, S.B., Natvig, D.O. and Kogoma, T. (1985) Toxicity and mutagenicity of plumbagin and the induction of a possible new DNA repair pathway in *Escherichia coli. J. Bacteriol.* **164**, 1309–1316.

Gardner, P. and Fridovich, I. (1991a) Superoxide sensitivity of the *Escherichia coli* 6-phosphogluconate dehydratase. *J. Biol. Chem.* **266**, 1478–1483.

Gardner, P. and Fridovich, I. (1991b) Superoxide sensitivity of the *Escherichia coli* aconitase. *J. Biol. Chem.* **266**, 19328–19333.

Greenberg, J.T., Chou, J.H., Monach, P.A. and Demple, B. Activation of oxidative stress genes by mutations at the *soxQ/cfxB/marA* locus of *Escherichia coli. J. Bacteriol.* **1733**, 4433–4439.

Greenberg, J.T. and Demple, B. (1988) Overproduction of peroxide-scavenging enzymes in *Escherichia coli* suppresses spontaneous mutagenesis and sensitivity to redox-cycling agents in *oxyR* mutants. *EMBO J.* **7**, 2611–2617.

Greenberg, J.T. and Demple, B. (1989) A global response induced in *Escherichia coli* by redox-cycling agents overlaps with that induced by peroxide stress. *J. Bacteriol.* **171**, 3933–3939.

Greenberg, J.T., Monach, P., Chou, J.H., Josephy, P.D. and Demple, B. (1990) Positive control of a global antioxidant defense regulon activated by superoxide-generating agents in *Escherichia coli. Proc. Natl. Acad. Sci. USA* **87**, 6181–6185.

Gregory, E.M. and Fridovich, I. (1973) Induction of superoxide dismutase by molecular oxygen. *J. Bacteriol.* **114**, 543–548.

Groat, R.G., Schultz, J.E., Zychlinsky, E., Bockman, A. and Matin, A. (1986) Starvation proteins in *Escherichia coli*: kinetics of synthesis and role in starvation survival. *J. Bacteriol.* **168**, 486–493.

Hassan, H.M. and Fridovich, I. (1977a) Enzymatic defenses against the toxicity of oxygen and of streptonigrin in *Escherichia coli. J. Bacteriol.* **129**, 1574–1583.

Hassan, H.M. and Fridovich, I. (1977b) Regulation of the synthesis of superoxide dismutase in *Escherichia coli. J. Biol. Chem.* **252**, 7667–7672.

Hassan, H.M. and Sun, H.-C.H. (1992) Regulatory roles of Fnr, Fur, and Arc in expression of manganese-containing superoxide dismutase in *Escherichia coli. Proc. Natl. Acad. Sci. USA* **89**, 3217–3221.

Imlay, J.A. and Linn, S. (1986) Bimodal pattern of killing of DNA-repair-defective or anoxically grown *Escherichia coli* by hydrogen peroxide. *J. Bacteriol.* **166**, 519–527.

Iuchi, S. and Lin, E.C.C. (1991) Adaptation of *Escherichia coli* to respiratory conditions: regulation of gene expression. *Cell* **66**, 5–7.

Jenkins, D.E., Schultz, J.E. and Matin, A. (1988) Starvation-induced cross protection against heat or H₂O₂ challenge in *Escherichia coli. J. Bacteriol.* **170**, 3910–3914.

Kao, S.M. and Hassan, H.M. (1985) Biochemical characterization of a paraquat-tolerant mutant of *Escherichia coli. J. Biol. Chem.* **260**, 10478–10481.

Kogoma, T., Farr, S.B., Joyce, K.M. and Natvig, D.O. (1988) Isolation of gene fusions (*soi*::*lacZ*) inducible by oxidative stress in *Escherichia coli. J. Bacteriol.* **85**, 4799–4803.

Lange, R. and Hengge-Aronis, R. (1991) Identification of a central regulator of stationary phase gene expression in *E. coli. Mol. Microbiol.* **5**, 49–59.

Liochev, S.I. and Fridovich, I. (1992) Fumarase C, the stable fumarase of *Escherichia coli*, is controlled by the *soxRS* regulon. *Proc. Natl. Acad. Sci. USA* **89**, 5892–5896.

Loewen, P.C. and Triggs, B.L. (1984) Genetic mapping of *katF*, a locus that with *katE* affects the synthesis of a second catalase species in *Escherichia coli. J. Bacteriol.* **160**, 668–675.

McCann, M.P., Kidwell, J.P. and Matin, A. (1991) The putative σ factor KatF has a central role in the development of starvation-mediated general resistance in *Escherichia coli. J. Bacteriol.* **1738**, 4188–4194.

Mito, S., Zhang, Q-M. and Yonei, S. (1993) Isolation and characterization of *Escherichia coli* strains containing new gene fusions (*soi*::*lacZ*) inducible by superoxide radicals. *J. Bacteriol.* **175**, 2645–2651.

Mulvey, M.R. and Loewen, P.C. (1989) Nucleotide sequence of *katF* of *Escherichia coli* suggests *katF* protein is a novel σ transcription factor. *Nucleic Acids Res.* **17**, 9979–9991.

Mulvey, M.R., Sorby, P.A., Triggs-Raine, B.L. and Loewen, P.C. (1988) Cloning and physical characterization of *katE* and *katF* required for catalase HPII expression in *Escherichia coli. Gene* **73**, 337–345.

Nunoshiba, T., Hidalgo, E., Amábile Cuevas, C.F. and Demple, B. (1992) Two-stage control of an oxidative stress regulon: the *Escherichia coli* SoxR protein triggers redox-inducible expression of the *soxS* regulatory gene. *J. Bacteriol.* **174**, 6054–6060.

Reeve, C.A., Bockman, A.T. and Matin, A. (1984) Role of protein degradation in the survival of carbon-starved *Escherichia coli* and *Salmonella typhimurium. J. Bacteriol.* **157**, 758–763.

Richter, H.E. and Loewen, P.C. (1981) Induction of catalase in *Escherichia coli* by ascorbic acid involves hydrogen peroxide. *Biochem. Biophys. Res. Commun.* **100**, 1039–1046.

Sak, B.D., Eisenstark, A. and Touati, D. (1989) Exonuclease III and the catalase hydroperoxidase II in *Escherichia coli* are both regulated by the *katF* gene product. *Proc. Natl. Acad. Sci. USA* **86**, 3271–3275.

Sammartano, L.J., Tuveson, R.W. and Davenport, R. (1986) Control of sensitivity to inactivation by H₂O₂ and broad-spectrum near-UV radiation by the *Escherichia coli katF* locus. *J. Bacteriol.* **168**, 13–21.

Spector, M.P., Aliabadi, Z., Gonzalez, T. and Foster, J.W. (1986) Global control in *Salmonella typhimurium*: two-dimensional electrophoretic analysis of starvation-, anaerobiosis-, and heat shock-inducible proteins. *J. Bacteriol.* **168**, 420–424.

Spiro, S. and Guest, J.R. (1991) Adaptive responses to oxygen limitation in *Escherichia coli. TIBS* **16**, 310–314.

Storz, G., Christman, M.F. and Ames, B.N. (1987) Spontaneous mutagenesis and oxidative damage to DNA in *Salmonella typhimurium*. *Proc. Natl. Acad. Sci. USA* **84**, 8917–8921.

Storz, G., Tartaglia, L.A. and Ames, B.N. (1990) Transcriptional regulator of oxidative stress-inducible genes: Direct activation by oxidation. *Science* **248**, 189–194.

Tao, K., Makino, K., Yonei, S. and Shinagawa, H. (1989) Molecular cloning and nucleotide sequencing of *oxyR*, the positive regulatory gene of a regulon for an adaptive response to oxidative stress in *Escherichia coli*: Homologies between OxyR protein and a family of bacterial activator proteins. *Mol. Gen. Genet.* **218**, 371–376.

Tardat, B. and Touati, D. (1991) Two global regulators repress anaerobic expression of MnSOD in *Escherichia coli*: Fur (ferric uptake regulation) and Arc (aerobic respiration control). *Mol. Microbiol.* **5**, 455–465.

Tartaglia, L.A., Gimeno, C.J., Storz, G. and Ames, B.D. (1992) Multidegenerate DNA recognition by the *oxyR* transcriptional regulator. *J. Biol. Chem.* **267**, 2038–2045.

Tsaneva, I.R. and Weiss, B. (1990) *soxR*, A locus governing a superoxide response regulon in *Escherichia coli* K-12. *J. Bacteriol.* **172**, 4197–4205.

Tuveson, R.W. (1980) Genetic control of near-UV sensitivity independent of excision deficiency (*uvrA6*) in *Escherichia coli* K12. *Photochem. Photobiol.* **32**, 703–705.

Tyrell, R.M. (1985) A common pathway for protection of bacteria against damage by solar UVA (334 nm, 365 nm) and an oxidising agent (H_2O_2). *Mutation Res.* **145**, 129–136.

Walkup, L.K.B. and Kogoma, T. (1989) *Escherichia coli* proteins inducible by oxidative stress mediated by the superoxide radical. *J. Bacteriol.* **171**, 1476–1484.

Wu, J. and Weiss, B. (1991) Two divergently transcribed genes, *soxR* and *soxS*, control a superoxide response regulon in *Escherichia coli*. *J. Bacteriol.* **173**, 2864–2871.

Wu, J. and Weiss, B. (1992) Two-stage induction of the *soxRS* (superoxide response) regulon of *Escherichia coli*. *J. Bacteriol.* **174**, 3915–3920.

Subject Index

448

456